T-Cell Dependent and Independent B-Cell Activation

Editor

E. Charles Snow

Associate Professor
Department of Microbiology and Immunology
University of Kentucky Medical Center
Lexington, Kentucky

CRC Press
Boca Raton Ann Arbor Boston

Library of Congress Cataloging-in-Publication Data

T-cell dependent and independent B-cell activation/editor, E.
Charles Snow.
p. cm.
Includes bibliographical references.
Includes index.
ISBN 0-8493-6819-7
1. B cells. 2. T cells. 3. Cellular immunity. I. Snow, E.
Charles.
[DNLM: 1. Antibody Formation--physiology. 2. Antigen-Antibody
Reactions--immunology. 3. B-Lymphocytes--immunology. 4. T
-Lymphocytes--immunology. 5. Thymus Gland--immunology. QW 575
T249]
QR185.8.B15T34 1991
616.07′93--dc20
DNLM/DLC
for Library of Congress 90-2375
CIP

Direct all inquiries to CRC Press, Inc., 2000 Corporate Blvd., N.W., Boca Raton, Florida 33431.

International Standard Book Number 0-8493-6819-7

Library of Congress Card Number 90-2375
Printed in the United States

PREFACE

We have known for over 100 years that mammals protect themselves from pathogenic organisms, at least in part, through the production and release into the circulatory system of specific antibody. During the first half of this century, immunochemically oriented studies dominated and provided a significant amount of evidence describing the physicochemical characteristics of the antibody molecule and its interaction with antigen. Interest in the biological characteristics of the humoral immune system has developed relatively recently. This new era in immunology was launched by Sir Macfarlane Burnet's theoretical considerations of antibody formation during the 1940s. These landmark papers were preceded by the identification of the lymphocyte as the mediator of immune responses. Elegant experiments from many laboratories during the 1960s and early 1970s detailed the requirements for cellular interactions during the development of a humoral response. At the same time, membrane-associated immunoglobulin was identified as the B cell's receptor for antigen. Experiments utilizing anti-immunoglobulin reagents and lipopolysaccharide as polyclonal stimulants for B cells provided the conceptual bases for the important one and two signal hypotheses which attempted to explain the mechanism by which B cells are activated to expand and ultimately differentiate into antibody-forming cells. Furthermore, our appreciation of the role played by soluble mediators, many of which are secreted by helper T cells, developed during the latter half of the 1970s and on into the 1980s. As so we learned that the antigen-mediated induction of B cell growth and differentiation involves the active participation of specific, cell surface receptors, direct cellular interactions between B and T cells, and the actions of hormones released by T cells and macrophages.

The purpose of this volume is to introduce the reader to the experimental and theoretical considerations currently employed to evoke explanations for the mechanism by which small, resting B cells become activated to proliferate and differentiate into cells capable of high-rate antibody synthesis and secretion. An attempt has been made to cover many of the relevant subdisciplines which have evolved over the past 10 or so years. I would like to point out that there does exist a certain amount of overlap between various contributions to this book. This has been my intention for the following reasons. I wanted to enlist contrasting views that more fully convey to the reader the differing points-of-view which serve to guide experimentation within this field. It is important to note that from this overlap one can gain an appreciation of how the individual subdisciplines meld together forming a more holistic view of the humoral immune response. All of the contributors have provided a brief historical perspective of their subdisciplines allowing the reader the opportunity to survey the current theories and model systems in the light of the experimental observations utilized during their development. We hope that this approach will provide a ready resource for any student of immunology interested in expanding their understanding of the humoral immune response.

THE EDITOR

E. Charles Snow, Ph.D., is Associate Professor of Microbiology and Immunology at the University of Kentucky Medical Center, Lexington, Kentucky.

Dr. Snow graduated in 1972 from Tulane University, New Orleans, Louisiana, with a B.S. degree in biology and obtained his Ph.D. degree in 1978 from the Department of Anatomy, University of Iowa Medical School in Iowa City. His postdoctoral training in cellular immunology was at the University of Texas Health Science Center at Dallas.

Dr. Snow is a member of the American Association for the Advancement of Science, American Association of Immunologists, Midwest Association of Immunologists, New York Academy of Science, and the Reticuloendothelial Society. He has chaired symposia at both the regional and national levels, and serves as an ad hoc reviewer for numerous journals and funding agencies. Dr. Snow has been the recipient of research grants from the National Institutes of Health and the Burroughs Wellcome Company, presented over 40 talks at national and international meetings and as a guest lecturer at Universities and Institutes, and he has published over 30 papers. His current research interests are in delineating the biochemical and molecular changes occurring in B cells which have received the major growth stimulus as a consequence of direct cellular contact with helper T cells, and understanding the reason(s) that lymphocytes express cell surface receptors for hormones, such as insulin, during the activation cascade.

CONTRIBUTORS

Marta Alarcon-Riquelme
Department of Immunology
The Arrhenius Laboratories for Natural Sciences
Stockholm University
Stockholm, Sweden

Arthur Baluyut
Department of Microbiology and Immunology and Sanders-Brown Center on Aging
University of Kentucky
Lexington, Kentucky

William C. Bartlett
Department of Microbiology and Immunology
Medical College of Virginia
Richmond, Virginia

Helen Braley-Mullen
Harry S. Truman Memorial VA Hospital
Departments of Medicine and Molecular Microbiology and Immunology
University of Missouri—Columbia
Columbia, Missouri

Robin E. Callard
Reader in Immunology
Institute of Child Health
University of London
London, England

D. Mark Estes
Department of Microbiology
University of Texas Health Science Center
San Antonio, Texas

Thomas L. Feldbush
Harry S. Truman Memorial VA Hospital
Departments of Medicine and Molecular Microbiology and Immunology
University of Missouri—Columbia
School of Medicine
Columbia, Missouri

Carmen Fernandez
Department of Immunology
The Arrhenius Laboratories for Natural Sciences
Stockholm University
Stockholm, Sweden

Suzanne Fish
Department of Biology
Princeton University
Princeton, New Jersey

Keith A. Krolick
Associate Professor
Department of Microbiology
University of Texas Health Science Center
San Antonio, Texas

Jonathan Lee
Department of Immunology
University of Toronto
Toronto, Ontario, Canada

David Leitenberg
Harry S. Truman Memorial VA Hospital
Departments of Medicine and Molecular Microbiology and Immunology
University of Missouri—Columbia
Columbia, Missouri

Tim Manser
Assistant Professor
Department of Biology
Princeton University
Princeton, New Jersey

Eleanor S. Metcalf
Professor
Department of Microbiology
U.S.U.H.S.
Bethesda, Maryland

Göran Möller
Professor
Department of Immunology
Stockholm University
Stockholm, Sweden

Natarajan Muthusamy
Department of Microbiology and Immunology and Sanders-Brown Center on Aging
University of Kentucky
Lexington, Kentucky

Randolph J. Noelle
Associate Professor
Department of Microbiology
Dartmouth Medical School
Hanover, New Hampshire

E. Charles Snow
Associate Professor
Department of Microbiology and Immunology
University of Kentucky Medical Center
Lexington, Kentucky

Yves St-Pierre
Department of Immunology
University of Toronto
Toronto, Ontario, Canada

Bondada Subbarao
Associate Professor
Department of Microbiology and Immunology and Sanders-Brown Center on Aging
University of Kentucky
Lexington, Kentucky

Judy M. Teale
Associate Professor
Department of Microbiology
University of Texas Health Science Center
San Antonio, Texas

Thomas Waldschmidt
Assistant Professor
Department of Pathology
University of Iowa College of Medicine
Iowa City, Iowa

Tania H. Watts
Assistant Professor
Department of Immunology
University of Toronto
Toronto, Ontario, Canada

TABLE OF CONTENTS

Chapter 1

THE HUMORAL IMMUNE RESPONSE—A HISTORICAL PERSPECTIVE

E. Charles Snow

TABLE OF CONTENTS

I. INTRODUCTION

Since Jenner's demonstration of vaccination against smallpox in the late eighteenth century, scientists have been interested in the mechanisms by which the vertebrate host protects itself from numerous pathogenic organisms through the production of humoral factors. Progress in this area of biology has followed a rather tortuous trail, due in part to the vagarious nature of research and the individuals who perform it. This book will provide the reader with a detailed analysis of the current thoughts concerning the many subdisciplines making up the humoral immune response. This chapter will attempt to provide a perspective for the later chapters, by considering some of the important experimental and theoretical considerations which, over the years since Jenner, have helped to delineate this exciting area of biology.

II. EARLY OBSERVATIONS OF HUMORAL IMMUNITY

Even before the experiments of Jenner, physicians realized that individuals who had survived contagious diseases could safely administer to patients afflicted with what seemed to be a similar disease. This was puzzling to physicians of that time, in part because they did not understand the causative nature of these deadly diseases. In the late nineteenth century, Pasteur initiated studies which led to an understanding of the principles governing the causative nature of infectious diseases. He demonstrated that chickens could be rendered resistant to chicken cholera if they were inoculated with the attenuated microorganism.[1] These results were utilized by Pasteur to develop his germ theory explaining the nature of acquired immunity.[2] This theory predicted that microorganisms were responsible for most, if not all, diseases which plagued mankind. Immunity must, therefore, represent a state in which the microorganisms are destroyed within the host. He first postulated that during the rapid growth phase of the microbe, a point might be reached in which the organisms would deplete the host of a nutrient important for the organism's growth.[1] At the same time, reports were appearing in the literature which suggested a more active role for a serum component during the host-mediated demise of the microbe. Nuttall was the first to demonstrate the presence, within the serum of normal animals, of substances with natural toxicity towards certain microorganisms.[3] von Behring and colleagues reported that protection against tetanus and diphtheria was mediated by serum factors which could likewise passively protect naive recipients.[4,5] These observations were followed several years later by a report by Pfeiffer showing that this serum factor actually lysed the bacteria.[6] In addition, he also documented the ability of this factor to be passively transferred from animal to animal. These experiments, along with others, served as the basis for the side-chain theory of Ehrlich.[7] This theory was actually an extension of Pasteur's nutrient depletion idea, except Ehrlich proposed that cell surface receptors for nutrients (the side-chains), and not the nutrients, were responsible for neutralizing the harmful microbes. He envisioned a crossreactivity existing between the pathogens or their natural products (such as toxins) and these nutrient transport receptors. In addition, he suggested that the occupancy of these receptors would induce the cell to overproduce the receptor resulting in the release of these molecules into bodily fluids. The demonstration that some bacterial products, such as toxins, displayed tissue selectivity, suggested to him that they bind to surface receptors upon the sensitive cells. The overproduction and release of these toxin receptors accounted for the serum factors exhibiting antimicrobial activities. As long as only bacteria, or their products, seemed to generate these serum factors, this side-chain theory was able to account for many of the experimental results. The demonstration by Bordet, in 1899, that antibodies could be formed against things other than microorganisms, in this case red blood cells, indicated the need for a modification of the original hypothesis.[8]

During the last part of the nineteenth century, a second school of thought concerning the mechanism by which vertebrates protected themselves from infectious diseases, was forcefully expounded by Metchnikoff.[9,10] This theory was based upon his study of phagocytic cells in the crustacean *Daphnia,* and suggested that phagocytic cells represented the first line of defense against pathogenic organisms. This point-of-view was widely denounced for several reasons. First, at the time Metchnikoff proposed it, most pathologists considered the inflammatory response to be detrimental to the host. The idea that the most prominent cell in the response, the phagocyte, was actually beneficial was too foreign a concept. In addition, as the reports continued to appear showing the ability of antibody to mediate devastating effects upon microbes, the sentiment in the field sided with humoral factors being of more importance than the phagocytes. In particular the German school of thought, led initially by Koch and later by Ehrlich, were vociferous in their defense of the humoral theory of immunity.[11] Their views were strengthened by each new demonstration that antibody could precipitate proteins or agglutinate microorganisms.[12,13] Even the demonstration, by Wright and Douglas, that antibody functioned as opsonins promoting the activity of phagocytes did little to alter the opinions of those propounding the chemical nature of the humoral response.[14] Many factors were involved in this victory of the humoralists over the cellularists, including such nonscientific concerns as nationalism (a wonderful and detailed account of this has recently been presented by Silverstein).[15] The result of this victory endured for four decades as immunochemical studies dominated over those concerned with understanding cellular activities. Until relatively recently, this retarded the development of our understanding the cellular components operative during immune responses. Much of what has been learned over the past 30 years will form the basis of this book.

III. THE FIRST THEORIES

The chemists maintained their domination over immunologic thought through the first five decades of the twentieth century.[15] During this period our knowledge of the physical characteristics of the antibody molecule, as well as many of the features of antigen-antibody interactions was expanded. When the immunologists of that time thought about the biological characteristics of the humoral immune response, they approached the topic as chemists, and tended not to incorporate biological considerations within the framework of their theories. While impressive progress was made during this period, these studies contributed little to understanding how the host made antibody.

Besides Ehrlich's sidechain theory,[11] which I will come back to, the first theories concerning the origin of antibody simply stated that antibodies were derivatives of the antigen.[16,17] In some manner, pieces of the immunizing agent served to bind and neutralize new antigen as it entered the system. For such a model to be tenable, the amount of antigen which entered the system had to equal the amount of antibody formed. It was soon realized that the average amount of antibody formed after the immunization of an animal far outweighed the amount of antigen inoculated.[18-20] Also contributing to the demise of this hypothesis was the demonstration that in many instances antibodies were structurally different from the antigens for which they were specific.[21,22] This left Ehrlich's side-chain theory as the sole explanation for how antibody formation occurred.

As I mentioned above, the side-chain theory gradually lost favor as the size of the antibody repertoire became apparent. One could, perhaps, imagine the existence of enough nutrient receptors to deal with the limited numbers of toxins, but not sufficiently diverse to contend with the limitless supply of synthetic haptens produced by Landsteiner and colleagues. It was, in fact, the necessity of dealing with the results reported by the immunochemist Landsteiner and his synthetic antigens which prompted other chemists to predict that the antigen itself must serve as the blueprint during antibody production. It seemed reasonable, in that time, to suggest that the antigen might alter the configuration of some serum component allowing it to effectively bind new antigen as it entered the body.[23-25] This new theory accounted for the quantitative differences between antigen and antibody, which compromised Buchner's theory, since one antigen molecule was speculated to be able to blueprint for more than one serum component. As knowledge concerning the structure of proteins advanced, Pauling suggested that antigen served as a template upon which the protein chain of antibody folded and that antigen, therefore, instructed the formation of antibody; this idea was favored by other investigators as well.[26-29] These instruction theories easily dealt with concerns of repertoire size and the fact that antibody levels always exceeded the levels of antigen to which the animal was exposed. These theories found tremendous support amongst the majority of immunologists prior to the start of the 1940s. However, those who formulated these theories never sought to accommodate known biological characteristics of humoral immunity.

IV. THE RETURN TO CELL BIOLOGY

In 1941, a book published by Sir MacFarlane Burnet reintroduced biological considerations to scientists interested in the humoral immune response.[30] When considering this event in immunology from our current vantage point, it must be kept in mind that at the time of its publication, lymphocytes were still thought to represent end stage cells without any apparent function. Therefore, the immunologists of that time, interested in the biology of the immune response, operated under the distinct disadvantage of knowing virtually nothing about the cell which might be mediating the activity in question. Burnet reasoned that antigen might induce some cell to produce antibody. Therefore, antigen must be able to change the biochemical capabilities of the selected cell. A popular concept in the early 1940s was that cells could be induced to synthesize a unique set of proteins called 'adoptive enzymes'. It was thought that cells produced these adoptive enzymes at crucial moments to mediate some new function required by the cell. Burnet suggested that antigen might induce such enzymes in the eventual antibody-forming cell. Included in this book was his suggestion that antigen would also induce the expansion of these antibody-forming cells, thus increasing the capability of the organism to mount a humoral defense against that particular offending agent. Burnet later modified his original theory based upon the newly recognized connection between genes and proteins.[31] Since the proteins expressed by a cell are controlled by genes, Burnet suggested that antigen must provide information directly to the genome of the antibody-forming cell so that it could subsequently synthesize the antibody. Jerne, in his natural selection theory, first attempted to explain how the antigen induced the formation of antibody.[32] He suggested that antibodies of differing specificities were continuously produced within the host. The antigen would bind to the antibody exhibiting the highest affinity and this complex taken up by the future antibody-forming cell. This internalized complex would imprint its information directly into the genome of the cell and thus induce the cell to a state of high-rate antibody synthesis. Although the mechanism turned out to be incorrect, this theory shaped immunologic thought by introducing the concept of antigen selecting the cells which participated in the generation of the antibody. The culmination of these thoughts was the publication by Burnet of the clonal selection theory.[33] This theory was derived from both Ehrlich's side-chain theory and Jerne's natural selection theory. Burnet correctly concluded that some lymphocytes expressed antibody upon their cell surfaces and any one lymphocyte would express antibody of only one specificity. Antigen would therefore bind directly upon the surface of only those cells expressing complementary antibody and the consequence of such an association would be the activation of the antigen-binding cells. The activated cells would subsequently proliferate to increase the overall numbers of the responding cells, and then finally differentiate into cells which would exhibit high-rate secretion of antibody. This important theory continues to serve as an impetus for experimentation into the humoral immune response.

The experiments which ultimately identified the cells participating in the humoral response came from many different directions. Harris et al. first provided evidence for lymphocytes serving as the antibody producing cells.[34] They analyzed the level of specific antibody associated with either lymphocytes or the fluid phase (lymph) recovered from the draining lymph node of an immunized animal. The antibody segregated with the recovered cells, particularly early in the response when antibody levels were peaking within the lymph node itself. The association of antibody with the cellular extracts was shown not to be due to passively absorbed antibody. This report suggested that the antigen-induced antibody-forming cell discussed by both Burnet and Jerne was most probably a lymphocyte. The first hint that lymphocytes might be divided into separate, and functionally distinct, subpopulations came from Bruton's description of a congenital agammaglobulinemia.[35] Young boys with this disease were repeatedly afflicted with pyrogenic bacterial infections as soon as the levels of maternal antibodies waned, and they were unable to mount a specific humoral response to any immunogen. The intriguing finding from these patients was that some lymphoid compartments were normal (such as the peripheral blood, thymus, paracortical areas of lymph nodes, and the periarteriolar sheath in the spleen), while other compartments were severely depleted of lymphocytes (manifested by an almost complete loss of all follicles and germinal centers). In addition, some functions, such as graft rejection and delayed type hypersensitivity (DTH), were normal in magnitude. Together with the observation by Chase that lymphocytes could transfer DTH, and those of Harris et al. indicating that lymphocytes might make antibody, this experiment of nature provided the first clue to the existence of lymphocyte subpopulations.[34,36]

Several years after Bruton's description of congenital agammaglobulinemia, Glick et al. reported their serendipitous finding that birds in which the bursa of Fabricius had been removed at birth, failed to make antibodies to the O antigen of *Salmonella typhimurium.*[37] They also reported that bursectomy of adult birds had little or no effect upon the humoral immune response. At the same time, Good and Varco described a clinical condition of acquired agammaglobulinemia in which there was also an occurrence of thymomas.[38] As seen in experiments with bursectomized birds, the removal of the thymus from adult animals did not appreciably influence the animals ability to mount an antibody response.[39,40] Animals neonatally thymectomized displayed a marked impairment in their ability to mount humoral immune responses.[41,42] A reasonable conclusion based upon these clinical observations and experimental findings was that a cell derived from the bursa of Fabricius in the chicken (or the hypothetical bursal equivalent in mammals) and one isolated from the thymus participated in the development of a humoral response. Whether these separate experimental approaches identified the same or different cells was not apparent from these early studies. It was noted, however, that the cell suspected of being the antibody secreting cell, the plasma cell, underwent at least one round of proliferation prior to assuming the structural features unique to such cells.[43-46] The precursor of plasma cells was not known at the time, nor was there any evidence suggesting how the antibody-forming cell precursor was driven through the differentiative process necessary for it to express its effector function.

Prior to the early 1960s, the importance of the lymphocyte within the immune system was suspected, but not understood. The eloquent experiments by Gowans and his colleagues demonstrated among other things, that the chronic depletion of thoracic duct lymphocytes compromised the ability of the animal to mount a humoral response, which could be restored through the infusion of lymphocytes back into the animal.[47] Nowell's work provided the exciting discovery that plant lectins, such as phytohemagglutinin, induced cultured lymphocytes to enter the proliferative cycle.[48] These studies indicated that lymphocytes were not the end-staged cells that everyone had thought they were, and in an extension of his clonal selection hypothesis, Burnet went on to suggest that the antigen-selected lymphocyte would be stimulated by the interaction with antigen to differentiate directly into a plasma cell.[49] This rather simplistic and attractive thought was shattered following the report of Claman et al. in 1966.[50] Utilizing the adoptive transfer technique, they demonstrated that both thymus-derived and bone marrow-derived cells were required in order to repopulate the irradiated animals humoral immune system. This was the first direct demonstration that more than one cell was involved during the generation of plasma cells. The next piece of the puzzle was provided by Davies et al., who utilized cells recovered from radiation chimeras for adoptive transfer experiments similar to those of Claman et al.[51] Several fundamentally important aspects of the humoral immune system were illuminated by this study. For example, although the thymus-derived cells did not produce any detectable antibody within the recipient animal, the cells proliferated following an antigenic challenge. The bone marrow-derived cells, on the other hand, made small levels of antibody when transferred alone, but did not exhibit signs of cellular expansion to the antigenic challenge. They confirmed Claman et al. by showing that a vigorous response occurred if both populations were utilized to repopulate the irradiated animals. Therefore, the thymus-derived cells most probably were not the cells destined to make antibody, and the bone marrow-derived cells seemed incapable of proliferating to antigen in the absence of thymus-derived cells. Although not understood at the time, this was the first hint that the binding of antigen to its specific receptor on B cells was not sufficient to activate B cells to expand and differentiate. Mitchell and Miller and colleagues employed a variety of approaches, such as chromosomally marked bone marrow- or thymus-derived donor cells in adoptive transfer experiments, to confirm that antibody-forming cells were derived from bone marrow-derived cells.[52-54] An additional level of complexity was exposed when Mosier demonstrated that a third cell type, the macrophage, was necessary for the elicitation of a humoral response *in vitro*.[55] He predicted that macrophages functioned to process and present antigen so that the lymphocyte could respond. However it was still not clear which of the responding cells in the humoral response required this intimate contact with the macrophage. Now that the cast of characters were somewhat identified, the mechanism by which they interacted to induce the differentiation of bone marrow-derived cells into plasma cells captured the interests of cellular immunologists.

V. THE IMPORTANCE OF CELLULAR INTERACTIONS

In 1963, Ovary and Benacerraf reported an experimental observation which intensified the search for the cellular mechanisms responsible for the initiation and regulation of humoral immune responses.[56] They were utilizing hapten-carrier conjugates as a simple tool for dissecting the amanestic antibody response. In the course of this study, they discovered that the induction of a secondary anti-hapten antibody response required that the animal be boosted with the hapten physically linked to the same carrier used during the primary immunization. They termed this result the 'carrier effect'. The first plausible explanation for this phenomenon suggested that the cell surface receptor recognizing hapten only did so in reference to its 'immediate environment'.[57] In other words, the receptor was, in part, recognizing the chemical bond between hapten and the carrier protein. An alternative hypothesis suggested that the carrier effect was best explained through considerations of cellular cooperation, with the receptor on one cell recognizing hapten, while the second cell's receptor recognized carrier.[58-60] These experiments prompted Mitchison to propose that the antigen bridged the hapten-binding bone marrow-derived cell to the carrier-binding, thymus-derived cell, thus allowing effective cooperation to occur.[61,62] Mitchison ruled out the environmental hypothesis by demonstrating that the insertion of spacer molecules between the hapten and the carrier did not interfere with the ability of hapten-carrier to induce a secondary anti-hapten response.[63] He also went on to duplicate the carrier effect in an adoptive transfer system providing additional support for the cellular collaboration model.[64] The demonstration, in an adoptive transfer system, by Raff that anti-theta positive cells (thymus-derived or T cells) recognized carrier while the theta negative cells (bone marrow-derived or B cells) recognized hapten was the final blow to the local environment theory.[65] Kisielow et al. extended this by fractionating T cells into Ly-1$^+$ and Ly-2$^+$,3$^+$ cells and demonstrating that the Ly-1$^+$ subpopulation provided help to B cells entering a thymus-dependent immune response.[66] These experimental results clearly supported Mitchison's antigen-bridging hypothesis, and provided focus to future studies intent upon elucidating the mechanisms mediating these important cellular interactions. Many of the contributors to this volume will address our current understanding of these mechanisms.

VI. THE MHC AND CELLULAR INTERACTIONS

Simultaneous with this revelation that cells of the immune system interacted with one another was the recognition, by several laboratories, of an underlying genetic component operative during these cellular interactions. The impetus for these genetic studies came from experiments performed in the 1940s and 1950s showing that the magnitude of the responsiveness of offspring to antigen could be predicted based upon the responsiveness exhibited by the parents.[67-69] These first studies were not able to document the existence of Mendelian genetic patterns within the response profiles. Levine et al. suggested that this may have been more a reflection of the complexity in terms of the number and diversity of antigenic determinants exhibited by the antigens utilized.[70-72] They simplified this through the use of inbred guinea pigs and defined hapten-carrier conjugates, and demonstrated that the ability of an animal to respond to a given antigen was transmitted as a unigenic Mendelian dominant trait. Importantly, they were able to identify the carrier, the part of the antigen recognized by the T cell, as the antigenic determinant mediating this genetic transmittance. Therefore, the mechanism by which this operated had to occur prior to the actual production of antibody. A similar example of genetic control was also reported in terms of induction of antibody synthesis in inbred mice.[73-76] McDevitt and colleagues, employing high and low responding inbred mice to synthetic polypeptide antigens, were the first to establish a correlation between an animal's ability to mount an antibody response and its major histocompatibility (MHC) locus.[77,78] This genetic trait was also found to be transmittable by the transfer of high responding spleen cells into a low responding animal, directly suggesting that the protein products of these genes were intimately involved with the events occurring during the genesis of a humoral response.[77] This linkage of immune responsiveness to the MHC haplotype was quickly extended to also include graft rejection and delayed-type hypersensitivity in inbred strains of mice, rats, and guinea pigs against a variety of antigenic determinants.[79] This work culminated with the identification of the Ir-1 (immune response) locus and its assignment to a position between H-2K and Ss-Slp in the mouse H-2 complex.[80]

Nossal et al. were the first to indicate the importance for transferring syngeneic B cells and T cells to obtain a successful adoptively transferred antibody response. This result was also confirmed and extended by Katz and colleagues.[54,81,82] Through the use of the Mishell-Dutton *in vitro* culture system, Katz et al. demonstrated that the use of populations of B and T cells which differed at H-2K and H-2D, but shared the I region, allowed for the appropriate cellular interactions for antibody formation.[83,84] They went on to hypothesize that this genetic restriction of cellular interactions was mediated through cell surface expressed molecules, called cell-interaction (CI) molecules.[85] These CI molecules, therefore, functioned as the means for T cells, B cells and macrophages to communicate with each other. The first plausible mechanism for how the cells of the immune system interacted with each other emanated from studies concerning the role of macrophages during the activation of T cells. Unanue and Askonas provided direct evidence for macrophages being able to transfer antigen from the peritoneal cavity of one animal into a recipient animal resulting in the induction of an immune response within the second animal.[86] This was consistent with the phagocytic role which had been assigned the macrophage. These, and future studies, were initiated in an attempt to delineate how T cells recognized antigen. While B cells had been shown to express cell surface immunoglobulin and to be able to bind soluble antigen, the nature of the antigen-specific receptor upon T cell surfaces was unclear.[87-94] Initially, it was shown that the induction of a T cell proliferative response required macrophages to take-up soluble antigen and present it to the antigen-reactive T cells.[94] A report by Rosenthal and Shevach demonstrated that the antigen presenting macrophage and the responding T cell had to be histocompatible at the I region of the MHC locus.[95] This MHC-restriction during the process of T cell activation was explained in terms of the expression upon the surface of the macrophage of processed antigen in association with I region molecules. Much later, studies by Chesnut and Grey and by Lanzavecchia concluded that B cells also present processed antigen, in association with surface I region products, to T cells.[96,97] To accommodate these new experimental findings, Mitchison's original antigen bridging hypothesis has been modified to postulate that antigen-mediated T cell-macrophage and T cell-B cell contact are mediated by similar mechanisms. Current thoughts about this will serve as the focus of several chapters in this book.

VII. EXTENSION OF THE CLONAL SELECTION HYPOTHESIS

Having reviewed some of the important experimental observations made during the period 1950 to 1975, I would like to return to the theoretical frameworks developed during this time. I have already discussed the basic concepts of the clonal selection hypothesis as expounded by both Burnet and Jerne.[30-33] This hypothesis suggested that lymphocytes expressed cell surface receptors for antigen and that it was the introduction of antigen into the system which resulted in the selection of a small percentage of cells expressing that particular antigen-specific receptor. Once selected, these lymphocyte clones would expand and differentiate into effector cells. This theory also maintained that a mechanism must exist for the efficient removal or inactivation of clones which responded to self. Burnet suggested that this occurred by a process in which self reactive B cells were deleted from the available repertoire.[33] Then, as now, the mechanism for this process was not apparent. Lederberg presented the first theoretical consideration of just how the immune system might effectively deal with clones of self-reactive lymphocytes.[98] His theory maintained that during the process of lymphocyte differentiation from the stem cell stage to the mature, peripheral lymphocyte, the cell passed through a stage during which the binding of antigen permanently inactivated it. This view rested upon the assumption that the only plausible difference between self and nonself antigens must be that unlike nonself, self-antigenic determinants were always present within the system.[99] This 'window in time' hypothesis was discounted by the identification in adult animals of B cell clones reactive to continuously expressed antigenic determinants.[100] However, Lederberg did pose important questions which needed to be addressed concerning the differences and/or similarities between positive and negative signals delivered to lymphocytes. In particular, Lederberg wondered about the possible role(s) of the antigen binding receptors during the induction of either activation or inactivation. Were these processes similar or entirely different mechanisms? The first suggestion that multiple points of binding by the B cell during the induction period might aid in the B cell's determination of self from nonself was made by Forsdyke.[101] In this hypothesis, he predicted that single point binding would only inactivate the cell, while two-point binding would result in cellular activation. Bretscher and Cohn extended this concept in the late 1960s in their associative recognition model of self-nonself discrimination.[102,103] They reasoned that lymphocytes must be able to easily make this discrimination and looked to the newly described carrier phenomenon for a clue to how this might work. Since the carrier effect suggested that two different determinants on an antigen were required for the successful induction of a thymus-dependent antibody response, they suggested that during B cell activation both determinants must be occupied. This would happen by one determinant binding the antigen receptor on the B cell, while the second determinant binding what they called associative antibody, which in their view was somehow derived from the carrier reactive T cell. In the absence of this duality of binding, the B cell could not be activated. Therefore, cells which received only signal 1, as a consequence of the occupancy of only the antigen binding receptor, would be inactivated. Two important principles of this hypothesis maintained that the occupancy of the antigen receptor could not deliver a growth stimulus to the B cell, and that the successful activation of B cells absolutely required the participation of T cells. If this model was correct, one supposition would be that the presence of a paralyzing signal should be countermanded through the artificial coupling of a T cell to the antigen-binding B cell. This was in fact shown experimentally by adding the appropriately responding allogeneic T cells (specific for one of the class II MHC antigens expressed by the B cells) to animals exposed to a tolerogen.[104-106]

The above hypothesis assigned a pivotal role for the T cell in regulating both B cell activation and inactivation. Chiller et al. provided the first direct evidence for tolerance occurring in both the T cell and the B cell compartments.[107,108] However, they also demonstrated that potentially important differences existed in the characteristics of T cell and B cell tolerance induction.[108] In reference to the present considerations, their key observations were that T cells became inactivated by lower concentrations of antigen, and that one could detect conditions in which the T cells were tolerant and the B cells were not. In addition, two other groups reported that the injection of a potentially immunogenic material into T cell deficient animals rendered the animals tolerant to the material.[109,110] This indicated that in the absence of T cell help, the interaction of only antigen with B cells resulted in B cell inactivation. This early information on the biological characteristics of the tolerant state were supportive of the two-signal model of B cell activation.

In 1972, Andersson et al. reported that the addition of lipopolysaccharide (LPS) to cultures of lymphocytes induced a polyclonal B cell response.[111] This result suggested to them that the interaction between antigen and the receptor for antigen upon B cell surface membranes was not an essential feature of the activation cascade for the simple reason that it could be by-passed. It was also shown that strains of mice whose B cells could not respond to the mitogenic actions of LPS were incapable of mounting a specific antibody response to hapten conjugated to LPS.[112-115] These, and many additional experimental observations, led Coutinho and Möller to propose the one signal hypothesis for B cell activation.[116,117] Simply stated, this hypothesis maintained that the surface immunoglobulin receptor on B cells served only as a focusing element during the activation process. The receptor did not function as a signal transducing element, and therefore could not transmit to the antigen-binding cell the signal 1 hypothesized by Bretscher and Cohn. The only signal necessary for the activation of B cells was delivered either by the mitogenic component of the antigen (in the case of a thymus-independent antigen) or by T cells or their products (during a thymus-dependent response). In both cases, the activation signal was mediated by the binding of the antigenic determinant onto an inert receptor. An important corollary of this hypothesis was that tolerant B cells did not exist and all of tolerance was explained by the inactivation of the helper T cell.[118]

The one- and two-signal hypotheses eventually spawned two separate and competing camps concerning the role, or lack of a role, for the surface immunoglobulin receptor during the activation of resting B cells. The controversy basically settled upon whether this surface protein was capable of delivering a biochemical signal to the cell. The proponents of the two-signal hypothesis ended up suggesting that this receptor in fact delivered the major growth stimulus to the resting cell. This stimulus would drive the resting cells into high-rate RNA synthesis, and promoted their responsiveness to progression factors. Those who continued to favor the one-signal hypothesis argued for a nonsignaling role for the receptor. As so often happens in research, what has turned out to be the truth borrows from some of the premises of both groups. This will be discussed in detail in later chapters.

VIII. SOME CONCLUDING THOUGHTS

Just over 100 years ago Pasteur first proposed the notion that many diseases were caused by pathogenic organisms overpowering a host organism. It made perfectly good sense to predict, therefore, that if the host could safely combat these harmful organisms, the host would be free from illness. Over the ensuing century, scientists have made giant leaps in our understanding of how mammals coup with an environment brimming with dangerous microbes. In this lead-off chapter, I have tried to provide the reader with a sense of how some of the early discoveries concerning the humoral immune system were uncovered, and a feel for the theoretical considerations the early studies provoked. I decided to limit this brief survey to papers published in 1975 and before, because in my opinion the level of our understanding of B cells and T cells by 1975 provided a strong foundation from which the more modern studies are based. It was at about this time that many of the subdisciplines which make up the study of the humoral system took form. These artificially circumscribed areas of study will be described in detail over the remaining chapters of this book. Through them, I hope that you gain some appreciation for the wonderous world of B cells, T cells, and macrophages, and how they interact with one another during the elicitation of an antibody response.

REFERENCES

1. **Pasteur, L.,** Sur les maladies virulentes, et en particulier sur la maladie appelee vulgairement cholera des poules, *C. R. Hebd. Seances Acad. Sci.,* 90, 239, 1880.
2. **Pasteur, L., Joubert, J., and Chamberland, C.,** La theorie des germes et ses applications a la medicine et a la chirurgie, *C. R. Hebd. Seances Acad. Sci.,* 86, 1037, 1878.
3. **Nuttall, R.,** Experimente uber die bacterienfeindlichen einflusse des thierischen korpers, *Z. Hyg.,* 4, 353, 1888.
4. **von Behring, E. and Kitasato, S.,** Ueber das zustandekommen der diphtherie-immunitat und der tetanus-immunitat bei thieren, *Dtsch. Med. Wochenschr.,* 16, 1113, 1890.
5. **von Behring, E. and Wernicke, E.,** Ueber immunisirung und heilung von versuchsthreien bei der diphtherie, *Z. Hyg.,* 12, 10, 1892.
6. **Pfeiffer, R.,** Weitere untersuchungen uber das wesen der cholera-immunitat und uber specitisch baktericide processe, *Z. Hyg.,* 18, 1, 1895.
7. **Ehrlich, P.,** *Experimental Researches on Specific Therapeutics,* Lewis, London, 1908.
8. **Bordet, J.,** Sur l'agglutination et la dissolution des globules ronges par le serum. D'animaux injectes de sang defribrine, *Ann. Inst. Pasteur Paris,* 12, 688, 1898.
9. **Metchnikoff, E.,** Ueber eine sprosspilzkrankheit der daphnien. Beitrag zur lehre uber den kampf der phagocyten gegen krankheitserreger, *Virchows Arch.,* 96, 177, 1884.
10. **Metchnikoff, E.,** *Lectures on the Comparative Pathology of Inflammation,* Keegan, Paul, Trench, Trubner, London, 1893.
11. **Ehrlich, P.,** The Croonian Lecture, *Proc. R. Soc. London Ser. B,* 66, 424, 1900.
12. **Kraus, R.,** Ueber specifische reactionen in keimfreien filtraten aus cholera, typhus und pestboiullon-culturen, erzeugt durch homologes serum, *Wien, Klin. Wochenschr.,* 10, 736, 1897.

13. **Gruber, M. and Durham, H. E.,** Eine neue methode zur raschen erkennung des choleravibrio und des typhusbacillus, *Muench. Med. Wochenschr.*, 43, 285, 1896.
14. **Wright, A. E. and Douglas, S. R.,** An experimental investigation of the role of the blood-fluids in connection with phagocytosis, *Proc. R. Soc. London Ser. B,* 72, 364, 1903.
15. **Silverstein, A. M.,** *A History of Immunology,* Academic Press, New York, 1989.
16. **Buchner, H.,** Ueber bacteriengifte und gegengifte, *Muench. Med. Wochenschr.*, 40, 449, 1893.
17. **Hertzfeld, E. and Klinger, R.,** Cemische studien zur physiologie und pathologie. II. Die immunitatsreaktienen, *Biochem. Z.*, 85, 1, 1918.
18. **Roux, E. and Vaillard, L.,** Contribution a L'etude du tetanos. Prevention et traitement par le serum antitoxique, *Ann. Inst. Pasteur Paris,* 7, 65, 1893.
19. **Knorr, A.,** Das tetanusgift und seine beziehungen zum thierischen organismus. Eine experimentelle studie uber krankliert und heilung, *Muench. Med. Wochenschr.*, 45, 321, 1898.
20. **Heidelberger, M. and Kendall, F. E.,** The amount of circulating precipitin following the injection of a soluble antigen, *Science,* 72, 23, 1930.
21. **Hooker. S. B. and Boyd, W. C.,** A quantitative aspect of the hypothetical incorporation of injected antigen in resulting antibody, *J. Immunol.*, 23, 465, 1932.
22. **Berger, E. and Erlenmeyer, H.,** Stoffliche beziehungen zwischen antigen und antikorper, *Hyg. Infektionskr.*, 113, 79, 1931.
23. **Bail, O. and Tsuda, K.,** Versuche uber bakteriolytische immunkorper mit besonderer berucksichtigung des normalen rinderserums, *Immunitaetsforsch.*, 1, 546, 1909.
24. **Bail, O.,** Untersuchungen uber normalhamolysine, *Immunitaetsforsch.*, 21, 202, 1914.
25. **Pauling, L. and Campbell, D. H.,** The production of antibodies *in vitro, Science,* 95, 440, 1942.
26. **Pauling, L. and Delbruck, M.,** The nature of the intermolecular forces operative in biological processes, *Science,* 92, 77, 1940.
27. **Topley, W. W. C.,** The role of the spleen in the production of antibodies, *J. Pathol. Bacteriol.*, 33, 341, 1930.
28. **Breinl, F. and Haurowitz, F.,** Untersuchung des praezipitates aus haemoglobin und aub-haemoglobinserum, *Z. Physiol. Chem.*, 192, 45, 1930.
29. **Rothen, A. and Landsteiner, K.,** Adsorption of antibodies by egg albumin films, *Science,* 90, 65, 1939.
30. **Burnet, F. M.,** *The Production of Antibodies,* Macmillan, New York, 1941.
31. **Burnet, F. M. and Fenner, F.,** *The Production of Antibodies,* 2nd ed., Macmillan, New York, 1949.
32. **Jerne, N. K.,** The natural-selection theory of antibody formation, *Proc. Natl. Acad. Sci. U.S.A.*, 41, 849, 1955.
33. **Burnet, F. M.,** A modification of Jerne's theory of antibody production using the concept of clonal selection, *Aust. J. Sci.*, 20, 67, 1957.
34. **Harris, T. N., Grimm, E., Mertens, E., and Ehrich, W. E.,** The role of the lymphocyte in antibody formation, *J. Exp. Med.*, 73, 1945.
35. **Bruton, O. C.,** Agamaglobulinemia, *Pediatrics,* 9, 722, 1952.
36. **Chase, M. W.,** The cellular transfer of cutaneous hypersensitivity to tuberculin, *Soc. Exp. Biol. Med. Proc.*, 59, 134, 1945.
37. **Glick, B., Chang, T. S., and Jaap, R. G.,** The bursa of Fabricius and antibody production, *Poult. Sci.*, 35, 224, 1956.
38. **Good, R. A. and Varco, R. L.,** A clinical and experimental study of agammaglobulinemia, *Lancet,* 75, 245, 1955.
39. **Harris, T. N., Rhoads, J., and Stokes, J.,** A study of the role of the thymus and spleen in the formation of antibodies in the rabbit, *J. Immunol.*, 58, 27, 1948.
40. **MacLean, L. D., Zak, S. L., Varco, R. L., and Good, R. A.,** The role of the thymus in antibody production: an experimental study of the immune response in thymectomized rabbits, *Transplant. Bull.*, 4, 21, 1957.
41. **Archer, O. and Pierce, J. C.,** Role of the thymus in development of the immune response, *Fed. Proc.*, 20, 26, 1961.
42. **Miller, J. F. A. P.,** Effect of neonatal thymectomy on the immunological responsiveness of the mouse, *Proc. R. Soc. London,* B156, 415, 1962.
43. **Fagraeus, A.,** Antibody production in relation to the development of plasma cells, *Acta Med. Scand.*, 130 (Suppl.), 204, 1948.
44. **Schooley, J. C.,** Autoradiographic observations of plasma cell formation, *J. Immunol.*, 86, 331, 1961.
45. **Nossal, G. and Makela, O.,** Autoradiographic studies of the immune response. I. The kinetics of plasma cell proliferation, *J. Exp. Med.*, 115, 209, 1962.
46. **Baney, R. N., Vazquez, J. J., and Dixon, F. J.,** Cellular proliferation in relation to antibody synthesis, *Soc. Exp. Biol. Med. Proc.*, 109, 1, 1962.
47. **Gowans, J. L. and McGregor, D. D.,** The immunological activities of lymphocytes, *Prog. Allergy,* 9, 1, 1965.

48. **Nowell, P. C.,** Phytohaemagglutinin: an initiator of mitosis in cultures of human leukocytes, *Cancer Res.*, 20, 462, 1960.
49. **Burnet, F. M.,** The immunological significance of the thymus: an extension of the clonal selection theory of immunity, *Australas. Ann. Med.*, 11, 79, 1962.
50. **Claman, H. N., Chaperon, E. A., and Triplett, R. F.,** Thymus-marrow cell combinations. Synergism in antibody production, *Soc. Exp. Biol. Med. Proc.*, 122, 1167, 1966.
51. **Davies, A. J. S., Leuchars, E., Wallis, V., Marchant, R., and Elliott, E. V.,** The failure of thymus-derived cells to produce antibody, *Transplantation*, 5, 222, 1967.
52. **Mitchell, G. F. and Miller, J. F. A. P.,** Immunological activity of thymus and thoracic-duct lymphocytes, *Proc. Natl. Acad. Sci. U.S.A.*, 59, 296, 1968.
53. **Miller, J. A. F. P. and Mitchell, G. F.,** Cell to cell interaction in the immune response. I. Hemolysin-forming thymus or thoracic duct lymphocytes, *J. Exp. Med.*, 128, 801, 1968.
54. **Nossal, G. J. V., Cunningham, A., Mitchell, G. F., and Miller, J. F. A. P.,** Cell to cell interaction in the immune response. III. Chromosomal marker analysis of single antibody-forming cells in reconstituted, irradiated or thymectomized mice, *J. Exp. Med.*, 128, 839, 1968.
55. **Mosier, D. E.,** A requirement for two cell types for antibody formation *in vitro*, *Science*, 158, 1573, 1967.
56. **Ovary, Z. and Benacerraf, B.,** Immunological specificity of the secondary response with dinitrophenylated proteins, *Soc. Exp. Biol. Med. Proc.*, 114, 72, 1963.
57. **Levine, B. B.,** Studies on delayed hypersensitivity. I. Inferences on the comparative binding affinities of antibodies mediating delayed and immediate hypersensitivity reactions in the guinea pig, *J. Exp. Med.*, 121, 873, 1965.
58. **Mitchison, N. A.,** Antigen recognition responsible for the induction *in vitro* of the secondary response, *Cold Spring Harbor Symp. Quant. Biol.*, 32, 431, 1967.
59. **Rajewsky, K. and Rottlander, E.,** Tolerance specificity and the immune response to lactic dehydrogenase isoenzymes, *Cold Spring Harbor Symp. Quant. Biol.*, 32, 547, 1967.
60. **Rajewsky, K., Schirrmacher, V., Nase, S., and Jerne, N. K.,** The requirement of one more antigenic determinant for immunogenicity, *J. Exp. Med.*, 129, 1131, 1969.
61. **Mitchison, N. A.,** Cell populations involved in immune responses, in *Immunological Tolerance*, Landy, M. and Braun, W., Eds., Academic Press, New York, 1969, 113.
62. **Mitchison, N. A., Rajewsky, K., and Taylor, R. B.,** Cooperation of antigenic determinants and of cells in the induction of antibodies, in *Proc. Prague Symp. Devel. Aspects Antibody Form. Struct.*, Vol. II, Sterzl, J. and Riha, I., Eds., Publishing House of the Czechoslovak Academy of Science, Prague, 1970, 547.
63. **Mitchison, N. A.,** The carrier effect in the secondary response to hapten-protein conjugates. I. Measurement of the effect with transferred cells and objections to the local environment hypothesis, *Eur. J. Immunol.*, 1, 10, 1971.
64. **Mitchison, N. A.,** The carrier effects in the secondary response to hapten-protein conjugates. II. Cellular cooperation, *Eur. J. Immunol.*, 1, 18, 1971.
65. **Raff, M. C.,** Role of thymus-derived lymphocytes in the secondary humoral immune response in mice, *Nature*, 226, 1257, 1970.
66. **Kisielow, P., Hirst, J. A., Shiku, H., Beverley, P. C. L., Hoffmann, M. K., Boyse, E. A., and Oettgen, H. F.,** Ly antigens as markers for functionally distinct subpopulations of thymus-derived lymphocytes of the mouse, *Nature*, 253, 219, 1975.
67. **Fjord-Scheibel, I.,** Hereditary differences in the capacity of guinea pigs for the production of diphtheria antitoxin, *Acta Pathol. Microbiol. Scand.*, 20, 464, 1943.
68. **Carlinfanti, E.,** The predisposition for immunity, *J. Immunol.*, 59, 1, 1948.
69. **Sang, J. H. and Sobey, W. R.,** The genetic control of response to antigenic stimuli, *J. Immunol.*, 72, 52, 1954.
70. **Levine, B. B., Ojeda, A., and Benacerraf, B.,** Studies on artificial antigens. III. The genetic control of the immune response to hapten-poly-L-lysine conjugates in guinea pigs, *J. Exp. Med.*, 118, 953, 1963.
71. **Levine, B. B., Ojeda, A., and Benacerraf, B.,** Basis for the antigenicity of hapten-poly-L-lysine conjugates in random bred guinea pigs, *Nature*, 200, 544, 1963.
72. **Levine, B. B. and Benacerraf, B.,** Genetic control in guinea pigs of immune response to conjugates of haptens and poly-l-lysine, *Science*, 147, 517, 1965.
73. **Gorer, P. A. and Schutze, H.,** Genetical studies on immunity in mice. II. Correlation between antibody formation and resistance, *J. Hyg.*, 38, 647, 1938.
74. **Fink, M. A. and Quinn, V. A.,** Antibody production in inbred strains of mice, *J. Immunol.*, 70, 61, 1953.
75. **Ipsen, J.,** Differences in primary and secondary immunizability of inbred mice strains, *J. Immunol.*, 83, 448, 1959.
76. **McDevitt, H. O. and Sela, M.,** Genetic control of the antibody response. I. Demonstration of determinant-specific differences in response to synthetic polypeptide antigens in two strains of inbred mice, *J. Exp. Med.*, 122, 517, 1965.

77. **McDevitt, H. O. and Tyan, M. L.,** Genetic control of the antibody response in inbred mice. Transfer of response by spleen cells and linkage to the major histocompatibility (H-2) locus, *J. Exp. Med.,* 128, 1, 1968.
78. **McDevitt, H. O. and Chinitz, A.,** Genetic control of the antibody response: relationship between immune response and histocompatibility (H-2) type, *Science,* 163, 1207, 1969.
79. **Benacerraf, B. and McDevitt, H. O.,** The histocompatibility-linked specific immune response genes, *Science,* 175, 273, 1972.
80. **McDevitt, H. O., Deak, B. D., Shreffler, D. C., Klein, J., Stimpfeling, J. H., and Snell, G. D.,** Genetic control of the immune response. Mapping of the Ir 1 locus, *J. Exp. Med.,* 135, 1259, 1972.
81. **Katz, D. H., Hamaoka, T., and Benacerraf, B.,** Cell interactions between histocompatible T and B lymphocytes. II. Failure of physiologic cooperative interactions between T and B lymphocytes from allogeneic donor strains in humoral response to hapten-protein conjugates, *J. Exp. Med.,* 137, 1405, 1973.
82. **Katz, D. H., Hamaoka, T., Dorf, M. E., and Benacerraf, B.,** Cell interactions between histocompatible T and B lymphocytes. III. Demonstration that the H-2 gene complex determines successful physiologic lymphocyte interactions, *Proc. Natl. Acad. Sci. U.S.A.,* 70, 2624, 1973.
83. **Mishell, R. L. and Dutton, R. W.,** Immunization of dissociated spleen cell cultures from normal mice, *J. Exp. Med.,* 126, 423, 1967.
84. **Katz, D. H., Graves, M., Dorf, M. E., Dimuzio, H., and Benacerraf, B.,** Cell interactions between histoincompatible T and B lymphocytes. VII. Cooperative responses between lymphocytes are controlled by genes in the I region of the H-2 complex, *J. Exp. Med.,* 141, 263, 1975.
85. **Katz, D. H. and Benacerraf, B.,** Hypothesis. The function and interrelationships of T cell receptors, Ir genes and other histocompatibility gene products, *Transplant Rev.,* 22, 135, 1974.
86. **Unanue, E. R. and Askonas, B. A.,** The immune response of mice to antigen in macrophages, *Immunology,* 15, 287, 1968.
87. **Sell, S. and Gell, P. G.,** Studies on rabbit lymphocytes *in vitro*. I. Stimulation of blast transformation with an anti-allotype serum, *J. Exp. Med.,* 122, 423, 1965.
88. **Naor, D. and Sulitzeanu, D.,** Binding of radioiodinated bovine serum albumin to mouse spleen cells, *Nature,* 214, 687, 1967.
89. **Pernis, B., Forni, L., and Amante, L.,** Immunoglobulin spots on the surface of rabbit lymphocytes, *J. Exp. Med.,* 132, 1001, 1970.
90. **Davie, J. M. and Paul, W. E.,** Receptors on immunocompetent cells. II. Specificity and nature of receptors on dinitrophenylated guinea pig albumin-^{125}-binding lymphocytes of normal guinea pigs, *J. Exp. Med.,* 134, 495, 1971.
91. **Hammerling, U. and Rajewsky, K.,** Evidence for surface associated immunoglobulin on T and B lymphocytes, *Eur. J. Immunol.,* 1, 447, 1971.
92. **Marchalonis, J. J., Cone, R. E., and Atwell, J. L.,** Isolation and partial characterization of lymphocyte surface immunoglobulin, *J. Exp. Med.,* 135, 956, 1972.
93. **Vitetta, E. S., Bianco, C., Nussenzweig, V., and Uhr, J. W.,** Cell surface immunoglobulin. IV. Distribution among thymocytes, bone marrow cells, and their derived populations, *J. Exp. Med.,* 136, 81, 1972.
94. **Waldron, J. A., Horn, R. G., and Rosenthal, A. S.,** Antigen induced proliferation of guinea pig lymphocytes *in vitro*: obligatory role of macrophages in the recognition of antigen by immune T-lymphocytes, *J. Immunol.,* 111, 58, 1973.
95. **Rosenthal, A. S. and Shevach, E. M.,** Function of macrophages in antigen recognition by guinea pig T lymphocytes. I. Requirement for histocompatible macrophages and lymphocytes, *J. Exp. Med.,* 138, 1194, 1973.
96. **Chesnut, R. W. and Grey, H. M.,** Studies on the capacity of B-cells to serve as antigen-presenting cells, *J. Immunol.,* 126, 1075, 1981.
97. **Lanzavecchia, A.,** Antigen-specific interaction between T- and B-cells, *Nature,* 314, 537, 1985.
98. **Lederberg, J.,** Genes and antibodies, *Science,* 129, 1649, 1959.
99. **Triplett, E. L.,** On the mechanism of immunologic self recognition, *J. Immunol.,* 89, 505, 1962.
100. **Boyse, E. A., Bressler, E., Iritani, C. A., and Lardis, M.,** Cytotoxic M autoantibodies in mouse alloantisera, *Transplantation,* 9, 339, 1970.
101. **Forsdyke, D. R.,** The liquid scintillation counter as an analogy for the distinction between "self" and "not-self" in immunological systems, *Lancet,* 1, 281, 1968.
102. **Bretscher, P. A. and Cohn, M.,** Minimal model for the mechanism of antibody induction and paralysis by antigen, *Nature (London),* 220, 444, 1968.
103. **Bretscher, P. A. and Cohn, M.,** A theory of self-nonself discrimination, *Science,* 169, 1042, 1970.
104. **McCullagh, P. J.,** The abrogation of sheep erythrocyte tolerance in rats by means of the transfer of allogeneic lymphocytes, *J. Exp. Med.,* 132, 916, 1970.
105. **Katz, D. H., Davie, J. M., Paul, W. E., and Benacerraf, B.,** Experimental conditions for the induction of hapten specific tolerance or the stimulation of anti-hapten amnestic responses by non-immunogenic polypeptide conjugates, *J. Exp. Med.,* 134, 201, 1971.

106. **Katz, D. H., Paul, W. E., and Benacerraf, B.,** Carrier function in anti-hapten antibody responses. V. Analysis of cellular events in the enhancement of antibody responses by the 'allogeneic effect' in DNP-OVA-primed guinea pigs challenged with a heterologous DNP-conjugate, *J. Immunol.,* 107, 1319, 1971.
107. **Chiller, J. M., Habicht, G. S., and Weigle, W. O.,** Cellular sites of immunologic unresponsiveness, *Proc. Natl. Acad. Sci. U.S.A.,* 65, 551, 1970.
108. **Chiller, J. M., Habicht, G. S., and Weigle, W. O.,** Kinetic differences in unresponsiveness of thymus and bone marrow cells, *Science,* 171, 813, 1971.
109. **Mitchell, G., Lafleur, L., and Andersson, K.,** Evidence for readily induced tolerance to heterologous erythrocytes in nude mice, *Scand. J. Immunol.,* 3, 39, 1974.
110. **Schrader, J. W.,** The mechanism of bone marrow (B) derived activation. II. A 'second signal' for antigen-specific activation provided by flagellin and lipopolysaccharide, *Eur. J. Immunol.,* 4, 20, 1974.
111. **Andersson, J., Sjoberg, O., and Möller, G.,** Induction of immunoglobulin and antibody synthesis *in vitro* by lipopolysaccharide, *Eur. J. Immunol.,* 2, 349, 1972.
112. **Watson, J. and Riblet, R.,** Genetical control of responses to bacterial lipopolysaccharide in mice. I. Evidence for a single gene that influences mitogenic and immunogenic responses to LPS, *J. Exp. Med.,* 140, 1147, 1974.
113. **Coutinho, A., Gronowicz, E., and Sultzer, B.,** Genetic control of B cell responses: selective unresponsiveness to lipopolysaccharide, *Scand. J. Immunol.,* 4, 138, 1975.
114. **Coutinho, A. and Gronowicz, E.,** Genetical control of B cell responses. III. Requirement for functional mitogenicity of the antigen in thymus-independent specific responses, *J. Exp. Med.,* 141, 753, 1975.
115. **Skidmore, B. J., Chiller, J. M., Morrison, D. C., and Weigle, W. O.,** Immunological properties of bacterial lipopolysaccharide (LPS). Correlation between the mitogenic adjuvant and immunogenic activities, *J. Immunol.,* 114, 770, 1975.
116. **Coutinho, A. and Möller, G.,** Immune activation of B cells: evidence for one nonspecific triggering signal not delivered by the Ig receptors, *Scand. J. Immunol.,* 3, 133, 1974.
117. **Coutinho, A. and Möller, G.,** Thumus independent B cell induction and paralysis, *Adv. Immunol.,* 21, 113, 1975.
118. **Möller, G.,** One non-specific signal triggers B lymphocytes, *Transplant. Rev.,* 23, 126, 1975.

Chapter 2

THE SURFACE IMMUNOGLOBULIN RECEPTOR

E. Charles Snow

TABLE OF CONTENTS

I. DISCOVERY OF THE B CELL'S RECEPTOR FOR ANTIGEN

The concept that lymphocytes express cell surface receptors for antigen was originally introduced by Ehrlich as an integral component of his 'nutrient side-chain' theory.[1] This hypothesis attempted to explain the ability of vertebrates to defend themselves against bacterial toxins, and was not perceived to be a viable alternative once the size of the antibody repertoire became apparent. However, Burnet incorporated the idea of lymphocytes expressing cell surface associated side chains into the clonal selection hypothesis, which was his extension of the natural selection hypothesis first postulated by Jerne.[2,3] The clonal selection hypothesis proposed that lymphocytes express antibody upon their surface membranes so that antigen might select for the expansion and differentiation of only those lymphocytes bearing antibody molecules complementary to the antigen. This hypothesis provided the theoretical framework for much of the subsequent experimentation into the humoral immune response.

While studying cell surface expressed major histocompatibility complex (MHC) proteins on mouse lymphocytes by indirect immunofluorescence, Möller noticed a consistent level of background staining when he utilized anti-mouse immunoglobulin reagents as the secondary antibody.[4] This suggested that some of the cells were expressing surface immunoglobulins. That such molecules not only existed upon lymphocyte surface membranes, but also exerted control over lymphocyte behavior, was shown experimentally in a number of different ways. The *in vitro* induction of antibody formation by the addition of antigen to lymphocytes recovered from antigen-primed animals was inhibited in the presence of anti-immunoglobulin antibodies.[5-7] Therefore, antibodies which bound to immunoglobulin molecules interfered with the ability of specific antigen to stimulated an antibody response, and this observation indirectly supported the notion that at least some lymphocytes employed immunoglobulin-like molecules as their receptors for antigen. A more direct demonstration came from the work of Gell and Sell who demonstrated that antibodies specific for allotypic determinants could induce lymphocyte blast transformation.[8,9] Their observation was replicated several years later with lymphocytes isolated from both humans and chickens.[10-12] These experimental results from lymphocytes isolated from three different species implicated an immunoglobulin-like molecule being functionally involved during the induction of lymphocyte proliferation. This conclusion was supported by immunofluorescent studies showing that some lymphocytes expressed cell surface associated immunoglobulin, referred to by this time as surface immunoglobulin (sIg) receptors.[13,14] The lymphocytes expressing such receptors were eventually identified as the bone marrow-derived or B lymphocyte which served as the precursor cell to the antibody secreting plasma cell.[15] Finally, radiolabeled antigen was shown to specifically bind to the surfaces of some lymphocytes.[16,17] Importantly, the binding of the antigen could be blocked by anti-immunoglobulin antisera, and the treatment of lymphocyte populations with radiolabeled antigen was shown to deplete all of the antigen reactive cells.[16,18] It was clear from these studies that B cells employed cell surface localized immunoglobulin as their specific receptors for antigen.

The next major question centered upon which immunoglobulin isotype functioned in this role as an antigen-binding receptor. Immunofluorescent analysis of human peripheral blood lymphocytes revealed the striking observation that most B cells expressed both IgM and IgD upon their surfaces.[19-23] This was also seen when murine lymphocytes were examined by immunofluorescent staining as well as by metabolic labeling and immunochemical analysis.[24-27] Both the IgM and IgD receptors present upon a single cell expressed the same antigen-binding capability, indicating that both molecules displayed the same variable regions.[28-31] During the ontogeny of the B cell, IgM appeared first upon the surface membrane, followed at a later time by the co-expression of the IgD receptor.[32-34] The observation that mature, immunocompetent B cells found localized within secondary lymphoid tissues expressed more sIgD than sIgM receptors lead to the suggestion that the receptor role of the two isotypes may vary.[35] The demonstrations that the sIgD receptors were more sensitive to proteolysis, and that the delta heavy chain possessed a greater degree of segmental flexibility due to its extended hinge region supported this notion.[27,36,37] Except in immature lymphocytes, both forms of the antigen receptor on mature B cells were capable of transmitting signals into resting cells, and the structural reason for this has been localized to the transmembrane region of the receptors.[38-43] Following the activation of resting, mature B cells, the level of detectable cell surface IgD decreased by about 80% and this was mediated, at least in part, by a block in transcription of the delta gene.[44-47]

Following the realization that B cells were able to simultaneously express two unrelated isotypes bearing the same antigen-binding specificity forced immunologists of the mid-1970s to focus their attention upon the regulatory mechanisms operative during the expression of this interesting gene locus. Many laboratories, for example, sought to understand how a B cell could express the same protein as both a cell surface receptor and a secretory product. The first clue came from the observation that membrane and secretory IgM differed in their buoyant densities and that membrane IgM exhibited a molecular weight shift in the presence of detergent.[48-50] These results led Melcher et al. to predict that sIgM was an integral membrane protein.[49] Their conclusion was further supported by the identification of three species of IgM heavy chains following the metabolic labeling of B cell tumors, by the carboxypeptidase-A release of residues from sIgM heavy chain not found in the C-terminus of the secreted heavy chain, and by the identification of a peptide difference between the tryptic digest of the membrane and secreted forms of IgM.[50-55] These data are consistent with the presence of a hydrophobic piece attached to the C-terminus of the cell surface expressed u heavy chain. This hypothesis was verified by comparing protein and nucleic acid sequences of the C-terminus from secretory and membrane u heavy chains.[56-59] An identical picture has emerged from the analysis of membrane-associated IgD.[60-62] The comparison of the sequences determined for sIgM, sIgD, and sIgG revealed the following information. All three receptors have a 12-25 amino acid acidic region abutting the external surface of the plasma membrane, and a highly conserved 26 amino acid transmembrane domain. All three receptors exhibit relatively short intracellular domains which are not particularly suited for signal delivery. How this shortcoming was overcome by this class of receptor will be discussed in a later section of this chapter. The expression of membrane vs. secretory form of immunoglobulins is regulated by differential splicing of the same primary RNA transcript generating two mRNAs with different 3′ ends.[56-58] This process of differential splicing either results in the attachment of a relatively hydrophilic carboxy terminus coded for by a secretory exon found contiguous to the final heavy chain exon, or in the attachment of a relatively hydrophobic carboxy terminus coded for by one of two membrane (M) exons located downstream from the final heavy chain exon.[57,58] The regulation of this interesting biological process was originally thought to be due to the choice of polyadenylation site usage.[63-64] More recent studies have suggested that the differential splicing of the immunoglobulin heavy chain message is actually more complicated.[65-68] One possibility is that the termination site found downstream from the first M exon may become altered during the transcription of this locus. A second possibility is that the splicing event may be regulated by the differential usage of particular polyadenylation and/or splice sites due to transient changes in the efficiency or accessibility of these sites. Likewise, the ability of the B cells to co-express two different isotypes bearing the same variable region has been shown to be regulated by the posttranscriptional modification of one primary RNA transcript.[68-72]

As I mentioned above, the membrane spanning domain of sIg molecules is only modestly hydrophobic because just under half of the 26 amino acids comprising this region are polar amino acids.[73] In fact, the first detectable biosynthetic form of the sIgM heavy chain actually partitions with the aqueous phase when B cells are disrupted in the presence of a detergent.[74] Pillai and Baltimore went on to show that the myristoylation of this protein converts it into a detergent soluble protein. Since both secretory IgM and immunoglobulin light chains are also myristoylated, the addition of acyl groups onto the protein seems not to be the sole reason for this rather important conversion. Pillai and Baltimore suggested that the acylation of the sIgM heavy chain might alter the folding pattern of the protein, revealing a more hydrophobic nature to the transmembrane region. This interesting proposal will need to be confirmed by a more detailed physical-chemical analysis of the various biosynthetic forms of the sIgM receptor.

II. THE BIOLOGICAL ROLE OF sIg

The only uncontested role for the sIg receptors resides in their binding of specific antigen. What happens following this remains a controversial topic. About the time that immunologists were identifying immunoglobulin molecules as the B cell receptor for antigen, Bretscher and Cohn proposed their associative recognition model for explaining how antigens stimulated an antibody response.[75] The essence of this hypothesis is that at least two signals must be delivered to B cells for them to become activated, with the first signal being delivered to the cell as a consequence of sIg occupancy. The delivery of only signal 1 to the resting cells would inactivate them. Bretscher and Cohn developed this hypothesis to provide an explanation for self-nonself discrimination, but this theory also provided the first clear-cut statement concerning the consequences of antigen binding onto sIg receptors. Bretscher and Cohn directly implicated signaling through these receptors as an integral component of the antigen-induced activation cascade for B cells. Several years later, Coutinho and Moller countered this with their one signal hypothesis.[76] Their theory was based upon results obtained following the stimulation of B cells with lipopolysaccharide (LPS). This polyclonal stimulant had been shown to effectively stimulate B cell expansion and differentiation in the absence of sIg involvement.[77] The one signal hypothesis stated that sIg receptors functioned only to focus either a mitogenic component of antigen or T cell-derived help onto the appropriate antigen-binding cell. The receptor itself did not participate during the initiation of the activation cascade. Work over the preceeding years have sought to discriminate between these two diametrically opposing hypotheses.

The first reported effect of occupying the sIg receptors with anti-Ig reagents was the capping of the receptors to one pole of the cell.[78] This process was shown to be temperature sensitive and inhibitable by chemical agents disruptive to cytoskeletal structures. Taylor et al. cited unpublished data showing that the binding of fluorescently labeled antigen to lymphocytes also resulted in the capping of the sIg receptors if the antigen-binding cells were cultured at 37°C. Both the anti-Ig reagents and the antigen utilized in these capping studies needed to be at least divalent. When sIg capped cells were analyzed by electron microscopy, the cap always segregated to the pole of the cell containing the Golgi apparatus and most of the cellular organelles, and frequently all of these structures were associated with a uropod-like appendage.[79] In addition, pinocytotic vesicles were found underneath the cap, suggesting that the material bound onto the sIg receptors was being actively transported into the cell. The receptors were eventually returned to the surface membranes, either due to the recycling or resynthesis of the receptors.[79,80] Later studies confirmed Raff and de Petris' assertion that the binding of specific antigen to the sIg receptors caused capping of the receptors. Also, Goding and Layton directly demonstrated that antigen binding resulted in the capping of both the sIgM and sIgD receptors.[81-83] These experimental results supported the emerging concept of the fluid nature of the plasma membrane, and predicted that the cross-linking of the sIg receptors promoted the physical association of the receptors with cytoskeletal components which then mediated this redistribution.[84-86] This latter prediction was verified by the demonstration by Braun et al. that cross-linked sIg receptors became associated with the detergent-insoluble cytoskeletal matrix.[87,88] This occurred following the linkage of the bound receptors to α-actinin with the subsequent involvement of actin, myosin, fodrin, and vimentin.[89-91] Recently, Albrecht and Noelle have suggested that a subpopulation of the sIgM receptors are permanently attached to the cytoskeleton (cyto IgM) and the cross-linking of the free floating receptors (both IgM and IgD) to the cyto IgM by antigen serves as one crucial feature of the capping process.[92,93] The role that capping plays during the activation process remains unanswered. Capping of occupied receptors may simply be an efficient mechanism by which the B cell protects itself from blockade of its antigen-specific receptors. On the other hand, during the process of initiating capping, or during the capping process itself, the receptors may assume the configuration required for the engagement of the signaling machinery.

Even prior to the formal identification of surface bound immunoglobulin as the antigen-sepcific receptor on B cells, Gell and Sell and others had shown that the addition of anti-Ig reagents to rabbit, chicken or human lymphocytes in culture induced some of the cells to undergo blast transformation.[9-12] This exciting finding was tempered by others demonstrating that such reagents appeared to block the proliferation of activated lymphocytes.[5-7] Also, many workers found that the anti-Ig stimulation of mouse lymphocytes *in vitro* inhibited, rather than promoted B cell proliferation.[94] This controversy widened the gap between those supporting the one signal hypothesis vs. those supporting the two signal hypothesis. The difficulty with stimulating murine B cells with anti-Ig reagents turned out to be due to anti-Ig antibodies simultaneously binding sIg and Fc receptors. The covalent attachment of the anti-Ig antibodies onto polyacrylamide beads converted nonstimulatory antibodies into ones which effectively induced murine B cell proliferation.[95] It was also reported that soluble rabbit anti-mouse Ig was more effective at stimulating murine B cells recovered from old mice. This was later determined to be due to a gradual loss in the ability of Fc receptor occupancy to down-regulate B cell proliferation with advancing age.[96,97] Goat antibodies bind less avidly than rabbit immunoglobulins to murine Fc receptors and were found to function efficiently as stimulants of murine B cell expansion when added to the cultures as soluble antibody.[98] Rabbit antibodies could be utilized in soluble form following the enzymatic removal of the Fc portion.[99] In agreement with the studies on sIg capping, anti-Ig-mediated B cell activation required the cross-linking of the receptors.[100-102]

Not all B cells respond equally well to the growth promoting actions of anti-Ig preparations. Initially, it was shown that B cells isolated from young mice (less than 6 weeks of age) responded poorly when challenged *in vitro* with anti-Ig.[103] This observation was extended through the demonstration that anti-Ig stimulated mainly B cells expressing high levels of sIgD and low levels of sIgM, a phenotype reflective of mature B cells.[104-107] Furthermore, although virtually all the B cells enlarge within hours of being stimulated by anti-Ig, only 30 to 60% of the cells actually enter the cell cycle.[108] The reason(s) for this differential reactivity of mature B cells to anti-Ig stimulation is not clear at this time. This may reflect differences in the activation or developmental stage of the individual B cells.[109]

It has been apparent for some time that the polyclonal stimulation of B cells with anti-Ig does not induce an increase in the amount of immunoglobulin secreted by the activated B cells.[95,110,111] This, of course, was in direct contrast with what had been seen following the stimulation of B cells with LPS.[77] These results were reminiscent of a model proposed in 1971 by Dutton and colleagues.[112] The model was compatible with the two-signal hypothesis and predicted that the activation of both B cell expansion and differentiation required not only occupancy of the sIg receptor, but also the involvement of nonspecific factors released from T cells. The first strong experimental evidence for this model was provided by the results from Kishimoto and Ishizaka.[113] These investigators demonstrated that hapten primed B cells differentiated into antibody forming cells *in vitro* when challenged by anti-

Ig covalently attached to a carrier protein as long as carrier-specific T cells were included within the cultures. Importantly, they went on to show that normal B cells were also induced to proliferate and differentiate in this experimental system. They concluded that the complete activation of resting B cells required both a signal through the sIg receptor along with some sort of help from an activated helper T cell. This observation was subsequently confirmed by numerous demonstrations that the addition of T cell-derived factors to cultures of anti-Ig stimulated B cells enhanced the level of B cell expansion, and induced B cell differentiation into antibody forming cells.[99,114-116] This important and interesting topic will be addressed in detail in other chapters in this book.

Many of these early studies supported the supposition that not only did the sIg receptors function in the selection of the appropriate B cell clones to respond, but also they seemed to serve an important role during the process of signaling resting cells to enter the early stages of the cell cycle. One could predict, therefore, that occupancy of these receptors delivered the major growth stimulus to the B cells. Such a stimulus can be defined as one which drives resting cells from G_0 into G_1, and promotes the acquisition of responsiveness of the cells to progression factors. In this case, the progression factors are represented by the soluble mediators released by activated T cells and macrophages. For sIg to deliver the major growth stimulus, it's occupancy must induce cellular changes commensurate with cells entering the cell cycle.

III. BIOCHEMICAL CONSEQUENCES FOLLOWING ANTI-IG OCCUPANCY OF sIg RECEPTORS

Braun et al. were the first to provide direct evidence for cross-linkage of sIg mediating important cellular biochemical changes.[117] The purpose for their study was to determine if receptor-mediated changes in cellular calcium might be associated with the capping phenomena. They found that the stimulation of $^{45}Ca^{++}$ labeled B cells with anti-Ig released approximately 30% of the label within two minutes of adding the stimulant. They went on to make the important observation that this calcium was mobilized from internal stores of the ion. These observations were verified and improved upon through the use of a fluoresceinated, highly selective calcium indicator known as quin2.[118] Quin2 loaded lymphocytes were monitored with a spectrofluorimeter during the time they were exposed to anti-Ig. This technique allowed the discrimination between internal and external movement of calcium, and the results revealed that the calcium movements reported by Braun et al. could be divided into two phases. The first phase consisted of calcium being released from a non-mitochondrial, intracellular source and accounted for only a small amount of the total level of calcium mobilized. The second phase could be accounted for in terms of transport into the cell of environmental calcium. The transport of external calcium is regulated in some manner by the amount of internally mobilized calcium.[119-122] For example, if the availability of internal calcium is compromised prior to signaling through sIg, the stimulated cells do not efficiently transport externally available calcium. Monroe and Cambier demonstrated that anti-Ig stimulation induced a state of membrane depolarization which peaked at around one hour post-stimulation.[123] They suggested that this depolarization might contribute to the second phase of the calcium response based upon results from other systems showing that membrane depolarization facilitates the opening of calcium channels.[124] More recent studies concluded that sIg occupancy in fact induces a state of hyperpolarization most probably due to K^+ channels being opened following the elevation of cytosolic levels of calcium.[125,126] To understand the details of how this process is regulated, I need to first discuss the biochemical connection between receptor occupancy and calcium mobilization.

The increased intracellular levels of calcium induced following the binding of ligand by various membrane receptors has been linked with changes in the turnover of membrane phosphoinositides.[127] The first evidence for this biochemical pathway being operative in lymphocytes was provided over 20 years ago when Fisher and Mueller demonstrated that the polyclonal stimulation of T cells resulted in the incorporation of ^{32}P into membrane-associated phosphatidylinositol (PI).[128] Maino et al. went on to confirm that the turnover of PI also occurred following the anti-Ig stimulation of B cells.[129] Interest in PI turnover was rejuvenated following the demonstration that the two breakdown products of PI turnover functioned as cellular second messengers.[130-134] One of the products was inositol 1,4,5-triphosphate (IP_3) which participated in the cellular changes in calcium always seen following ligand binding to these types of receptors. The second product was diacylglycerol (DAG) and it was shown to help up-regulate the activity of an interesting membrane-associated kinase known as protein kinase C (PKC). The experimental results of Cambier and colleagues refocused attention back to the original observation made by Maino et al.[129,135-137] They were able to mimic phenotypic changes known to be induced following the occupancy of sIg (such as hyper-Ia expression) by artificially stimulating PKC by supplying the cells with phorbol diesters or analogs of DAG. Also, anti-Ig treatment was found to induce the turnover of only PI and did not promote the *de novo* synthesis of phospholipids in general during the first several hours following the addition of the stimulus. Subsequently, anti-Ig treated cells were shown chemically to release both IP_3 and DAG, confirming that sIg occupancy signaled the cell through the release of these two second messengers.[138,139]

The sIg-mediated PI turnover in resting B cells suggested that IP_3 mediated the release of calcium from intracellular stores and accounted for the first phase of the receptor-mediated calcium response. The results of Ransom et al. supported this conclusion by demonstrating that the addition of IP_3 to lightly saponized B cells induced the release of sequestered $^{45}Ca^{2+}$.[140] IP_3 has been shown to mediate this action by binding to specific receptors localized to certain areas of the endoplasmic reticulum, particularly those close to the nucleus, and the nuclear envelop.[141-143] These receptors have also been found on small membraneous vesicles, called calciosomes, which seem to be integral components of calcium partitioning within mammalian cells.[144] So, the two phase calcium response induced by anti-Ig stimulation of resting B cells appears to be regulated in the following manner. The rapid, early peak of calcium is released from certain areas of the endoplasmic reticulum and calciosomes by the interaction of IP_3 with specific receptors. This early peak of calcium gradually diminishes, but may play some role in regulating calcium influx from the environment. The second phase is lesser in intensity but of longer duration and is mediated by a conductive calcium channel which resembles voltage-gated calcium channels found on excitable cells.[145] The regulation of these plasma membrane localized voltage-gated channels is still unclear, with current findings suggesting that the channels are regulated either by intracellular concentrations of calcium, or through the actions of IP_3.[119-122,145-147]

As I mentioned above, DAG is also released during the active turnover of PI and has been shown to serve as a cellular second messenger participating in the activation and membrane localization of PKC.[131-134] The important role that PKC plays during the process of B cell activation was initially shown through phorbol diester stimulation or by inhibiting the enzyme prior to stimulating the cells polyclonally.[148,149] The stimulation of B cells with the phorbol esters was shown to induce some of the phenotypic changes seen when B cells were stimulated with anti-Ig reagents.[135,136,139] Chen et al. went on to formally show that anti-Ig stimulation induced the translocation of PKC from the cytosol into the plasma membrane.[150]

Activated and membrane-associated PKC influences cellular physiology by phosphorylating, and thus altering the functional capability of a variety of membrane and cytosolic proteins on seryl and threonyl residues.[133,150] It should be kept in mind that the consequence of an active PKC depends on the developmental and activational stage of the cell.[151-153] During the process of B cell activation, numerous substrates for PKC have been identified and the physiologic significance of phosphorylating them is still under intense investigation.[154-157] For example, PKC has been shown to phosphorylate α-actinin, thus engaging the association of this important cytoskeletal component with membrane-associated proteins.[155,158] As discussed earlier, α-actinin has been shown to directly interact with sIg molecules, and the activation of PKC can at least be suggested to be functionally linked to the initiation of the capping process.[89]

Considerable experimental evidence has shown that activated PKC participates during the up-regulation of numerous mammalian genes.[159,160] An interesting example of this resides with the ability of an induced PKC to alter the transcription rate for the Ig heavy chain locus.[161] This study by Hogbom et al. linked PKC to the process by which the transcription of sIgD decreases early during the activation of B cells through the sIg receptor.[44-47] Also, PKC has been implicated in the up-regulation of a number of immediate/early genes, many of which code for proteins involved in transcriptional regulation.[159-167] Signaling through the sIg receptors results in the appearance of a host of proteins which serve active roles in regulating many of the later events required for the activated cells to traverse the cell cycle. Two of the best characterized immediate/early genes are c-*myc* and c-*fos*. Both of these genes are induced either following anti-Ig or phorbol diester and ionomycin stimulation of resting B cells.[168-171] One of the cellular actions of PKC appears to function as the connection between the receptor-mediated PI turnover and the enhanced transcription of these protooncogenes.[172-174] An important consideration, which I will return to later in this chapter, is that even though these genes are expressed early in the activation cascade, their expression does not necessarily reflect the acquisition of competence of a cell to enter the cell cycle.[170-172] In addition, these same genes are transcribed in additional stages of the cell cycle requiring the investigator to fully understand the cell cycle positioning of the cells under consideration.

IV. PHENOTYPIC CHANGES IN B CELLS STIMULATED THROUGH THE sIg RECEPTOR

The sIg receptor-mediated induction of the PI cycle, along with the release into the B cells of IP_3 and DAG, results in important phenotypic changes within the cell. In this section I will discuss what is known about these cellular changes, but will wait until later in the chapter to suggest how such changes are involved during the early stages of B cell activation.

B cells signaled through sIg display an increased expression of cell surface MHC class II proteins.[175] Subsequent studies have implicated a role for both the increased levels of calcium and the activated PKC in the link between the receptor occupancy and the increased expression of these surface proteins.[135,136,176,177] This phenotypic change is potentially quite important to B cells which have encountered antigen because it allows the antigen binding cells (ABC) to more efficiently interact with the helper T cell. Experimentally this can be shown by the enhanced ability of such B cells to present antigen to T cell lines and clones.[178-180] Such B-T cell interactions play an intimate role in the successful thymus-dependent induction of B cell activation and will be covered in detail in several chapters of this book.

The cross-linking of sIg has also been reported to induce the desialylation of B cell surface proteins.[181] Based on these results, Kearse et al. suggested that MHC class II proteins were desialylated by this process, thus promoting the direct cellular contact between antigen-binding B cells and T cells. This result was reminiscent of the findings from Frohman and Cowings that the glycosylation state of the class II molecules was important in determining how well cells interacted.[182] However, Kreiger et al. have clearly shown that the presence or absence of sialic acid residues on class II proteins does not influence the capacity of the proteins to present antigen to T cells when incorporated into artificial planar membranes.[183] So the enhancement of B cell antigen presentation following the occupancy of sIg receptors appears to be mediated at some level other than physically altering the class II proteins.

Provocative findings from Dustin and Springer have demonstrated that the affinity of interaction between cell surface adhesion molecules was modulated following the stimulation of T cells through their receptor for antigen.[184] The adhesion molecules were transiently phosphorylated by an as yet undetermined kinase. This study clearly opens the possibility that similar regulatory mechanisms may be operative during the early stages of B cell activation. Such a mechanism, possibly implemented following the occupancy of sIg, would prepare the B cell to form the tight contacts with helper T cells required to fully induce a thymus-dependent antibody response.

V. THE COUPLING OF sIg TO THE SIGNALING PATHWAY

The sIg receptor is apparently devoid of any intrinsic signaling capability. The 26 amino acid transmembrane domains of sIgM and sIgD are approximately 60% homologous, and both receptors have identical three amino acid (Lys-Val-Lys) cytoplasmic tails.[57,62,185] Cambier and colleagues have been suggesting for some time that sIg-mediated hydrolysis resembles that seen following the occupancy of the α- and β-adrenergic receptors.[186] The adrenergicreceptors are coupled to increased PI turnover through the actions of guanine nucleotide regulatory (G) proteins (for review see References 187 and 188). Basically, once occupied by ligand, these receptors interact directly with a G protein probably utilizing one of the conserved regions within the transmembrane domain shared by all G proteins.[189-191] Prior to this interaction, the G protein is associated with GDP and, thus, in an inactive state. Following the interaction with the receptor, the G protein dissociates from GDP and then associates with GTP. The GTP-bound protein represents the activated form of the G protein. The activated form of the G protein directly modifies the activity of select membrane-associated enzymes, such as phospholipase C (PLC). Once activated, PLC cleaves PI into DAG and IP_3. G proteins possess intrinsic GTPase activity which eventually converts the GTP to GDP, and in the process inactivates the G protein itself. There are a family of G proteins functioning to couple receptors to PLC, with each individual G protein selectively coupling only one type of receptor.[192] Just about all the members of this family of G proteins (referred to as G_p) are ribosylated, and thus inactivated, by pertussis toxin, but vary in their abilities to potentiate PI turnover following receptor coupling. Obviously, much remains to be learned concerning the biological properties of these interesting proteins.

Mizuguchi et al. indirectly suggested a role for G proteins during the signaling cascade induced through sIg by showing that some stages of B cell activation were sensitive to the actions of NaF, an inhibitor of some G proteins.[193] Gold et al. utilized lightly saponized WEHI-231 cells to show that the inclusion of a nonhydrolyzable GTP analog during the anti-Ig stimulation of B cells greatly enhanced the level of PI turnover. Harnett and Klaus confirmed this observation by using permeabilized normal murine B cells and both anti-IgM and anti-IgD as the co-stimulant.[195] Both of these groups reported that the sIg associated G protein was insensitive to the actions of pertussis toxin. Monroe and Haldar have recently extended these initial findings by demonstrating that the anti-Ig stimulation of B cell membrane preparations increased the level of GTP binding by the membranes.[196] They also demonstrated that the pretreatment of the membranes with pertussis toxin uncoupled the sIg receptors from the signaling pathway leading to increase PI turnover. This discrepancy with the early two studies may reflect an inherent difficulty in effectively neutralizing G proteins in some whole cell systems.[197] I feel that the main conclusion to be drawn from these studies is that B cells couple their sIg receptors to the signaling pathway controlling PLC via a pertussis toxin-sensitive G protein.

The main glitch to this entire scenario rests in determining how the sIg receptors interact with the G proteins. Receptors which are coupled to PLC regulation via a pertussis toxin-sensitive G protein characteristically span the membrane seven times and have three intracellular loops. It is within some of the transmembrane and intracellular regions that investigators have localized conserved regions which might serve as contact points with the G proteins following the occupancy of the receptor.[188-191,198] The transmembrane region and cytoplasmic tail of sIg molecules do not possess the appropriate consensus sequences to allow the receptors to directly associate with G proteins. Therefore, sIg molecules must utilize intermediary proteins to couple with and activate G proteins. There is a precedence for coupling a receptor in this manner. The T cells receptor for antigen (TCR), which also lacks these conserved regions, has been shown to couple to the appropriate G proteins through a complex of receptor-associated proteins collectively referred to as the CD3 complex.[199-201]

It turned out not to be a simple matter to identify CD3-like proteins associated with sIg receptors. Over the years there have been several reports showing the possible association of proteins with sIg molecules. Sidman et al. developed an experimental system designed to identify integral membrane proteins displayed on the plasma membranes of B cells.[202] They found a 28 kDa integral membrane protein which routinely immunoprecipitated along with sIg. Rosenspire and Choi reported that a 34 kDa cytosolic protein and a 55 kDa integral membrane protein were both attached to sIg.[203] Koch and Haustein did a 'close-neighbor' analysis by cross-linking iodinated B cell surface proteins with a bifunctional crosslinker.[204] They found two proteins of 46 and 56 kDa which associated exclusively with sIgM. Haustein and Von der Ahe were able to identify a 30 kDa protein apparently disulfide linked to sIgM on the surfaces of both normal and neoplastic B cells.[205] The basic problem with all of these studies stemmed from the inability of most laboratories to reproducibly identify proteins associated with the sIg receptors. As in many such problems encountered in research, the answer to the dilemma was a rather simple one once the systems were better understood. Sakaguchi et al., in a study designed to clone genes expressed at prominent stages during B cell development, identified a gene they called *mb-1*.[206] This gene was expressed in pre-, mature, and activated B cells, but was not found in fully differentiated plasmacytomas. The deduced sequence of the MB-1 protein was consistent with MB-1 being a membrane-bound glycoprotein of 26 to 35 kDa in size exhibiting extracellular, transmembrane, and intracytoplasmic domains. Sakaguchi and colleagues concluded that MP-1 looked very much like the γ-chain of the CD3 complex. Hombach et al. reported that transfected sIgM was expressed upon the plasma membranes of lymphoma, but not plasmacytoma, cells.[207] They extended this study by isolating a sIgM-transfected plasmacytoma variant which did express sIgM.[208] Their results were consistent with the involvement of a 34 kDa (B34) glycoprotein in the successful transport of sIgM onto B cell plasma membranes. The B34 protein appeared upon the surface as a disulfide-linked homodimer and displayed a similar distribution pattern as Sakaguchi et al. reported for MP-1. The chemical identification of the CD3 complex required that the T cells be solubilized in a mild, nonionic detergent, like digitonin.[209] The use of more harsh detergents, such as the commonly used NP-40, dissociated the CD3 protein from the TCR. Hombach et al. found this was also the case in their demonstration that B34 associated noncovalently with sIgM.[208] Campbell and Cambier verified this by utilizing digitonin to identify a complex of three proteins which assembled together to form the putative CD3-like complex associated with sIg receptors.[210] Their results demonstrated that this complex consisted of three proteins (pp33, pp34, and pp37) associated with IgD and three proteins (pp32, pp34, and pp37) linked to IgM. The two isotype specific proteins (pp32 and pp33) were found to be disulfide-linked to either pp34 or pp37 and the pp32 and pp33 were separate proteins. At this time, the data is consistent with pp32 being identical to B34 and MB-1. Wienands et al. have recently identified a 35 kDa protein (B35) which associated only with sIgD and may be identical to pp33.[211] All four of these proteins are phosphorylated following sIg cross-linking, but the kinase responsible has not been determined. It is also unclear what changes in receptor function occur following the phosphorylation of these four proteins. Reth has recently pointed out that the cytoplasmic domains of all three proteins forming the CD3 complex and the pp32/B34 sIgM-associated protein share a common consensus sequence.[212] He suggested that this region of the cytoplasmic domain might form an alpha-helical barrel which binds to proteins. One possibility, of course, would be that this consensus sequence represents the site at which the G protein binds following the occupancy of either the TCR or sIg receptor. Finally, Pesando et al. have recently reported that in human B cells, the CD19 surface protein is both functionally and physically associated with sIg receptors.[213] What relationship the human CD19 protein has to any of the three murine proteins described above remains to be determined. These preliminary studies certainly indicate that we are approaching a better understanding concerning the mechanism by which the sIg receptors initiates the signaling cascade resulting ultimately in the enhanced turnover of PI.

VI. THE BIOLOGICAL RELEVANCE OF SPECIFIC ANTIGEN BINDING TO sIg RECEPTORS

It is easy to see how many have come to the conclusion that occupancy of the sIg receptor delivers the major growth stimulus to resting B cells.[214] The stimulation of B cells with anti-Ig results in the movement of the cells into the early stages of the cell cycle and promotes the responsiveness of the cells to lymphokines which represent the progression factors operative in this cellular system. However, such a point of view relegates T helper activity to simply releasing nonspecific factors upon demand. This sort of a situation would seriously reduce the ability of the T cell to regulate only those B cell clones which express sIg receptors specific for the invading antigen. In the absence of a more intimate role for T cells during a thymus-dependent antibody response, the system would be seriously hindered in its ability to distinguish self from nonself.[75]

The binding of specific antigen by the sIg receptors induces the clearance of the bound material through the capping process.[6,16-18,215] These results suggested that there must be some sort of signal delivered through the receptor following antigen occupancy to initiate the associative biochemical processes required for capping and internalization of the receptors. The results did not, however, address whether or not these receptor-induced changes participated in the delivery of the major growth stimulus to the resting B cells.

The ability to experimentally determine if ligand association with the sIg receptors delivers the major growth stimulus to resting B cells awaited the development of procedures designed to isolate populations of hapten-specific B cells.[216-219] These methodologies prepare populations of cells which are at least 60% specific for a given antigenic reactivity. The use of such populations allowed investigators to study B cells whose sIg receptors have been bound in a physiologically relevant manner. It is important to point out that the cells isolated by these various procedures were capable of responding to various challenges which would cause any B cell to proliferate and/or differentiate *in vitro*.[217-223] This is more than just a trivial point due to the somewhat heroic efforts required to isolate these antigen-binding cells. Also, since the sIg receptors have to be occupied to identify the cells to be isolated, efforts needed to be made to ensure that the cells were not activated during the selection procedures. Taking into account all of the possible variables, I feel that these systems have been useful in furthering our understanding of the B cell activation process. I base this conclusion mainly upon the reproducibility, regardless of which selection procedure utilized.

Kenny et al. provided the first evidence that the binding of specific antigen to sIg initiated some sort of a signaling mechanism.[224] They isolated sheep red blood cell (SRBC)-specific B cells from both immune and nonimmune animals and demonstrated that the ABC exhibited a significant increase in their synthesis of cellular phospholipids. This study was completed prior to the current appreciation concerning the importance of the PI cycle during receptor signaling, but did provide a biochemical basis for the existence of an antigen-induced signaling pathway. Cambier et al. demonstrated by acridine orange analysis that antigen-specific cells stimulated with thymus-dependent antigens did not exit the G_0 stage of the cell cycle.[225] They also found that the addition of lymphokines simultaneously with the antigenic challenge was still inefficient in moving resting B cells into the cell cycle. We were able to substantiate their conclusion by showing that thymus-dependent antigenic challenge of resting, antigen-specific B cells did not promote thymidine incorporation.[215] We went to great lengths to rule out trivial explanations for our inability to show the induction of B cell growth by specific antigen. In short, these cells were not activated by increasing the concentration of antigen, by increasing the amount of hapten conjugated onto the carrier protein, by attaching the hapten onto SRBC or Sepharose beads, or by purposefully cross-linking cell bound hapten-carrier conjugate with anti-hapten antibody. This latter experiment dealt with the question of low affinity receptors not being appropriately cross-linked by antigen. We were able to confirm that specific antigen cross-linked sIg receptors by showing that the ABC efficiently capped their bound antigen.[215] Alderson et al. monitored the delivery of a growth stimulus to hapten-specific B cells cultured as single cells by measuring changes in cell size.[226] Their results demonstrated that the binding of specific antigen by the hapten-specific B cells did not induce the ABC to enlarge. Finally, the stimulation of antigen-specific B cells with a variety of thymus-dependent antigens did not promote the responsiveness of the ABC to either the proliferative or the differentiative properties of the soluble mediators which regulate B cell biology.[215,227-229] The only conclusion one can make upon reviewing these results is that the binding of specific antigen to the sIg receptors displayed upon the surfaces of resting B cells does not deliver the major growth stimulus.

Several experimental observations mitigate against the sIg receptors functioning solely as focusing molecules with absolutely no physiologic role during the process of B cell activation. First, as mentioned previously, Kenny et al. have shown that the binding of SRBC to SRBC-specific B cells induced enhanced synthesis of phospholipids.[224] Second, LoCascio et al., utilizing a B cell lymphoma, bearing sIg receptors specific for SRBC, demonstrated that antigen binding to the lymphoma cells significantly increased the number of antibody-forming cells induced by the artificial coupling of B and T cells.[230] We were able to confirm this observation by following the impact of specific antigen addition to hapten-specific B cells stimulated by autoreactive T cell clones.[231] Third, we documented that the exposure of antigen-specific B cells to specific antigen enhanced the sensitivity of the B cells to lipopolysaccharide by at least a thousand fold.[215] These results suggested that antigen association with sIg altered the biological characteristics of the cell, and therefore, the receptor seemed to be initiating a signaling cascade within ABC.

Since anti-Ig stimulation of resting B cells induced a dramatic increase in the turnover of membrane-associated PI, we decided to determine if the occupancy by antigen initiated this same signaling pathway.[135-141] We first looked to see if antigenic stimulation of hapten-specific B cells induced enhanced expression of c-*myc* mRNA, since this increase has been shown by others to correlate with active PI turnover.[172,173] As seen when normal B cells were stimulated with anti-Ig, antigenic challenge of hapten-specific B cells induced elevated levels of c-*myc* mRNA within two hours.[168-170,232] We went on to formally show that antigenic occupancy of sIg induced PI turnover, a result verified by others.[233-235] These results demonstrated that consistent with the anti-Ig stimulation of B cells, antigenic challenge of hapten-specific B cells induced a robust PI cycle. Therefore, one can conclude that the sIg receptor does, in fact, function as a signal transducing element for resting B cells. However, the consequences to resting B cells of these elicited signals are not sufficient to induce cellular movement into the cell cycle. The occupancy of the sIg receptors does not deliver the major growth stimulus to resting B cells. Recent information implicates the direct cellular contact between B and T cells as delivering this growth signal to the resting B cell. This concept has been addressed in other places and will be discussed by Noelle and Bartlett in Chapter 6.[236,237]

These experimental results lead to an obvious question. Why are the biologic consequences to B cells different if the sIg receptors are occupied by specific antigen rather than by anti-Ig? Unfortunately, there aren't any adequate answers at this time. This is an important consideration because so many experimental systems utilize anti-Ig stimulation as a model for the normal, antigen-mediated activation of resting B cells. I will discuss one possible reason for the discrepancy between anti-Ig and antigen.

As I mentioned earlier, the stimulation of antigen-specific B cells with specific antigen results in the ABC becoming responsive to a 1000-fold lower concentration of the polyclonal stimulant, LPS.[215] A report by Wilder et al. helped to illuminate the relevance of our observation in terms of anti-Ig stimulation of B cells.[238] They acquired five commercial batches of anti-Ig and measured the LPS content in each. They found that the higher the inherent LPS concentration, the 'higher quality' the preparation was in terms of activating resting B cells. Some preparations contained as much as 25 ng/ml of LPS. The addition of that amount of LPS, in the absence of anti-Ig, was found to induce a significant amount of B cell proliferation, which could be enhanced by supplementing the culture medium with 2-mercaptoethanol. It has been recognized for sometime that anti-Ig reagents induce higher levels of B cell proliferation when the cells were cultured in media supplemented with 2-mercaptoethanol.[239] Finally, Wilder et al. found that the depletion of the contaminating LPS by incubating the five anti-Ig reagents with agarose beads conjugated with polymyxin B was found to change a high quality anti-Ig reagent into one not nearly as good in terms of B cell stimulatory capability.

VII. A WORKING MODEL FOR THE BIOLOGIC ROLE OF THE sIg RECEPTORS

I would suggest that the role played by the sIg receptors during the process of antigen-driven B cell activation lies somewhere between that proposed by the one- and two-signal hypotheses.[75,76] The biochemical and molecular studies have clearly demonstrated that the occupancy of sIg, by either specific antigen or by anti-Ig, induces the turnover of membrane PI and the release into the cell of second messengers. These results certainly refute the notion that the sIg receptors serve only to focus antigen onto the correct clones of reactive B cells. On the other hand, the two signal hypothesis indirectly implied that the occupancy of the antigen binding receptor delivered to the B cell an important growth promoting signal. In other words, the hypothesis suggested that the occupancy of the sIg receptors induced the responsiveness of the ABC to either the T cells or their soluble factors. In my opinion, this clearly does not happen.

The occupancy of surface immunoglobulin by a thymus-dependent antigen initiates a sequence of events which results in the expression upon the cell's surface membrane of antigenic fragments in associated with the MHC class II proteins. What role, if any, the PI cycle and its by-products play during antigen processing and presentation is completely unknown. The involvement of the sIg receptor does fulfill the main tenet, however, of the one signal hypothesis, since antigen-presentation by B cells allows the direct contact of the ABC with a helper T cell which is an absolute requirement for the full activation of the ABC. I also propose that the biochemical pathways initiated following antigen binding to sIg up-regulate a series of important growth related genes called immediate/early genes. These genes code mainly for transacting factors, many of which play a prominent role during the activation of resting B cells. However, the expression of these proteins does not signify that the cell has reached competence for cell cycle progression. This implies that some of the cellular changes induced by the major growth stimulus may interface with the changes induced by signaling through sIg. Its possible, for example, that the activity of some of the sIg-induced transacting factors is enhanced following their phosphorylation by a kinase induced as a consequence of the cell receiving the major growth stimulus. In the absence of such interactions, the biochemical and molecular changes induced following sIg occupancy do not promote B cell cycle entrance, and may, in fact, function to inactivate the ABC. The simple fact that antigen-binding by sIg results in the important process of selecting the clones of B cells to become activated by the helper T cells highlights the importance of these clonotypically expressed receptors. The actual role that the biochemical changes induced through sIg play during B cell activation and/or inactivation remains a fertile area for future work.

REFERENCES

1. **Ehrlich, P.,** *Experimental Researches on Specific Therapeutics,* Lewis, London, 1908.
2. **Jerne, N. K.,** The natural-selection theory of antibody formation, *Proc. Natl. Acad. Sci. U.S.A.,* 41, 849, 1955.
3. **Burnet, F. M.,** A modification of Jerne's theory of antibody production using the concept of clonal selection, *Aust. J. Sci.,* 20, 67, 1957.
4. **Möller, G.,** Demonstration of mouse isoantigens at the cellular level by the fluorescent antibody technique, *J. Exp. Med.,* 114, 415, 1961.
5. **Dray, S.,** Effect of maternal isoantibodies on the quantitative expression of two allelic genes controlling the γ-globulin allotypic specificities, *Nature,* 195, 677, 1962.
6. **Mitchison, N. A.,** Antigen recognition responsible for the induction *in vitro* of the secondary response, *Cold Spring Harbor Symp. Quant. Biol.,* 32, 431, 1967.
7. **Fuji, H. and Jerne, N. K.,** Primary immune response *in vitro:* reversible suppression by anti globulin antibodies, *Ann. Inst. Pasteur,* 117, 801, 1969.
8. **Sell, S. and Gell, P. G. H.,** Studies on rabbit lymphocytes *in vitro.* I. Stimulation of blast transformation with an antiallotype serum, *J. Exp. Med.,* 122, 423, 1965.
9. **Gell, P. G. H. and Sell, S.,** Studies on rabbit lymphocytes. II. Induction of blast transformation with antisera to six Ig allotypes and summation with mixtures of antisera to different allotypes, *J. Exp. Med.,* 112, 813, 1965.
10. **King, R. A., Messner, R. P., and Williams, R. C.,** Human lymphocyte transformation induced by anti-gamma globulin factors, *Arthritis Rheum.,* 12, 597, 1969.
11. **Oppenheim, J. J., Rogentine, G. N., and Terry, W. D.,** The transformation of human lymphocytes by monkey antisera to human immunoglobulin, *Immunology,* 16, 123, 1969.
12. **Ivanyi, J., Marvanova, H., and Skamene, E.,** Immunoglobulin synthesis and lymphocyte transformation by anti-immunoglobulin sera in bursectomized chickens, *Immunology,* 17, 325, 1969.
13. **Raff, M. C., Sternberg, M., and Taylor, R. B.,** Immunoglobulin determinants on the surface of mouse lymphoid cells, *Nature,* 225, 553, 1970.
14. **Pernis, B., Forni, L., and Amante, L.,** Immunoglobulin spots on the surface of rabbit lymphocytes, *J. Exp. Med.,* 132, 1001, 1970.
15. **Kincade, P. W. and Cooper, M. D.,** Development and distribution of immunoglobulin-containing cells in the chicken, *J. Immunol.,* 106, 371, 1972.
16. **Ada, G. L. and Byrt, P.,** Specific inactivation of antigen-reactive cells with ^{125}I-labeled antigen, *Nature (London),* 222, 1291, 1969.
17. **Davie, J. M. and Paul, W. E.,** Receptors on immunocompetent cells. II. Specificity and nature of receptors on dinitrophenylated guinea pig albumin-^{125}I-binding lymphocytes of normal guinea pigs, *J. Exp. Med.,* 134, 495, 1971.
18. **Warner, N. L., Byrt, P., and Ada, G. L.,** Blocking of the lymphocyte antigen receptor site with anti-Ig sera *in vitro, Nature (London),* 226, 942, 1970.
19. **van Boxel, J. A., Paul, W. E., Terry, W. D., and Greene, I.,** IgD-bearing human lymphocytes, *J. Immunol.,* 109, 648, 1972.

20. **Knapp, W., Bolhuis, R. L., Radl, J., and Hijmans, W. J.,** Independent movement of IgD and IgM molecules on the surface of individual lymphocytes, *J. Immunol.,* 111, 1295, 1973.
21. **Rowe, D. S., Hug, K., Forni, L., and Pernis, B.,** IgD as a lymphocyte receptor, *J. Exp. Med.,* 138, 965, 1973.
22. **Fu, S. M., Winchester, R. J., and Kunkel, H. G.,** Occurrence of surface IgM, IgD and free L chains on human lymphocytes, *J. Exp. Med.,* 139, 451, 1974.
23. **Kubo, R. T., Grey, H. M., and Pirofsky, B.,** IgD: a major Ig on the surface of lymphocytes from patients with CLL, *J. Immunol.,* 112, 1952, 1974.
24. **Melcher, U., Vitetta, E. S., McWilliams, M., Lamm, M. E., Phillips-Quagliata, J., and Uhr, J. W.,** Cell surface immunoglobulin. X. Identification of an IgD-like molecule on the surface of murine splenocytes, *J. Exp. Med.,* 140, 1427, 1974.
25. **Parkhouse, R. M. E. and Abney, E. R.,** Heterogeneity of surface Igs of murine B lymphocytes, in *Membrane Receptors of Lymphocytes,* Vol. 1, Seligmann, M., et al., Eds., North-Holland, Amsterdam, 1975, 51.
26. **Lisowska-Bernstein, B. and Vassalli, P.,** The surface Igs of mouse spleen B lymphocytes: radioiodination, biosynthetic labeling and immunofluorescent studies, in *Membrane Receptors of Lymphocytes,* Vol. 1, Seligmann, M. et al., Eds., North-Holland, Amsterdam, 1975, 39.
27. **Vitetta, E. S., Cambier, J., Forman, J., Kettman, J. R., Yuan, D., and Uhr, J. W.,** Immunoglobulin receptors on murine B lymphocytes, *Cold Spring Harbor Symp. Quant. Biol.,* 41, 185, 1976.
28. **Pernis, B., Brouet, J. C., and Seligmann, M.,** IgM and IgD on the membrane of lymphoid cells in macroglobulinemia, *Eur. J. Immunol.,* 4, 776, 1974.
29. **Salsano, P., Froland, S. S., Natvig, J. B., and Michaelsen, T. E.,** Same idiotype of lymphocyte membrane IgD and IgM. Formal evidence for monoclonality of chronic lymphatic leukaemia cells, *Scand. J. Immunol.,* 3, 841, 1974.
30. **Fu, S. M., Winchester, R. J., and Kunkel, H. G.,** Similar idiotypic specificity for the membrane IgD and IgM of human lymphocytes, *J. Immunol.,* 114, 250, 1975.
31. **Stern. C. and McConnell, I.,** IgM and IgD as antigen-binding receptors on the same cell with shared specificity, *Eur. J. Immunol.,* 6, 225, 1976.
32. **Abney, E. R. and Parkhouse, R. M. E.,** Candidate for immunoglobulin D present on murine B lymphocytes, *Nature,* 252, 600, 1974.
33. **Vitetta, E. S., Melcher, U., McWilliams, M., Lamm, M. E., Philips-Quagliata, J. M., and Uhr, J. W.,** Cell surface immunoglobulin. XI. The appearance of an IgD-like molecule on murine lymphoid cells during ontogeny, *J. Exp. Med.,* 141, 206, 1975.
34. **Vossen, J. M. and Hijmans, S. W.,** Membrane associated immunoglobulin determinants on bone-marrow and blood lymphocytes in the paediatric age group and on foetal tissues, *Ann. N.Y. Acad. Sci.,* 254, 262, 1975.
35. **Vitetta, E. S. and Uhr, J. W.,** Immunoglobulin receptors revisited, *Science,* 189, 964, 1975.
36. **Vitetta, E. S. and Uhr, J. W.,** IgD and B cell differentiation, *Immunol. Rev.,* 37, 50, 1977.
37. **Tucker, P. W., Liu, C. P., Mushinski, J. F., and Blattner, F. R.,** Mouse immunoglobulin D: messenger RNA and genomic DNA sequences, *Science,* 209, 1353, 1980.
38. **Tisch, R., Roifman, C. M., and Hozumi, N.,** Functional differences between immunoglobulins M and D expressed on the surface of an immature B-cell line, *Proc. Natl. Acad. Sci. U.S.A.,* 85, 6914, 1988.
39. **Ales-Martinez, J. E., Warner, G. L., and Scott, D. W.,** Immunoglobulins D and M mediate signals that are qualitatively different in B cells with an immature phenotype, *Proc. Natl. Acad. Sci. U.S.A.,* 85, 6919, 1988.
40. **Mizuguchi, J., Tsang, W., Morrison, S. L., Beaven, M. A., and Paul, W. E.,** Membrane IgM, IgD, and IgG act as signal transmission molecules in a series of B lymphomas, *J. Immunol.,* 137, 2162, 1986.
41. **Tisch, R., Watanabe, M., Letarte, M., and Hozumi, N.,** Assessment of antigen-specific receptor function of surface immunoglobulin M and D with identical hapten specificity, *Proc. Natl. Acad. Sci. U.S.A.,* 84, 3831, 1988.
42. **Monroe, J. G., Blank, K. J., Owen, C. S., and Sykes, N.,** Signalling through sIgA on a novel murine B lymphoma, *Mol. Immunol.,* 26, 593, 1989.
43. **Webb, C. F., Nakai, C., and Tucker, P. W.,** Immunoglobulin receptor signaling depends on the carboxyl terminus but not the heavy-chain class, *Proc. Natl. Acad. Sci. U.S.A.,* 86, 1977, 1989.
44. **Bourgois, A., Kitajima, K., Hunter, I. R., and Askonas, B. A.,** Surface immunoglobulin of lipopolysaccharide-stimulated cells. The behavior of IgM, IgD, and IgG, *Eur. J. Immunol.,* 7, 151, 1977.
45. **Monroe, J. G., Havran, W. L., and Cambier, J. C.,** B lymphocyte activation: entry into the cell cycle is accompanied by decreased expression of IgD but not IgM, *Eur. J. Immunol.,* 13, 208, 1983.
46. **Yuan, D. and Tucker, P. W.,** Regulation of IgM and IgD synthesis in B lymphocytes. I. Changes in biosynthesis of mRNA for μ and J chains, *J. Immunol.,* 132, 1561, 1984.
47. **Yuan, D.,** Regulation of IgM and IgD synthesis in B lymphocytes. II. Translational and post-translational events, *J. Immunol.,* 132, 1566, 1984.

48. **Melcher, U. and Uhr, J. W.,** An electrophoretic difference between surface and secreted IgM of murine splenocytes, *J. Exp. Med.*, 138, 1282, 1973.
49. **Melcher, U., Eidels, L., and Uhr, J. W.,** Are Igs membrane proteins?, *Nature,* 258, 434, 1975.
50. **Melcher, U. and Uhr, J. W.,** Density differences between membrane and secreted immunoglobulins of murine splenocytes, *Biochemistry,* 16, 145, 1977.
51. **Bergman, Y. and Haimovich, J.,** B lymphocytes contain three species of μ chains, *Eur. J. Immunol.*, 8, 876, 1978.
52. **Goding, J. W., Black, S. J., and Herzenberg, L. A.,** Biosynthesis and two-dimensional gel analysis of IgM and IgD receptors, in *B Lymphocytes in the Immune Response,* Cooper, M., Mosier, D. E., Scher, I., and Vitetta, E. S., Eds., Elsevier North Holland, Amsterdam, 1979, 9.
53. **Williams, P. B., Kubo, R. T., and Grey, H. M.,** μ-Chains from a nonsecretor B cell line differ from secreted μ-chains at the C-terminal end, *J. Immunol.*, 121, 2435, 1978.
54. **Vassalli, P., Tedghi, K., Lisowski-Bernstein, B., Tartakoff, A., and Jaton, J. C.,** Evidence for hydrophobic region within heavy chains of mouse B lymphocyte membrane-bound IgM, *Proc. Natl. Acad. Sci. U.S.A.*, 76, 5515, 1979.
55. **Yuan, D., Uhr, J. W., and Vitetta, E. S.,** A peptide difference between the μ-chains from cell-associated and secreted IgM of the BCL_1 tumor, *J. Immunol.*, 125, 40, 1980.
56. **Alt, F. W., Bothwell, A. L. M., Knapp, M., Siden, E., Mather, E., Koshland, M., and Baltimore, D.,** Synthesis of secreted and membrane-bound immunoglobulin, μ heavy chains is directed by mRNAs that differ at their 3′ ends, *Cell,* 20, 293, 1980.
57. **Rogers, J., Early, P., Carter, C., Calame, K., Bond, M., Hood, L., and Wall, R.,** Two mRNAs with different 3′ ends encode membrane-bound and secreted forms of immunoglobulin μ chain, *Cell,* 20, 303, 1980.
58. **Early, P., Rogers, J., Davis, M., Calame, K., Bond, M., Wall, R., and Hood, L.,** Two mRNAs can be produced from a single immunoglobulin μ gene by alternative RNA processing pathways, *Cell,* 20, 313, 1980.
59. **Kehry, M., Ewald, S., Douglas, R., Sibley, C., Raschke, W., Fambrough, D., and Hood, L.,** The immunoglobulin μ chains of membrane-bound and secreted IgM molecules differ in their C-terminal segments, *Cell,* 21, 393, 1980.
60. **Goding, J. W.,** Structural studies of murine lymphocyte surface IgD, *J. Immunol.*, 124, 2082, 1980.
61. **Goding, J. and Herzenberg, L. A.,** Biosynthesis of lymphocyte surface IgD in the mouse, *J. Immunol.*, 124, 2540, 1980.
62. **Cheng, H. L., Blattner, F. R., Fitzmaurice, L., Mushinski, J. F., and Tucker, P. W.,** Structure of genes for membrane and secreted murine IgD heavy chains, *Nature,* 296, 410, 1982.
63. **Peterson, M. L. and Perry, R. P.,** Regulated production of μ_m and μ_s mRNA requires linkage of the poly(A) addition sites and is dependent on the length of the μ_s-μ_m intron, *Proc. Natl. Acad. Sci. U.S.A.*, 83, 8883, 1986.
64. **Law, R., Kuwabara, M. D., Briskin, M., Fasel, N., Hermanson, G., Sigman, D. S., and Wall, R.,** Protein-binding site at the immunoglobulin $\mu_{membrane}$ polyadenylation signal: possible role in transcriptional termination, *Proc. Natl. Acad. Sci. U.S.A.*, 84, 9160, 1987.
65. **Galli, G., Guise, J., Tucker, P. W., and Nevins, J. R.,** Poly(A) site choice rather than splice site choice governs the regulated production of IgM heavy-chain RNAs, *Proc. Natl. Acad. Sci. U.S.A.*, 85, 2439, 1988.
66. **Peterson, M. L. and Perry, R. P.,** The regulated production of μ_m and μ_s mRNA is dependent on the relative efficiencies of μ_s poly(A) site usage and the C_μ4-to-M1 splice, *Mol. Cell. Biol.*, 9, 726, 1989.
67. **Guise, J. W., Lim, P. L., Yuan, D., and Tucker, P. W.,** Alternative expression of secreted and membrane forms of immunoglobulin μ-chain is regulated by transcriptional termination in stable plasmacytoma transfectants, *J. Immunol.*, 140, 3988, 1988.
68. **Weiss, E. A., Michael, A., and Yuan, D.,** Role of transcriptional termination in the regulation of μ mRNA expression in B lymphocytes, *J. Immunol.*, 143, 1046, 1989.
69. **Gilbert, W.,** Why genes in pieces?, *Nature,* 271, 501, 1978.
70. **Wall, R., Choi, E., Carter, C., Kuehl, M., and Rogers, J.,** RNA processing in immunoglobulin gene expression, *Cold Spring Harbor Symp. Quant. Biol.*, 45, 879, 1980.
71. **Yuan, D. and Tucker, P. W.,** Transcriptional regulation of the μ-J gene in normal murine B lymphocytes, *J. Exp. Med.*, 160, 564, 1984.
72. **Yuan, D. and Tucker, P. W.,** Regulation of IgM and IgD synthesis in B lymphocytes. I. Changes in biosynthesis of mRNA for μ- and J-chains, *J. Immunol.*, 132, 1561, 1984.
73. **Tyler, B. M., Courman, A. F., Gerondakis, S. D., Adams, J. M., and Bernard, O.,** mRNA for surface immunoglobulin γ chains encodes a highly conserved transmembrane sequence and a 28-residue intracellular domain, *Proc. Natl. Acad. Sci. U.S.A.*, 79, 2008, 1982.

74. **Pillai, S. and Baltimore, D.,** Myristoylation and the post-translational acquisition of hydrophobicity by the membrane immunoglobulin heavy-chain polypeptide in B lymphocytes, *Proc. Natl. Acad. Sci. U.S.A.*, 84, 7654, 1987.
75. **Bretscher, P. A. and Cohn, M.,** A theory of self-nonself discrimination, *Science,* 169, 1042, 1970.
76. **Coutinho, A. and Möller, G.,** Immune activation of B cells: evidence for one nonspecific triggering signal not delivered by the Ig receptors, *Scand. J. Immunol.*, 3, 133, 1974.
77. **Andersson, J., Sjoberg, O., and Möller, G.,** Induction of immunoglobulin and antibody synthesis *in vitro* by lipopolysaccharide, *Eur. J. Immunol.*, 2, 349, 1972.
78. **Taylor, R. B., Duffus, W. P. H., Raff, M. C., and de Petris, S.,** Redistribution and pinocytosis of lymphocyte surface immunoglobulin molecules induced by anti-immunoglobulin antibody, *Nature New Biol.*, 233, 225, 1971.
79. **Raff, M. C. and de Petris, S.,** Movement of lymphocyte surface antigens and receptors: the fluid nature of the lymphocyte plasma membrane and its immunological significance, *Fed. Proc.*, 32, 48, 1973.
80. **Loor, F., Forni, L., and Pernis, B.,** The dynamic state of the lymphocyte membrane. Factors affecting the distribution and turnover of surface immunoglobulin, *Eur. J. Immunol.*, 2, 203, 1972.
81. **Goding, J. W. and Layton, J. E.,** Antigen-induced co-capping of IgM and IgD-like receptors on murine B cells, *J. Exp. Med.*, 144, 852, 1976.
82. **Diener, E. and Paetkau, V. H.,** Antigen recognition: early surface-receptor phenomena induced by binding of a tritium-labeled antigen, *Proc. Natl. Acad. Sci. U.S.A.*, 69, 2364, 1972.
83. **Ashman, R. F.,** Lymphocyte receptor movement induced by sheep erythrocyte binding, *J. Immunol.*, 111, 212, 1973.
84. **Frye, L. D. and Edidin, M.,** The rapid intramixing of cell surface antigens after formation of mouse-human heterokaryons, *J. Cell Sci.*, 7, 319, 1970.
85. **Blaisie, J. K. and Worthington, C. R.,** Molecular localization of frog retinal receptor photopigments by electron microscopy and low-angle X-ray diffraction, *J. Mol. Biol.*, 38, 407, 1969.
86. **Singer, S. J. and Nicolson, R. L.,** The fluid mosaic model of the structure of cell membranes, *Science,* 175, 720, 1972.
87. **Braun, J., Hochman, P. S., and Unanue, E. R.,** Ligand-induced association of surface immunoglobulin with the detergent-insoluble cytoskeletal matrix of the B lymphocyte, *J. Immunol.*, 128, 1198, 1982.
88. **Woda, B. A. and McFadden, M. L.,** Ligand-induced association of rat lymphocyte membrane proteins with the detergent insoluble lymphocyte cytoskeletal matrix, *J. Immunol.*, 131, 1917, 1983.
89. **Gupta, S. K. and Woda, B. A.,** Ligand-induced association of surface immunoglobulin with the detergent insoluble cytoskeleton may involve α-actinin, *J. Immunol.*, 140, 176, 1988.
90. **Gabbiani, G., Chaponnier, C., Zumbe, A., and Vassalli, P.,** Actin and tubulin co-cap with surface immunoglobulin in mouse B lymphocytes, *Nature,* 269, 697, 1977.
91. **Dellagi, K. and Brouet, J. C.,** Redistribution of intermediate filaments during capping of lymphocyte surface molecules, *Nature,* 298, 284, 1982.
92. **Albrecht, D. L. and Noelle, R. J.,** Membrane Ig-cytoskeletal interactions. I. Flow cytofluorometric and biochemical analysis of membrane IgM-cytoskeletal interactions, *J. Immunol.*, 141, 3915, 1988.
93. **Albrecht, D. L. and Noelle, R. J.,** Membrane Ig-cytoskeletal interactions. II. Biochemical analysis of cytoskeletally associated IgM, *Mol. Immunol.*, 26, 575, 1989.
94. **Greaves, M., Janossy, G., Feldmann, M., and Doenhoff, M.,** Polyclonal mitogens and the nature of B lymphocyte activation mechanisms, in *The Immune System,* Sercarz, E. E., Williamson, A. R., and Fox, C. F., Eds., Academic Press, New York, 1974, 271.
95. **Parker, D. C.,** Stimulation of mouse lymphocytes by insoluble anti-mouse immunoglobulin, *Nature,* 258, 361, 1975.
96. **Weiner, H. L., Moorhead, J. W., and Claman, H. N.,** Anti-immunoglobulin stimulation of murine lymphocytes. I. Age dependency of the proliferative response, *J. Immunol.*, 116, 1656, 1976.
97. **Scribner, D. J., Weiner, H. L., and Moorhead, J. W.,** Anti-immunoglobulin stimulation of murine lymphocytes. V. Age-related decline in Fc receptor-mediated immunoregulation, *J. Immunol.*, 121, 377, 1978.
98. **Sieckmann, D. G., Asofsky, R., Mosier, D. E., Zitron, I. M., and Paul, W. E.,** Activation of mouse lymphocytes by anti-immunoglobulin. I. Parameters of the proliferative response, *J. Exp. Med.*, 147, 814, 1978.
99. **Parker, D. C., Wadsworth, D. C., and Schneider, G. B.,** Activation of murine B lymphocytes by anti-immunoglobulin is an inductive signal leading to immunoglobulin secretion, *J. Exp. Med.*, 152, 138, 1980.
100. **Fanger, M. W., Hart, N. A., Wells, J. V., and Nisonoff, A.,** Requirement for cross-linkage in the stimulation of transformation of rabbit peripheral lymphocytes by antiglobulin reagents, *J. Immunol.*, 105, 1484, 1970.
101. **Weiner, H. L., Moorhead, J. W., Yamaga, K., and Kubo, R. T.,** Anti-immunoglobulin stimulation of murine lymphocytes. II. Identification of cell surface target molecules and requirements for cross linkage, *J. Immunol.*, 117, 1527, 1976.

102. **Sidman, C. L. and Unanue, E. R.,** Requirements for mitogenic stimulation of murine B cells by soluble anti-IgM antibodies, *J. Immunol.*, 122, 406, 1979.
103. **Sieckmann, D. G., Scher, I., Asofsky, R., Mosier, D. E., and Paul, W. E.,** Activation of mouse lymphocytes by anti-immunoglobulin. II. A thymus-independent response by a mature subset of B lymphocytes, *J. Exp. Med.*, 148, 1628, 1978.
104. **Scher, I., Sharrow, S. O., Wistar, R., Asofsky, R., and Paul, W. E.,** B-lymphocyte heterogeneity: ontogenetic development and organ distribution of B-lymphocyte populations defined by their density of surface immunoglobulin, *J. Exp. Med.*, 144, 494, 1976.
105. **Scher, I., Ahmed, A., and Sharrow, S. O.,** Murine B lymphocyte heterogeneity: distribution of complement receptor-bearing and minor lymphocyte-stimulating B lymphocytes among cells with different densities of total surface Ig and IgM, *J. Immunol.*, 119, 1938, 1977.
106. **Vitetta, E. S. and Uhr, J. W.,** IgD and B cell differentiation, *Immunol. Rev.*, 37, 50, 1978.
107. **Sieckmann, D. G., Habbersett, R., Scher, I., and Paul, W. E.,** Activation of mouse lymphocytes by anti-immunoglobulin. III. Analysis of responding B lymphocytes by flow cytometry and cell sorting, *J. Immunol.*, 127, 205, 1981.
108. **DeFranco, A. L., Raveche, E. S., Asofsky, R., and Paul, W. E.,** Frequency of B lymphocytes responsive to anti-immunoglobulin, *J. Exp. Med.*, 155, 1523, 1982.
109. **Andersson, J., Schreier, M. H., and Melchers, F.,** T-cell-dependent B-cell stimulation is H-2 restricted and antigen dependent only at the resting B-cell level, *Proc. Natl. Acad. Sci. U.S.A.*, 77, 1612, 1980.
110. **Sell, S., Rowe, D. S., and Gell, P. G. H.,** Studies on rabbit lymphocytes in vitro. III. Protein, RNA, and DNA synthesis by lymphocyte cultures after stimulation with phytohaemagglutinin, with staphylococcal filtrate, with anti-allotypic serum, and with heterologous antiserum to rabbit whole serum, *J. Exp. Med.*, 122, 823, 1965.
111. **Sidman, C. L. and Unanue, E. R.,** Control of proliferation and differentiation in B lymphocytes by anti-Ig antibodies and a serum-derived co-factor, *Proc. Natl. Acad. Sci. U.S.A.*, 75, 2401, 1978.
112. **Dutton, D. W., Falkoff, R., Hirst, J. A., Hoffman, M., Kappler, J. W., Kettman, J. R., Lesley, J. F., and Vann, D.,** Is there evidence for a non-antigen specific diffusable chemical mediator from the thymus-derived cell in the initiation of the immune response?, in *Progress in Immunology,* Amos, B., Ed., Academic Press, New York, 1971, 355.
113. **Kishimoto, T. and Ishizaka, K.,** Regulation of antibody response *in vitro*. IX. Induction of secondary anti-hapten IgG antibody response by anti-immunoglobulin and enhancing soluble factor, *J. Immunol.*, 114, 585, 1975.
114. **Kishimoto, T., Miyake, T., Nishizawa, Y., Watanabe, T., and Yamamura, Y.,** Triggering mechanisms of B lymphocytes. I. Effect of anti-immunoglobulin and enhancing soluble factor on differentiation and proliferation of B cells, *J. Immunol.*, 115, 1179, 1975.
115. **Parker, D. C., Fothergill, J. J., and Wadsworth, N. C.,** B lymphocyte activation by insoluble anti-immunoglobulin: induction of immunoglobulin secretion by a T cell-dependent soluble factor, *J. Immunol.*, 123, 931, 1979.
116. **Pure, E., Isakson, P. C., Takatsu, K., Hamaoka, T., Swain, S. L., Dutton, R. W., Dennert, G., Uhr, J. W., and Vitetta, E. S.,** Induction of B cell differentiation by T cell factors. I. Stimulation of IgM secretion by products of a T cell hybridoma and a T cell line, *J. Immunol.*, 127, 1953, 1981.
117. **Braun, J., Sha'Afi, R. I., and Unanue, E. R.,** Crosslinking by ligands to surface immunoglobulin triggers mobilization of intracellular $^{45}Ca^{2+}$ in B lymphocytes, *J. Cell Biol.*, 82, 755, 1979.
118. **Pozzan, T., Arslan, P., Tsien, R. Y., and Rink, T. J.,** Anti-immunoglobulin, cytoplasmic free calcium, and capping in B lymphocytes, *J. Cell Biol.*, 94, 335, 1982.
119. **Bijsterbosch, M. K., Rigley, K. P., and Klaus, G. G. B.,** Crosslinking of surface immunoglobulin on B lymphocytes induces both intracellular Ca^{2+} release and Ca^{2+} influx: analysis with indo-1, *Biochem. Biophys. Res. Commun.*, 137, 500, 1986.
120. **Dugas, B., Calenda, A., Delfraissy, J. F., Vazquez, A., Bach, J. F., and Galanaud, P.,** The cytosolic free calcium in anti-μ-stimulated human B cells is derived partly from extracellular medium and partly from intracellular stores, *Eur. J. Immunol.*, 17, 1323, 1987.
121. **Ransom, J. T., Digiusto, D. L., and Cambier, J. C.,** Single cell analysis of calcium mobilization in anti-immunoglobulin-stimulated B lymphocytes, *J. Immunol.*, 136, 54, 1986.
122. **Ransom, J. T., Chen, M., Sandoval, V. M., Pasternak, J. A., Digiusto, D., and Cambier, J. C.,** Increased plasma membrane permeability to Ca^{2+} in anti-Ig-stimulated B lymphocytes is dependent on activation of phosphoinositide hydrolysis, *J. Immunol.*, 140, 3150, 1988.
123. **Monroe, J. G. and Cambier, J. C.,** B cell activation. I. Anti-immunoglobulin induced receptor cross-linking results in a decrease in the plasma membrane potential of murine B lymphocytes, *J. Exp. Med.*, 157, 2073, 1983.
124. **Conn, P. M., Marian, J., McMillian, M., Stern, J., Rogers, D., Hamby, M., Penna, A., and Grant, E.,** Gonadotropin-releasing hormone action in the pituitary: a three step mechanism, *Endocr. Rev.*, 2, 171, 1981.

125. **LaBaer, J., Tsien, R. Y., Fahey, K. H., and DeFranco, A. L.**, Stimulation of the antigen receptor on WEHI-231 B lymphoma cells results in a voltage-independent increase in cytoplasmic calcium, *J. Immunol.*, 137, 1836, 1986.
126. **MacDougall, S. L., Grinstein, S., and Gelfand, E. W.**, Activation of Ca^{2+}-dependent K^+ channels in human B lymphocytes by anti-immunoglobulin, *J. Clin. Invest.*, 81, 449, 1988.
127. **Michell, R. H.**, Inositol phospholipids and cell surface receptor function, *Biochem. Biophys. Acta,* 415, 81, 1975.
128. **Fisher, D. B. and Mueller, G. C.**, An early alteration in the phospholipid metabolism of lymphocytes by phytohaemagglutinin, *Proc. Natl. Acad. Sci. U.S.A.*, 60, 1396, 1968.
129. **Maino, V. C., Hayman, M. J., and Crumpton, M. J.**, Relationship between enhanced turnover of phosphatidylinositol and lymphocyte activation by mitogens, *Biochem. J.*, 146, 247, 1975.
130. **Berridge, M. and Irvine, R.**, Inositol triphosphate, a novel second messenger in cellular signal transduction, *Nature,* 312, 315, 1984.
131. **Kishimoto, A., Takai, Y., Mori, T., Kikkawa, U., and Nishizuka, Y.**, Activation of calcium and phospholipid dependent protein kinase by diacylglycerol, its possible relationship to phosphatidylinositol turnover, *J. Biol. Chem.*, 255, 2273, 1980.
132. **Ku, Y., Kishimoto, A., Takai, Y., Ogawa, Y., Kimura, S., and Nishizuka, Y.**, A new possible regulatory system for protein phosphorylation in human peripheral lymphocytes. II. Possible relation to phosphatidylinositol turnover induced by mitogens, *J. Immunol.*, 127, 1375, 1981.
133. **Nishizuka, Y.**, The role of protein kinase C in cell signal transduction and tumor promotion, *Nature,* 308, 693, 1984.
134. **Berridge, M.**, Inositol trisphosphate and diacylglycerol as second messengers, *Biochem. J.*, 220, 345, 1984.
135. **Monroe, J. G., Niedel, J. E., and Cambier, J. C.**, B cell activation. IV. Induction of cell membrane depolarization and hyper-I-A expression by phorbol diesters suggests a role for protein kinase C in murine B lymphocyte activation, *J. Immunol.*, 132, 1472, 1984.
136. **Coggeshall, K. M. and Cambier, J. C.**, B cell activation. VI. Effects of exogenous diglyceride and modulators of phospholipid metabolism suggest a central role for diacylglycerol generation in transmembrane signaling by mIg, *J. Immunol.*, 134, 101, 1985.
137. **Coggeshall, K. M. and Cambier, J. C.**, B cell activation. VIII. Membrane immunoglobulins transduce signals via activation of phosphatidylinositol hydrolysis, *J. Immunol.*, 133, 3382, 1984.
138. **Grupp, S. A. and Harmony, J. A. K.**, Increased phosphatidylinositol metabolism is an important but not an obligatory early event in B lymphocyte activation, *J. Immunol.*, 134, 4087, 1985.
139. **Bijsterbosch, M. K., Meade, C. J., Turner, G. A., and Klaus, G. G. B.**, B lymphocyte receptors and polyphosphoinositide degradation, *Cell,* 41, 999, 1985.
140. **Ransom, J. T., Harris, L. K., and Cambier, J. C.**, Anti-Ig induces release of inositol 1,4,5-triphosphate which mediates mobilization of intracellular Ca^{++} stores in B lymphocytes, *J. Immunol.*, 137, 708, 1986.
141. **Ross, C. A., Meldolesi, J., Milner, T. A., Satoh, T., Supattapone, S., and Synder, S. H.**, Inositol 1,4,5-trisphosphate receptor localized to endoplasmic reticulum in cerebellar Purkinje neurons, *Nature,* 339, 468, 1989.
142. **Furuichi, T., Yoshikawa, S., Miyawaki, A., Wada, K., Maeda, N., and Mikoshiba, K.**, Primary structure and functional expression of the inositol 1,4,5,-trisphosphate-binding protein P_{400}, Nature, 342, 32, 1989.
143. **Ferris, C. D., Huganir, R. L., Supattapone, S., and Snyder, S.**, Purified inositol 1,4,5,-trisphosphate receptor mediates calcium flux in reconstituted lipid vesicles, *Nature,* 342, 87, 1989.
144. **Berridge, M. J. and Irvine, R. F.**, Inositol phosphates and cell signalling, *Nature,* 341, 197, 1989.
145. **MacDougall, S. L., Grinstein, S., and Gelfand, E. W.**, Detection of ligand-activated conductive Ca^{2+} channels in human B lymphocytes, *Cell,* 54, 229, 1988.
146. **Kuno, M. and Gardner, P.**, Ion channels activated by inositol 1,4,5,-trisphosphate in plasma membranes of human T lymphocytes, *Nature,* 326, 301, 1987.
147. **Gelfand, E. W., MacDougall, S. L., Cheung, R. K., and Grinstein, S.**, Independent regulation of Ca^{2+} entry and release from internal stores in activated B cells, *J. Exp. Med.*, 170, 315, 1989.
148. **Nel, A. E., Wooten, M. W., Goldschmidt-Clermont, P. J., Miller, P. J., Stevenson, H. C., and Galbraith, R. M.**, Polymyxin B causes coordinate inhibition of phorbol ester-induced C-kinase activity and proliferation of B lymphocytes, *Biochem. Biophys. Res. Commun.*, 128, 1364, 1985.
149. **Guy, G. R., Gordon, J., Walker, L., Michell, R. H., and Brown, G.**, Redistribution of protein kinase C during mitogenesis of human B lymphocytes, *Biochem. Biophys. Res. Commun.*, 135, 146, 1986.
150. **Kikkawa, U., Kishimoto, A., and Nishizuka, Y.**, The protein kinase C family: heterogeneity and its implications, *Annu. Rev. Biochem.*, 58, 31, 1989.
151. **Scott, D. W., Livnat, D., Pennell, C. A., and Keng, P.**, Lymphoma models for B cell activation and tolerance. III. Cell cycle dependence for negative signalling of WEHI-231 B lymphoma cells by anti-μ, *J. Exp. Med.*, 164, 156, 1986.

152. **Gold, M. R. and DeFranco, A. L.,** Phorbol esters and dioctanoylglycerol block anti-IgM-stimulated phosphoinositide hydrolysis in the murine B cell lymphoma WEHI-231, *J. Immunol.*, 138, 868, 1987.
153. **Ingles, J., Engel, P., De La Calle, O., and Gallart, T.,** Differential responsiveness of human B lymphocytes to phorbol ester and calcium ionophore based on their state of activation, *Immunology,* 67, 359, 1989.
154. **Monroe, J. G. and Gaulton, G. N.,** Comparison of protein phosphorylation induced by mitogen and phorbol diester of murine T and B lymphocytes, *Serv. Immunol. Rev.*, 4, 192, 1985.
155. **Hornbeck, P. and Paul, W. E.,** Anti-immunoglobulin and phorbol ester induced phosphorylation of proteins associated with the plasma membrane and cytoskeleton in murine B lymphocytes, *J. Biol. Chem.*, 261, 14817, 1986.
156. **Hornbeck, P., Nakabayashi, H., Fowlkes, B. J., Paul, W. E., and Kligman, D.,** A major myristoylated substrate of protein kinase C and protein kinase C itself are differentially regulated during murine B- and T-lymphocyte development and activation, *Mol. Cell. Biol.*, 9, 3727, 1989.
157. **Burke, T., Pollok, K., Cushley, W., and Snow, E. C.,** Phosphorylation of class I but not class II MHC molecules by membrane-localized protein kinase C, *Mol. Immunol.*, 26, 1095, 1989.
158. **Burn, P.,** Phosphatidylinositol cycle and its possible involvement in the regulation of cytoskeleton-membrane interactions, *J. Cell. Biochem.*, 36, 15, 1988.
159. **Rabin, M. S., Doherty, P. J., and Gottesman, M. M.,** The tumor promoter phorbol 12-myristate 13-acetate induces a program of altered gene expression similar to that induced by platelet-derived growth factor and transforming oncogenes, *Proc. Natl. Acad. Sci. U.S.A.*, 83, 357, 1987.
160. **Comb, M., Birnberg, N. C., Seasholtz, A., Herbert, E., and Goodman, H. M.,** A cyclic AMP- and phorbol ester-inducible DNA element, *Nature,* 323, 353, 1986.
161. **Hogbom, E., Martensson, I. L., and Leanderson, T.,** Regulation of immunoglobulin transcription rates and mRNA processing in proliferating normal B lymphocytes by activators of protein kinase C, *Proc. Natl. Acad. Sci. U.S.A.*, 84, 9135, 1987.
162. **Shirakawa, F. and Mizel, S. B.,** In vitro activation and nuclear translocation of NF-kB catalyzed by cyclic AMP-dependent protein kinase and protein kinase C, *Mol. Cell. Biol.*, 9, 2424, 1989.
163. **Denhardt, D. T., Edwards, D. R., and Parfett, C. L.,** Gene expression during the mammalian cell cycle, *Biochim. Biophys. Acta,* 865, 83, 1986.
164. **Lau, L. F. and Nathans, D.,** Expression of a set of growth-related immediated early genes in Balb/c3T3 cells: coordinate regulation with c-*fos* and c-*myc*, *Proc. Natl. Acad. Sci. U.S.A.*, 84, 1182, 1987.
165. **Lim, R. W., Varnum, B. C., and Herschman, H. R.,** Cloning of tetradecanoyl phorbol ester-induced 'primary response' sequences and their expression in density-arrested Swiss 3T3 cells and a TPA non-proliferative variant, *Oncogene,* 1, 263, 1987.
166. **Almendral, J. M., Sommer, D., MacDonald-Bravo, H., Burckhardt, J., Perera, J., and Bravo, R.,** Complexity of the early genetic response to growth factors in mouse fibroblasts, *Mol. Cell. Biol.*, 8, 2140, 1988.
167. **Seyfert, V. L., Sukhatme, V. P., and Monroe, J. G.,** Differential expression of a zinc finger-encoding gene in response to positive versus negative signaling through receptor immunoglobulin in murine B lymphocytes, *Mol. Cell. Biol.*, 9, 2083, 1989.
168. **Smeland, E., Godal, T., Ruud, E., Beiske, K., Funderud, S., Clark, E. A., Pfeifer-Ohlsson, S., and Ohlsson, R.,** The specific induction of *myc* protooncogene expression in normal human B cells is not a sufficient event for acquisition of competence to proliferate, *Proc. Natl. Acad. Sci. U.S.A.*, 82, 6255, 1985.
169. **Lacy, J., Sarkar, S. N., and Summers, W. C.,** Induction of c-*myc* expression in human B lymphocytes by B-cell growth factor and anti-immunoglobulin, *Proc. Natl. Acad. Sci. U.S.A.*, 83, 1458, 1986.
170. **Smeland, E. B., Beiske, K., Ek, B., Watt, R., Pfeifer-Ohlsson, S., Blomhoff, H. K., Godal, T., and Ohlsson, R.,** Regulation of c-*myc* transcription and protein expression during activation of normal human B cells, *Exp. Cell Res.*, 172, 101, 1987.
171. **Monroe, J. G.,** Up-regulation of c-*fos* expression is a component of the mIg signal transduction mechanism but not indicative of competence for proliferation, *J. Immunol.*, 140, 1454, 1988.
172. **Conghlin, S. R., Lee, W. M. F., Williams, P. W., Giels, G. M., and Williams, L. T.,** c-*myc* gene expression is stimulated by agents that activate protein kinase C and does not account for the mitogenic effect of PDGF, *Cell,* 43, 243, 1985.
173. **Faletto, D. L., Arrow, A. S., and Macara, I. G.,** An early decrease in phosphatidylinositol turnover occurs on induction of Friend cell differentiation and proceeds the decrease in c-*myc* expression, *Cell,* 43, 315, 1985.
174. **Wright, T. C., Pukac, L. A., Castellot, J. J., Karnovsky, M. J., Levine, R. A., Kim-Park, H. Y., and Campisi, J.,** Heparin suppresses the induction of c-*fos* and c-*myc* mRNA in murine fibroblasts by selective inhibition of a protein kinase C-dependent pathway, *Proc. Natl. Acad. Sci. U.S.A.*, 86, 3199, 1989.

175. **Mond, J. J., Seghal, E., Kung, J., and Finkelman, F. D.,** Increased expression of I-region-associated antigen (Ia) on B cells after crosslinking of surface immunoglobulin, *J. Immunol.*, 127, 881, 1981.
176. **Klaus, G. G. B., Bijsterbosch, M. K., and Holman, M.,** Activation and proliferation signals in mouse B cells. VII. Calcium ionophores are non-mitogenic polyclonal B-cell activators, *Immunology*, 56, 321, 1985.
177. **Ransom, J. T. and Cambier, J. C.,** B cell activation. VII. Independent and synergistic effects of mobilized calcium and diacylglycerol on membrane potential and I-A expression, *J. Immunol.*, 136, 66, 1986.
178. **Roehm, N. W., Leibson, H. J., Zlotnick, A., Kappler, J., Marrack, P., and Cambier, J. C.,** Interleukin-induced increase in Ia expression by normal mouse B cells, *J. Exp. Med.*, 160, 679, 1984.
179. **Bottomly, K., Jones, B., Kaye, J., and Jones, F.,** Subpopulations of B cells distinguished by cell surface expression of Ia antigens: correlation of Ia and idiotype during activation by cloned Ia-restricted T cells, *J. Exp. Med.*, 158, 265, 1983.
180. **Sanders, V. M., Fernandez-Botran, R., Uhr, J. W., and Vitetta, E. S.,** Interleukin 4 enhances the ability of antigen-specific B cells to form conjugates with T cells, *J. Immunol.*, 139, 2349, 1987.
181. **Kearse, K. P., Cassatt, D. R., Kaplan, A. M., and Cohen, D. A.,** The requirement for surface Ig signaling as a prerequisite for T cell:B cell interactions, *J. Immunol.*, 140, 1770, 1988.
182. **Frohman, M. and Cowing, C.,** Presentation of antigen by B cells: functional dependence on radiation dose, interleukins, cellular activation, and differential glycosylation, *J. Immunol.*, 134, 2269, 1985.
183. **Kreiger, J., Jenis, D. M., Chesnut, R. W., and Grey, H. M.,** Studies on the capacity of intact cells and purified Ia from different B cell sources to function in antigen presentation to cells, *J. Immunol.*, 140, 388, 1988.
184. **Dustin, M. L. and Springer, T. A.,** T-cell receptor cross-linking transiently stimulates adhesiveness through LFA-1, *Nature*, 341, 619, 1989.
185. **Tucker, P. W., Cheng, H. L., Richards, J. E., Fitzmaurice, L., Mushinski, J. F., and Blattner, F. R.,** Genetic aspects of IgD expression. III. Functional implications of the sequence and organization of the C gene, *Ann. N.Y. Acad. Sci.*, 399, 266, 1982.
186. **Cambier, J. C., Justement, L. B., Newell, M. K., Chen, Z. Z., Harris, L. K., Sandoval, V. M., Klemsz, M. J., and Ransom, J. T.,** Transmembrane signals and intracellular "second messengers" in the regulation of quiescent B-lymphocyte activation, *Immunol. Rev.*, 95, 37, 1987.
187. **Stryer, L. and Bourne, H. R.,** G proteins: a family of signal transducers, *Annu. Rev. Cell Biol.*, 2, 391, 1986.
188. **Gilman, A. G.,** G proteins: transducers of receptor-generated signals, *Annu. Rev. Biochem.*, 56, 615, 1987.
189. **Strader, C. D., Dixon, R. A. F., Cheung, A. H., Candelore, M. R., Blake, A. D., and Sigal, I. S.,** Mutations that uncouple the β-adrenergic receptor from G_s and increase agonist affinity, *J. Biol. Chem.*, 262, 16439, 1987.
190. **O'Dowd, B. F., Hnatowich, M., Regan, J. W., Leader, W. M., Caron, M. G., and Lefkowitz, R. J.,** Site-directed mutagenesis of the cytoplasmic domains of the human α_2-adrenergic receptor, *J. Biol. Chem.*, 263, 15985, 1988.
191. **Weiss, E. R., Kelleher, D. J., Woon, C. W., Soparkar, S., Osawa, S., Heasley, L. E., and Johnson, G. L.,** Receptor activation of G proteins, *FASEB J.*, 2, 2841, 1988.
192. **Ashkenazi, A., Petalta, E. G., Winslow, J. W., Ramachandran, J., and Capon, D. J.,** Functionally distinct G proteins selectively couple different receptors to PI hydrolysis in the same cell, *Cell*, 56, 487, 1989.
193. **Mizuguchi, J., Yong-Yong, J., Nakabayaschi, H., Huang, K. P., Beaven, M. A., Chused, T., and Paul, W. E.,** Protein kinase C activation blocks anti-IgM-mediated signaling in Ba117 B lymphoma cells, *J. Immunol.*, 139, 1054, 1987.
194. **Gold, M. R., Jakway, J. P., and DeFranco, A. L.,** Involvement of a guanine nucleotide-binding component in membrane IgM-stimulated phosphoinositide breakdown, *J. Immunol.*, 139, 3604, 1987.
195. **Harnett, M. M. and Klaus, G. G. B.,** G protein coupling of antigen receptor-stimulated polyphosphoinositide hydrolysis in B cells, *J. Immunol.*, 140, 3139, 1988.
196. **Monroe, J. G. and Haldar, S.,** Involvement of a specific guanine nucleotide binding protein in receptor immunoglobulin stimulated inositol phospholipid hydrolysis, *Biochim. Biophys. Acta*, 1013, 273, 1989.
197. **Katada, T. and Ui, M.,** Slow interaction of islet-activating protein with pancreatic islets during primary culture to cause reversal of α-adrenergic inhibition of insulin secretion, *J. Biol. Chem.*, 255, 9580, 1980.
198. **Sullivan, K. A., Miller, R. T., Masters, S. B., Beiderman, B., Heideman, W., and Bourne, H. R.,** Identification of receptor contact site involved in receptor-G protein coupling, *Nature*, 330, 758, 1987.
199. **Reinherz, E. L., Meuer, S., Fitzgerald, K. A., Hussey, R. E., Levine, H., and Schlossman, S. F.,** Antigen recognition by human T lymphocytes is linked to surface expression of the T3 molecular complex, *Cell*, 30, 735, 1982.

200. **Meuer, S. C., Fitzgerald, K. A., Hussey, R. E., Hodgdon, J. C., Schlossman, S. F., and Reinherz, E. L.,** Clonotypic structures involved in antigen-specific human T cell function: relationship to the T3 molecular complex, *J. Exp. Med.,* 157, 705, 1983.
201. **Clevers, H., Alarcon, B., Wileman, T., and Terhorst, C.,** The T cell receptor/CD3 complex: a dynamic protein ensemble, *Annu. Rev. Immunol.,* 6, 629, 1988.
202. **Sidman, C., Bercovici, T., and Gitler, C.,** Membrane insertion of lymphocyte surface molecules, *Mol. Immunol.,* 17, 1575, 1980.
203. **Rosenspire, A. and Choi, Y. S.,** Relation between actin-associated proteins and membrane immunoglobulin in B-cells, *Mol. Immunol.,* 19, 1515, 1982.
204. **Koch, N. and Haustein, D.,** Association of surface IgM with two membrane proteins on murine B lymphocytes detected by chemical crosslinking, *Mol. Immunol.,* 20, 33, 1983.
205. **Haustein, D. and Von der Ahe, D.,** A 30-kDa protein is disulfide linked to IgM on normal and neoplastic murine B lymphocytes, *Eur. J. Immunol.,* 16, 113, 1986.
206. **Sakaguchi, N., Kashiwamura, S. I., Kimoto, M., Thalmann, P., and Melchers, F.,** B lymphocyte lineage-restricted expression of *mb*-1, a gene with CD3-like structural properties, *EMBO J.,* 7, 3457, 1988.
207. **Hombach, J., Sablitzky, F., Rajewsky, K., and Reth, M.,** Transfected plasmacytoma cells do not transport the membrane form of IgM to the cell surface, *J. Exp. Med.,* 167, 652, 1988.
208. **Hombach, J., Leclercq, L., Radbruch, A., Rajewsky, K., and Reth, M.,** A novel 34-kd protein co-isolated with the IgM molecule in surface IgM-expressing cells, *EMBO J.,* 7, 3451, 1988.
209. **Oettgen, H. C., Pettey, C. L., Maloy, W. L., and Terhorst, C.,** A T3-like protein complex associated with the antigen receptor on murine T cells, *Nature,* 320, 272, 1986.
210. **Campbell, K. S. and Cambier, J. C.,** B lymphocyte antigen receptors (mIg) are non-covalently associated with a disulfide-linked, inducibly phosphorylated glycoprotein complex, *EMBO J.,* 9, 441, 1990.
211. **Weinands, J., Hombach, J., Radbruch, A., Riesterer, C., and Reth, M.,** Molecular components of the B cell antigen receptor complex of class IgD differs partly from that of IgM, *EMBO J.,* 9, 441, 1990.
212. **Reth, M.,** Antigen receptor tail clue, *Nature,* 338, 383, 1989.
213. **Pesando, J. M., Bouchard, L. S., and McMaster, B. E.,** CD19 is functionally and physically associated with surface immunoglobulin, *J. Exp. Med.,* 170, 2159, 1989.
214. **Monroe, J. G. and Kass, M. J.,** Molecular events in B cell activation. I. Signals required to stimulate G_0 to G_1 transition of resting B lymphocytes, *J. Immunol.,* 135, 1674, 1985.
215. **Snow, E. C., Noelle, R. J., Uhr, J. W., and Vitetta, E. S.,** Activation of antigen-enriched B cells. II. Role of linked recognition in B cell proliferation to thymus-dependent antigens, *J. Immunol.,* 130, 614, 1983.
216. **Haas, W. and Layton, J. E.,** Separation of antigen-specific lymphocytes. I. Enrichment of antigen-binding cells, *J. Exp. Med.,* 141, 1004, 1975.
217. **Kenny, J. J., Merrill, J. E., and Ashman, R. F.,** A two-step centrifugation procedure for the purification of sheep erythrocyte antigen-binding cells, *J. Immunol.,* 120, 1233, 1978.
218. **Cambier, J. C. and Neale, M. J.,** Isolated phosphorylcholine binding lymphocytes. I. Use of a cleavable crosslinking reagent for solid phase adsorbent isolation of functional antigen binding cells, *J. Immunol. Methods,* 51, 209, 1982.
219. **Snow, E. C., Vitetta, E. S., and Uhr, J. W.,** Activation of antigen-enriched B-cells. I. Purification and response to thymus-independent antigens, *J. Immunol.,* 130, 607, 1983.
220. **Julius, M. H., Masuda, T., and Herzenberg, L. A.,** Demonstration that antigen-binding cells are precursors of antibody-producing cells after purification with a FACS, *Proc. Natl. Acad. Sci. U.S.A.,* 69, 1934, 1972.
221. **Haas, W.,** Separation of antigen-specific lymphocytes. II. Enrichment of hapten specific antibody-forming cell presursors, *J. Exp. Med.,* 141, 1015, 1975.
222. **Venkataraman, M. and Scott, D. W.,** Cellular events in tolerance. VIII. Analysis of the proliferative capacity of antigen-binding cells isolated from normal or tolerant spleens, *J. Immunol.,* 124, 607, 1980.
223. **Veaux, D. L., Pike, B. L., and Nossal, G. J. V.,** Antibody production by single, hapten-specific B lymphocytes: an antigen-driven cloning system free of filler or accessory cells, *Proc. Natl. Acad. Sci. U.S.A.,* 78, 7702, 1981.
224. **Kenny, J. J., Martinez-Maza, O., Fehniger, T., and Ashman, R. F.,** Lipid synthesis: an indicator of antigen-induced signal transduction in antigen-binding cells, *J. Immunol.,* 122, 1278, 1979.
225. **Cambier, J. C., Monroe, J. G., and Neale, M. J.,** Definition of conditions that enable antigen-specific activation of the majority of isolated trinitrophenol-binding B cells, *J. Exp. Med.,* 156, 1635, 1982.
226. **Alderson, M. R., Pike, B. L., and Nossal, G. J. V.,** Effects of antigens and lymphokines on early activation of single hapten-specific B lymphocytes, *J. Immunol.,* 138, 1056, 1987.
227. **Noelle, R. J., Snow, E. C., Uhr, J. W., and Vitetta, E. S.,** Activation of antigen-specific B-cells: role of T-cells, cytokines and antigen in induction of growth and differentiation, *Proc. Natl. Acad. Sci. U.S.A.,* 80, 6628, 1983.

228. **Pike, B. L. and Nossal, G. J. V.,** A reappraisal of ''T-independent'' antigens. I. Effect of lymphokines on the response of single adult hapten-specific B lymphocytes, *J. Immunol.*, 132, 1687, 1984.
229. **Krusemeier, M. and Snow, E. C.,** Induction of lymphokine responsiveness of hapten-specific B lymphocytes promoted through an antigen-mediated T helper lymphocyte interaction, *J. Immunol.*, 140, 367, 1988.
230. **LoCascio, N. J., Haughton, G., Arnold, L. W., and Corley, R. B.,** Role of cell surface immunoglobulin in B-lymphocyte activation, *Proc. Natl. Acad. Sci. U.S.A.*, 81, 2466, 1984.
231. **Nagarkatti, P. S., Snow, E. C., and Kaplan, A. M.,** Characterization and function of autoreactive T-lymphocyte clones isolated from normal, unprimed mice, *Cell. Immunol.*, 94, 32, 1985.
232. **Snow, E. C., Fetherston, J. D., and Zimmer, S.,** Induction of the c-*myc* protooncogene after antigen binding to hapten-specific B cells, *J. Exp. Med.*, 164, 944, 1986.
233. **Grupp, J. A., Snow, E. C., and Harmony, J. A. K.,** Early events in B cell activation: binding of antigen to antigen-specific cells results in a phosphatidylinositol (PI) response, *Ann. N.Y. Acad. Sci.*, 494, 211, 1987.
234. **Grupp, S. A., Snow, E. C., and Harmony, J. A. K.,** The phosphatidylinositol response is an early event in the physiologically relevant activation of antigen-specific, *Cell. Immunol.*, 109, 181, 1987.
235. **Myers, C. D., Kriz, M. K., Sullivan, T. J., and Vitetta, E. S.,** Antigen-induced changes in phospholipid metabolism in antigen-binding B lymphocytes, *J. Immunol.*, 138, 1705, 1987.
236. **Snow, E. C. and Noelle, R. J.,** Thymus-dependent antigenic stimulation of hapten-specific B lymphocytes, *Immunol. Rev.*, 99, 173, 1987.
237. **Noelle, R. J. and Bartlett, W. C.,** T cell-dependent B cell activation, in *T-Cell Dependent and Independent B-Cell Activation,* Snow, E. C., Ed., CRC Press, Boca Raton, FL, 1990.
238. **Wilder, J. A., Cowdery, J. S., and Ashman, R. F.,** The influence of lipopolysaccharide content on the apparent B cell stimulating activity of anti-u preparations, *J. Immunol. Methods,* 110, 63, 1988.
239. **Sieckmann, D. G., Asofsky, R., Mosier, D. E., Zitron, I. M., and Paul, W. E.,** Activation of mouse lymphocytes by anti-immunoglobulin. I. Parameters of the proliferative response, *J. Exp. Med.*, 147, 814, 1978.

Chapter 3

FUNCTIONALLY IMPORTANT B CELL SURFACE MOLECULES

Bondada Subbarao, Arthur R. Baluyut, and Natarajan Muthusamy

TABLE OF CONTENTS

I. INTRODUCTION

Study of cell surface molecules has played a major role in understanding the functioning of the immune system. Early studies with conventional antisera led to the definition of cell lineage specific molecules which were also known as differentiation antigens. The nomenclature of Lyb and Lyt antigens was introduced to distinguish between the molecules expressed uniquely on B or T cells. We had earlier proposed that these differentiation antigens specific to T or B cells may also have a role in the functioning of the lymphocyte, especially with regard to its ability to perceive extracellular signals and to interact with other cells of the lymphoid system.[1] This has been very well demonstrated by the specific roles of CD4, CD8 and LFA molecules in T cell selection, activation and in their interaction with other accessory cells.[2,3] Similarly, Lyb-2, CD23, CD20, CD19, and CD39 molecules on murine and human B cells appear to function in the activation and progression of resting B cells from quiescence into cell cycle and finally into antibody secreting cells.[1,4]

Monoclonal antibody technology has revolutionized the field of the identification of cell surface molecules and is reflected in the development of the CD nomenclature. The description of murine B cell specific molecules has clearly lagged behind the human system though several new ones have been described. A critical new family of molecules consists of the ones that are identified as receptors for various cytokines that act on B cells and thus far none of these receptors have been correlated unequivocally with previously defined differentiation antigens. In this review we will discuss the new knowledge gained about a selected set of interesting cell surface molecules associated with B cell function. For more extensive description of the murine as well as human B cell markers we refer the reader to several reviews that have appeared recently[1,4,5,7] and to the other articles in this volume. Recently a unique B cell subpopulation that expresses the T cell lineage specific Ly1 antigen has been described. The $Ly1^+$ B cells, also known as $CD5^+$ B cells, are enriched in the peritoneal cavity and appear to consist of a unique B cell lineage. These B cells are self-renewing and have a tendency to produce autoantibodies.[6] This subject has been reviewed on several occasions.[6]

B lymphocytes are generated from stem cells and pass through a poorly characterized stage called pro B cell, in which the genes for immunoglobulin loci are in germ line configuration. The pro-B cells become committed to B cell lineage when they undergo productive rearrangements of heavy chain gene segments V, D, J and enter the pre-B cell stage.[5] This process takes place in two steps: the D and J segments rearrange first which is then followed by the joining of the V_H segments and expression of the cytoplasmic μ heavy chains. These pre-B cells undergo rearrangement in the light chain loci and become mature resting B cells that express IgM on the cell surface.[5] The process is antigen independent until the acquisition of the cell surface IgM receptors. The IgM^+ B cells express a second isotype, IgD, on their cell membranes and constitute the majority of the resting B cells. The IgM^+ IgD^+ B cells appear to participate in antigen specific responses. The antigen drives the clonal expansion of the mature resting B cells and their differentiation into Ig secreting cells or into memory cells. The events involved in B cell activation have been the subject of intense study and has led to the identification of specific biochemical pathways that mediate the transition of resting B cells into the G_1 state of cell cycle.[8] It appears to involve the increase in the inositol phosphates (IP), mobilization of intracellular calcium, translocation of protein kinase C (PKC) and transcriptional activation of oncogenes all of which are also reviewed elsewhere in this volume. In the presence of various cytokines or helper T cells these activated B cells enter into the S phase and later differentiate into IgM secreting cells or switch into cells that secrete other isotypes. In this review, we will first describe the important cytokine receptors on B cells, then describe the role of a select group of B cell surface antigens in B cell activation and finally a collection of new B cell surface

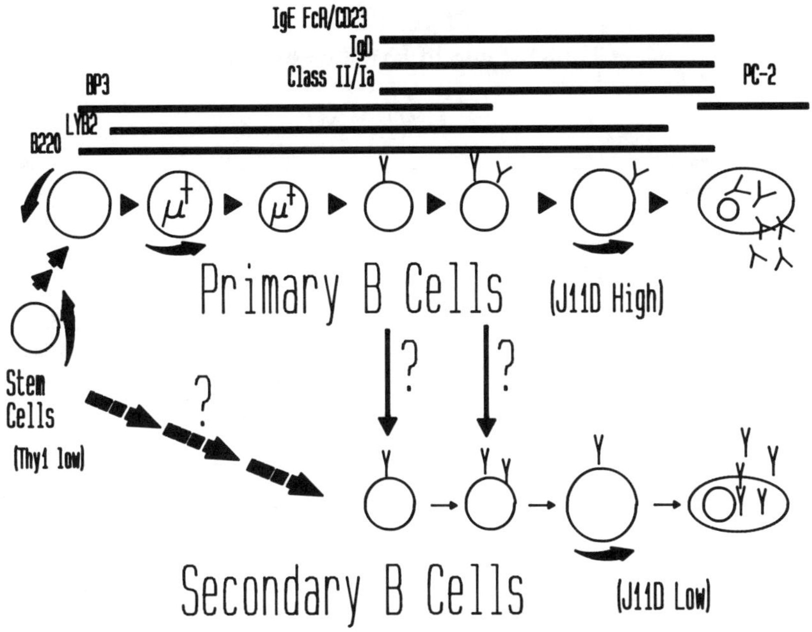

FIGURE 1. Stage-specific expression of B cell surface molecules. See text and References 4 and 5 for further details.

antigens that have interesting implications for B cell differentiation. Some of the steps in B cell differentation are the cell surface antigens associated with various stages of B cell development are summarized in Figure 1.

II. RECEPTORS FOR CYTOKINES

Availability of ultrapure lymphokines and monokines has allowed the identification of their respective receptors on the target cells, though very few are so far defined by monoclonal antibodies. Analysis of the structure and function of these receptors is very important in the light of the proven role of various cytokines in the induction of B cell proliferation and maturation. Since some of these lymphokines such as interleukins (IL) 4 and 5 have demonstrable effects both in the early stages related to growth and in the late stages related to antibody secretion and isotype switch, study of these receptors will be crucial to understand how the same agent is able to induce distinct effects at different stages of B cell activation. In the following we will focus on only those receptors that have been characterized at the biochemical level. Some important characteristics of these receptors are summarized in Table 1.

A. INTERLEUKIN-2 RECEPTORS

Although IL-2 is the major T cell growth factor, it is now becoming clear that activated B cells express IL-2 receptors.[9,10] In the human system, IL-2 has a profound role in induction of antibody secretion and the IL-2 specific receptors are easily detected.[9] In the murine system IL-2 receptors were better expressed on B cells activated with both LPS and anti-Ig antibodies.[10] The role of IL-2 in B cell differentiation was elegantly demonstrated in the murine B cell lymphoma (BCL_1) model system by Koshland and colleagues.[11,12] BCL_1 cells expressed both high and low affinity IL-2 receptors just like the T cells and the dissociation constants were of same magnitude for the binding of IL-2 to both B and T cells. The IL-2 receptor on the B cells was also composed of the 55 kDa α chain (Tac antigen) and the 75

TABLE 1
Characteristics of Cytokine Receptors

Receptor[a]	Molecular weight	Binding sites of: High affinity K_d	High affinity No./cell	Low affinity K_d	Low affinity No./cell	Mechanism of signal transduction
IL-2R	α-55 kDa, β-75 kDa	13 p*M* 14 p*M*	800 (BCL_1) 2,100 (T cell)	2.0 n*M* 14 n*M*	125,000 500,000	c-*myc* ↑ PKC activity is not essential
IL-4R	140 kDa (75 kDa fragment seen)	30—100 p*M*	300	30 n*M* (human B cells)	1,000	Induction of protein phosphorylation
IL-5R	47 kDa	55 p*M*	250	9.4 n*M*	6,200	Not known
IL-6R	80 kDa	10 p*M*	100—300	1 n*M*	24,000	Associates with gp130
IL-1-R[b]	80 kDa in T cells, 68 kDa in B cells	8 p*M* (pre-B cells)	4,000	0.4—2.1 n*M* (pre-B cells)	900—4,000	PKC activation, c-*myc*, Ha-*ras*
IFN-γ-R[c]	90 kDa (50 kDa fragment seen)	100—400 p*M*	1,360—11,700	—	—	cAMP ↑, Ca^{++} ↑ (slow increase)

[a] cDNAs for all the receptors (except IL-5R) have been cloned. Except in case of IL-5-R, all other receptors do not mediate increase in IP_3.
[b] IL-1 has only one class of binding sites in most cells.
[c] IFN-γ has only one class of binding sites.

kDa β chain which has been shown to be involved in signal transduction in T cells. Addition of IL-2 to the BCL_1 cells specifically induced synthesis of the J chain which is essential for the assembly and the secretion of the pentameric IgM. The IL-2 mediated BCL_1 differentiation also required the heterodimer of the α and β chains of the receptor since blocking the Tac antigen also inhibited the induction of J chains.

Studies with T cells showed that the α and β chains may bind different epitopes on the IL-2 molecule and that the β chain is important for signal transduction.[13] The genes for both α and β chains have been cloned. The detailed structure of these genes and regulation has already been reviewed.[14] The expression of IL-2 receptors on B cells is also modulated by IL-4 and IL-5. Loughnan et al. showed that IL-5 induced expression of the 55 kDa α chain (Tac) on the membrane as well as release of soluble Tac protein from murine B cells.[15] They suggested that the soluble receptor may be released by proteolytic cleavage of the surface receptors.[15] In addition, IL-4 induced expression of the β chain of IL-2-R as well as proliferative response in B cells to high doses of IL-2. In the presence of both IL-4 and IL-5 murine B cells responded to low doses of IL-2 and this was shown to be due to expression of both α and β chains which associate to form the high affinity receptors. Thus the expression of the two chains of the IL-2-R on B cells is uniquely regulated by the IL-4 and IL-5. In contrast, the expression of high affinity receptors on the HT-2 T cells and the BCL_1 B cells was partially down-regulated by IL-4.[16] There appears to be a tight and complex regulation of the biological response and the expression of interleukin-2 receptors which may be important for optimal beneficial response *in vivo*.

The mechanism by which the IL-2 receptors on B cells transduce the differentiation signal is not known but it did not involve the generation of IP_3, increase in calcium or cyclic nucleotides or translocation of protein kinase C.[11] IL-2 mediated T cell activation also does not appear to require protein kinase C.[17] The differentiative effects of IL-2 on BCL_1 cells were inhibited by IL-4. The inhibition appeared to be at the level of signal transduction since IL-4 did not affect the binding of IL-2 to its receptor.[11]

B. INTERLEUKIN-4 RECEPTORS

IL-4 is an important lymphokine that affects B cell activation at several stages.[18] IL-4 activates resting B cells to leave the G_o stage and to express high levels of class II molecules and receptors for IgE. Induction of B cell proliferation with antigen or with low doses of anti-Ig antibodies is critically dependent on IL-4. In mitogen activated B cells low doses of IL-4 induce IgG_1 secretion and inhibit IgG_{2a} synthesis. At high doses, IL-4 induces IgE secretion and is shown to be the essential factor required for the high levels of IgE found in helminth infected animals.[19a] IL-4 also induces proliferation of T cells and mast cells and activation of certain macrophage populations. More recently IL-4 was shown to have anti-tumor activities.[20]

Resting murine B cells had ~300 receptors/cell which bound IL-4 with an equilibrium constant of $33 \pm 0.7 \times 10^{10}\ M^{-1}$. More recently studies with human B cells suggested that human IL-4 may also bind to a much larger number of low affinity (K_d of ~30 nM) binding sites.[21] In addition to B cells, IL-4 receptors (IL-4-R) were also found on T cell, macrophage and mast cell lines at higher densities. Activated B cells had fivefold more receptors than resting B cells and IL-4, upregulated its own receptor expression. Cross-linking studies showed that IL-4 bound to a protein of 60 to 75 kDa when low to optimal doses of radiolabeled IL-4 were employed.[18,19] At higher concentrations of IL-4, both a 60 to 75 kDa and a 105 kDa molecule were identified on B cells as well as on T cells.[21,22]

The gene for the murine IL-4 receptor was cloned by subtractive hybridization technique using T cell lines that were selected for high density of IL-4-R expression.[23] Expression of one of these cDNA clones in COS-7 cells resulted in an IL-4 binding protein of molecular weight 140,000. This agrees well with the more recent experiments which employed a cocktail of protease inhibitors and showed that the receptor must be ~145 kDa molecule though fragments of 60 to 75 kDa were also observed as in the original studies.[24] The analysis of the cDNA sequence shows that the extracellular domain consists of 208 amino acids including seven cysteines and five N-linked glycosylation sites and a large intracellular domain of 553 amino acids in length. One interesting property of the receptor molecule was a relationship between the p75 component of the IL-2 receptor and the extracellular domain of the IL-4-R. The N-terminal segments had identical residues in 40 positions, and 40 conservative substitutions. Also, there was a preponderance of acidic residues in the cyto-plasmic domains of the IL-4 receptor and the IL-2 receptor β chain. These relationships may be biologically relevant if they share same signal transduction mechanisms and may explain why IL-4 inhibits the function of IL-2 in certain systems.

Another interesting feature of the IL-4-R cDNA clones was a 114 bp insertion at nu-cleotide 598 in two of the clones.[23] The insertion results in the production of a soluble form of the IL-4 receptor that is 230 residues in length and diverges nine amino acids from the transmembrane region adding seven new residues before termination. The soluble form also binds IL-4 with similar apparent Ka values as the membrane form and even inhibits the biological activity of IL-4 in inducing T cell proliferation. It remains to be seen if the B cell form of the IL-4 receptor is identical to the T cell derived molecule.

In working with human peripheral blood B cells, Foxwell et al. found that IL-4 receptor might consist of a 65 to 70 kDa doublet and a 110 kDa polypeptide.[21] They proposed that the association status of these three chains may determine the high and low affinity binding sites. It is conceivable that the 110 kDa molecule is similar to the 140 kDa polypeptide identified in the murine system. The requirement for additional chains is not clear since the transfection studies with the cDNA for the membrane or soluble form of the murine IL-4-R reproduced the basic binding properties of the natural receptor. Is this due to difference in the murine and human receptors or due to an artifact of proteolytic degradation? The former is a distinct possibility since the human IL-4 has more inhibitory than activating affects on human B cell function.

The mechanisms by which the IL-4 receptor transduces signals are not known. However, as in the case of IL-2, IL-4 also does not induce any increase in PI metabolism or calcium mobilization or increase in cyclic nucleotides or translocation of PKC.[8,18,19] IL-4 was shown to induce phosphorylation of a 42 kDa protein. The cDNA sequence for the IL-4-R indicates the presence of target sites for its phosphorylation but no kinase like domain.[23]

C. INTERLEUKIN-5 RECEPTORS

In the earlier studies IL-5 was known as B cell growth factor II (BCGF II) or B151K12-T cell-replacing factor (B151TRF).[19,25] Interleukin-5 has also pleiotropic properties in that it induces proliferation of the BCL_1 cells or of dextran sulfate activated B cells, antibody secretion from resting B cells as well as differentiation of T cells and eosinophils. IL-5 was originally characterized as a factor produced by the T cell hybridoma B151K12 or by a long term T cell line. cDNA for IL-5 has been cloned and recombinant IL-5 is now available. Binding studies with radiolabeled IL-5 demonstrated both low (K_d = 9.4 n*M*, 6200 sites/cell) and high (K_d = 55 p*M*, 250 sites/cell) affinity binding sites on the BCL_1 cells. The high affinity binding sites were undetectable on resting B cells, T cells, thymocytes or Con A activated blasts. LPS activated B cells exhibited about 40 high affinity binding sites per cell.[26] Transformed cell lines of both B and T cell lineage expressed tenfold more receptors than normal B cells. Recently two monoclonal antibodies (R52-120, R52-625) that react with the IL-5-R were developed.[27] Immunoprecipitation studies with the monoclonals and the cross-linking studies with the radiolabeled IL-5 gave discrepant results about the biochemical nature of the receptor. IL-5 was bound to a 47 kDa molecule upon reaction with bifunctional cross-linking reagents. On the other hand, mAbs immunoprecipitated three polypeptide chains of M_r 46, 130, and 140 kDa which are linked to each other by noncovalent interactions.[27] One possible explanation of these results suggested by Rolink et al. is based on analogy to the IL-2 receptors. Thus, binding of IL-5 to the 45 kDa molecule may generate only the low affinity binding sites whereas interaction with the 130 and/or 140 kDa chains may give rise to high affinity binding sites. Though the antibodies reacted with only a minor population of B cells (1 to 5% in immunofluorescence assays) they inhibited the IL-5 induced proliferation of the IL-5 dependent pro-B cell lines.[27]

Earlier studies with conventional antisera and T-cell replacing factor containing crude supernatants showed that IL-5 receptor was absent or nonfunctional on B cells from DBA/2 Ha mice which have an x-linked immune deficiency.[25] However, such receptors on B cells from CBA/N mice which have another x-linked immune defect were functionally normal. It will be interesting to know if this x-linked control of IL-5-R expression can be confirmed with the monoclonal antibodies and the recombinant IL-5. At present, there is no information about the nature of any second messengers that are generated in IL-5 stimulated lymphocytes.

D. INTERLEUKIN-6 RECEPTORS

The multifunctional cytokine IL-6 is made by a variety of cell lines and has important regulatory functions in many distinct biological systems.[28] For cells of B lineage, IL-6 induces Ig secretion from preactivated B cells and supports growth of hybridomas and plasmacytomas. IL-6 induces proliferation and IL-2 secretion in T cells, synthesis of acute phase proteins in hepatocytes and inhibits the growth and differentiation of myeloid leukemia cell lines. In the human system IL-4 induces resting B cells to produce IL-6 whereas in both murine and human systems autocrine production of the IL-6 was observed with several transformed cells of B lineage.[28] The receptor for human IL-6 was characterized extensively by Kishimoto and colleagues.[28,29] Resting B cells do not express any receptors for IL-6 whereas activated B cells expressed a few hundred receptors with a dissociation constant of $\sim 10^{-11}$ *M*. Once again, in analogy with the IL-2 system, low affinity binding sites with a K_d of 10^{-9} *M* were also present on B cells in much greater number.

Employing the expression vectors CDM8, Yamasaki et al. cloned the cDNA encoding the receptor protein for human IL-6.[29] This receptor consisted of a single polypeptide with 468 amino acid residues of which ~82 residues were present in the cytoplasmic domain. The intracellular domain had no similarity to any known tyrosine kinase and thus IL-6 differs from other growth factors, that signal cells by phosphorylation of tyrosine residues. In the external segment of the receptor, a region between residues 20 and 110 exhibits homology to the Ig-like domain of the constant 2 (C2) set of the members of the Ig superfamily.[29]

Biochemical studies employing radiolabeled IL-6 and a bifunctional cross-linker or a mAb MT18 identified an 80 kDa molecule on the myeloma cell line V66.[28-30] Since the cDNA codes for only a 50 kDa polypeptide, the remainder of the protein must be made up of carbohydrate moieties which was shown to be the case with endoglycosidase-F digestion. The most interesting aspect of this receptor is that the 80 kDa polypeptide was found to be associated with a 130 kDa cell surface glycoprotein (gp 130).[30] The association between the 80 kDa and the gp 130 polypeptides can be observed only when the cells are cultured with IL-6 at 37°C and this heterodimer can be detected as soon as 5 min after the addition of IL-6. The human IL-6-R was able to associate with murine gp 130 when a murine B cell line (M12) was transfected with the human IL-6-R cDNA. Two mutant cDNAs that either lacked only the cytoplasmic domain or both the transmembrane and cytoplasmic domain, were constructed and were transfected into the myeloid leukemia cell line M1. The cDNA lacking the cytoplasmic domain specified a membrane bound receptor whose binding properties were unaltered. Further the growth regulatory signals were transmitted perfectly well in the absence of the large cytoplasmic domain. The mutant cDNA from which both the transmembrane and the cytoplasmic domain were deleted specified a soluble form of the IL-6 receptor which was secreted into the medium. Furthermore, the soluble form of IL-6-R was able to associate with the gp 130 on M1 cells in presence of IL-6 and transmitted the growth inhibitory signals of IL-6 to these cells. Thus IL-6 receptors appear to be unique in that they make use of only the external domain and the gp 130 but not the cytoplasmic domain for signal transduction. The gp 130 molecule appears to be conserved in man and mouse. However, since IL-6 can mediate a diversity of effects in a number of distinct cell types, it is conceivable that the cytoplasmic domain and gp 130 may have functions in certain situations but not others.

E. INTERLEUKIN-1 RECEPTORS

IL-1 is produced primarily by the cells of the monocytic lineage. This cytokine exhibits a wide spectrum of biological activities on a diverse number of cell types.[31] In the two signal model of T cell activation, IL-1 probably provides the second signal required to stimulate T cell proliferation. IL-1 stimulates hematopoietic cell growth and differentiation, induces synthesis of acute phase proteins, pyrogenicity and synthesis of mediators like prostaglandins from fibroblasts and chondrocytes. IL-1 has been shown to enhance B cell proliferation and differentiation when used in concert with other mitogens and lymphokines.[19,32] The murine pre-B cell line 70Z/3 is induced by IL-1 to differentiate into a surface Ig expressing mature B cell.[33] Therefore, it will be important to study the IL-1 receptors on B cells to better define the role of IL-1 in B cell activation.

Most cells exhibit only one class of IL-1 receptors, whose affinity varies depending on whether IL-1α or IL-1β is employed.[33-35] EL4 cells, a thymoma, express about ~4000 IL-1β receptors/cell with a k_d of 0.4 n*M* whereas the K_d for the IL-1β receptors on the B cell line, RAJ1, is 2.1 n*M*. The differing K_d values of the IL-1-βR from these two cell lines appears to be due to changes both in the association and in the dissociation rates. The receptors on the T cells were made of an 80 kDa polypeptide whereas a 68 kDa receptor was found on the B cells. The differences in the M_r are not due to changes in glycosylation since treatment with neuraminidase or endoglycosidase F affected similarly the size of both the B and T cell forms of the receptor. Certain variants of the EL-4 cells, Con A activated spleen cells and the murine pre-B cell line 70Z/3 possess a second class of binding sites that are of high affinity (K_d = 63 p*M*, 112 sites/cell).[33] Thus far there is no evidence for a second chain associated with the IL-1 receptor.

Recently cDNA for the T cell form of the IL-1 receptor from the T cell line EL4.6.1C10 was cloned using the expression vector pDC201.[36] Expression of the IL-1-R in COS cells by transfection with the cDNA for the IL-1-R yielded an 80 kDa polypeptide that has IL-1 binding activity. The IL-1-R was found to be made up of 576 amino acids with a putative transmembrane segment between the residues 320 and 340. The amino-terminal domain consisted of 319 amino acids which contained all of the seven N-linked glycosylation sites and is likely to be the external domain. The cytoplasmic domain was 217 amino acid residues long. The predicted molecular weight is 64,598 suggesting that the remainder of the 80 kDa chain must be made up of carbohydrates. Since the cDNA transfected COS cells express IL-1 receptors which are of same affinity as the receptors on the EL4 cell line from which the mRNA was isolated, it is likely that the 80 kDa molecule is the sole receptor that gives rise to the low affinity binding sites. This was further confirmed by a recent experiment in which a deletion mutant cDNA that lacked the transmembrane and cytoplasmic domains was constructed and was employed to produce soluble IL-1 receptors. Such a soluble IL-1-R bound IL-1 with same affinity as the membrane IL-1. Therefore, it is not clear if the differences in sizes of B and T cell receptors are due to the existence of another gene or due to post-transcriptional or translational modifications or due to differences in the murine (EL4 cells) and the human (RAJI cells) IL-1-R genes.

The external segment of the IL-1 receptor is organized into three domains which are related to each other and to the members of the Ig gene superfamily. The relationships based on percentage amino acid identity are domain 1 with rat CD8 chain II, 31.6%, domain 2 with human IgCγ 1, 31.6% and domain 3 with human T cell receptor Cβ, 33.3%. The cytoplasmic domain has no tyrosine kinase like structure or tyrosine phosphorylation site but has a PKC phosphorylation site. In contrast to the other lymphokine receptors reviewed thus far, IL-1 was shown to activate PKC in the pre-B cells and production of diacylglycerol in two T cell lines. In addition, IL-1 induced an increase in expression of the oncogenes c-*myc* and Ha-*ras* but not of c-*fms* in the RAJ1 cells.[34]

F. INTERFERON-γ (IFN-γ) RECEPTORS

IFN-γ is produced by the activated T cells and in the murine system it is restricted to the T_h1 subset of T cells.[37,38] B lymphocytes are uniquely regulated by IFN-γ. The activation of murine B cells with IL-4 or anti-Ig but not with anti-Lyb-2 antibodies is inhibited by IFN-γ and the inhibition occurs during early stages of activation (18 and our unpublished observations). IFN-γ inhibits IL-4 induced class II gene expression in B cells but enhances class I gene expression in monocytic cells.[18,19,37,38] During late stages of B cell activation IFN-γ inhibits IL-4 induced IgG_1 secretion but promotes IgG_{2a} synthesis in mitogen activated B cells. IFN-γ has also been shown to act as T cell replacing factor in antibody responses to sheep erythrocytes.[39] In the pre-B cell line 70Z/3, which expresses only cytoplasmic μ chains, IFN-γ induces κ chain synthesis and surface expression of IgM.[33] In contrast to its inhibitory effects on murine B cells, IFN-γ enhances the proliferative response of human B cells to anti-Ig stimulation. IFN-γ synergizes with IL-2 to enhance Ig secretion from human B cells activated with *Staphylococcus aureus* Cowan I strain (SAC).

There is only one class of binding sites for IFN-γ on most cells which is unlike the other cytokines discussed here.[33,37,40] The actual affinity of the IFN-γ receptor does vary depending on the cell type. Thus the K_d values range from 100 to 400 p*M* and the number of binding sites range from ~1360 on resting B cells to 11,700 on the RAJI and CB cell lines. Like the RAJI cells, the murine pre-B cell line also has a large number of IFN-γ binding sites. Interestingly, phosphorylated IFN-γ is employed for the binding studies whereas most studies with other cytokines employed radioiodinated products.[37] The ^{32}P labeled IFN-γ offers better specific activity (10- to 100-fold) and biological activity than the iodinated molecule.

Cross-linking experiments as well as affinity purification with mAB demonstrated the existence of 90 and 50 kDa molecules on human B cells and B cell lines.[37,40,41] Aguet and colleagues isolated a cDNA clone for the human IFN-γ-R by screening λgt11 expression libraries with a rabbit antiserum to the receptor.[41] With the help of the cDNA clone a genomic clone was isolated from a cosmid library. Transfection of mouse L1210 cells with this gene yielded IFN-γ receptor positive cell lines. Such transfectants expressed receptor specific mRNA species which was of same size as found in the RAJI cells. However, only some of the transfectants appeared to respond to IFN-γ with an increase in class I gene expression.

The cDNA encodes a protein of 489 amino acids with a transmembrane region that spans from position 279 to the position 299. The extracellular domain consists of five potential N-linked glycosylation sites and several serines and threonines (12% of the amino acids were serines) which can harbor O-linked sugars. Therefore, the difference between the predicted M_r of 54,000 and the observed value of 90 kDa is likely to be due to glycosylation. The 50 kDa polypeptide appears to be a degradation product of the 90 kDa molecule. The cDNA does not reveal any obvious clues about the signal transduction mechanism. In the pre-B cells IFN-γ induces PKC translocation and Na/H^+ exchange though these events do not appear to be required for the induction of surface Ig expression.[43] In normal B cells IFN-γ enhances intracellular cyclic AMP whereas in macrophages it induces a slow rise in calcium and increase in PKC activity without affecting PI turnover.[37] Because of this diversity of the signaling mechanisms, it will be interesting to know if the receptor structure in different cell types is the same.

III. OTHER FUNCTIONALLY RELEVANT MURINE B CELL SURFACE MOLECULES

In this section we will review the functionally important B cell surface molecules which are not shown to be receptors for any growth or differentiation factors. These will include Lyb-2, Ia antigens, Ly-5, IgE and C3 receptors, and Mb-1 molecules which have all been shown to affect B cell function. We will concentrate on the murine system since the human CD antigens were reviewed recently.

A. LYB-2

Sato and Boyse defined the Lyb-2 antigen by means of a conventional allo antiserum prepared against the I-29 ascites tumor cells.[44] Subsequently the monoclonal antibodies 10.1.D2, 9-6.1, K24.67 recognizing the Lyb-2.1 allele and the clone NEI-7 reacting with the Lyb-2.3 allele were produced.[1,45] Four alleles and five specificities of this antigen can be distinguished with a combination of alloantisera and mAb.[45] It is specified by a gene linked to the MUP-1, Lyb-4, Lyb-6 loci on chromosome 4 and was shown to be a glycoprotein of M_r 45 kDa.[1,45] Lyb-2 antigen is expressed very early during B cell differentiation. It is present on pro-B cells, pre-B cells as well as on mature B cells but not on macrophages, T cells or any other tissues that were tested. Lyb-2 antigen is found on bone marrow but not on fetal liver derived pre-B cells. Earlier we showed that mAB against Lyb-2 induced blast transformation and proliferation of resting pre-B lymphocytes.[46] The anti-Lyb-2 activated B cells were induced to secrete Ig in the presence of T cell derived factors.[46,56] Two interesting properties of the Lyb-2 mediated B cell activation are: first, immobilized anti-Lyb-2 antibodies were unable to induce B cell proliferation suggesting that Lyb-2 molecules may have to be internalized for eliciting a response.[47] Alternately, the very short external domain of

Lyb-2 molecule may not have been easily accessible to the immobilized antibody. Second, B cell activation with anti-Lyb-2 mAb did not require cross-linking of the Lyb-2 molecules since monovalent Fab′ fragments were as effective as $(Fab')_2$ antibodies in inducing B cell proliferation.[48] Although such a phenomenon is rare, an anti-idiotype antibody to the T cell receptor on the D10 T cell line also induced T cell activation in the absence of cross-linking. Since most growth factors are monomeric, bind single chain molecules and may mediate their effects without cross-linking the receptors, we had suggested that Lyb-2 molecules may be receptors for a B cell growth factor. We and others showed that anti-Lyb-2 induced an increase in B cell Ia-expression which is a characteristic property of the lymphokine, IL-4.[49-51] Also, in some studies but not others, anti-Lyb-2 antibodies enhanced anti-μ induced B cell proliferation on the basis of which it was suggested that Lyb-2 molecules may be the receptors for IL-4.[49,51,52] However, by performing absorption experiments to remove IL-4 in the presence and absence of anti-Lyb-2 mAb we showed that Lyb-2 was not the receptor for IL-4. This was also confirmed in the direct binding experiments with radiolabeled IL-4 and by the recent cloning of the gene for the IL-4 receptor (see above).[18,23]

Nakayama et al. cloned the cDNA specifying the Lyb-2 molecules using the CDM8 expression vector.[53] It was a protein of 354 amino acids with a core M_r of 40,496. The only N-glycosylation site was on the region C-terminal to the predicted transmembrane domain between the residues 96 to 116. There was no traditional signal peptide. Therefore it appeared that Lyb-2 may have an inverted orientation with a cytoplasmic N-terminal domain. Indeed, the carbohydrates were shown to be N-linked by demonstrating their susceptibility to endoglycosidase F. Lyb-2 is thus a type II membrane protein with a 96 amino acid long N-terminal domain inside the cell. A partial relationship was found between the Lyb-2 protein and the human CD23 (low affinity Fc receptor for IgE) and both rat and human asialoglycoprotein receptors. These sequences were, respectively, 29.1, 28.3, and 27.5% identical to Lyb-2.1 when comparisons were made over restricted regions. We have analyzed the sequence of the cytoplasmic domain for the presence of a tyrosine kinase like structure and found no evidence.

Ashida and colleagues reported that Lyb-2 may not be a single polypeptide but that it may be disulfide bonded to a 105 kDa chain.[54] On the other hand, Nakamura et al. found that Lyb-2 was made of 45 kDa and 95 kDa polypeptide chains and that it was related to the alloantigens Ly-19 and Ly-32.[45] The genes specifying Ly-19 and Ly-32 were shown to be closely linked to the Lyb-2 locus on chromosome 4. Unlike Lyb-2 which is B cell specific, Ly-19 and Ly-32 antigens were found on B cells as well as on a subpopulation of T cells. Sequential immunoprecipitation experiments showed that clearing of the lysates with one of the two antibodies to Ly-19 or Lyb-2 removed all the molecules reactive with the other mAb suggesting that these epitopes are shared on a heterodimer. In contrast to these studies, Nakayama et al. found that COS cells transfected with the Lyb-2 cDNA bound the anti-Lyb-2 mAb and that immunoprecipitation analysis detected only a single 45 kDa polypeptide chain.[53] Therefore, the additional molecules seen in some experiments may be loosely associated with Lyb-2 or an artifact.

Perturbation of the Lyb-2 molecules with the mAb induces an increase in PI metabolism and elevation of IP_3 and cytosolic calcium.[49,55] Recently we showed that pertussis toxin, cholera toxin, forskolin and dibutyryl cAMP inhibited anti-μ but not anti-Lyb-2 induced B cell proliferation (Figure 2). This difference did not appear to be due to the involvement of different B cell subsets, since the number of B cells entering the G_1 phase of cell cycle with anti-μ was not further increased by the addition of anti-Lyb-2 (Muthusamy and Subbarao, manuscript in preparation). The selective inhibition of Ig but not Lyb-2 mediated activation can also be accomplished with a physiological ligand such as IFN-γ which is shown to elevate intracellular cAMP. These agents blocked the entry of B cells into the G_1 phase of cell cycle as measured by changes in cell size and transferrin receptor expression. Indeed,

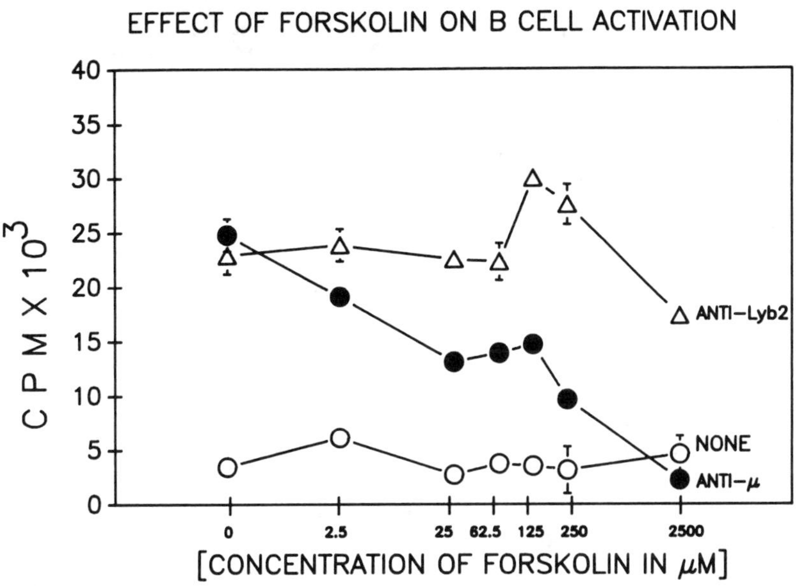

FIGURE 2. Effect of forskolin on anti-IgM or anti-Lyb-2 induced B cell proliferation. 5 $\times$ 10^5 resting B cells were cultured in serum free IF12 medium for 48 h with goat anti-IgM (50 μg/ml) or anti-Lyb-2.1 (10 μg/ml) antibodies in the presence of indicated concentrations of forskolin.

forskolin blocked anti-μ but not anti-Lyb-2 induced IP_3 generation as well as calcium mobilization (Figures 3 and 4). Thus Ig and Lyb-2 appear to employ different pathways to generate the second messengers IP_3 and Ca^{++} of which only one pathway is susceptible to inhibition by cAMP. It is conceivable that these two receptors are employing different transducer proteins such as the postulated G proteins to activate the phospholipase C or that they are stimulating distinct phospholipases which differ in their susceptibility to cAMP mediated feedback inhibition. Thus phospholipase C isozyme γ but not the isozymes β and δ were shown to be phosphorylated by cAMP dependent PKA in a rat glioma cell line.[57] Since the Lyb-2 molecules have a large cytoplasmic domain unlike the IgM molecules, Lyb-2 may transduce activation signals directly. Thus Lyb-2 but not Ig mediated phospholipase C activation may be resistant to cAMP induced phosphorylation events. Thus this system of stimulation of B cells through the surface Lyb-2 provides evidence for the first time in favor of the functioning of multiple pathways to generate IP_3 in B cells. Further elucidation of these mechanisms will definitely unravel the components of the signaling apparatus in B cells.

B. LY-5 OR B-220 ANTIGEN

Allo- as well as xeno-specific antibodies define the Ly-5 antigens which are present on both T and B lymphocytes.[58-61] Ly-5 antigens were made up of 200 kDa single polypeptide chains and hence are also referred to as T200 molecules or CALLA for common acute lymphocytic leukemia antigen or CD10 in the human.[4,59] However, one form of the T200 molecule consists of a 220 kDa polypeptide which is uniquely expressed in B cells and is known as B220 antigen.[5,60] It is expressed on pre-B cells and mature B lymphocytes and is widely used to quantitate B cell precursors.[5] cDNA cloning data showed that Ly-5 molecules have a large external domain of 402 amino acids, a cytoplasmic domain of 705 amino acids and a 22 amino acid transmembrane segment.[61] The multiple forms of Ly-5 molecule in different cell types appear to be due to the presence of various species of mRNA that have arisen by alternate splicing.[62]

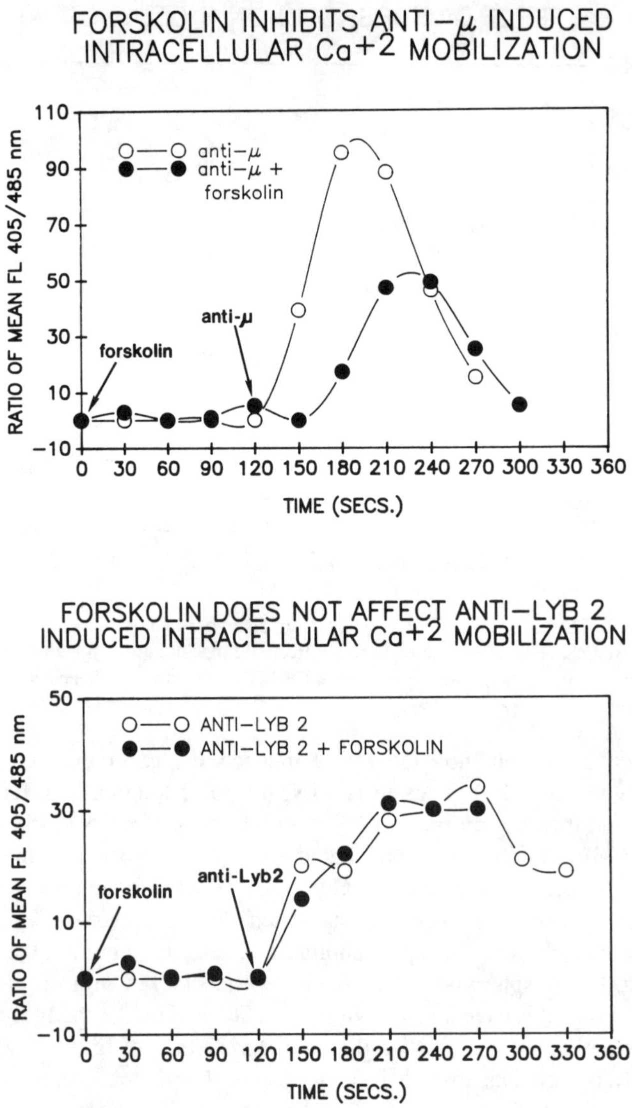

FIGURE 3. Effect of forskolin on anti-IgM or anti-Lyb-2 induced IP3 generation. 2×10^6 small resting B cells labeled with ^{3}H myoinositol were preincubated with 100 μM of forskolin at 37°C for 5 min wherever indicated. The cells were then stimulated with either goat anti-mouse IgM or anti-Lyb-2.1 antibodies for 3 min. IP_3 was separated on ion exchange columns. The results are expressed as counts per minute/2×10^6 cells. The results are mean of four independent experiments with SE bars.

Anti-Ly-5 mAb inhibited antibody responses to the TI-2 antigen TNP-Ficoll and to the TD antigen SRBC but not to the TI-1 antigens TNP-BA or TNP-LPS.[63] Since Ly-5 is present on B cells, T cells and macrophages, inhibition studies were performed with B cells from Ly-5.1 and T cells or macrophages from Ly-5.2 congenic strains of mice. Use of allele specific Ly-5 antibodies in this system demonstrated that inhibition of antibody response was obtained only when the antibodies were directed against the Ly-5.1 allele present on the B cells. Both TI-2 and TD responses require accessory cell function whereas the TI-1 antigen responses are much less dependent on macrophages suggesting that anti-Ly-5 may block an accessory cell signal.[63,64] This appears to be the case since a macrophage but not T cell derived factor overcame the inhibitory effects of the anti-Ly-5 mAb.[63]

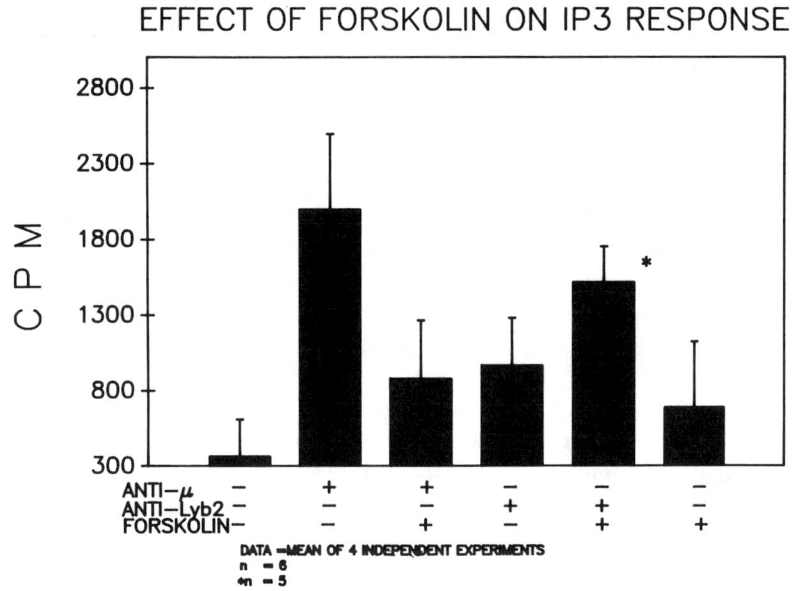

FIGURE 4. Effect of forskolin on anti-IgM or anti-Lyb-2 induced increase in intracellular Ca^{++} levels. 1×10^6 B cells loaded with Indo-1 AM were stimulated with anti-IgM or anti-Lyb-2 antibodies in the presence or absence of 100 μ*M* forskolin as indicated. Changes in the fluorescence emission at 405 nm (FL2) and 485 nm (FL1) were measured on a FACSTAR PLUS machine. The values of ratio of fluorescence are plotted as a function of time.

Yakura et al. have recently shown that anti-Ly-5 antibodies inhibited LPS induced IgG but not IgM responses.[65] All IgG subclasses were equally affected and the inhibitory effects were primarily due to a reduction in the precursor frequency. It was suggested that Ly-5 may have a role in the generation of B cells that can switch to other isotypes. In these studies antibodies that react with Ly-5 epitopes on all lymphoid cells were employed. It will be interesting to know if antibodies recognizing the B cell form of Ly-5, i.e., B-220, will have similar effects.

Recent studies of a lymphocyte derived protein tyrosine phosphatase (PTPase) revealed a surprising similarity of the PTPase sequence to that of two cytoplasmic domains of the CD45 molecule.[97] On the basis of these studies CD45 was hypothesized to dephosphorylate proteins. The hypothesis was proved to be correct when such an intrinsic PTPase activity was demonstrated for the CD45 molecule. CD45 was found to have two cytoplasmic domains that have PTPase-like sequences though it is not clear if both have PTPase activity. B220 and CD45 belong to the same family of 180 to 220 kDa molecules present on lymphocytes. Studies on the B220 molecule in mouse will have to be performed to confirm these PTPase-like properties which might explain the inhibitory effects of anti-Ly-5 antibodies on B cell function.

C. RECEPTOR FOR COMPLEMENT COMPONENT C3

B cells have been known to possess receptors for the C3d, C3b which are cleavage products of the complement component C3. These receptors have long been implicated in B cell responses.[66-69] In the human system C3d receptor is known as CR2 or CD21 antigen and was found to serve as Epstein-Barr virus (EBV) receptor.[4,70] EBV is known to be a potent polyclonal B cell activator for human B cells.[4,71] Some antibodies to CR2 induce proliferation of human B cells.[4,71] In the murine system Lernhardt et al. showed that immobilized C3 or C3d or C3b induce proliferation of preactivated B cells but had no effect on resting B cells.[72] Soluble C3 was without any effect. However, it was noted that a T cell derived soluble factor known as α-BCGF also promoted growth of activated B cells and the effect of α-BCGF was inhibited by a peptide derived from C3. Probing of the T cell mRNA with an oligonucleotide probe from murine C3 cDNA identified 4.2 and 1.9 kb mRNA species. Since the C3 producing cells contain 5.4 kb message, the α-BCGF is related to C3 but is distinct from C3. Since C3 appears to act on preactivated B cells which are in the G_1 phase and facilitated their entry into the S phase of cell cycle, it was considered to be a progression factor.[72]

The C3d receptor in humans is a 140 kDa glycoprotein whereas the murine homolog may exist in multiple species of M_r 145, 150, and 190 kDa.[73-75] The genes for these receptors have been cloned and were found to be homologous.[73,74] The murine C3d receptor is a transmembrane protein with a small cytoplasmic domain of 35 amino acids. The C3b receptors (CR) are 210 to 290 kDa in humans but only 75 kDa in the mouse. It is thought that the human CR1 may have evolved due to duplication of murine CR1 and inheritance of three additional 60 amino acid repeats as evidenced by the homologies in the gene sequences for the two proteins.[75] Currently the mechanisms by which C3 fragments transmit activation signals to B cells are unknown. Since C3 behaves like α-BCGF, further analysis of C3 receptor genes may help in the definition of the receptor for α-BCGF.

D. IgE Fc RECEPTOR

The low affinity receptor for IgE ($FC_\epsilon RII$) is expressed on several types of hematopoietic cells including B lymphocytes.[4,76,77] The importance of the $FC_\epsilon RII$ for B cell activation was suggested by several findings. First, the human B cell specific CD23 antigens were the same as $FC_\epsilon RII$. Second, IL-4 enhanced the expression of the CD23 molecules on B cells. Third, some anti-CD23 antibodies synergized with PMA to induce B cell proliferation. Fourth, the CD23 antigen was shown to be cleaved from the cell surface, which acted as a growth factor for activated B cells.[77]

The cDNAs encoding human CD23 were isolated by several groups a few years ago and the cDNA specific for murine $FC_\epsilon R$ was isolated recently.[78-81] In both systems the receptor is encoded by a gene that lacks any signal sequence and has its transmembrane segment closer to the NH_2-terminal. This finding together with the observation that the only N-glycosylation site is C-terminal to the transmembrane domain suggested that the IgE receptors are inversely oriented with the N-terminal segment inside the cytoplasm. Thus the CD23 antigen and the $FC_\epsilon RII$ are similar to the human CD20 antigen and the murine Lyb-2 molecule in being type II membrane proteins.[4,53,78-81] The cytoplasmic segments are 21 and 25 amino acids long, respectively, in the human and in the mouse. The two molecules share a 57% identity (identical residues) or a 74% identity if conservative substitutions are considered. Both proteins exhibit extensive homology with C-type animal lectins. In both the systems transfection experiments demonstrated that the proteins coded by the cDNAs bind IgE. Further, COS cells transfected with the cDNA for the murine $FC_\epsilon RII$ bind rodent but not human IgE which is also the property of the natural receptor. Both the proteins have a core of M_r ~37 kDa whereas the glycosylated molecules were 45 kDa in M_r.

Immunologically, murine CD23 homolog was defined with the rat monoclonal antibody B3B4.[76] B-cell expression of murine $FCR_\epsilon II$ is also enhanced with IL-4 but the $FCR_\epsilon II$ was super-induced in the presence of IL-4 plus LPS.[82] Anti-Ig was unable to synergize with IL-4 in enhancing the expression of this receptor. Murine B cells resembled human B cells in shedding the fragments of the receptor generated by a proteolytic cleavage.[76,82] In contrast to the soluble human CD23 fragments, the soluble form of the murine receptor did not induce B cell proliferation.[76] In this context it is interesting to note that the soluble CD23 molecules prepared by transfection with mutant cDNA did not exhibit the growth-promoting effects of the natural fragments.[83] These mutant soluble CD23 molecules were of M_r 25 to

27 kDa which are not glycosylated but bind IgE. In the original experiments of Swendemann and Thorley-Lawson soluble CD23 consisted of 25 and 12 kDa fragments.[84] It is not clear if the 12 kDa fragment is required for detecting the autocrine properties of the 25 kDa fragment. More recently, Delespesse reported that another soluble CD23 created by recombinant DNA technology enhanced IgE synthesis.[77] These differences in the various soluble CD23 molecules may provide an explanation for the inability of mouse $FC_{\epsilon}RII$ to induce B cell growth. Study of these soluble receptors is important since for a long time T cells have been reported to make IgE binding factors that regulate IgE synthesis.[76,77] Thus far there is no evidence for the existence of $FC_{\epsilon}RII$ specific mRNA in murine or human T cells.[77,81]

Very recently a novel role for $FC_{\epsilon}RII$ receptors was proposed. Kehry and Yamashita suggested that the B cell Fc receptors for IgE may help focus the antigen onto B cells and its subsequent degradation and presentation to T cells.[85] B cells were loaded with a DNP-specific IgE so that the $FC_{\epsilon}RII$ was occupied. Such B cells presented DNP conjugated rabbit Ig to T cells specific for rabbit Ig better than the unconjugated normal rabbit Ig molecule. The antigen presentation in this system was as efficient as the one mediated by surface Ig as evidenced by the presentation of rabbit anti-Ig with the same B cell population. Thus $FC_{\epsilon}RII$ may act as an adjuvant to increase specific T cell responses to small doses of antigen.

E. Ia ANTIGENS

Class II molecules, specified by MHC linked genes, play a major role in the activation of T lymphocytes and in the selection of the T cell repertoire. Ia antigens, also known as class II molecules, are not usually discussed in reviews on B cell differentiation antigens. They are viewed as passive recognition structures on antigen presenting cells that facilitate cell-cell interactions. However, among all cells of the body B lymphocytes express a maximum number of Ia antigens per cell in the absence of ongoing immune response or inflammation. Conceivably Ia antigens on B cells may have a functional role in B cell activation and currently there is extensive evidence in favor of such a notion (reviewed in Reference 86).

The observations supporting a functional role for B cell Ia can be summarized as:

1. IL-4, anti-Ig antibodies as well as several B cell stimulants increase the expression of class II molecules on B cells.[87,88]
2. Antibodies to HLA-DR in man and to class II molecules in the mouse were shown to replace T helper cell function.[89]
3. Overwhelming the B cells with a number of anti-Ia antibodies simultaneously inhibited proliferation and differentiation response of B cells to the powerful B cell-mitogen LPS.[90,90a]
4. Perturbation of the Ia antigens on some CH series of B cell lymphomas induced Ig secretion only in presence of antigen.[91] Anti-Ia antibodies induced Ig secretion from the BKS series of lymphomas in the absence of antigen.[92]
5. Allo- or self-Ia reactive T cells induced proliferation and differentiation of the appropriate Ia^+ B cells.[93,94]

Most of the systems were subject to the criticism that the effects on B cells are not due to a signaling role for Ia but are due to indirect effects mediated by other factors generated in the system. This issue was addressed in two model systems. First, helper T cells were incapacitated such that they do not secrete any soluble factors. Interaction of such T cells with B cells was shown to be essential to activate resting B cells.[95,96] This approach is reviewed elsewhere in this volume.

In the second system developed by us, purified B cells were cultured with anti-μ and anti-Ia antibodies to evaluate the ability of Ia molecules to function as signaling molecules.[86]

In serum free cultures suboptimal doses of anti-μ antibodies induced modest B cell proliferation. Addition of purified IA or IE specific monoclonal antibodies significantly enhanced the proliferative response of resting B cells to anti-μ. In the absence of anti-μ, anti-Ia antibodies had no effect on B cell response. The effects of the anti-Ia mAb were shown not to be due to production of cytokines. In this system, interaction of Ia with the anti-Ia mAb resulted in an activation signal that augmented B cell response to anti-μ. Also, the addition of anti-Ia but not of anti-μ can be delayed up to 30 h to observe the synergistic effects. The anti-Ia antibodies did not affect the entry of anti-Ig activated B cells to enter the G_1 phase of the cell cycle and therefore, must be facilitating the transition of cells from the G_1 into the S phase of cell cycle. Consistent with this, anti-Ia antibodies did not induce or augment anti-Ig mediated PI turnover or calcium mobilization which are early events in B cell activation.[8,55] However, anti-Ia antibodies enhanced anti-μ induced increases in the levels of the mRNA for the oncogene c-*myc* (our unpublished observations). Inhibition of c-*myc* expression by treatment with anti-sense oligonucleotides blocked T cells and fibroblasts in the G_1 phase of cell cycle.[98] Thus c-*myc* oncogene may have a role in the movement of cells from G_1 into the S phase of cell cycle, which is the step facilitated by the perturbation of the B cell Ia antigens. Consistent with these findings Cambier et al. also found that anti-Ia antibodies and lymphokines induced growth and differentiation of B cells.[99] In addition, Pereira et al. demonstrated that anti-Ia antibodies significantly enhanced B cell proliferation induced by anti-μ-LPS conjugates.[100] We are proposing that perturbation of B cell-Ia during T cell-B cell interactions will lead to activation signals that facilitate the entry of B cells into the cell cycle. Such growth signals delivered to B cells via the Ia antigens may be as important as the ones generated by the surface Ig and the lymphokine receptors.

What are the biochemical mechanisms by which class II molecules transduce activation signals? $(Fab')_2$ Anti-Ia antibodies alone did not induce elevation of IP_3 but increased the intracellular levels of cAMP.[101] Delayed addition of cAMP agonists was found to enhance B cell proliferation induced by PMA and calcium ionophore, ionomycin. Anti-Ia mAbs were also shown to induce translocation of PKC to the nucleus.[102] In the human B cells, anti-DR antibodies were shown to induce an increase in cytosolic calcium.[103,103a] In murine B cells, anti-Ia antibodies had to be cross-linked with an anti-IgG antibody to observe such increases in cytosolic calcium (Figure 5). The increase in Ca^{++} appeared to be mainly due to influx from outside the cell. These observations definitively establish the concept that class II molecules deliver activation signals to B cells which leads to generation of some of the well known second messengers. It remains to be established that these same signaling events take place in B cells during cognate T cell-B cell interactions involving the recognition of B cell Ia and antigenic peptides.

F. Mb-1/B34 PROTEINS

Two novel approaches to the study of B cell specific surface antigens led to the discovery of the antigens known as Mb-1 and B34. The discovery of these structures will revolutionize the understanding of the putative signaling structures associated with surface IgM molecules present on mature B cells. The role of surface Ig in signaling resting B cells to enter growth cycle is now established unequivocally. Despite this, it was noticed that the μ_m heavy chain has only a three amino acid long cytoplasmic tail which is too short to permit specific association with any structures on the inner membrane or in the cytoplasm. In addition, the transmembrane domain of the μ heavy chain is highly conserved between different species and Ig classes. It was postulated that there might be interaction between the transmembrane part of IgM with other membrane molecules. Such an association of T cell receptor with CD3 complex is well documented in T cells. Two candidates for the Ig-associated structures are the B34 and Mb-1 proteins, respectively, discovered by Hombach et al. and Sakaguchi et al.[104,105]

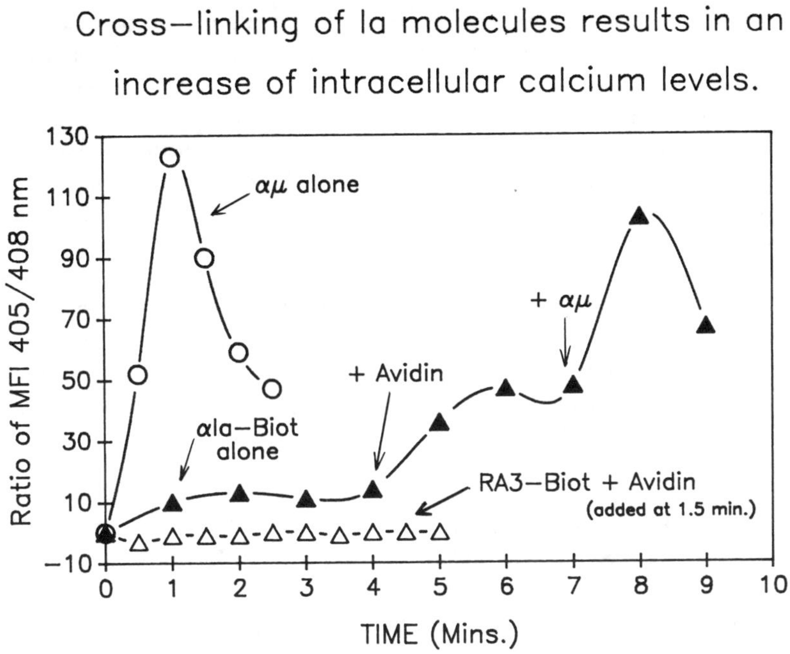

FIGURE 5. Effect of cross-linking of Ia molecules on increase in intracellular Ca^{++} levels. Indo-1 AM loaded B cells were stimulated with biotinylated anti-Ia (MKD6) antibodies followed by avidin at indicated time points. As a control the biotinylated anti-B220 antibody (RA3-3A1) was used. The increase in intracellular calcium was measured as described in Figure 4.

Hombach et al. found that transfection of expression vectors, coding for μ_m protein into lymphoma and plasmacytoma cells resulted in fully assembled membrane IgM in lymphomas but not in plasmacytomas.[104] An expression vector (psvμm) that codes for the membrane bound form of μ heavy chain was constructed and transfected into the 558L plasmacytoma cells. The transfectants were called 558Lμm. The transfected plasmacytoma cells did not express surface IgM. A variant (558LμM3) that expresses surface IgM was selected by flow cytometry techniques. Immunoprecipitation analysis of the surface IgM in the lysates of 558LμM3 cells, generated with the mild detergent, digitonin, revealed the existence of a 34 kDa molecule (B34) that was associated with the surface IgM. SDS PAGE analysis of these lysates was aided by the fact that the transfected μ_m gene was specific to the hapten, NP, which permitted the affinity purification of the IgM complex on NP-coupled Sephadex. There was very little nonspecific reactivity under these conditions. B34 was found to be a glycosylated and disulfide bonded protein. It was present in the IgM^+ transfectant but not in the parental cell line that did not express sIgM. It is not clear at present if B34 is associated with surface IgD. B34 may function to transport the assembled IgM to the cell surface and/or in Ig mediated signal transduction in B cells.

Sakaguchi et al. constructed a pre-B cell specific library of cDNA clones by subtractive hybridization with the mRNA from a T cell hybridoma.[105] One of these clones named m-mb-1 was uniquely expressed in pre-B cells and mature B cells but not in plasma cells. The cDNA clone identified transcripts of ~1 kb in length. Mb-1 transcripts were detected in resting B cells, LPS blasts and fetal thymocytes and were absent in liver cells.

Sequence analysis of a full length cDNA clone showed that Mb-1 protein is made up of 220 amino acids with two N-glycosylation sites. A 22 amino acid long transmembrane domain and a 61 residue long C-terminal cytoplasmic domain are predicted to be present in mb-1. It was not related to any known protein. A human homolog was isolated from the Burkit lymphoma cell line Daudi. In the segment that was sequenced, the human mb-1 gene showed strong homology (~90% in nucleotide sequence) to the murine m-mb-1 gene. Antibodies raised to a synthetic peptide corresponding to the intracellular region of mb-1 detected this protein in the cytoplasm of pre-B and B cells. Antibodies to a fusion protein with protein A gene detected surface expression of this protein on B cells. It is of interest that the mb-1 transcripts were present in the membrane IgM expressing 558 Lμm3 cells of Hombach et al. The parental cells did not have Mb-1 transcripts suggesting that mb-1 gene may be related to B34.

In a third approach Cambier and colleagues found that a brief exposure of B cells to anti-Ig antibodies results in the generation of ~30 to 34 kDa phosphoprotein associated with IgM.[106,106a] Its relationship to mb-1 or B34 is not yet established. It is clear from these studies that IgM associated structures exist which may form the B cell-Ig complex in analogy with the T cell receptor-CD3 complex in T cells. Such a complex may play a vital role in signal transduction.

IV. B CELL SURFACE MOLECULES IMPORTANT FOR B CELL FUNCTION AND DIFFERENTIATION

In this section we will briefly describe three sets of markers that may be important for B cell differentiation and function. These are the cell surface antigens BP-1, BP-3, J11d, ThB and CD2.

A. BP-1, BP-3

In a novel approach to identify interesting cell surface molecules, Cooper and colleagues immunized wild mice with the pre-B cell line 18.81. Two antibodies BP-1, BP-3 were isolated.[107,108] BP-1 antibody reacted with pre-B and immature B cells in the bone marrow. A rat mAb, 6C3, made against Abelson murine leukemia virus transformed pre-B cells reacted with a different epitope on the BP-1 defined molecule.[109] It is a 140 kDa glycoprotein with a restricted distribution. Thymocytes, T cells and mature B cells in peripheral lymphoid tissues did not express BP-1/6C3. Pre-B cells generated in long-term bone marrow cultures express high levels of BP-1 molecules which are phosphorylated. Absence of BP-1 molecules in mature B cells appears to be due the action of a transacting gene in mature B cells, since BP-1 expression is extinguished in pre-B-plasmacytoma hybrids.

BP-3 is another antibody that is produced in a similar fusion experiment that gave rise to BP-1. Expression of BP-3 antigen closely parallels BP-1 with two exceptions. BP-3 is present on immature B cells found in the periphery and thus its expression lasts longer than BP-1 during B cell development. BP-3 but not BP-1 is also present on myeloid cells and its expression increases as myeloid cells differentiate. BP-3 antigen is found to be a cell surface glycoprotein of 36 to 48 kDa. Specific absence of BP-1, BP-3 molecules in mature B cells suggests that they may have an important role in pre-B cell development.

B. J11d/T_hB

Bruce et al. described a rat anti-mouse B cell monoclonal antibody, called J11d.[110] The J11d defined antigen is present on most B cells, thymocytes and erythrocytes but not on mature T cells in the periphery. Expression of J11d distinguishes mature and immature thymocytes and was used extensively to define thymocyte development.[111] Pre-B cells, immature and mature B cells express J11d. The biochemical nature of the J11d antigen is unknown. A series of monoclonals prepared by Eckhardt and Herzenberg also define an antigen shared by thymocytes and splenic B cells.[112] This antigen was described as T_hB in accordance with a similar antigen defined previously by a goat antiserum raised against

MOPC 104E cells.[113] Immunoprecipitation studies were unsuccessful in defining the biochemical nature of the T_hB molecule. The characteristics of tissue specific expression suggest a relationship between T_hB and J11d specificities.

It was reported that most primary B cells that respond to TI antigens and SRBC expressed J11d antigen in high levels and were killed by anti-J11d + complement. However, B cells activated previously with a TD antigen, influenza hemagglutinin or a TI antigen, fluoresceinated Ficoll expressed very low levels of J11d and were resistant to lysis with J11d antibody plus complement.[114] Thus antigen activated cells down-regulate the expression of J11d molecules and this offers a potential marker to separate virgin and memory B cells. In fact, splenic B cells can be divided into J11d low and J11d high phenotypes by flow cytometry techniques.[115]

$J11d^{low}$ B cells constitute only 10 to 14% of surface Ig-positive cells in unimmunized mice. Linton et al. transferred $J11d^{low}$ or $J11d^{high}$ B cells and carrier (hemocyanin) primed helper T cells into SCID mice, which express the severe combined immunodeficiency defect. Upon challenge with DNP-hemocyanin, SCID mice repopulated with $J11d^{high}$ B cells mounted a good primary response whereas the mice receiving $J11d^{low}$ B cells mounted a small response which decayed quickly. Interestingly, boosting with a second dose of antigen induced very high titers (tenfold more) of anti-DNP antibody from the SCID mice hosting the $J11d^{low}$ B cells but not from the mice reconstituted with $J11d^{high}$ population. Analysis of $J11d^{high}$ and $J11d^{low}$ B cell responses to the hapten, (4-hydroxy-3-nitrophenyl) acetyl (NP) revealed that the characteristics of the antibodies made by $J11d^{low}$ B cells resembled those from secondary or memory B cells in light chain composition and in accumulation of somatic mutations. These studies led Linton et al. to suggest that the secondary B cells arise from a distinct subpopulation of precursor B cells rather than from unequal division of a single B cell population.[115]

Another rat anti-mouse mAb, NIM-R2, also behaves similar to J11d in distinguishing virgin and primed memory B cells.[116]

C. CD2

Human T cells express CD2 molecules which were also known as T11 previously.[117] The receptor for SRBC was shown to be made up of the CD2 molecule. Combinations of antibodies to CD2 induce T cell activation. CD2 molecules transduce activation signals on their own or in association with the T cell receptor complex.[117] The cell contact between T cells and accessory cells required for T cell activation is mediated in part by the CD2 molecule.[118] A natural ligand binding to CD2 was identified to be the LFA-3 antigen present on accessory cells.[119] Recently the mouse homolog of CD2 was first identified by molecular cloning of the cDNA and later by means of monoclonal antibody.[120,121] The CD2 structure was well conserved between mouse, rat and human suggesting a common functional role. The mature mouse polypeptide consisted of 327 amino acids which included an N-terminal region of 186 amino acids, a 25 residue long transmembrane and a cytoplasmic domain of 116 amino acids.[120] Recent experiments with mutant human CD2 molecules demonstrated that the cytoplasmic domain was essential for signaling the T cells but not for binding to the LFA-3 structures.[122]

Monoclonal anti-CD2 antibodies revealed its surprising presence on murine B cells.[121,123] The B cell expression of CD2 molecule was confirmed in immunoprecipitation studies as well as by Northern blot analysis. The size of the transcript and the molecule were identical to that in the T cells. CD2 antigen was expressed during all stages of B cell development. Thus, 65% of IgM-$B220^+$ pre-B cells in the bone marrow were CD2 positive. Mature B cells and plasmacytomas also expressed CD2 molecules. CD2 was not expressed on B cells in the rat and the human. The function of CD2 on B cells is at present unknown.

V. FUTURE PERSPECTIVES

The new knowledge on lymphokine receptors and Ig-associated molecules will further the understanding of signaling events and differentiation steps in B cells. The molecules like BP-1 may facilitate the isolation of normal pre-B cell populations and the study of factors that control programmed gene rearrangements during B cell development. Alternate pathways to generate known second messengers will be revealed by further study of signaling through B cell surface molecules like Lyb-2 and Ig. Since most lymphokine receptors fail to employ currently known signal transduction mechanisms such as tyrosine phosphorylation, elevation of known several second messengers, etc., new mechanisms of signaling cells may be revealed by studying the B cell differentiation events mediated by these lymphokine receptors. Exploitation of J11d-like markers will finally permit B cell immunologists to isolate memory B cell subsets and study the factors that govern the development of memory B cells. The availability of monoclonal antibodies and cDNA probes to various B cell surface structures will most definitely promote rapid progress in defining the steps in B cell generation, growth, and development. The exquisite sensitivity and specificity of measuring antibody production and the modern tools of mAbs and cDNA probes to various B cell receptors offer B cells as excellent models to study the events that control cell growth and differentiation.

ADDENDUM

Due to the rapid progress in this field several new and important pieces of information emerged since the original draft of this review. Receptor for another cytokine, interleukin-7 has been characterized and is shown to be present on cells of lymphoid and myeloid origin. In the B cell lineage only pre-B cells respond to IL-7 and express low- and high-affinity receptors for IL-7. Recently, cDNAs that specify IL-7 and IL-7 receptor have been cloned.[1,2] cDNA clones that specify only the soluble form of the human IL-7R have also been identified. IL-7R sequence exhibits significant homology to several cytokine receptors such as IL-4R, IL-6R, receptors for erythropoietin, GM-CSF, growth hormone and prolactin. It appears that there is a receptor superfamily characterized by the occurrence of two to four cysteines and the sequence Trp-Ser-X-Trp-Ser in the external domain.[3,4] Even though the external domains of these receptors are conserved with a few exceptions, the cytoplasmic domains exhibit very little similarity to each other or to protein tyrosine kinases.

Another B cell surface molecule that will be important in the future is the receptor for the cytokine interleukin-10 but it has not yet been characterized biochemically. IL-10 appears to prolong the survival of murine B cells and it also induces an increase in the cell surface Ia expression on mature B cells. IL-10 is made by TH_2 type of T cells, inhibits cytokine secretion by the TH_1 clones and is also known as cytokine synthesis inhibitory factor (CSIF).[5,6]

A human homolog of Lyb-2 has been identified by cross-species hybridization of a human cDNA clone.[7] The murine and human Lyb-2 molecules exhibit extensive similarity in the protein sequence with 63.8, 76.2 and 48.9% identify respectively in the cytoplasmic, transmembrane and the external domains. The human molecule is also a type 2 transmembrane protein.

The protein tyrosine phosphatase activity of the B220 molecule may have an important role in signaling lymphocytes based on several recent studies on T cells. Koretzky et al. have isolated CD45 deficient variants of the human leukemic T cell line HPB-ALL, which fails to produce phosphatidyl inositol derived second messengers upon stimulation through the T cell receptor.[8] The early signaling events were restored when CD45 expression was reconstituted by transfection with a cDNA specifying the CD45 molecule. It is interesting to note that plasma cell transfectants expressing membrane IgM or IgD also fail to generate second messengers like calcium when stimulated with anti-receptor antibodies.[9] The lack of signal generation may also be related to the level of B220 (the B cell form of CD45 expression on these cells.

Recent studies have clearly established that the surface Ig receptors on B cells also exist as a molecular complex analogous to the CD3-T cell receptor complex on T cells. The B cell surface IgM and IgD are associated noncovalently with phosphoproteins that exist as a heterodimer and a homodimer. The polypeptide chain pp32 is unique to IgM whereas the species pp33 is found in association with IgD. Mb-1 molecule discussed in this review appears to be the same as pp32 (References 10 and 11 and References 106 and 106a in the review). One of the polypeptides found in association with IgD appears to be specific to the tissues from which the B cells are derived whereas the IgM associated chains are similar in all the tissues examined. Interestingly, one polypeptide in the IgM complex identified by Chen at all to be of 42 kDa in size, is missing from B cells derived from CBA/N mice that exhibit the X-linked immunodeficiency.[11]

ADDENDUM REFERENCES

1. **Goodwin, R. G., Lupton, S., Schmierer, A., Hjerrild, K. A., Jerzy, R., Clevenger, W., Gillis, S., Cosman, D., and Namen, A. E.,** Human interleukin 7: molecular cloning and growth factor activity on human and murine B-lineage cells, *Proc. Natl. Acad. Sci. U.S.A.*, 86, 302, 1989.
2. **Goodwin, R. G., Friend, D., Ziegler, S. F., Jerzy, R., Falk, B. A., Gimpel, S., Cosman, D., Dower, S. K., March, C. J., Namen, A. E., and Park, L. S.,** Cloning of the human and murine interleukin-7 receptors: demonstration of a soluble form and homology to a new receptor superfamily, *Cell*, 60, 941, 1990.
3. **Idzerda, R. L., March, C. J., Mosley, B., Lyman, S. D., Bos, T. V., Gimpel, S. D., Din, W. S., Grabstein, K. H., Widmer, M. B., Park, L. S., Cosman, D., and Beckman, P.,** Human interleukin 4 receptor confers biological responsiveness and defines a novel receptor superfamily, *J. Exp. Med.*, 171, 961, 1990.
4. **D'Andrea, A. D., Fasman, G. D., and Lodish, H. F.,** Erythropoietin receptor and interleukin-2 receptor β chain: a new receptor family, *Cell*, 58, 1023, 1989.
5. **Fiorentino, D. F., Bond, M. W., and Mosmann, T. R.,** Two types of mouse T helper cells. IV. Th2 clones secrete a factor that inhibits cytokine production by Th1 clones, *J. Exp. Med.*, 170, 2081, 1989.
6. **Howard, M.,** personal communication.
7. **Von Hoegen, I., Nakayama, E., and Parnes, J. R.,** Identification of a human protein homologous to the mouse Lyb-2 B cell differentiation antigen and sequence of the corresponding cDNA, *J. Immunol.*, 144, 4870, 1990.
8. **Koretzky, G. A., Picus, J., Thomas, M. L., and Weiss, A.,** Tyrosine phosphatase CD45 is essential for coupling T-cell antigen receptor to the phosphatidyl inositol pathway, *Nature*, 346, 66, 1990.
9. **Justement, L. B., Wienands, J., Hombach, J., Reth, M., and Cambier, J. C.,** Membrane IgM and IgD molecules fail to transduce Ca^{2+} mobilizing signals when expressed in differentiated B lineage cells, *J. Immunol.*, 144, 3272, 1990.
10. **Wienands, J., Hombach, J., Radbruch, A., Riesterer, C., and Reth, M.,** Molecular components of the B cell antigen receptor complex of class IgD differ partly from those of IgM, *EMBO J.*, 9, 449, 1990.
11. **Chen, J., Stall, A. M., Herzenberg, L. A., and Herzenberg, L. A.,** Differences in glycoprotein complexes associated with IgM and IgD on normal murine B cells potentially enable transduction of different signals, *EMBO J.*, 9, 2117, 1990.

ACKNOWLEDGMENT

Our thanks are due to Ms. C. Burchett and W. Smith for excellent secretarial assistance.

This work was supported in part by the NIH grants to B.S. A.R.B. is supported by a fellowship from the NIH training grant.

REFERENCES

1. **Subbarao, B. and Mosier, D. E.,** Lyb antigens and their role in B lymphocyte activation, *Immunol. Rev.,* 69, 81, 1982.
2. **Moller, G., Ed.,** Function of CD structures on lymphocytes, *Immunol. Rev.,* 109, 1, 1989.
3. **Springer, T. A., Dustin, M. L., Kishimoto, T., and Marlin, S. D.,** The lymphocyte function associated LFA-1, CD2 and LFA-3 molecules: cell adhesion receptors of the immune system, *Annu. Rev. Immunol.,* 5, 223, 1987.
4. **Clark, E. A. and Ledbetter, J. A.,** Structure, function and genetics of human B cell-associated surface molecules, *Adv. Cancer Res.,* 52, 81, 1989.
5. **Kincade, P. W., Lee, G., Pietrangeli, C. E., Hayashi, S., and Gimble, J. M.,** Cells and molecules that regulate B lymphopoiesis in bone marrow, *Annu. Rev. Immunol.,* 7, 111, 1989.
6. **Hayakawa, K. and Hardy, R. R.,** Normal, autoimmune, and malignant $CD5^+$ B cells: the LY-1 B cell lineage, *Annu. Rev. Immunol.,* 6, 197, 1988.
7. **Morse, H. C., III, Shen, F. W., and Hämmerling, U.,** Genetic nomenclature for loci controlling lymphocyte antigens, *Immunogenetics,* 25, 71, 1987.
8. **Cambier, J. C. and Ransom, J. T.,** Molecular mechanisms of transmembrane signaling in B lymphocytes, *Annu. Rev. Immunol.,* 5, 175, 1987.
9. **Kehrl, J. H., Muraguchi, A., Goldsmith, P. K., and Fauci, A. S.,** The direct effects of interleukin 1, interleukin 2, interferon α, interferon γ, B-cell growth factor, and a B cell differentiation factor on resting and activated human B cells, *Cell. Immunol.,* 96, 38, 1985.
10. **Zubler, R. H., Lowenthal, J. W., Erard, E., Hashimoto, N., Devos, R., and MacDonald, H. R.,** Activated B cells express receptors for, and proliferate in response to, pure interleukin 2, *J. Exp. Med.,* 160, 1170, 1984.
11. **Tigges, M. A., Casey, L. S., and Koshland, M. E.,** Mechanism of interleukin-2 signaling: mediation of different outcomes by a single receptor and transduction pathway, *Science,* 243, 781, 1989.
12. **Blackman, M. A., Tigges, M. A., Minie, M. E., and Koshland, M. E.,** A model system for peptide hormone action in differentiation: interleukin 2 induces a B lymphoma to transcribe the J chain gene, *Cell,* 47, 609, 1986.
13. **Wang, H. and Smith, K. A.,** The interleukin 2 receptor. Functional consequences of its bimolecular structure, *J. Exp. Med.,* 166, 1055, 1987.
14. **Waldmann, T. A.,** The multi-subunit interleukin-2 receptor, *Annu. Rev. Biochem.,* 58, 875, 1989.
15. **Loughnan, M. S. and Nossal, G. J.,** Interleukins 4 and 5 control expression of IL-2 receptor on murine B cells through independent induction of its two chains, *Nature,* 340, 76, 1989.
16. **Fernandez-Botran, R., Sanders, V. M., and Vitetta, E. S.,** Interactions between receptors for interleukin 2 and interleukin 4 on lines of helper T cells (HT-2) and B lymphoma cells (BCL_1), *J. Exp. Med.,* 169, 379, 1989.
17. **Valge, J., Wong, G. P., Datlof, B. M., Sinskey, A. J., and Rao, A.,** Protein kinase C is required for responses to T cell receptor ligands but not to interleukin-2 in T cells, *Cell,* 55, 101, 1988.
18. **Paul, W. E. and Ohara, J.,** B cell stimulatory factor-1/interleukin 4, *Annu. Rev. Immunol.,* 5, 429, 1987.
19. **Möller, G., Ed.,** IL-4 and IL-5: biology and genetics, *Immunol. Rev.,* Vol. 102, 1988.
19a. **Finkleman, F. D., Holmes, J., Katona, I. M., Urban, J. F., Beckmann, M., Park, L. S., Schooley, K. A., Coffman, R. L., Mosmann, T. R., and Paul, W. E.,** Lymphokine control of in vivo immunoglobulin isotype selection, *Annu. Rev. Immunol.,* 8, 303, 1990.
20. **Tepper, R. J., Pattengale, P. K., and Leder, P.,** Murine Interleukin-4 displays potent anti-tumor activity in vivo, *Cell,* 57, 503, 1989.
21. **Foxwell, B. M., Woerly, G., and Ryffel, B.,** Identification of interleukin-4 receptor associated proteins and expression of both high- and low-affinity binding on human lymphoid cells, *Eur. J. Immunol.,* 19, 1637, 1989.
22. **Fernandez-Botran, R., Uhr, J. W., and Vitetta, E. S.,** Cross-linking of interleukin-4 to surface molecules on murine T and B lymphocytes, *Proc. Natl. Acad. Sci. U.S.A.,* 86, 4235, 1989.
23. **Mosley, B., Beckmann, M. P., March, C. J., Idzerda, R. L., Gimpel, S. D., VandenBos, T., Friend, D., Alpert, A., Anderson, D., Jackson, J., Wignall, J. M., Smith, C., Gallis, B., Sims, J. E., Urdal, D., Widmer, M. B., Cosman, D., and Park, L. S.,** The murine interleukin-4 receptor: molecular cloning and characterization of secreted and membrane bound forms, *Cell,* 59, 335, 1989.
24. **Park, L. S., Tushinski, R. J., Mochizuki, D. Y., and Urdal, D. L.,** Interleukin-4 binds to murine fibroblasts: receptor characterization and induction of colony stimulating activity production, *J. Cell. Biochem.,* Suppl. 12A, 111, 1988.
25. **Hamaoka, T. and Ono, S.,** Regulation of B-cell differentiation: interactions of factors and corresponding receptors, *Annu. Rev. Immunol.,* 4, 167, 1986.
26. **Mita, S., Tominaga, A., Hitoshi, Y., Sakamoto, K. et al.,** Characterization of high-affinity receptors for interleukin-5 on interleukin 5-dependent cell lines, *Proc. Natl. Acad. Sci. U.S.A.,* 86, 2311, 1989.

27. **Rolink, A., Melchers, F., and Palacios, R.,** Monoclonal antibodies reactive with the mouse interleukin 5 receptor, *J. Exp. Med.,* 169, 1693, 1989.
28. **Kishimoto, T. and Hirano, T.,** Molecular regulation of B lymphocyte response, *Annu. Rev. Immunol.,* 6, 485, 1988.
29. **Yamasaki, K., Taga, T., Hirata, Y., Yawata, H., Kawanishi, Y., Seed, B., Taniguchi, T., Hirano, T., and Kishimoto, T.,** Cloning and expression of the human interleukin-6 (BSF-2) receptor, *Science,* 241, 825, 1984.
30. **Taga, T., Hibi, M., Hirata, Y., Yamasaki, K., Yasukawa, K., Matsuda, T., Hirano, T., and Kishimoto, T.,** Interleukin-6 triggers the association of its receptor with a possible signal transducer, gp130, *Cell,* 58, 573, 1989.
31. **Dinarello, C. A. and Savage, N.,** Interleukin-1 and its receptor, *CRC Crit. Rev. Immunol.,* 9, 1, 1989.
32. **Booth, R. J., Prestidge, R. L., and Watson, J. D.,** Interleukin 1 and B cell responsiveness, *Lymphokines,* 12, 74, 1985.
33. **Bomsztyk, K., Stanton, T. H., Smith, L. L., Rachie, N. A., and Dower, S. K.,** Properties of interleukin-1 and interferon-gamma receptors in B lymphoid cell line, *J. Biol. Chem.,* 264, 6052, 1989.
34. **Horuk, R. and McCubrey, J. A.,** The interleukin-1 receptor in Raji human B-lymphoma cells. Molecular characterization and evidence for receptor-mediated activation of gene expression, *Biochem. J.,* 260, 657, 1989.
35. **Scapigliati, G., Ghiara, P., Bartalini, M., Tagliabue, A., and Boraschi, D.,** Differential binding of IL-1 alpha and IL-1 beta to receptors on B and T cells, *FEBS Lett.,* 243, 394, 1989.
36. **Sims, J. E., March, C. J., Cosman, O., Widmer, M. B., MacDonald, H. R., McMahan, C. J., Grubin, C. E., Wignall, J. M., Jackson, J. L., Call, S. M., Friend, D., Alpert, A. R., Gillis, S., Urdal, D. L., and Dower, S. K.,** cDNA expression cloning of the IL-1 receptor, a member of the immunoglobulin superfamily, *Science,* 241, 585, 1988.
37. **Möller, G., Ed.,** Gamma interferon, *Immunol. Rev.,* Vol. 97, 1987.
38. **Mosmann, T. R. and Coffman, R. L.,** TH1 and TH2 cells: different patterns of lymphokine secretion lead to different functional properties, *Annu. Rev. Immunol.,* 7, 145, 1989.
39. **Leibson, H. J., Gefter, M., Zlotnik, A., Marrack, P., and Kappler, J. W.,** The role of γ-interferon in antibody-producing responses, *Nature,* 309, 799, 1984.
40. **Nakagawa, T., Nakagawa, N., Delsing, G. A., Volkman, D., and Kehrl, J. H.,** Demonstration and partial characterization of the interferon-gamma receptor on human B lymphocytes, *J. Cell. Biochem.,* 40, 417, 1989.
41. **Aguet, M., Dembić, Z., and Merlin, G.,** Molecular cloning and expression of the human interferon-γ receptor, *Cell,* 55, 273, 1988.
42. **Kumar, C. S., Muthukumaran, G., Frost, L. J., Noe, M., Ahn, Y. H., Mariano, T. M., and Pestka, S.,** Molecular characterization of the murine interferon γ receptor cDNA, *J. Biol. Chem.,* 264, 17939, 1989.
43. **Smith, L. L., Stanton, T. H., Calalb, M. B., and Bomsztyk, K.,** Recombinant murine interferon-γ-induced differentiation of pre-B lymphocytes is associated with Na^+/H^+ exchange-dependent and independent cytoplasmic alkalinization, *J. Biol. Chem.,* 263, 7359, 1988.
44. **Sato, H. and Boyse, E. A.,** A new alloantigen expressed selectively on B cells: the Lyb-2 system, *Immunogenetics,* 3, 565, 1976.
45. **Nakamura, S., Ikegami, S., Dairiki, K., Kochiya, M., Fujimori, T., Tamaoki, N., and Tada, N.,** Biochemical similarity of Ly-19, Ly-32, and Lyb-2 alloantigens encoded in the gene cluster on mouse chromosome 4, *Immunogenetics,* 28, 314, 1988.
46. **Subbarao, B. and Mosier, D. E.,** Induction of B lymphocyte proliferation by monoclonal anti-Lyb-2 antibody, *J. Immunol.,* 130, 2033, 1983.
47. **Subbarao, B. and Melchers, F.,** The action of an Lyb 2.1-specific monoclonal antibody in soluble or immobilized form on resting and activated B cells, *Curr. Topics Microbiol. Immunol.,* 113, 72, 1984.
48. **Subbarao, B. and Mosier, D. E.,** Activation of B lymphocytes by monovalent anti-Lyb-2 antibodies, *J. Exp. Med.,* 159, 1796, 1984.
49. **Subbarao, B., Morris, J., and Baluyut, A. R.,** Properties of Lyb-2 mediated B-cell activation and the relationship between Lyb-2 molecule and receptors for B-cell stimulatory factor-1 on murine B lymphocytes, *Cell. Immunol.,* 112, 329, 1988.
50. **Polla, B. S., Ohara, J., Paul, W. E., Nabavi, N., Meyer, A., Liou, H. C., Shen, F. N., Gillis, S., Bonventre, J. V., and Glimcher, L. H.,** Differential induction of class II gene expression in murine pre B-cell lines by B-cell stimulatory factor-1 and by antibodies to B-cell surface antigens, *J. Mol. Cell Immunol.,* 3, 363, 1988.
51. **Yakura, H., Kawabata, I., Ashida, T., Shen, F. W., and Katagiri, M.,** A role for Lyb-2 in B cell activation mediated by a B cell stimulatory factor, *J. Immunol.,* 137, 1475, 1986.
52. **Laurindo, I., Finkelman, F. D., Mizuguchi, J., and Mond, J. J.,** B-cell activation stimulated by monoclonal anti-Lyb-2.1 antibody is mediated through a receptor distinct from the BSF-1 receptor, *Cell. Immunol.,* 108, 473, 1987.

53. **Nakayama, E., Hoegen, I. V., and Parnes, J. R.,** Sequence of the Lyb-2 B cell differentiation antigen defines a gene superfamily of receptors with inverted membrane orientation, *Proc. Natl. Acad. Sci. U.S.A.*, 86, 1352, 1989.
54. **Ashida, T., Yakura, H., Kawabata, I., Murakami, M., and Katagiri, M.,** Structural features of the Lyb-2 molecule, *Immunogenetics*, 28, 286, 1988.
55. **Grupp, S., Harmony, J. A. K., Baluyut, A. R., and Subbarao, B.,** Early events in B cell activation: anti-Lyb-2, but not BSF-1 induces a phosphatidylinositol response in murine B cells, *Cell. Immunol.*, 110, 131, 1987.
56. **Snow, E. C., Mond, J. J., and Subbarao, B.,** Enhancement by monoclonal anti-Lyb-2 antibody of antigen-specific B lymphocyte expansion stimulated by TNP-Ficoll and T lymphocyte-derived factors, *J. Immunol.*, 137, 1793, 1986.
57. **Kim, U., Kim, J. W., and Rhee, S. G.,** Phosphorylation of phospholipase C- by cAMP-dependent protein kinase, *J. Biol. Chem.*, 264, 20167, 1989.
58. **Komuro, K., Itakura, K., Boyse, E. A., and John, M.,** Ly-5: a new T-lymphocyte antigen system, *Immunogenetics*, 1, 452, 1975.
59. **Omary, M. B., Trowbridge, I. S., and Scheid, M. P.,** T200 cell surface glycoprotein of the mouse: polymorphism defined by the Ly-5 system of alloantigens, *J. Exp. Med.*, 151, 1311, 1980.
60. **Coffmann, R. L.,** Surface antigen expression and immunoglobulin gene rearrangement during mouse pre-B cell development, *Immunol. Rev.*, 69, 5, 1983.
61. **Saga, Y., Tung, J. S., Shen, F. W., and Boyse, E. A.,** Sequence of Ly-5 cDNA: isoform-related diversity of Ly-5 mRNA, *Proc. Natl. Acad. Sci. U.S.A.*, 83, 6940, 1986.
62. **Streuli, M., Hall, L. R., Saga, Y., Schlossman, S. F., and Saito, H.,** Differential usage of three exons generates at least five different mRNAs encoding of human leukocyte common antigens, *J. Exp. Med.*, 166, 1548, 1989.
63. **Yakura, H., Shen, F. W., Bourcet, E., and Boyse, E. A.,** On the function of Ly-5 in the regulation of antigen-driven B cell differentiation, *J. Exp. Med.*, 157, 1077, 1983.
64. **Mosier, D. E. and Subbarao, B.,** Thymus-independent antigens—complexity of B lymphocyte activation revealed, *Immunol. Today*, 3, 217, 1982.
65. **Yakura, H., Kawabata, I., Ashida, T., and Katagiri, M.,** Differential regulation by Ly-5 and Lyb-2 of IgG production induced by lipopolysaccharide and B cell stimulatory factor-1 (IL-4), *J. Immunol.*, 141, 875, 1988.
66. **Fearon, D. T.,** Cellular receptors for fragments of the third component of complement, *Immunol. Today*, 5, 105, 1984.
67. **Ross, G. D.,** Structure and function of membrane complement receptors, *Fed. Proc.*, 41, 3089, 1982.
68. **Klaus, G. G. B. and Humphrey, J. H.,** The generation of memory cells. I. The role of C3 in the generation of B memory cells, *Immunology*, 33, 31, 1977.
69. **Pepys, M. B.,** Role of complement in induction of the allergic response, *Nature New Biol.*, 237, 157, 1972.
70. **Fingeroth, J. D., Weis, J. J., Tedder, T. F., Strominger, J. L., Biro, P. A., and Fearon, D. T.,** Epstein-Barr virus receptor of human B lymphocytes is the C3d receptor CR2, *Proc. Natl. Acad. Sci. U.S.A.*, 81, 4510, 1984.
71. **Frade, R., Crevon, M. C., Barel, M., Vazquez, A., Krikorian, L., Charriaut, C., and Galanaud, P.,** Enhancement of human B cell proliferation by an antibody to the C3d receptor, the gp140 molecule, *Eur. J. Immunol.*, 15, 73, 1985.
72. **Lernhardt, W., Karasuyama, H., Rolink, A., and Melchers, F.,** Control of cell cycle of murine B lymphocytes: The nature of α- and β-B-cell growth factors and of B-cell maturation factors, *Immunol. Rev.*, 99, 241, 1987.
73. **Weis, J. J., Toothaker, L. E., Smith, J. A., Weis, J. H., and Fearon, D. T.,** Structure of the human B lymphocyte receptor for C3d and the Epstein Barr virus and relatedness to other members of the family of C3/C4 binding proteins, *J. Exp. Med.*, 167, 1047, 1988.
74. **Kingsmore, S. F., Vik, D. P., Kurtz, C. B., Leroy, P., et al.,** Genetic organization of complement receptor-related genes in the mouse, *J. Exp. Med.*, 169, 1479, 1989.
75. **Kurtz, C. B., Paul, M. S., Aegerter, M., Weis, J. J., and Weis, J. H.,** Murine complement receptor gene family. II. Identification and characterization of the murine homolog (Cr2) to human CR2 and its molecular linkage to Crry, *J. Immunol.*, 143, 2058, 1989.
76. **Conrad, D. H.,** Low affinity IgE receptors (Fc epsilon RII), *Clin. Rev. Allergy*, 7, 165, 1989.
77. **Delespesse, G., Sarfati, M., and Hofstetter, R. H.,** Human IgE-binding factors, *Immunol. Today*, 10, 159, 1989.
78. **Kikutani, H., Inui, S., Sato, R., Barsumian, E. L., Owaki, H., Yamasaki, K., Kaisho, T., Uchibayashi, N., Hardy, R. R., Hirano, T., Tsunasawa, S., Sakiyama, F., Suemura, M., and Kishimoto, T.,** Molecular structure of human lymphocyte receptor for immunoglobulin E, *Cell*, 47, 657, 1986.

79. **Lüdin, C., Hofstetter, H., Sarfati, M., Levy, C. A., Suter, U., Alaimo, D., Kilchherr, E., Frost, H., and Delespesse, G.,** Cloning and expression of the cDNA coding for a human lymphocyte IgE receptor, *EMBO J.*, 6, 109, 1987.
80. **Ikuta, K., Takami, M., Kim, W., Honjo, T., Miyoshi, T., Tagaya, Y., Kawabe, T., and Yodoi, J.,** Human lymphocyte Fc receptor for IgE: sequence homology of its cloned cDNA with animal lectins, *Proc. Natl. Acad. Sci. U.S.A.*, 84, 819, 1987.
81. **Bettler, B., Hofstetter, H., Rao, M., Yokoyama, W. M., Kilchherr, F., and Conrad, D. H.,** Molecular structure and expression of the murine lymphocyte low-affinity receptor for IgE ($FC_{\epsilon}RII$), *Proc. Natl. Acad. Sci. U.S.A.*, 86, 7566, 1989.
82. **Conrad, D. H., Keegan, A. D., Kalli, K. R., Van Dusen, R., et al.,** Superinduction of low affinity IgE receptors on murine B lymphocytes by lipopolysaccharide and IL-4, *J. Immunol.*, 141, 1091, 1988.
83. **Uchibashij, N., Kikutani, H., et al.,** Recombinant soluble Fc_{ϵ} receptor II ($FC_{\epsilon}RII$/CD23) has IgE binding activity but no B cell growth promoting activity, *J. Immunol.*, 142, 3901, 1989.
84. **Swendeman, S. and Thorley-Lawson, D. A.,** The activation antigen BLAST-2, when shed is an autocrine BCGF for normal and transformed B cells, EMBO J., 6, 1637, 1987.
85. **Kehry, M. and Yamashita, L.,** Low-affinity IgE receptor (CD 23) function on mouse B cells. Role in IgE-dependent antigen focusing, *Proc. Natl. Acad. Sci. U.S.A.*, 86, 7556, 1989.
86. **Baluyut, A. R. and Subbarao, B.,** The synergistic effects of anti-IgM and monoclonal anti-Ia antibodies in induction of murine B lymphocyte activation, *J. Mol. Cell. Immunol.*, 4, 45, 1988.
87. **Mond, J. J., Seghal, E., Kung, J., and Finkelman, F. D.,** Increased expression of I-region associated antigens (Ia) on B cells after cross-linking of surface Ig, *J. Immunol.*, 127, 881, 1981.
88. **Noelle, R., Krammer, P. H., Ohara, J., Uhr, J. W., and Vitteta, E. S.,** Increased expression of Ia antigens on resting B cells: an additional role for B cell growth factor, *Proc. Natl. Acad. Sci. U.S.A.*, 81, 6149, 1984.
89. **Palacios, R., Martinez-Maza, O., and Guy, K.,** Monoclonal antibodies against HLA-DR antigens replace T helper cells in activation of B lymphocytes, *Proc. Natl. Acad. Sci. U.S.A.*, 80, 3456, 1983.
90. **Forsgren, S., Pobor, G., Coutinho, A., and Pierres, M.,** The role of I-A/E molecules in activation: inhibition of lipopolysaccharide-induced responses by monoclonal antibodies, *J. Immunol.*, 133, 2104, 1984.
90a. **Forsgren, S., Martinez-A., C., and Coutinho, A.,** The role of I-A/E molecules in B-lymphocyte activation, *Scand, J. Immunol.*, 25, 225, 1987.
91. **LoCascio, N. J., Haughton, G., Arnold, L. W., and Corley, R. B.,** Role of cell surface immunoglobulin in B lymphocyte activation, *Proc. Natl. Acad. Sci. U.S.A.*, 81, 2466, 1984.
92. **Baluyut, A. R., Udhaykumar, V., Morris, J., and Subbarao, B.,** A role for Ia antigens in B cell activation—studies with normal B cells and B cell lymphoma, in *Histocompatibility Antigens Structure and Function,* David, C. S., Ed., Plenum Press, New York, 1987, 501.
93. **Nagarkatti, P. A., Seth, M., Nagarkatti, M., Muthusamy, N., Rychlik, B., and Subbarao, B.,** A specific defect in the proliferative capacity of B cells from young mice stimulated with an autoreactive T cell clone, *Cell Immunol.*, 120, 102, 1989.
94. **Pobor, G., Pettersson, S., Bandeira, A., Martinez, A. C., and Coutinho, A.,** B lymphocyte activation upon exclusive recognition of major histocompatibility antigens by T helper cells, *Eur. J. Immunol.*, 14, 222, 1984.
95. **Swain, S. L. and Dutton, R. W.,** Consequences of the direct interaction of helper T cells with B cells presenting antigen, *Immunol. Rev.*, 99, 263, 1987.
96. **Krusemeier, M. and Snow, E. C.,** Induction of lymphokine responsiveness of hapten-specific B lymphocytes promoted through an antigen-mediated T helper lymphocyte interaction, *J. Immunol.*, 140, 367, 1988.
97. **Tonks, N. K. and Charbonneau, H.,** Protein tyrosine dephosphorylation and signal transduction, *TIBS*, 14, 497, 1989.
98. **Heikkila, R., Schwab, R., Wickstrom, E., Loke, S. L., Pluznick, D. H., Watt, R., and Neckers, L. M.,** A c-myc antisense oligonucleotide inhibits entry into S phase but not progress from G_0 to G_1, *Nature (London),* 328, 445, 1987.
99. **Cambier, J. C. and Lehmann, K. R.,** Ia-mediated signal transduction leads to proliferation of primed B lymphocytes, *J. Exp. Med.*, 170, 877, 1989.
100. **Pereira, P., Bandeira, A., Kleine, B., Marquez, C., Countinho, A., and Martinez-A. C.,** A model system for the analysis of B-cell activation and effector T-cell functions. T cell-dependent B-cell responses facilitated by anti-I-A antibodies, *Scand. J. Immunol.*, 29, 49, 1989.
101. **Cambier, J. C., Newell, M. K., Justement, L. B., McGuire, J. C., Leach, K. L., and Chen, Z. Z.,** Ia binding ligands and cAMP stimulate nuclear translocation of PKC in B lymphocytes, *Nature (London),* 327, 629, 1987.
102. **Chen, Z. Z., McGuire, J. C., Leach, K. L., and Cambier, J. C.,** Transmembrane signaling through B-cell MHC class II molecules: anti-Ia antibodies induce protein kinase C translocation to the nuclear fraction, *J. Immunol.*, 138, 2345, 1987.

103. **Yano, A., Yamashita, K., Yui, K., Aosai, F., Sugane, K., Fujishita, J., Ohno, S., and Suzuki, K.,** Antigen presentation and signal transduction of antigen presenting cells infected by toxaplasma, 7th Int. Congr. of Immunology, Berlin, 1989, 371.

103a. **Lane, P. J. L., McConnell, F. M., Schieven, H. L., Clark, E. A., and Ledbetter, J. A.,** The role of class II molecules in human B cell activation: association with phosphatidyl inositol turnover, protein tyrosine phosphorylation, and proliferation, *J. Immunol.,* 144, 3864, 1990.

104. **Hombach, J., Leclercq, L., Radbruch, A., Rajewsky, K., and Reth, M.,** A Novel 34-Kd protein co-isolated with the IgM molecule in surface IgM-expressing cells, *EMBO J.,* 7, 3451, 1988.

105. **Sakaguchi, N., Kashiwamura, S., Kimoto, M., Thalmann, P., and Melchers, F.,** B lymphocyte lineage-restricted expression of mb-1, a gene with CD3-like structural properties, *EMBO J.,* 7, 3457, 1988.

106. **Campbell, K. S. and Cambier, J. C.,** Lymphocyte antigen receptors (mIg) are non-covalently associated with a disulfide linked, inducibly phosphorylated glycoprotein complex, *EMBO J.* 9, 441, 1990.

106a. **Hombach, J., Tsubata, T., Leclercq, L., Stappert, H., and Reth, M.,** Molecular components of the B-cell antigen receptor complex of th IgM class, *Nature,* 343, 760, 1990.

107. **Cooper, M. D., Mulvaney, D., Coutinho, A., and Cazenave, P.,** A novel cell surface molecule on early B lineage cells, *Nature,* 321, 616, 1986.

108. **McNagy, K. M., Cazenae, P., and Cooper, M. D.,** BP-3 alloantigen: a cell surface glycoprotein that marks early B lineage cells and mature myeloid lineage cells in mice, *J. Immunol.,* 141, 2551, 1988.

109. **Tidmarsh, G. F., Dailey, M. O., and Weissman, I. L.,** Expression of a monoclonal antibody-defined, B-lineage transformation antigen specifically identifies Abelson-diseased animals. Genetically determined resistance to Abelson murine leukemia virus acts before induction of gp160 (6C3), *J. Exp. Med.,* 164, 1356, 1986.

110. **Bruce, J., Symington, F. W., McKearn, J., and Sprent, J.,** A monoclonal antibody discriminating between subsets of T and B cells, *J. Immunol.,* 127, 2496, 1981.

111. **Von Boehmer, H.,** The developmental biology of T lymphocytes, *Annu. Rev. Immunol.,* 6, 309, 1988.

112. **Eckhardt, L. A. and Herzenberg, L. A.,** Monoclonal antibodies to ThB detect close linkage of Ly-6 and a gene regulating ThB expression, *Immunogenetics,* 11, 275, 1980.

113. **Stout, R. D., Yutoku, M., Grossberg, A., Pressman, D., and Herzenberg, L. A.,** A surface membrane determinant shared by subpopulations of thymocytes and B lymphocytes, *J. Immunol.,* 115, 508, 1975.

114. **Pillai, P. S., Reynolds, S. V., Chace, J. H., and Scott, D. W.,** Differential expression of a surface antigen recognized by a monoclonal antibody, J11d, on unprimed and primed B cells, *J. Immunol.,* 137, 791, 1986.

115. **Linton, P. J., Decker, D. J., and Klinman, N. R.,** Primary antibody-forming cells and secondary B cells are generated from separate precursor cell subpopulations, *Cell,* 59, 1049, 1989.

116. **Andrew, E. M., Marshall-Clarke, S., and Parkhouse, R. M.,** Association of cell function with quantitative variation in cell surface markers, *Ann. Immunol. (Paris),* 135D, 204, 1984.

117. **Meuer, S. C., Hussey, R. E., Fabbi, M., Fox, D., Acuto, O., Fitzgerald, K. A., Hodgdon, J. C., Protentis, J. P., Schlossman, S. F., and Reinherz, E. L.,** An alternative pathway of T cell activation: a functional role for the 50kd T11 sheep erythrocyte receptor protein, *Cell,* 36, 897, 1984.

118. **Brottier, P., Boumell, L., Gehn, C., and Bernard, A.,** T cell activation via CD2 (T,gp50) molecules: accessory cells are required to trigger T cell activation via CD2-D66 plus CD2-9.6/T11 epitopes, *J. Immunol.,* 135, 1624, 1985.

119. **Selvaraj, P., Plunkett, M. L., Dustin, M., Sanders, M. E., Shaw, S., and Springer, T. A.,** The T lymphocyte glycoprotein CD2 binds the cell surface ligand LFA-3, *Nature (London),* 326, 400, 1987.

120. **Yagita, H., Okumura, K., and Nakauchi, H.,** Molecular cloning of the murine homologue of CD2: homology of the molecule to its human counterpart T11, *J. Immunol.,* 140, 1321, 1988.

121. **Yagita, H., Nakamura, T., Karasuyama, H., and Okumura, K.,** Monoclonal antibodies specific for murine CD2 reveal its presence on B as well as T cells, *Proc. Natl. Acad. Sci. U.S.A.,* 86, 645, 1984.

122. **Chang, H. C., Moingeon, P., Lopez, P., Krasnow, H., Stebbins, C., and Reinherz, E. L.,** Dissection of the human CD2 intracellular domain. Identification of a segment required for signal transduction and interleukin 2 production, *J. Exp. Med.,* 169, 2073, 1989.

123. **Yagita, H., Nakamura, T., Asakawa, J., Matsuda, H., Tansyo, S., Yutaka, L., and Okumura, K.,** CD2 expression in murine B cell lineage, *Int. Immunol.,* 1, 94, 1989.

Chapter 4

THYMUS-INDEPENDENT ANTIGENS

Göran Möller, Carmen Fernandez, and Marta Alarcon-Riquelme

TABLE OF CONTENTS

I. INTRODUCTION

Long before T and B lymphocytes were discovered, it was known that certain antigens, usually of bacterial origin, were fundamentally different from protein or particulate antigens. A typical member of these "different" antigens was lipopolysaccharide (LPS) from Gram negative bacteria. LPS was found to be immunogenic at very low concentrations and even 1000 molecules could induce a detectable immune response.[1] The immune response to LPS was dominated by 19S antibodies (IgM),[2] and the response was characterized by regular cyclical peaks of IgM antibodies.[3] It was shown that LPS persisted for prolonged time periods in an immunogenic form in the animals.[4] The cyclical appearance of the immune response to LPS was explained as the result of antibody feed-back suppression. It was shown that the IgM produced in the first peak suppressed further antibody synthesis by blocking the antigenic determinants. The IgM antibodies already produced then declined rapidly and the antigenic determinants became free to induce a new wave of antibody synthesis and this process was repeated. When the IgM antibodies then rapidly declined, antibody suppression disappeared and the immunogenic LPS could induce a new peak.

It was also shown that specific and very long-lasting immunological tolerance could be induced to LPS by injecting larger doses of detoxified LPS.[5] Although the tolerant animals did not produce antibodies to LPS:[6] their spleens contained lymphocytes that specifically bound LPS.[5] The presence of antigen-binding cells in tolerant mice was characteristic for LPS and was not seen in mice tolerant to protein antigens.

II. THYMUS-INDEPENDENT ANTIGENS

One of the important findings leading to the discovery of T and B cells was the realization that most antigens required both cell types for induction of an antibody response (for review see Reference 7). It was soon found that this was not the case with all antigens. Thus, LPS[8-9] and polyvinylpyrrolidone[10] could induce an immune response without the help of T cells. Other antigens, such as Ficoll, dextran, pneumococcal polysaccharide SIII, levan and polymerized flagellin from Salmonella bacteria were later added to the list.[11-13] A common characteristic of these molecules was that they consisted of repeated antigenic determinants. This property was considered to explain their T cell independency: the repeated antigenic determinants should result in an efficient crosslinking of the immunoglobulin membrane receptors on the specific B cells and this should give activation signals to the B cells. This view is still held by many immunologists. Two other views on the nature of thymus-independent antigens have also been proposed. One is simply to deny their existence: thymus-independent antigens are only less T cell dependent than other antigens.[14] The other view is that all thymus-independent antigens are polyclonal B cell activators and this is what triggers B cells that have bound the antigen by their specific immunoglobulin receptors.[15]

Thymus-independent antigens are now grouped in two types, depending on their ability to induce an immune response in the CBA/N mouse strain.[16] This strain is incapable of responding to certain thymus-independent antigens, called type II antigens, due to an immunodeficiency caused by gene(s) on the X-chromosome. Dextran, Ficoll, and PVP are all thymus-independent type II antigens. Other thymus-independent antigens, such as LPS, *Brucella abortus* bacteria and lipoprotein are type I antigens and can induce an immune response in CBA/N mice.

III. THYMUS-INDEPENDENT ANTIGENS AND POLYCLONAL B CELL ACTIVATION

It was known before that LPS could induce DNA synthesis in B cells.[17] We showed that LPS also could induce antibody synthesis in B cells.[18] We found that LPS activated B cells produced antibodies against any antigen we tested and coined the phenomenon polyclonal B cell activation. We postulated the existence of an activation receptor for LPS on B cells. After activation with LPS, the B cells produced the antibody they were genetically programmed to synthesize. Thus, LPS would be competent to induce the synthesis of any antibody specificity. Initially, this raised serious criticism. The most common argument was that LPS crossreacted with the particular antigen we happened to study. The development of the protein A plaque assay[19] and the demonstration that 30% of B cells were activated by LPS to antibody synthesis[20] weakened these arguments.

When we found that five additional thymus-independent antigens were polyclonal B cell activators,[21] the crossreactive argument became untenable. These five antigens had no common structures or antigenic determinants, but they had one common property: they were polyclonal B cell activators.

LPS, pneumococcal polysaccharide and polymerized flagellin were all classified as TI type I antigens. It is now generally accepted that TI type I antigens are polyclonal B cell activators and that TI type II antigens are not. However, this is not correct. Both dextran and levan are TI type II antigens and they are both potent polyclonal B cell activators for induction of antibody synthesis.[22-23] Ficoll and PVP are not polyclonal B cell activators, but as will be shown later, Ficoll can be made to function as a polyclonal B cell activator.

In order to induce polyclonal B cell activation by LPS it is necessary to add about a 1,000-fold higher concentration *in vitro* than is required for the induction of a specific immune response.[24] The one nonspecific signal hypothesis states that a specific immune response to a determinant on a TI antigen is the result of binding of the determinant to the B cell by their immunoglobulin receptors and thereafter the B cells are activated by inherent activation properties of the TI antigen. In this model, the immunoglobulin receptors transmitted no signals, they only passively focussed the antigen to the B cells. In the case of LPS, the activation signals are transmitted to the B cells by Lipid A,[25] most likely via genetically determined LPS receptors on the B cells.

IV. GENETIC DETERMINATION OF THE IMMUNE RESPONSE TO THYMUS-INDEPENDENT ANTIGENS

A. LPS

It was first shown that a substrain of C3H mice, the C3H/HeJ strain, was resistant to the toxic effects of LPS.[26] Later it was demonstrated that B cells from C3H/HeJ mice could not be polyclonally activated by LPS.[27-28] Finally, it was found that C3H/HeJ could not produce antibodies to any antigenic determinant conjugated to LPS as a carrier,[29] whereas they responded normally to these antigenic determinants conjugated to other carriers. Thus, the immunological repertoire of the C3H/HeJ mice was normal. Their immunological defect was an inability to respond to LPS and all antigenic determinants present on LPS. These findings have had a profound effect on our understanding of the mechanism of B cell activation to IgM synthesis.

The genetic defect of the C3H/HeJ mice was localized to the 4th chromosome.[27] The gene behaved as recessive trait and consequently F_1 hybrids between C3H/HeJ and LPS responder strains are competent to respond to LPS, although to a smaller extent.

The gene product of the LPS non-responder gene on the 4th chromosome is still not known. Antibodies have been produced in responder C3H strains against B cells from C3H/HeJ mice and these antibodies were found to react with the surface of about 30% of the B cells from C3H/HeJ mice.[30] Selective binding of LPS to B cells from normal C3H mice as compared to cells from C3H/HeJ has been reported, but contradictory findings have also been published (for review see Reference 31). Another strain with a similar gene defect (C57BL/10ScSr) was later identified.[32]

B. IMPLICATIONS OF THE C3H/HeJ STRAIN FOR THE MECHANISM OF B CELL ACTIVATION

The fact that C3H/HeJ and C57BL/10 ScSr mice can never be made to produce antibodies to any antigenic determinant present on LPS, but respond normally when the same antigenic determinants are conjugated to other thymus-dependent or -independent carrier molecules has important implications for the mechanism of B cell activation to IgM synthesis. Since the immunoglobulin repertorie is normal in C3H/HeJ mice and their B cells can bind antigen, it follows that the immunoglobulin receptors cannot mediate activation signals to the B cells. The defect in C3H/HeJ is that B cells from this strain cannot be activated by LPS to polyclonal antibody synthesis. Any model of B cell activation ascribing a triggering role of the immunoglobulin receptors fails to account for the inability of C3H/HeJ to produce antibodies to antigenic determinants present on LPS. However, hypotheses based on the existence of genetically determined activation receptors that receive activation signals from a carrier, such as LPS, can easily account for the findings with C3H/HeJ mice.

C. THYMUS-INDEPENDENT TYPE II ANTIGENS

Another mouse strain, CBA/N, has a genetic defect on the X-chromosome that makes the mice unresponsive to some (Ficoll, PVP, and dextran), but not other (LPS) thymus-independent antigens, as discussed above. The nature of the genetic defect in CBA/N mice is not known. It seems probable that B cell differentiation or maturation is affected in CBA/N mice and they have fewer B cells than normal mice. It is possible that they lack T cells producing an interleukin, since the defect can be reconstituted by injection of this yet unnamed switch interleukin.[33] Male CBA/N mice respond to type II antigens, suggesting that the defect is limited to the X-chromosome, but other findings suggest that the genetics may be more complicated. Thus, female (CBA/N × C57BL/10ScCr)F_1 hybrids are low responders to dextran, although this cross should lead to genetic complementation. This suggests that the only defect in CBA/N mice cannot be on the X-chromosome and the only defect in C57BL/10ScCr cannot be on the 4th chromosome.

D. DEXTRAN

Dextran B512 is a polysaccharide antigen extracted from *L. mesenteroides* and has been classified as a TI type II antigen. The immune response to dextran B512 (and also other dextrans) is regulated by several different genes. We will limit the discussion to dextran B512.

Dextran has certain unique immunological properties. It is made up of repeating glucose molecules in α 1-6 linkages. The immune response is very homogeneous with regard to antibody affinity, at least when plaque-forming cells are studied.[34]

E. A GENE CODING FOR ANTI-DEXTRAN ANTIBODIES IN THE Igh LOCUS

Some mouse strains respond to dextran and others do not. When the PFC response is studied the difference between responder and non-responder strains is nearly absolute. When serum antibodies are analyzed, PFC non-responder strains usually produce lower levels of low affinity antibodies, but the difference between responder and low-responder strains is less clearcut compared to the results obtained with PFC. We will mainly discuss PFC results for the sake of clarity.

Typical responder strains are C3H and C57BL and typical non-responders are A, AT.H, A.SW. Actually most strains are non-responders. Unresponsiveness to dextran B512 is linked to allotypes in the immunoglobulin heavy chain locus.[35] Responder strains have allotypes j and k. Non-responder strains cannot be made to respond to dextran, whether immunized with thymus-independent or thymus-dependent forms of dextran. Polyclonally activated B cells from non-responder strains (activated by, e.g., LPS) produce antibodies of all specificities tested for, except anti-dextran. Suppressor cells were not responsible for the lack of response.

Dextran non-responder strains can be activated by dextran to polyclonal antibody synthesis. This is an important finding, since it demonstrates that non-responder strains possess activation receptors for dextran as a polyclonal B cell activator.

Presumably the non-responder strains lack the gene(s) coding for anti-dextran antibodies. Alternatively, they have these genes, but cannot express them.

F. YOUNG MICE

Young mice from the dextran high-responder strains C3H and C57BL cannot produce antibodies to dextran.[36-37] It takes 3 months after birth for the mice to reach maximal immunocompetence. Dextran was found to be a strong polyclonal B cell activator for B cells from young mice. Activation of B cells from young mice with LPS as a polyclonal B cell activator induced antibodies of all specificities tested for, except anti-dextran antibodies. Thus, young mice from dextran responder strain can be activated by dextran to polyclonal antibody synthesis and they have the genes coding for anti-dextran antibodies. They do not produce anti-dextran antibodies after immunization with thymus-dependent or -independent forms of dextran, but they can produce antibodies to haptens conjugated to dextran. Presumably there exists some mechanism to specifically suppress or delay the expression of genes coding for anti-dextran antibodies. No suppressor cells were responsible for the late maturation of the anti-dextran response.[36]

G. ANTI-DEXTRAN ANTIBODIES OF NON-IgM ISOTYPE

C3H and CBA mice consistently produced IgM PFC after immunization with dextran, whether given in thymus-dependent or -independent form.

In contrast, C57BL mice were found to produce PFC detected after development with an anti-IgG antiserum.[38] These PFC will be referred to as IgG PFC. C57BL mice consistently produced IgG antibodies to dextran irrespective of the type of antigen: thymus-dependent or -independent. The ability of C57BL B cells to produce IgG antibodies was an inherent property of the B cells and not influenced by the environment. Thus, B cells from C57BL mice produced IgG antibodies after transfer to x-irradiated CBA mice, whereas B cells from CBA mice could not produce IgG anti-dextran antibodies in irradiated C57BL mice.

Later studies showed that C57BL mice can produce IgG3 and IgA antibodies to dextran, whereas CBA mice are very poor in this respect.[39] It was clearly shown that the IgG3 responses were thymus-independent. It must be concluded, therefore, that some mechanisms exist in C57BL mice to switch from IgM to IgG3 and IgA, respectively, that are T cell independent and do not exist in CBA mice. The molecular basis for these mechanisms remains unknown.

H. B CELL ACTIVATION BY TI TYPE II ANTIGENS

The mechanism of B cell activation by TII antigens can be adequately explained in terms of the one non-specific signal hypothesis. However, this is not the case with B cell activation in immune responses to TI type II antigens. Most of these antigens are not polyclonal B cell activators (with the exception of dextran and levan). Therefore, attention has been focussed on the role of the immunoglobulin receptors in immune responses against TI antigens of type II. The molecular characteristics of these antigens—in most cases being made up of repeated identical antigenic determinants—have suggested to many scientists that they are uniquely capable of crosslinking the immunoglobulin receptors on the B cells. Sufficient crosslinking of immunoglobulin receptors would result in activation signals to the B cells. By building various molecules with different hapten density Dintzis and collaborators could actually define the "immunon": the minimum requirement for epitope density necessary to give one activation signal to B cells.[40] It has also been shown by Brunnswick et al.[41] that anti-IgM and anti-IgD conjugated to Ficoll and LPS could activate B cells at very low antibody concentrations, suggesting that their crosslinking capacity was dramatically increased by conjugation to long carrier molecules, such as Ficoll.

It has actually been suggested that not only endogeneous immunoglobulin receptors, but also passively membrane incorporated antibody molecules could mediate activation signals to B cells. Thus, Peacock et al.[42] showed that B cells "decorated" with lipid conjugated anti-DNP monoclonal antibodies and exposed to DNP-polymerized flagellin were activated into polyclonal antibody synthesis. Such B cells were shown to express up to 10^5 antibody molecules/cells and were found to bind TNP conjugated type II thymus-independent antigens well.

Since we do not think that immunoglobulin receptors, whether endogenous or passively membrane "decorated", can transmit any activation signals, we decided to critically test their hypothesis. As Peacock et al.[42] we decorated B cells with anti-TNP monoclonal antibodies and stimulated them with TNP-LPS. However, we used B cells from two strains: the normal C3H strain and the LPS non-responder C3H/HeJ strain. The results were crisp: only B cells from the normal C3H strain were induced to polyclonal antibody synthesis and there was no activation of B cells from the C3H/HeJ strain.[43] This excludes any signal transmission from passively membrane incorporated monoclonal antibodies and demonstrates that the only variable determining the outcome of the experiment is whether the B cells can respond to activation signals delivered by LPS.

V. THYMUS-INDEPENDENT TYPE II ANTIGENS ARE POLYCLONAL B CELL ACTIVATORS

With the exception of dextran, thymus-independent type II antigens are not polyclonal B cell activators. We have previously suggested that their interaction with the postulated B cell activation receptors is of too low affinity to permit activation and, consequently, that these antigens required the focussing role of immunoglobulin receptors in order to reach sufficient concentration on the membrane of the B cells to activate the cells.

The use of passively membrane "decorated" antibodies on B cells as developed by Peacock et al.[42] allowed a direct test of our hypothesis. As shown above, such antibodies cannot transmit activation signals. However, they can function to focus the corresponding antigen to the B cells.

Thus, we have decorated B cells with monoclonal antibodies against the hapten TNP. In this way all B cells have been given the same immunological specificity. These B cells were then exposed to TNP-Ficoll *in vitro*. We found that they activated to polyclonal B cell antibody synthesis, whereas B cells "decorated" with antibodies of another specificity remained resting and "non-decorated" B cells exposed to TNP-Ficoll could not be activated or were activated by much higher concentrations.[44]

These experiments show that Ficoll is a polyclonal B cell activator, provided it can be focussed to the B cells in a sufficiently high concentration. In our experiments this was achieved by the use of passively membrane incorporated monoclonals, previously shown to be unable to deliver activation signals to B cells.

VI. MATURATION OF THE IMMUNE RESPONSE TO TI ANTIGENS

The immune response to TD antigens is characterized by an increased affinity of the antibodies produced with time after immunization. Several mechanisms have been suggested to account for the increase of antibody affinity, such as antibody feed-back inhibition, accumulation of somatic mutations in B cells and decreased antigen concentration in the animals.

As mentioned before, immune responses to TI antigens differ from those to TD antigens. This is particularly pronounced with regard to the secondary immune responses: usually only a new primary response develops, but the secondary responses may be decreased or even abolished.[44-45] Suppressed secondary responses are typical for the immune response to dextran B512 and have been ascribed to the spontaneous synthesis of autoanti-idiotypic antibodies.[45-46]

We have produced a collection of anti-dextran producing hybridomas in C57BL mice and have analyzed the monoclonals for Ig class, light chain type, idiotypes and affinity for dextran. IgM, IgG3, and IgA producing hybridomas were found and IgG3 synthesizing hybridomas were obtained from nude mice, indicating the T cell independence of this IgG subclass. Most monoclonals expressed κ light chains, but also λ light chains were found. The latter antibodies possessed the same idiotype as λ bearing antibodies, suggesting that the heavy chain was responsible for idiotype expression.[47]

The secondary response to dextran was suppressed also at the hybridoma cell level. However, the affinity of anti-dextran antibodies produced by hybridomas collected from mice making a secondary immune response was drastically increased compared to antibodies from primary immune responses. This was the case, whether the animals had been immunized with TI or TD forms of dextran. The antibody affinity was increased to the same extent in normal and nude mice, indicating that T cells did not participate in the process of affinity maturation.

A. IMMUNOLOGICAL TOLERANCE TO TI ANTIGENS

It is not possible to review the field of tolerance to TI antigens, because nearly every known phenomenon in immunology has been used to explain tolerance to TI antigens. Thus, tolerance to TI antigens has been ascribed to clonal elimination, blocking of Ig receptors, neutralization of antibodies by stable circulating TI antigens, suppressor T cells and anti-idiotypic antibodies. Tolerance to TI antigens has been partly reviewed elsewhere.[48-49] Instead of attempting to review this controversial field, we have decided to summarize our own findings on tolerance induction to dextran B512 as the TI antigen.

Specific and long-lasting immunological tolerance to dextran is induced in adult mice by the injection of a 1,000-fold higher antigen dose than required for induction of immunity.[50] Tolerance was not caused by suppressor cells. When spleen cells from tolerant mice were activated by polyclonal concentrations of LPS *in vitro* immediately after removal, there was induction of a polyclonal antibody response, but no anti-dextran antibodies were induced. However, if the spleen cells were incubated *in vitro* for 24 h before the addition of LPS, then anti-dextran antibodies were induced. This suggested the possibility that dextran was initially bound to the Ig receptors of the anti-dextran specific B cells and this binding prevented their activation by LPS. The *in vitro* incubation of the B cells for 24 h would allow shedding of the Ig receptors and allow them to be activated by LPS. This possibility was tested by treating spleen cells from tolerant mice with dextranase *in vitro* prior to activation by LPS. Such dextranase treated cells from tolerant mice were activated by LPS to the synthesis of anti-dextran antibodies. We showed that dextranase injected into mice was capable of removing dextran from mice previously injected with dextran. Therefore, we tried to break tolerance to dextran in mice by injecting them with dextranase, followed at various time periods by immunogenic doses of dextran. This never resulted in an immune response to dextran.

However, when dextran tolerant mice were treated with dextranase and thereafter immunized with TD forms of dextran, an anti-dextran response was regularly induced.[51-52] Pretreatment with dextranase was a prerequisite for breaking of tolerance with TD forms of dextran. Tolerance could never be broken by immunization with TD forms of dextran if the mice were not treated with dextranase.

These findings suggested that there were immunocompetent anti-dextran B cells in tolerant mice and that these cells could be activated by TD forms of dextran provided dextran had been eliminated by dextranase treatment. However, it was never possible to induce an immune response to dextran itself (the TI form) in tolerant animals, whether treated with dextranase or not. Thus, part of the anti-dextran specific B cell clone was tolerized, whereas another part was intact and could be activated into antibody synthesis.

We interpreted these findings in the following way: those B cells that could be activated by the TI form of dextran could also be tolerized by dextran. Other B cells with identical Ig receptors could not be activated or tolerized by dextran. However, these B cells could be activated by TD forms of dextran.

These findings show that tolerance to dextran only affects part of the anti-dextran clone. The cells that can be activated by the TI form of dextran can also be made tolerant to dextran. These B cells must have two properties, which are required for activation and tolerance induction: they must have both Ig and activation receptors for dextran. Those B cells that only have Ig receptors for dextran will bind the antigen, but are neither activated nor tolerized. Such B cells are ''blocked'' and cannot be activated by TD forms of dextran, unless dextran is removed by treating them with dextranase.

These findings again emphasize the passive, non-signaling role of Ig receptors.

REFERENCES

1. **Landy, M. and Baker, P. J.,** Cytodynamics of the distinctive immune response produced in regional lymph nodes by Salmonella somatic polysaccharide, *J. Immunol.*, 97, 670, 1966.
2. **Möller, G.,** 19S antibody production against soluble lipopolysaccharide antigens by individual lymphoid cells in vitro, *Nature*, 207, 1166, 1965.
3. **Britton, S. and Möller, G.,** Regulation of antibody synthesis against *E. coli* endotoxin. I. Suppressive effect of endogenously produced and passively transferred antibodies, *J. Immunol.*, 100, 1326, 1968.
4. **Britton, S., Wepsic, T., and Möller, G.,** Persistence of immunogenicity of two complex antigens retained in vivo, *Immunology*, 14, 49, 1968.
5. **Britton, S.,** Regulation of antibody synthesis against *E. coli* endotoxin. II. Specificity, dose requirements and duration of paralysis induced in adult mice, *Immunology*, 16, 513, 1969.
6. **Sjöberg, O. and Möller, E.,** Antigen-binding cells in tolerant animals, *Nature*, 228, 5273, 1970.
7. Antigen-sensitive cells, *Transplant. Rev.*, 1, 4, 1969.
8. **Möller, G.,** Triggering mechanisms for cellular recognition in *Immune Surveillance*, Smith, R. T. and Landy, M., Eds., Academic Press, New York, 1970.
9. **Möller, G. and Michael, G.,** Frequency of antigen-sensitive cells to thymus-independent antigens, *Cell Immunol.*, 2, 309, 1971.
10. **Andersson, B. and Blomgren, H.,** Evidence for thymus-independent humoral antibody production in mice against Polyvinylpyrrolidone and *E. coli* lipopolysaccharide, *Cell. Immunol.*, 2, 411, 1971.
11. **Howard, J., Christie, G. H., Courtenay, B. M., Leuchars, E., and Davies, A. J. S.,** Studies on immunological paralysis. VI. Thymic independence of tolerance and immunity to type 3 pneumococcal polysaccharide, *Cell. Immunol.*, 2, 614, 1971.
12. **Feldman, M. and Basten, A.,** The relationship between antigenic structure and the requirement for thymus-derived cells in the immune response, *J. Exp. Med.*, 134, 103, 1971.
13. **Miranda, J. J.,** Studies on immunological paralysis. IX. The immunogenicity and tolerogenicity of levan in mice, *Immunology*, 23, 829, 1972.
14. **Cohn, M. and Blomberg, B.,** The self-nonself discrimination: a one or two signal mechanism?, *Scand. J. Immunol.*, 4, 1, 1975.

15. **Coutinho, A. and Möller, G.,** Immune activation of B cells: evidence for the one nonspecific triggering signal not delivered by the Ig receptors, *Scand. J. Immunol.,* 3, 133, 1974.
16. **Mosier, D. E., Scher, I., Ruhl, H., Cohen, P. L., Zitron, I., and Paul, W. E.,** Activation of normal and defective B lymphocytes by thymus-independent antigens, in *Mitogens in Immunolobiology,* Oppenheim, J. J. and Rosenstreich, D. L., Eds., Academic Press, New York, 1976, 313.
17. **Peavy, D. L., Adler, W. H., and Smith, R. T.,** The mitogenic effect of endotoxin and staphylococcal endotoxin B on mouse spleen cells and human peripheral lymphocytes, *J. Immunol.,* 105, 1453, 1973.
18. **Andersson, J., Sjöberg, O., and Möller, G.,** Mitogens as probes for immunocyte activation and cellular cooperation, *Transplant Rev.,* 11, 131, 1972.
19. **Gronowicz, E., Coutinho, A., and Melchers, F.,** A plaque assay for all cells secreting Ig of a given type rr class, *Eur. J. Immunol.,* 6, 588, 1976.
20. **Andersson, J., Coutinho, A., Lernhardt, W., and Melchers, F.,** Clonal growth and maturation to immunoglobulin secretion in vitro of every growth-inducible B-lymphocyte, *Cell,* 10, 27, 1977.
21. **Coutinho, A. and Möller, G.,** B cell mitogenic properties of thymus-independent antigens, *Nature New Biology,* 245, 12, 1973.
22. **Coutinho, A. and Möller, G.,** Thymus-independent B cell induction and paralysis, *Adv. Immunol.,* 21, 114, 1975.
23. **Coutinho, A., Möller, G., and Richter, W.,** Molecular basis of B cell activation. I. Mitogenicity of native and substituted dextrans, *Scand. J. Immunol.,* 3, 321, 1974.
24. **Coutinho, A., Gronowicz, E., Bullock, W. W., and Möller, G.,** Mechanism of thymus-independent immunocyte triggering: mitogenic activation of B cells result in specific immune response, *J. Exp. Med.,* 139, 74, 1974.
25. **Andersson, J., Melchers, F., Galanos, C., and Luderitz, O.,** The mitogenic effect of lipopolysaccharide on bone marrow derived lymphocytes. Lipid A as the mitogenic part of the molecule, *J. Exp. Med.,* 137, 943, 1973.
26. **Sultzer, B. M.,** Genetic control of host responses to endotoxin, *Infect. Immun.,* 6, 107, 1972.
27. **Watson, J. and Riblet, R.,** Genetic control of responses to lipopolysaccharides in mice. Evidence for a single gene that influences mitogenic and immunogenic responses to lipopolysaccharides, *J. Exp. Med.,* 140, 1147, 1974.
28. **Coutinho, A., Möller, G., and Gronowicz, E.,** Genetic control of B cell responses. IV. Inheritance of unresponsiveness to lipopolysaccharides, *J. Exp. Med.,* 142, 253, 1975.
29. **Coutinho, A. and Gronowicz, E.,** Genetical control of B cell responses. III. Requirement for functional mitogenicity of the antigen in thymus-independent specific responses, *J. Exp. Med.,* 141, 753, 1975.
30. **Forni, L. and Coutinho, A.,** An antiserum that recognizes lipopolysaccharide B cells in the mouse, *Eur. J. Immunol.,* 8, 56, 1978.
31. **Morrison, D. C. and Ryan, J. L.,** Bacterial endotoxins and host immune responses, *Adv. Immunol.,* 28, 293, 1979.
32. **Coutinho, A. and Meo, T.,** Genetic basis for unresponsiveness to lipopolysaccharide in C57BL/10Cr mice, *Immunogenetics,* 7, 17, 1978.
33. **Ridderstad, A., Valugerdi, A., Ström, H., Möller, G., and Möller, E.,** Rheumatoid synovial fluid reconstitutes the B cell defect in CBA/N mice, *Scand. J. Immunol.,* in press.
34. **Fernandez, C. and Möller, G.,** The immune response against two epitopes on the same thymus-independent polysaccharide carrier. I. Role of epitope density in carrier dependent immunity and tolerance, *Immunology,* 33, 59, 1977.
35. **Fernandez, C., Liebermann, R., and Möller, G.,** The immune response to the alpha 1-6 epitope of dextran is determined by a gene linked to the IgCH locus, *Scand. J. Immunol.,* 10, 77, 1979.
36. **Fernandez C. and Möller, G.,** Immunological response to native dextran B512 in young animals of dextran high responder strains is due to lack of Ig expression. Evidence for a non-random expression on V_H genes, *J. Exp. Med.,* 147, 645, 1978.
37. **Howard, J. G. and Hale, C.,** Lack of neonatal susceptibility to induction of tolerance to polysaccharide antigen, *Eur. J. Immunol.,* 6, 486, 1977.
38. **Fernandez, C. and Möller, G.,** A thymus-independent IgG response against dextran B512 can be induced in C57BL, but not in CBA mice, even though both strains possess a V_H dextran gene, *Scand. J. Immunol.,* 10, 465, 1979.
39. **Fernandez, C. and Möller, G.,** Serum antibody and cellular immune response in mice to dextran B512, *Scand. J. Immunol.,* submitted.
40. **Dintzis, H. M., Dintzis, R. L., and Vogelstein, B.,** Molecular determinants of immunogenicity: the immunon model of immune response, *Proc. Natl. Acad. Sci. U.S.A.,* 73, 3771, 1976.
41. **Brunnswick, M., Finkelman, F. D., Highet, P. F., Inman, J. K., Dintzis, H. M., and Mond, J. J.,** Picogram quantities of anti-Ig antibodies coupled to dextran induce B cell proliferation, *J. Immunol.,* 140, 3364, 1988.

42. **Peacock, J. S., Lando, T. R., Roes, D. A., and Barisas, G.,** Biologic activity of antigen receptors artificially incorporated onto B lymphocytes, *J. Immunol.*, 137, 1916, 1986.
43. **Alarcon-Riquelme, M. and Möller, G.,** Thymus-independent B cell activation. Passively incorporated membrane antibodies serve to focus the antigen, but cannot transmit activation signals, *Scand. J. Immunol.*, submitted.
44. **Fernandez, C. and Möller, G.,** A primary immune response to dextran B512 is followed by a period of antigen-specific suppression caused by autoanti-idiotypic antibodies, *Scand. J. Immunol.*, 11, 53, 1980.
45. **Kluskens, L. and Köhler, H.,** Regulation of immune response by autogeneous antibody against receptor, *Proc. Natl. Acad. Sci. U.S.A.*, 71, 5083, 1974.
46. **Fernandez, C. and Möller, G.,** Antigen-induced strain-specific autoantiidiotypic antibodies modulate the immune response to dextran B512, *Proc. Natl. Acad. Sci. U.S.A.*, 76, 5944, 1979.
47. **Fernandez, C. and Möller, G.,** The role of T cells in the maturation of the immune response to dextran B512 tested at the hybridoma cell level, submitted.
48. Mechanism of B lymphocyte tolerance, *Immunol. Rev.*, 43, 1979.
49. **Nossal, G. J. V.,** Cellular mechanisms of immunological tolerance, *Annu. Rev. Immunol.*, 1, 33, 1983.
50. **Fernandez, C. and Möller, G.,** Induction of immunological tolerance requires that the B cells respond to the polyclonal B cell activation properties of the thymus-independent antigens, *J. Exp. Med.*, 146, 308, 1977.
51. **Fernandez, C. and Möller, G.,** Irreversible immunological tolerance to thymus-independent antigens is restricted to the clone of B cells having both Ig and PBA receptors for the tolerogen, *Scand. J. Immunol.*, 7, 137, 1978.
52. **Fernandez, C., Hammarström, L., Möller, G., Primi, D., and Smith, C. I. E.,** Immunological tolerance affects only a subpopulation of the antigen-specific B lymphocytes: evidence against clonal deletion as the mechanism of tolerance induction, *Immunol. Rev.*, 43, 3, 1979.

Chapter 5

ANTIGEN PRESENTATION BY B CELLS

Tania H. Watts, Yves St-Pierre, and Jonathan M. Lee

TABLE OF CONTENTS

I. INTRODUCTION

In contrast to B lymphocytes which can recognize soluble antigens, T lymphocyte recognition of foreign antigen requires that the antigen be recognized on the surface of an antigen presenting cell (APC) in association with gene products of the major histocompatibility complex (MHC). Antigen presentation refers to the process by which APC display foreign antigen for recognition by T lymphocytes. It is now generally accepted that protein antigens are first taken up by the antigen presenting cell, processed in some way in an intracellular acidic compartment and subsequently displayed on the APC surface as a complex with an MHC class II protein. CD4-positive T lymphocytes, via their specific, clonally distributed cell surface receptors, then recognize the complex between self-MHC class II protein and processed foreign antigen. Antigen presentation can thus be divided into several distinct phases: antigen uptake and processing; antigen-MHC II association and transport; and recognition by T lymphocytes.

Although it was initially thought that antigen presenting cells were primarily macrophages, it later became apparent that MHC class II-bearing dendritic cells and Langerhans cells could also function as APC (for review see Reference 1). Subsequently, several investigators found that B cell tumor lines[2-6] as well as normal B cells[1,7-10] could activate antigen-specific T cell lines and hybridomas. In this review we will discuss our current view of each phase of antigen presentation with particular emphasis on the properties which are unique to B cells as antigen presenting cells. In addition, we will review recent findings suggesting that an additional step should be added to the list above: namely that the initial recognition process between B cells and T cells leads to changes in the properties of both cells, which in turn lead to improved cell-cell interaction. Because of the broad scope of the topic, we will not attempt to be comprehensive in listing all the contributors to these areas, but will refer to several excellent reviews which have appeared in the last few years covering a number of these issues. As discussed elsewhere in this volume,[11-14] the ability of specific B cells to present antigen to specific T cells in a highly efficient manner is important in that it allows the B cell to receive from the T cell the signals required for differentiation into antibody secreting plasma cells.

II. BACKGROUND: REVIEW OF BASIC CONCEPTS OF MHC-RESTRICTED RECOGNITION

MHC proteins are highly polymorphic cell surface glycoproteins involved in the binding of peptide antigens and their presentation to T lymphocytes. There are two main classes of MHC protein involved in T cell activation, the class I MHC proteins (mouse H-2K, D and L; human HLA A,B,C) and the class II MHC proteins (mouse Ia antigens: I-A, I-E; human HLA-DR,DP,DQ) The class I MHC proteins are widely distributed on almost all nucleated cells of the body and consist of a 45 kDa polymorphic heavy chain encoded in the MHC locus and a non-covalently attached 12 kDa invariant polypeptide, β-2 microglobulin. The class II MHC proteins are much more restricted in their expression. They are found primarily on macrophages, dendritic cells, Langerhans cells of the skin, B cells, thymic epithelial cells and, in humans, on activated T cells. MHC class II proteins are heterodimeric proteins consisting of an α chain and a β chain of, respectively, 34 and 29 kDa. Each of the MHC class II chains has two extracellular domains, a transmembrane region and a short cytoplasmic domain. The N-terminal domains are most distal from the cell surface and contain most of the polymorphic amino acids. The polymorphic amino acids are thought to influence foreign antigen binding. The membrane proximal domains of MHC proteins are immunoglobulin-like and are more conserved in sequence (for review see References 15 through 17). MHC class II proteins are found associated intracellularly with a 31 kDa polypeptide known as the invariant chain. The invariant chain dissociates from the class II molecule before it reaches the cell surface.[18-20] CD4-positive T cells, which are primarily of the helper subset, preferentially recognize antigen in association with MHC class II proteins, while CD8-positive lymphocytes, which are primarily of the cytotoxic subset, recognize foreign antigen in association with MHC class I proteins.[21] Antigen presentation generally refers to the recognition of antigen with MHC class II proteins by CD4-positive T lymphocytes. MHC restriction refers to the observation that T lymphocytes only recognize foreign antigen when it is displayed on the surface of an APC that expresses an MHC molecule genetically identical to the MHC of the strain of origin of the responding T lymphocyte.[22,23] We now understand this restricted recognition to reflect a requirement that T cells recognize antigenic fragments only when they are embedded in the binding site of an MHC molecule.

III. ANTIGEN PROCESSING

The first stage in antigen presentation is the uptake and delivery of antigen to an internal compartment for antigen processing. Antigen presenting cells can take up antigen nonspecifically via fluid phase pinocytosis or more specifically via receptor mediated endocytosis. A currently accepted view of antigen processing is that antigen first enters the endocytic pathway where it becomes exposed to an acidic environment. Protein antigens are subsequently converted to smaller fragments by proteolysis and a subset of these fragments form complexes with MHC class II molecules and are thereby transported to the cell surface for recognition. As will be discussed below, there is no definitive evidence as to the cellular location of the processing or the MHC class II binding events. Here we will describe some of the key findings relating to the nature of the processing event. For a more comprehensive treatment of some of the earlier work in this area, the reader is referred elsewhere.[24]

A. EVIDENCE FOR AN INTRACELLULAR PROCESSING EVENT IN ANTIGEN PRESENTATION

The work of Gell and Benaceraff[25] showed that, in contrast to antibody mediated recognition which is often sensitive to the conformation of antigenic proteins, T cells can recognize denatured proteins. By comparing families of proteins from different species, and by using proteolytic fragments as well as synthetic peptides, a large number of laboratories have subsequently mapped the epitopes recognized by many antigen-specific T cells to linear sequences of about 10 to 15 amino acids (for review see Reference 26). Early work on the role of macrophages in T lymphocyte activation established that one could pulse macrophages with antigen prior to addition of T lymphocytes and thereby define a 30 to 60 min handling step that depended on cellular energy (reviewed in Reference 22). Zeigler and Unanue[27] found that the presentation of *Listeria monocytogenes* by macrophages involved a lag of 45 to 60 min before T cell recognition could be demonstrated. Subsequent studies by the same group showed that weak bases such as chloroquine or ammonium chloride could block this "handling" step.[28] Chloroquine rapidly penetrates cells and accumulates in acidic vacuoles (such as endosomes and lysosomes) resulting in an increase of the pH of those compartments of 1 to 2 pH units (for review see Reference 29). This led to the suggestion that antigen processing involves alteration, perhaps fragmentation, in an acidic compartment. However, it should be pointed out, as recently reviewed by Mellman et al.,[29] that treatment of cells with chloroquine has a number of cellular consequences: for example, inhibition of lysosomal degradation by acid hydrolases, inhibition of intracellular sorting and exocytic mechanisms, swelling of the intracellular vacuoles and direct inhibition of the enzymes transglutaminase and cathepsin D.

Further evidence for an essential processing step in antigen presentation came from the work of Shimonkevitz et al.,[30] which showed that B lymphoma cells, when lightly fixed with glutaraldehyde, were able to present tryptic or cyanogen bromide-cleaved fragments of ovalbumin but not the native protein antigen to ovalbumin-specific T cell hybrids. These results substantiated the idea that the processing even involves proteolysis of the antigen. The glutaraldehyde fixation experiments were further supported by experiments using purified MHC class II molecules in lipid bilayers to replace the antigen presenting cell.[31] Planar membrane bound I-A^d (described in Section VI) could present a peptide digest of ovalbumin but not the native protein to an I-A^d-restricted T cell hybrid.

Additional evidence that proteolysis plays an important role in the antigen processing event comes from the observation that protease inhibitors such as leupeptin can block the presentation of native, but not peptide, antigens.[32-34] However, as will be discussed below, Blum and Cresswell[35] have shown that leupeptin can also influence the intracellular transport of the MHC class II molecule by preventing its dissociation from the invariant chain.

B. NATURE OF THE PROCESSING EVENTS

Although antigen fragmentation appears to be an essential step in the generation of some T cell epitopes, work from several laboratories has shown that some proteins only require unfolding rather than actual fragmentation. For example, Streicher et al.[32] showed that, in the presence of leupeptin or chloroquine, an intact but denatured form of myoglobin (*s*-methylmyoglobin) could be presented to T cell clones. Based on these observations, Streicher et al.[32] suggested that the critical step in antigen processing was unfolding to expose buried hydrophobic areas which might be important for interaction with the antigen presenting cell membrane or with the MHC class II protein itself. Similar observations were described by Allen et al.[36-38] In mapping the region of hen egg lysozyme (HEL) recognized by two I-A^k-restricted T cell clones, they found that both clones recognized a region contained within amino acids 46 to 61 of HEL.[37] One clone (2A11) was able to recognize unfolded protein (carboxymethyl-HEL) presented by chloroquine-treated APC while the other clone (3A9) required further fragmentation of the antigen.[36] Further analysis of these clones showed differential sensitivity to truncation of the antigenic peptide. 2A11 showed a decreased responsiveness to antigen when HEL46-61 was further truncated by six amino acids at the N-terminus, while the response of 3A9 improved with truncation.[38] These data suggest that the degree of processing required for optimal antigenic stimulation will depend on the particular T cell clone. One criticism of these studies, however, is that partial unfolding of these globular proteins renders them extremely susceptible to proteolysis so that surface enzymes resistant to chloroquine could potentially process the antigen further. On the other hand, Sette et al.[39] have recently shown that intact unfolded antigens can bind to purified MHC class II molecules in a cell free system. Moreover, Donermeyer et al.[40] have used a minimal HEL peptide extended at each end with segments of 12 D amino acids and shown that this 34-amino acid peptide can be presented by fixed APC. Since the peptide bonds of D amino acid-peptides are protease resistant, this rules out the criticism raised above that further processing might take place due to the action of surface proteases. Another aspect of the study by Donermeyer et al. was the demonstration that binding of peptide to the MHC molecule protects the core region from the action of proteolytic enzymes. This observation lends support to a hypothesis put forward by Sercarz and collaborators[41] that the binding of partially unfolded proteins to MHC influences the outcome of the processing and presentation events and ultimately the emergence of dominant T cell epitopes.

It is conceivable that different APC produce different processed products, and that this differential processing might influence the immune repertoire. Indirect evidence that the processed form of antigen differs from cell type to cell type has been obtained in several laboratories. For example, Shastri et al.[42] found that L cells expressing transfected MHC class II molecules could present lysozyme peptides but not native lysozyme to specific T cells. Since other L cell transfectants had been shown to present a variety of protein antigens,

Shastri et al. suggested that the defect in this cell was not a failure to process but perhaps the nature of the processed product. However, in a system as complicated as antigen presentation, other explanations can be suggested to explain the data of Shastri et al. For example, Stockinger et al.[43] have suggested that the defect in the L cells used in Shastri's experiments was a deficiency in the invariant chain (see Section V.D.).

Michalek et al.[44] have obtained indirect evidence that two B lymphomas from the same genetic background (Balb/c) process the protein ovalbumin differently. They compared the ability of the B lymphomas A20 and M12 to present native ovalbumin (OVA) or processed peptides to the T cell hybridomas DO-11.10 and 3DO-54.8. Both DO-11.10 and 3DO-54.8 had been shown to respond to the peptide OVA323-339 together with I-A^d, but with different peptide fine specificity. 3DO-54.8 will respond to a minimal peptide defined by residue 326-336 of chicken ovalbumin while DO-11.10 is sensitive to truncation of the C-terminal three amino acids.[45] Michalek et al. found that M12 and A20 could present OVA323-339 equally well to the T hybrids but that M12 presented native ovalbumin only to 3DO-54.8. The A20 cell line presented native ovalbumin to both T hybrids. The simplest interpretation of these data is that A20 and M12 process ovalbumin differently, M12 producing less of the full length OVA323-339 determinant required for DO-11.10 activation. We have found that the Balb/c macrophage-like cell line P388D1 and the B lymphoma A20 also show differences in presentation of native ovalbumin to DO-11.10 (Y.S.P. and T.H.W., unpublished). By treatment with γ-interferon, P388D1 can be induced to express high levels of I-A^d and thereby become an effective APC.[46] We found that γ-IFN-treated P388D1 was much more effective in native ovalbumin presentation to DO-11.10 than was A20, although both lines were comparable in peptide (OVA323-339) presentation. Taken together, the above findings suggest that the three Balb/c APC lines M12, A20 and P388D1 produce the optimal peptide for DO-11.10 activation at very different levels.

Recently, Demotz et al.[47] have described the isolation of a naturally processed form of hen egg lysozyme. They pulsed antigen-presenting B lymphoma cells with native HEL, purified the I-E molecules and, using the supported planar membrane method (described in Section VI), showed that the purified I-E was effective in activation of an HEL-specific T cell without further addition of peptide. Acid elution of the peptide from the I-E followed by HPLC separation of the eluted peptide showed that the fractions which were active in T cell activation were similar in size and hydrophobicity to synthetic peptides representing the minimal epitope for that T cell and were therefore unlikely to have undergone any major chemical modifications besides fragmentation. Since HEL is a small protein (about 14 kDa), it is not unexpected that proteolytic cuts, which we assume are made at exposed turns of the folded polypeptide chain, would generate relatively short processed fragments. This does not rule out, however, that for other proteins, larger fragments might be produced. Taken together, the data described above suggest that the final form of processed antigen is indeed a peptide of size greater than or equal to 10 to 20 amino acids. However, these data do not completely rule out the possibility that, during the processing step in an APC, the antigen undergoes other modifications which facilitate the processing, transport, or MHC binding events.

Falo et al.[48] have shown that antigen associated with the cell surface of antigen-pulsed APC can be removed by the action of phospholipases. Phospholipase A2 or phospholipase C treatment of insulin-pulsed APC abrogated the activation of insulin-specific T hybrids but not the activation of alloreactive hybrids. This led them to suggest that the antigen may be attached to the cell surface in a lipid linkage, perhaps providing a pool of antigen for interaction with MHC class II. Since alloreactive hybrids may also be specific for peptide/MHC complexes, the peptides being derived from endogenous cellular proteins or serum proteins[49-50] (see Section V.A.2), it is not clear why the removal of antigen would be specific to the exogenously added antigen in these experiments. However, further work by the same group[51] showed that cerulenin, an inhibitor of sterol synthesis and lipid metabolism, is also a potent inhibitor of native but not peptide antigen presentation when added prior to but not after antigen pulsing. This experiment lends support to the idea that lipid molecules are somehow involved in some aspect of antigen processing or transport, prior to the formation of the antigen-MHC complex.

Pierce and collaborators[52-53] have described a peptide-binding protein on the surface of splenic B cells. This protein appears to play a role in antigen presentation, since a specific antiserum raised against it blocks T cell responses to cytochrome *c*. This peptide binding protein was subsequently shown to be a member of the heat shock protein (hsp 70) family.[53] As reviewed by Rothman,[54] heat shock proteins are known to bind to denatured or misfolded proteins as well as to newly synthesized proteins. Heat shock proteins have been implicated in transport of proteins through the endoplasmic reticulum[54] as well as the targeting of intracellular proteins for lysosomal degradation.[55] This raises the possibility that the protein identified by Pierce and collaborators is involved in targeting processed antigen to the site of MHC-peptide interaction. Although it is not clear how these data or those of Falo et al.[48,51] fit into our current model of antigen processing and presentation, they should certainly be kept in mind when considering possible active cellular mechanisms for facilitating antigen processing, intracellular transport of peptides or peptide-MHC association and dissociation.

C. CELLULAR LOCATION OF THE ANTIGEN PROCESSING EVENTS

Despite the criticisms mentioned above on the lack of specificity of both chloroquine and leupeptin, the simplest interpretation of the current data is that antigens taken up by pinocytosis enter the mildly acidic endosomal compartment and are partly degraded by endosomal proteases. In addition to the endosomal compartment, which is generally agreed to have a pH of between 5 and 6,[29] the cell contains other acidic compartments that could potentially be the site of antigen processing. Lysosomes have a pH 4.6 to 5 and coated vesicles as well as the trans-Golgi are also mildly acidic (for review see Reference 29). As has been pointed out by others[56] it is unlikely than antigen processing takes place in lysosomes since the extensive set of acid hydrolases present in the lysosomes would result in more or less complete degradation of the protein. In support of the role of endosomes in antigen processing, Stahl and collaborators[57-58] have shown that isolated endosomes contain acid hydrolases, particularly a membrane-associated cathepsin D-like protease.[58] Chain et al.[56] have also found neutral and acidic endopeptidase activity associated with the low density (endosomal or plasma membrane) fractions of B lymphoma membranes.

Further support that endosomes and not lysosomes are the relevant site for antigen processing comes from the recent work of McCoy et al.[59] These studies made use of temperature sensitive mutants of a Chinese hamster ovary (CHO) cell line that had been shown to be defective in early endosome acidification. The mutant CHO cells were transfected with cDNAs for the murine MHC class II proteins I-E^k or I-A^d. At the permissive temperature, the CHO transfectants could present native antigens as well as peptides to peptide-specific T cell clones. At the nonpermissive temperature, however, peptides but not native antigen could be presented. Although these data are suggestive that endosomes are the site of processing, the original mutants were also shown to have changes in Golgi and transport functions.[60] As McCoy et al. point out, it is possible that altered transport of the MHC class II molecule or of the processed fragments prevents the formation of the antigen-MHC complex, rather than the defect being in the processing compartment per se. One anticipates that further characterization of these cell lines will help resolve some of these issues.

D. TIME COURSE OF ANTIGEN PROCESSING

As mentioned above, the studies of Unanue and others[22,27,28] established that antigen processing as monitored by T cell recognition involves an initial lag of at least 45 to 60 min. More recently, Roosnek et al.[61] have used a rapid assay of T cell activation to investigate this lag period further. This study took advantage of calcium-selective fluorescent probes developed by Tsien and collaborators[62] which allow changes in intracellular free calcium to be measured within minutes. Using increased cytoplasmic free calcium as an indication of the T cell recognition event, Roosnek et al.[61] found that pre-processed antigens added to B cells at high doses show virtually no lag in T cell activation. In contrast, antigens which required intracellular processing showed a lag of about 1 h. This lag was not particularly dependent on antigen concentration, suggesting that a processing or transport step was the limiting step for the appearance of the cell surface antigen-MHC complex. By using antibodies against the transferrin receptor as the antigen (and mouse immunoglobulin-specific human T cells as the responders) Roosnek et al. further defined the rate-limiting step for appearance of the MHC-peptide complex on the cell surface. Since internalization of the transferrin receptor occurs within 10 min,[63] rate limiting steps accounting for the one hour lag time must take place after internalization of the antigen. Diment and Stahl[57] have shown that endocytosed proteins begin to be degraded within minutes of uptake by macrophages. This led Chain et al.[56] to suggest that the rate-limiting step for antigen processing is intracellular transport of the processed product to the site where it interacts with the MHC class II protein (see Section V.B).

IV. THE ROLE OF ANTIGEN-SPECIFIC B CELLS IN ANTIGEN DELIVERY TO THE PROCESSING COMPARTMENT

As discussed above, the internalization of antigen by the APC is a prerequisite for the presentation of most protein antigens examined to date. Macrophages, in addition to their nonspecific pinocytic and phagocytic capabilities, possess Fc receptors and C3b receptors which are thought to facilitate the internalization of opsonized antigens.[64] B cells possess a unique mechanism, by virtue of their antigen-specific immunoglobulin receptors, which allows them to bind specific antigens selectively and efficiently deliver them to an internal, chloroquine-sensitive compartment for processing (for review see Reference 65). The evidence for this model has come from several studies which showed that antigen presentation was effective at much lower doses of antigen when B cells specific for the antigen were present in the culture.[7,66-69] Critical evidence that the immunoglobulin receptor was involved in antigen presentation by B lymphocytes came from the work of Chesnut and Grey,[7] who showed that rabbit-anti-mouse immunoglobulin, but not normal rabbit serum, could be presented efficiently to rabbit-immunoglobulin-specific T cells by macrophage-depleted splenocytes. Subsequently, Rock et al.[67] demonstrated a similar phenomenon with hapten-specific B lymphocytes.

In a particularly elegant study, Lanzavecchia[68] showed that tetanus toxoid (TT)-specific EBV-transformed human B cells could present TT to TT-specific T cell clones at 10^4-fold lower antigen doses than nonspecific B cells. The fact that denatured TT was presented equally well by both specific and nonspecific B cells and that pre-addition of anti-immunoglobulin to the B cells blocked the enhanced antigen presentation indicated that the mechanism of enhanced presentation involved binding of the antigen to the B cell surface Ig. Experiments with chloroquine or glutaraldehyde added before or after pulsing with TT, showed that the enhanced antigen presentation by specific B cells involved processing of the antigen in an acidic compartment. The concentration of antigen used for these experiments was as low as 10^{-12} M, while the affinity of the sIg for TT is about 10^8 M^{-1}. Lanzavecchia[65] calculated from these data that there could be an average of only 10 to 50 TT molecules bound per APC in the culture, suggesting that the process of internalization and delivery of antigen to the MHC class II binding site must be highly efficient in these cells. He proposed that, by continually internalizing the antigen via the sIg, the B cell can accumulate sufficient

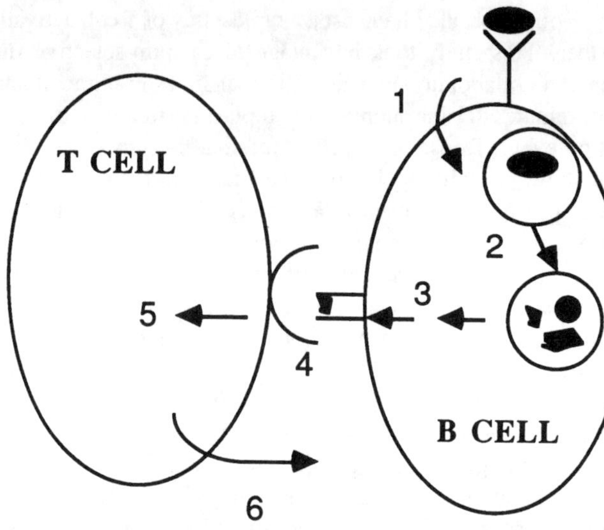

FIGURE 1. Antigen presentation by specific B cells. 1. Antigen bound specifically to the surface immunoglobulin receptor is taken up by receptor-mediated endocytosis and enters the early endosomal compartment. 2. Antigen is partially degraded by the action of proteases in a chloroquine-sensitive compartment. 3. At an as yet unidentified site, processed antigen binds to the MHC class II molecule and is ultimately transported to the cell surface. 4. The MHC class II protein/processed antigen complex is recognized on the cell surface by a specific T cell. 5. As a consequence of the recognition event, the T cell is activated. 6. The activated T cell delivers "help" to the B cell. (Adapted from Howard, J. C., *Nature (London),* 314, 494, 1985.)

antigen for T cell activation even at very low antigen concentrations. Lanzavecchia[65] has referred to this process as the "Vacuum Cleaner Model" of antigen presentation (Figure 1).

As illustrated in Figure 1 and reviewed by Howard,[70] the understanding of the role of antigen-specific B cells in antigen presentation to T cells, culminating in the work of Lanzavecchia, has greatly clarified our understanding of how T cell help can be delivered selectively to a B cell specific for the same antigen as that recognized by the T cell. This is important in that it allows the T cell to regulate the B cell response and, at limiting concentrations of antigen, bystander activation of B cells specific for other antigens is avoided.

Using human T cells specific for mouse immunoglobulins (obtained from patients undergoing immunotherapy with mouse monoclonal antibodies), Lanzavecchia et al.[71] have recently compared a number of surface structures for their ability to deliver antigen to the processing compartment for effective presentation. After normalization of the data to account for different levels of binding of the ligands to the cell surface (measured at 4°C) it was found that anti-immunoglobulin was presented about ten times more efficiently that anti-class II or anti-transferrin receptor and 100-fold more efficiently than anti-class I antibodies. As pointed out by Lanzavecchia et al.,[71] the differences between various surface proteins as delivery vehicles for antigen presentation, may be due to their different rates of internalization or due to their entrance into distinct cellular compartments.

Snider and Segal[72] have carried out a similar study of selective antigen presentation using hen egg lysozyme (HEL) bound to HEL-specific antibodies in heteroconjugates with other monoclonal antibodies specific for various B cell surface structures. In this study, anti-IgM and anti-IgD worked best, about tenfold better than anti-class I or anti-class II which were equally effective. Targeting to the B cell surface antigen B220 or to the FcγRII was completely ineffective. The rate of internalization was found to be substantially faster for the anti-IgD than the anti-class I heteroconjugates and the anti-IgM fell between the two. Anti-B220 heteroconjugates did not internalize at all, correlating with the failure of anti-B220 to deliver antigen for processing. Casten et al.[73] have carried out a similar study using cytochrome *c* peptides coupled to antibodies specific for various cell surface molecules. While they found a similar enhancement of antigen presentation by targetting to surface Ig or MHC molecules, their results are somewhat puzzling because in this case the enhanced presentation persisted even when the cells had been pre-fixed with glutaraldehyde, suggesting that antigen coupled by a short cross-linker to an immunoglobulin and attached to an irrelevant surface protein, can somehow reach the MHC class II binding site and activate a T cell response.

V. MHC-PEPTIDE INTERACTION

A. MHC CLASS II PROTEINS BIND PEPTIDES

There is now an extensive literature documenting the binding of small synthetic peptide antigens to purified MHC class II proteins. Prior to the first direct evidence for this interaction, a number of indirect pieces of evidence strongly pointed in that direction. First, the T cell response to foreign antigen was strongly dependent on the specificity of both the antigen and the MHC molecule.[74] Second, in dually specific T cells produced by fusion of two antigen-specific MHC-restricted hybrids, there was never independent sorting of the antigen and MHC specificities, strongly suggesting that a single T cell receptor was jointly responsible for both the antigen and the MHC specificity of the T cell.[75] Therefore, the simplest model was that the T cell receptor was specific for a complex between antigen and MHC. The first evidence that antigen binds to MHC class II molecules came from competition experiments which indicated that antigens whose immune response was restricted to the same MHC class II gene product could compete with one another for antigen presentation.[76-77] Subsequently, Watts et al.[31] showed that highly purified MHC class II molecules when reconstituted into lipid bilayers on glass supports, could present processed antigenic peptides to T cell hybridomas, indicating that no other APC surface proteins were required for peptide-MHC-T cell interaction.

The first direct evidence that a purified MHC class II protein could bind an antigenic peptide came from Babbitt et al.[78] who used equilibrium dialysis in detergent to show that the peptide HEL46-61 binds to I-A^k with a 2 μM K_d and that the I-A^k molecule possessed a single saturable site for binding this peptide. Shortly thereafter, these data were confirmed by Buus et al.[79] who showed that the peptide OVA323-339 could bind to I-A^d and not to I-A^k, while the peptide HEL46-61 could bind to I-A^k but not to I-A^d. This provided experimental evidence supporting the determinant selection hypothesis for MHC-restricted recognition: the observation that the ability to mount an immune response to a particular antigen in a particular strain of mouse maps to the class II region of the MHC locus could now be explained at the level of antigen-MHC binding. These observations were subsequently extended by Buus et al.[80] Twelve different antigenic peptides were analyzed for binding to four different MHC class II molecules in a gel filtration assay both by direct binding and competition assays. In all cases peptides could be demonstrated to bind to the MHC molecules that were known to act as restriction elements for those antigens. Several of the peptides could bind to more than one MHC class II protein, but in all but one case, the highest binding obtained was to the MHC class II restriction element dominantly used by that antigen.[80] It is now clear that many peptide antigens can bind to more than one haplotype or more than one isotype of MHC antigen,[81-83] although fine mapping of the determinants has shown that the amino acids required for binding to one isotype of MHC molecule (e.g.,

I-A) do not necessarily overlap with those required for binding to another isotype (e.g., I-E).[82]

In addition to the studies of peptide binding to MHC class II proteins in detergent using equilibrium dialysis, a number of other approaches have been used to demonstrate specific peptide-MHC interaction: gel filtration in detergent,[84] fluorescent peptide binding to purified MHC class II in lipid bilayers on glass supports,[85,86] photoaffinity cross-linking to MHC class II molecules on live cells[87] or on purified MHC molecules,[88,89] competition assays on live or fixed cells[90-93] as well direct binding to cell surfaces using radioiodinated peptides.[94]

Our initial attempts to show fluorescent peptide binding to purified MHC class II molecules in lipid bilayers on glass supports were not successful.[95] Fluorescent peptides were briefly incubated with MHC class II-containing planar membranes and then subjected to extensive washing. We now know that the on-rate for peptide-MHC association is very low,[84] explaining our initial failure to demonstrate binding by this method. However, we knew that an MHC class II-peptide complex must form in the presence of specific T cells since T cell activation could be demonstrated using only purified MHC class II plus peptide. This led us to speculate that the presence of the T cell could stabilize the interaction between the MHC class II molecule and the peptide. Evidence that this could occur was obtained using the technique of resonance energy transfer. It was found that resonance energy transfer between Texas red-labeled Ia and fluorescein-labeled peptide was induced by the binding of a specific T cell.[95] Since peptide-MHC interaction can occur in the absence of the T cell we interpret the former result to imply either that the presence of the specific T cell receptor made the complex between MHC and peptide more rigid and therefore led to increased energy transfer allowing us to detect a very low number of MHC-peptide complexes, or that the presence of the T cell receptor altered the kinetics of the interaction. In the latter case, the experiment is not physiologically relevant since the available evidence strongly supports a model whereby peptide-MHC interaction takes place intracellularly before the T cell recognition event.

1. Unusual Stability of the MHC Class II-Peptide Complex

Our first hint that the antigen-MHC complex was remarkably stable, came when we observed that antigen presenting B lymphomas (A20) could be pulsed with antigen, their plasma membranes isolated, solubilized in the detergent deoxycholate and dialyzed exhaustively without loss of T cell stimulatory ability.[85] Using fluorescein-labeled OVA323-339, we subsequently found that this stable association of antigen with the APC membrane could be explained at the level of antigen-MHC interaction. Fluorescent peptide could be incubated with purified I-A^d in lipid bilayers and the lipid vesicles dialyzed exhaustively, while retaining specific fluorescence and T cell stimulatory capacity. Binding of the peptide to the lipid vesicles specifically required the presence of the I-A^d molecule, pure lipid bilayers or glycophorin-containing membranes did not retain peptide. Moreover, the dialyzed, pre-formed I-A^d/OVA323-339 complexes were completely sufficient for T cell activation when converted to supported planar membranes (Section VI.A).[85] At about the same time as we reported these results, Buus et al.[84] also provided evidence that the complex between MHC class II and peptide was remarkably stable. They found that the complexes could be isolated by gel filtration chromatography in detergent and that the isolated pre-formed complexes, when reconstituted with lipid and coated on glass beads could stimulate T cells. The rapid gel filtration assay devised by Buus et al. was particularly useful because it was amenable to kinetic measurements. It was found that the on-rate for MHC-peptide interaction was remarkably slow (ka = 10^0 s^{-1} M^{-1}) in contrast to antibody-antigen interactions which are typically diffusion limited (10^5 s^{-1} M^{-1}). By the same method, it was found that the I-A^d/OVA323-339 complex dissociated with a kd of about 10^{-6} s^{-1}. At neutral pH, this results in a half-life for the antigen-MHC complex in the detergent NP40 of about 310 min, which is decreased to 160 min at pH 5.6 and to 29 min at pH 4.6.[84]

In contrast to the results described by Buus et al.,[84] we have recently found that at neutral pH the complex between I-A^d and OVA323-339 is stable for weeks at 4°C.[96] Jensen[97] has recently reported that the antigenic determinants associated with crude membrane preparations from antigen-pulsed-I-A-containing B lymphomas are remarkably stable to acid pH as monitored in a functional assay for T cell activation. We have come to similar conclusions using purified I-A preparations. The I-A^d/OVA323-339 complex was stable to prolonged exposure to pH 5, and required several hours at lower pHs for full dissociation.[98] The data from these studies suggest that brief transit through an endosomal compartment would be unlikely to release bound peptide from the MHC binding site. The difference between the studies of Buus et al.[84] and the more recent studies[97-98] are twofold. First, in the latter two studies, the MHC class II molecules were in lipid bilayers and not in detergent. Second, Buus et al. used iodinated peptides, while our experiments and those of Jensen made use of unlabeled peptide or fluorescein-labeled peptide. We have found that heavily radioiodinated OVA323-339 breaks down rapidly, as monitored by thin layer chromatography. This might explain the higher apparent dissociation rates observed by Buus et al.[84]

Lanzavecchia[65] has found that the half-life of antigen in/on antigen-pulsed specific B cells is about 24 h, as measured using T cell activation. Although this is a rather indirect calculation and may not strictly reflect the half-life of the actual antigen-MHC complex in an APC, the data are consistent with the complex being relatively stable in an APC as well, The finding that a high proportion of I-E molecules isolated from lysozyme-pulsed APC retain the antigenic peptide, also supports this view.[47]

The unusual stability of the antigen-MHC class II complex, led to speculation[84,85] that peptide binding and/or dissociation from MHC class II molecules may involve conformational changes in either the MHC or the peptide or both. Data consistent with such a structural change come from recent studies by Sadegh-Nasseri and McConnell.[86] Kinetic analysis of complex formation between I-E^k and pigeon cytochrome *c* 88-104 revealed the presence of an intermediate complex of lower stability that slowly converts into the final stable complex recognized by T cells.

2. The MHC Class II Molecule Does Not Discriminate Between Self and Nonself

It is now clear that the MHC class II protein does not discriminate between the binding of foreign antigens vs. self-peptides.[99] The first evidence that this was the case came from studies by Babbitt et al.[91] in the lysozyme system. They found that mouse lysozyme 46-61 could compete with HEL46-61 for presentation by the I-A^k molecule. A similar result was also obtained in the cytochrome *c* system by Lakey et al.[100] Subsequently, Lorenz and Allen[101] extended these observations to show that self-hemoglobin can bind to MHC class II molecules *in vivo* and *in vitro*. This observation and the observation that the MHC-peptide complex is so stable, have led to the suggestion that at any one time a large proportion of the binding sites of the MHC class II proteins are occupied with peptide fragments produced by fragmentation of serum proteins or endogenous cellular proteins. Indeed, Buus et al.[102] have acid eluted peptide-like material from purified MHC class II molecules and shown that the eluted material competes for ovalbumin peptide binding to I-A^d. One possibility for the slow on-rate observed for MHC class II interaction is therefore the need to displace endogenous peptides from the site.

When the number of MHC class II binding sites that can be occupied with peptide are analyzed in purified preparations of MHC proteins, only a subfraction of the molecules can be shown to bind peptide. In the case of I-A^d only about 10 to 20% of the binding sites are available for peptide binding.[96,103] Because of the remarkable stability of the complex, it is difficult to dissociate peptide from the MHC binding site without some loss of serological determinants.[98,102] For this reason it has been difficult to obtain pure populations of MHC-peptide complexes for structural analysis. Interestingly, in the study of Demotz et al.,[47] described in Section III.B above, estimates of the proportion of I-E occupied with a naturally processed HEL peptide ranged from 10 to 38%, suggesting that in a B cell, the process of MHC interaction with an immunodominant peptide is quite efficient.

The fact that the MHC class II molecules on the surface of APC are likely to be occupied with a broad array of self-peptides has important implications for allorecognition. In 1977, Matzinger and Bevan[49] suggested that alloreactive T cells (T cells which respond to MHC from another individual in an apparently antigen-independent manner) might be specific for both self-antigens on the APC together with self-MHC protein. The recent data suggesting that MHC class II molecules are constitutively occupied with endogenous peptides strongly supports such a model. Recently, Kappler and Marrack[50] have demonstrated that some alloreactive I-E-restricted-T cells are B-cell specific and fail to respond to the same MHC molecule on macrophage lines. Their favored interpretation of these data is that these T cells are specific for a B cell-specific peptide plus I-E.

3. X-Ray Crystallographic Analysis of MHC Molecules

The first high resolution X-ray derived model of an MHC molecule, reported by Bjorkman, Wiley, and collaborators in 1987,[15,16] represented a major step forward in the study of MHC-restricted recognition by T cells. X-ray crystallographic analysis of the human MHC class I protein, HLA-A2 revealed that the N-terminal domains of the protein, containing most of the polymorphic amino acids, consist of a platform of 8-stranded β sheet upon which rest two α-helical regions, which together form a groove of 25 Å length by 10 Å wide. This groove is thought to be the site for foreign antigen binding. A high proportion of the polymorphic amino acids in HLA-A2 are found either on the β sheet pointing up into the groove or on the inner faces of the α-helices pointing in toward the groove. This suggests that changes at these positions from one MHC haplotype to another could greatly influence antigen binding. In the HLA-A2 electron density map, the putative peptide binding groove is not empty but is occupied with unidentified density that cannot be traced to the HLA-A2 protein. Bjorkman et al.[16] have suggested that this material represents bound peptide that has co-purified and co-crystallized with the MHC protein. Brown et al.,[17] have proposed a model for MHC class II proteins based upon similarities to MHC class I and suggest that the overall three-dimensional conformation of the class II molecules will be quite similar to MHC class I with respect to the folding of the domains and the putative peptide-binding groove. This model is now providing a structural framework for site directed mutagenesis of the MHC class I[104] and II antigens,[105-108] for the analysis of the importance of the polymorphic and constant amino acids in peptide binding. So far, these studies have consistently supported a model in which amino acids lining the putative-peptide binding groove exert a major influence on T cell recognition.

4. Elucidation of the MHC and T Cell Contact Sites on Antigenic Peptides: Models for MHC Class II-Peptide Interaction

For a peptide antigen to be recognized by a T cell, it must have two functional domains, a region which interacts with the T cell receptor (the epitope) and a region which interacts with the MHC protein. Schwartz and collaborators coined the term aggretope to refer to the part of the peptide which contacts MHC (for review see Reference 74). In order to map the regions of the peptide recognized by T cells and those involved in contact with the MHC, the approach used by a number of groups has been to make single amino acid substitutions in the peptide antigen. If a substituted peptide fails to activate the T cell it is then used either in direct binding studies or in competition assays to determine whether the failure of the T cell response is due to loss of binding to the MHC protein or due to a loss of T cell recognition of the peptide/MHC complex.

Two of the best studied systems in this area to date have been the interaction of HEL46-

61 with I-A^k, by Allen and collaborators,[109,110] and the interaction of OVA323-339 with I-A^d, by the laboratory of Howard Grey.[103,111] The overall conclusion from these studies has been that while T cells are highly sensitive to amino acid substitutions at many positions in the peptide antigens which they recognize, MHC class II molecules are remarkably permissive in tolerating amino acid substitutions in the peptide antigen. When conservative and non-conservative single amino acid substitutions are made in OVA323-339 or HEL46-61, only three or four amino acid side chains seem important for binding to Ia.

The data of Allen et al.[109] were most readily interpreted in terms of segregation of the aggretope and epitope portions of HEL 52-61 when the peptide was modeled as an α-helix. Brown et al.[17] have fit the α-helical model of the HEL peptide into their hypothetical model for the class II protein, I-A^k. Although a unique fit cannot be determined for HEL52-61 binding to the MHC class II groove, the model at least appears to be able to accommodate the helical HEL peptide. The data of Sette et al.[111] on the analysis of OVA323-339 binding to I-A^d suggest a more complicated model. There was no clear segregation of amino acids into an aggretope and an epitope region and indeed some amino acids appeared to be important for both T cell contact and MHC contact. The data were consistent with neither a fully extended or a fully helical conformation of OVA323-339 in the I-A^d binding site.

While singly substituted peptides are certainly useful in formulating models for peptide-MHC interaction, it should be kept in mind that the interaction between the peptide and the MHC protein may be conformationally complex and could depend critically on the fold of the peptide backbone. It is therefore possible that multiple substitutions within a peptide will give different results than those obtained with the singly substituted peptides. In fact, in an earlier analysis of the T cell response to OVA323-339 we had found that valine 327 and histidine 328 could be individually substituted with alanine with only a marginal effect on T cell activation but that the two substitutions together had a synergistic effect.[112] Because of this difficulty, it is still not obvious how one can synthesize a peptide de novo based on analysis of the Ia contact sites with any assurance that it will bind to the MHC molecule of interest. On the other hand, one can construct hybrid peptides containing the T cell contact residues from one peptide and the remaining residues from another peptide which is known to compete for binding to the same restriction element. In one case, Guillet et al.[92] were able to transfer three amino acids from the λ repressor peptide cI 12-24 onto the backbone of the I-A^d-binding antigen OVA323-339 (at position 331, 333, and 336). The hybrid peptide could stimulate a cI-specific-I-A^d-restricted T cell. A similar approach was recently taken by Lorenz et al.[113] They transferred the critical residues required for RNase recognition into an influenza hemagglutinin peptide. In the latter experiment, one additional amino acid besides the four amino acids identified as T cell contact sites was required, again suggesting a possible conformational effect on the T cell determinant.

It has been known for some time that a single MHC class II protein must be able to bind a broad array of peptide antigens, and, based on the recent data, it must be able to bind them with reasonable affinity to form a stable complex. Therefore, considerable effort has been made to find a common structural motif in peptides which bind MHC proteins.[114-118] One of the first suggestions was that peptides bind to MHC class II molecules in an α-helical conformation which allows the segregation of the aggretope and epitope residues of the peptide to opposite sides of the helix.[74,112,114,119]

Bersofsky and collaborators[114-116] have championed the view that the tendency to form amphipathic structures (particularly α-helices) is a general property of most peptides which bind to MHC class II proteins. Indeed, some antigenic peptides have been shown to come from α-helical regions of proteins or to have a helical propensity as measured by the ability to form an α-helix in the helix promoting solvent trifluoroethanol (TFE).[119-121] To test the Bersofsky model, several groups have synthesized peptide analogs with substitutions designed to stabilize or destabilize helical structures. Carbone et al.[122] concluded from such studies that the T cell response to pigeon cytochrome *c* is improved when hydrophobic leader sequences are added to the N-terminus of a minimal peptide but that no difference was obtained when these regions contained helix-breaking amino acids. Allen et al.[121] found that by adding charged amino acids to the N- and C-terminus of HEL46-61, to stabilize the α-helical conformation by neutralizing the dipole of the α-helix, one could improve the antigenicity of the peptide. In addition, at limiting concentrations of the helix promoting solvent TFE, they could demonstrate, using circular dichroism, a modest increase in helix propensity of the modified peptide. However, no improved binding of the peptide to I-A was observed as a result of these changes, indicating that the improved response does not take place at the level of binding to the I-A molecule but may reflect an increased resistance of the peptide to proteolysis or an increased ability to interact with a lipid bilayer.

Sette et al.[111] have shown that the most critical region of OVA323-339 binding to I-A^d is contained within a six amino acid core (VHAAHA, in one-letter code). This core is remarkably similar in sequence to a core region found in other I-A^d binding peptides.[111,123] The core region, however, does not have the capability of binding to I-A^d alone but requires flanking residues on each side. Sette et al.[124] have analyzed a variety of flanking sequences for their ability to confer peptide binding to the OVA 327-332 core. The results were quite complex and no simple correlation could be made between the propensity to form helices or β-sheets and the ability to bind to I-A^d.

Taylor and Rothbard[117] have identified a linear sequence motif of four or five amino acids found in many peptide antigens known to associate with class I or class II MHC proteins. Their motif is as follows. Position 1 has a high probability of being charged or a glycine residue. Position 2 and 3 are hydrophobic amino acids in 45 to 50% of epitopes looked at, position 4 is primarily polar or glycine. This motif is relatively similar to the motif proposed by the Grey lab. However, Sette et al. have found a difference in the motif used for peptides that bind to I-A^d vs. I-E. As defined by Sette et al.,[124] the I-A^d binding peptides use a six amino acid motif: 1. hydrophobic, 2. positively charged or polar, 3. hydrophobic, 4. hydrophobic, 5. polar or charged, and 6. short side chain, while the I-E binding peptides tended to have the motif: basic, X, basic, where X is any amino acid. Clearly, these motifs are too broad to give a clear understanding of the molecular basis of the peptide-MHC class II interaction and the X-ray structure of an MHC class II-peptide complex is eagerly awaited. At the present time a major limitation in this area, besides the crystallization and structural analysis, will be the isolation of a pure MHC-peptide complex, free from the presence of co-purifying endogenous peptides.

B. CELL BIOLOGY OF MHC CLASS II-PEPTIDE INTERACTION

Relatively little is known about the cell biology of antigen-MHC class II interaction. The unusual stability of the MHC-peptide complex raises the possibility that peptides remain associated with MHC for the life-time of the MHC molecule and that only newly synthesized MHC class II molecules can bind and present antigen efficiently. On the other hand, Harding and Unanue[125] have recently reported that cycloheximide-treated B cells can present antigen to T cells, suggesting that MHC class II-peptide association can take place in the absence of newly synthesized MHC class II molecules. However, this point remains controversial especially in view of a recent report by Davis and Cresswell[126] demonstrating that human B cells cannot endocytose their MHC class II molecules to any significant extent. The experiments of Davis and Cresswell used protection from neuraminidase as an assay for internalization and were sensitive enough to readily detect a 10% neuraminidase-resistant population. In contrast, in the study of Harding and Unanue,[125] 15 to 35% of the surface MHC class II protein on the murine B lymphoma TA3 internalized within 10 min at 37°C as measured using the resistance of ^{125}I-labeled Fab fragments of anti-MHC class II antibodies to acid stripping from the surface. As discussed above, we and others have recently obtained evidence that peptide does not dissociate from MHC class II proteins at a pH equivalent to that found in endosomes.[97,98] This led us to suggest that if MHC recycling is important in antigen presentation then there may be active cellular processes involved in the regeneration of the peptide binding site.

We have recently done experiments similar to those reported by Harding and Unanue using murine B lymphomas transfected with I-A^k as the antigen presenting cells (Y.S.P. and T.H.W., unpublished). In contrast to the results described by Harding and Unanue, we found that presentation of native lysozyme but not HEL46-61 to an HEL-specific T cell hybrid was inhibited by cycloheximide treatment of B cells. Jensen[127] has shown that cycloheximide blocks the presentation of native antigens by macrophages. As Jensen has pointed out, it is possible to interpret such data in two ways: the cycloheximide may be inhibiting a step in the processing pathway or, antigen processed in the cell associates preferentially with newly synthesized MHC class II.

Recent data reported by Ceppellini et al.[94] show that the kinetics of peptide-MHC class II association are more rapid on live B cells than on fixed cells. This raises the possibility that there are active cellular mechanisms for promoting MHC-peptide association. An alternate interpretation is that if the MHC class II binding site on mature MHC class II molecules is already occupied with peptide, then the slow on-rate could reflect a requirement for peptide exchange in the binding site. On live cells, where a newly synthesized pool of empty MHC class II molecules is available, the on-rate should be much faster.

Cresswell[128] has shown that proteins which enter the cell by the endocytic route ultimately intersect a compartment where they meet with MHC class II proteins which are on their way out to the cell surface via the exocytic route. In an elegant series of experiments, Cresswell prepared conjugates between neuraminidase and Fe-transferrin. The conjugates were incubated with cells that had been pulsed with ^{35}S-methionine to label newly synthesized proteins. MHC class II molecules were immunoprecipitated from the cells and analyzed by 2D electrophoresis. As discussed below, newly synthesized MHC class II proteins are found associated with a 31 kDa invariant polypeptide that remains with the MHC class II protein until it is removed by proteolysis in a late or post-Golgi compartment. The invariant chain is not found associated with cell surface Ia. In Cresswell's experiments, sialic acid loss from the invariant chain could be demonstrated when neuraminidase was added as a conjugate with Fe-transferrin but not when neuraminidase was added alone. These data indicated that proteins which enter the cell via the endocytic route can reach an intracellular compartment where newly synthesized MHC class II molecules are found. The fact that the endocytosed neuraminidase met the MHC class II-invariant chain complexes before the dissociation of the invariant chain suggested that the compartment may be the trans-Golgi or a transport vesicle immediately post-Golgi.

C. THE ROLE OF THE INVARIANT CHAIN IN MHC-PEPTIDE INTERACTION

Intracellular MHC class II molecules associate with a non-polymorphic Type II membrane glycoprotein known as the invariant chain.[18-20] The MHC class II protein-invariant chain association initially takes place in the endoplasmic reticulum[19] and the complex remains associated until it reaches an acidic late or post-Golgi compartment where the invariant chain is removed by proteolytic digestion.[35] The murine invariant chain has two forms—a 31 and a 41 kDa form produced by alternate RNA splicing.[129] The significance of these two forms is not known.[43] In pulse-chase experiments, the invariant chain can be shown to rapidly dissociate from the MHC class II molecule starting at about 2 h after the chase.[130] Recently, Stockinger et al.[43] have shown that MHC class II-expressing L cell transfectants which are low in invariant chain are defective in presentation of protein antigens compared to L cells that have been supertransfected with invariant chain. As recently reviewed by Long,[20] several roles have been proposed for the invariant chain in antigen presentation. Blum and Cresswell[35] have suggested that the invariant chain serves to delay the MHC class II proteins in an acidic compartment to facilitate the interaction with newly processed antigen. The invariant chain could play a targeting role: in the absence of invariant chain, the MHC class II proteins may exit to the cell surface without meeting the endosomes containing processed antigen. It has also been proposed[20] that the binding of invariant chain to the MHC class II protein serves to prevent binding of peptides to MHC II until it reaches the, as yet unidentified, compartment where it meets newly processed endocytosed antigen. Alternatively, the invariant chain might facilitate peptide exchange in the MHC class II binding site.

VI. THE RECOGNITION PROCESS

A. THE MINIMAL SIGNAL FOR ACTIVATION OF T CELL HYBRIDS

For peptide-specific T cell hybridomas, all that is required for activation is a pre-formed complex between processed antigen and purified MHC class II protein in a supported lipid bilayer.[31,84,85] Supported planar membranes (developed in the laboratory of Harden McConnell, Reference 131) are prepared by allowing pre-formed phospholipid vesicles to interact with clean glass surfaces, followed by washing to remove the unbound vesicles. The vesicles appear to fuse on the surface to form a continuous membrane. By incorporating fluorescent lipid molecules into the bilayers and showing that they are uniformly distributed and diffuse over large areas it was established that these are continuous membranes, and not just collections of vesicles stuck to the glass.[132] Membrane proteins can be incorporated into the supported bilayers by first reconstituting them into the phospholipid vesicles. The incorporated proteins can be visualized either by labeling them directly with fluorophores prior to reconstitution or by binding specific antibodies to the bilayers after formation of the planar membranes.

When purified I-A^d was incorporated into the supported membranes it was quite effective in peptide presentation to a peptide-specific hybrid, 3DO-54.8. The same material in the form of small unilamellar liposomes was completely ineffective.[31] The ability of the supported planar membranes to present antigen may be due to the planar surface providing a larger surface area for contact with the T cells. Additionally, small unilamellar vesicles remain in suspension while the T cells rapidly settle to the bottom of the culture dish. Therefore, planar membranes may be more efficient than phospholipid vesicles in antigen presentation because the T cells settle onto the glass support and thereby make better contact with the stimulating surface. Perhaps for this reason, antigen presentation by supported planar membranes appears to be adhesion molecule independent.

Walden et al.[133,134] have successfully used liposome-bound Ia for antigen presentation. In their experiments, intact protein antigens were covalently coupled to lipid molecules in the Ia-containing liposomes. Although the physical details of how this sytem stimulates T cells are not understood, one explanation for the data is that the cross-linking process unfolds the antigen, allowing part of the lipid-linked protein to interact with the MHC molecule and stimulate T cells.

Planar membranes can be prepared on coverglasses[31] or glass beads,[135] for use in T cell activation assays, or in glass flow cells on quartz slides for analysis of fluorescent peptide binding by fluorescence microscopy.[85,86,96] T cell activation by purified MHC class II molecules and peptide antigen has now been demonstrated for a number of different peptide-specific T cell hybrids.[31,91,136] The MHC class II molecules in the lipid bilayer are immobile[31,132] and therefore the distance between them can be controlled by varying the lipid to protein ratio. The T cell response to planar membranes varies with Ia density, and at high densities of Ia, planar membranes can be as effective as APC in the activation of peptide-specific T cell hybrids. By looking at T cell responses to pre-formed complexes of I-A^d and fluorescein-labeled OVA323-339, where the distance between complexes was varied by changing the

MHC density in the lipid bilayer or the degree of saturation of the Ia with peptide, it was found that optimal T cell activation by the planar membranes required that the stimulating MHC-peptide complexes be close-packed.[96] This supports a model of T cell activation in which cross-linking of the T cell receptor by clusters of MHC-peptide complexes leads to signal transduction. Unfortunately, a minimum estimate of the number of MHC peptide complexes required to stimulate a T cell could not be obtained from this data, since the fact that the Ia is immobile in the supported bilayers means that an excess of MHC-peptide complexes is required in order to achieve a critical density for T cell activation. In live cells, where one expects both the T cell receptor molecules and the MHC class II molecules to cluster to the APC-T cell interface[137] one would expect this critical density could be achieved at relatively low antigen density. At the present time, the best estimate of the number of MHC-peptide complexes required for T cell activation comes from the work of Lanzavecchia on the highly efficient presentation of antigen by specific B cells (see Section IV). Lanzavecchia[65] calculates that after 24 h of pumping tetanus toxoid into cells, the TT-specific B cells can accumulate an average of no more than 2000 specific complexes. This suggests that the minimal number of complexes required for a detectable T cell response could be much less than 2000/APC.

B. THE ROLE OF ACCESSORY/ADHESION MOLECULES IN ANTIGEN PRESENTATION

It is now well known that T cell recognition of antigen/MHC is mediated by a specific receptor, an α/β heterodimer whose specificity is generated by gene rearrangement and varies from clone to clone (for review see Reference 138). Through the identification of monoclonal antibodies which can inhibit or activate T cell responses it has become apparent that a number of other proteins are involved in the recognition events that go on between T cells and antigen presenting cells during T cell activation.

The presence of the CD4 molecule on a T cell correlates with the restriction of that T cell to recognize antigen in association with an MHC class II molecule.[139] Functional data has shown that the CD4 molecule can augment MHC-restricted T cell responses by increasing cell to cell adhesion[140] and evidence for direct binding between CD4 and class II MHC proteins has recently been obtained.[141] In addition, there is an increasing body of evidence that CD4 and the T cell receptor co-cluster during the recognition event,[142-147] and that T cell activation leads to the activation of a CD4-associated tyrosine kinase.[148-150] These observations have led to a model of MHC class II-restricted recognition of antigen in which, in the nomenclature of Charles Janeway Jr.,[151] CD4 functions as a co-receptor with the T cell receptor for T cell activation. Janeway proposes that by binding to a constant region of the same MHC class II molecule that the T cell recognizes, the CD4 molecule and the T cell receptor are brought into close proximity, thereby generating critical signals for T cell activation.

Antigen nonspecific adhesion events between complementary pairs of cell surface adhesion proteins are thought to precede the antigen specific interaction between the T cell receptor and the antigen/MHC complex. Two such adhesion systems are the LFA-1-ICAM 1 (or ICAM-2) interaction and the CD2-LFA-3 interaction.[152-154] The CD2 molecule is primarily found on T cells while LFA-3 is broadly distributed and constitutive in its expression. LFA-1 is expressed widely on cells of hematopoietic lineage, while ICAM-1 has a more limited distribution and is inducible by lymphokines (for review see Reference 154). Recently it has been shown that L cells transfected with human MHC class II molecules are defective in antigen presentation unless they are supertransfected with the ICAM cDNA.[155]

Kupfer and collaborators[137] have done extensive work demonstrating receptor redistribution to the contact sites between T cells and antigen presenting cells as a consequence of specific recognition. Using antigen-specific mouse B cell-T cell conjugates, they have shown that CD4, T cell receptor and LFA-1 on the T cell cluster at the T cell-B cell interface during recognition in the presence of specific antigen but not in its absence. The redistribution of receptors to the cell-cell contact site during recognition is thought to be important in cell activation because it provides a mechanism of aggregating receptors which interact with monovalent ligands.

Recently, Dustin and Springer[156] have shown that activation of the T cell (via the T cell receptor/CD3 complex) can modulate the adhesive interaction between LFA-1 on the T cell and ICAM-1 on the APC. Previous work by this group had established that LFA-1-bearing cells could bind to purified ICAM-1 on solid supports in a temperature- and magnesium-dependent manner.[157] By contrast, the LFA-3-CD2 adhesion system operates at 4°C.[152-154] In the recent report,[156] Dustin and Springer showed that activation of the T cell with anti-CD3 or with the phorbol ester, phorbol 12-myristate 13-acetate (PMA), rapidly improved the adhesion of LFA-1 bearing lymphocytes to purified ICAM on solid supports. On the other hand, treatment of ICAM-1-bearing cells with PMA had no effect on binding of ICAM-bearing cells to purified LFA-1. Furthermore, the increased adhesiveness was transient, providing a mechanism for de-adhesion of the lymphocytes after the recognition event. PMA activates protein kinase C and, in combination with calcium ionophores, can mimic signaling through the T cell receptor (for review see Reference 158).

In their recent review,[137] Kupfer and Singer described an experiment in which pre-treatment of mouse T cells with PMA prior to induction of LFA-1 capping by anti-LFA, induced the association of the cytoskeletal protein talin with the LFA-1 clusters, suggesting a role for the cytoskeleton in PMA-induced receptor clustering. Taken together, the above findings suggest that the recognition events that go on between B cells and T cells during antigen presentation involve an initial contact between the cells during which cell surface molecules begin to cluster at the cell-cell interface. Subsequent signals induced via activation through the T cell receptor can then modulate the interactions leading to enhanced cell contact.[137]

C. THE ROLE OF COSTIMULATORY SIGNALS FOR T CELL CLONES

Although the minimum signal of an antigen-MHC complex is sufficient for the activation of a number of peptide-specific T cell hybrids, T cell clones appear to require additional signals (for review see Reference 159). It has recently been reported that T cell clones cannot respond to antigen presented by aldehyde or carbodiimide-fixed APC[160,161] or to antigen plus purified MHC class II molecules in planar membranes.[162] Furthermore, after exposure to purified antigen/MHC or to antigen/MHC on fixed cells, these clones become refractory to subsequent stimulation by antigen and live APC. The induction of this nonresponsive state is abrogated if the culture also contains live allogeneic accessory cells, which presumably act to provide a necessary costimulatory molecule which prevents the induction of the nonresponsive state.[163] In the studies of Schwartz and collaborators,[159] using T cell clones specific for cytochrome *c* and I-E^k, this costimulatory signal could be provided by low density allogeneic splenic accessory cells, to a lesser extent by resting B cells and by some (activated) T cells. Fibroblast lines (CHO cells or L cells) transfected with MHC class II molecules also stimulate T cell clones and therefore presumably can provide this accessory signal.[59] These recent studies have generated a revived interest in a two-signal model for T cell activation originally put forward by Lafferty and collaborators.[164] The current model suggested by the work of Schwartz and collaborators[159] is that exposure of T cell clones to antigen/MHC plus a costimulatory ligand leads to activation, while exposure to antigen and MHC in the absence of the costimulator leads to functional inactivation. Further studies by this group have led them to suggest that second messengers produced by ligation of the T cell receptor, induce the nonresponsive state unless additional signals provided by the costimulatory signal are present at the same time.[165] Mueller et al.[159] have tested IL-1,

IL-2, IL-3, IL-4, IL-6, IFN-γ, GM-CSF, TGF-β and TNF-α, none of which can function as a costimulatory ligand for activation of their cytochrome *c*-specific clones.

In other systems, IL-1 has been shown to act as a costimulatory molecule for T cell activation (for review see References 166 and 167). IL-1-like mediators are produced in soluble form by a wide-variety of cells including macrophages, dendritic cells, B cells, keratinocytes, and fibroblasts.[166] In addition, IL-1 has been shown to exist in a membrane-associated form, in both macrophages and B cells.[168,169] In macrophages, IL-1 can be induced by a variety of stimuli, including microorganisms and T cell contact.[166] As reviewed by Mizel,[167] IL-1 appears to act in concert with stimulation via the T cell receptor, to augment both IL-2 and IL-2 receptor gene transcription in T cells. Recent molecular evidence[170] has shown that in the murine IL-1-responsive line LBRM, IL-1 treatment leads to enhanced expression of c-*jun* mRNA while signaling through the T cell receptor enhances expression of the c-*fos* message. These two proteins then combine to form AP-1, a factor which binds to an IL-1-responsive regulatory element in the IL-2 promoter to promote IL-2 expression.

As reviewed elsewhere in this volume,[13] CD 4^+ T cells have recently been divided into two subsets, T_h1 and T_h2 on the basis of differential lymphokine secretion.[171] A major difference between these subsets are that activated T_h1 cells produce IL-2 and IFN-γ, while T_h2 cells produce IL-4 but not IL-2 or IFN-γ. In several recent studies it has been shown that IL-1 can provide the necessary costimulatory activity together with antigen/MHC for the proliferative response of T_h2 but not T_h1 cells.[172-175] In a recent study, Weaver et al.[174] compared macrophages and B cells for their ability to deliver accessory signals to T cell clones representing the two subsets. Macrophages which had been induced to express membrane IL-1, could activate T_h2 cells before or after fixation, while purified resting B cells, which appeared to express lower levels of membrane IL-1, could activate the T_h2 clones when unfixed but not after fixation with 1% paraformaldehyde. In contrast, the T_h1 clones could not use IL-1 as a costimulator but could respond to B cells which had been pre-treated with anti-immunoglobulin and IFN-γ and then fixed with 1% paraformaldehyde.[174,176] Based on the above discussion, it appears that both lineages of APC (macrophages and B cells) can provide costimulatory ligands for both T cell subsets (T_h1 and T_h2).

D. THE ROLE OF SIGNALING THROUGH THE MHC CLASS II MOLECULE IN ANTIGEN PRESENTATION BY B CELLS

In addition to their well-documented role as peptide binding proteins, it has been suggested that MHC class II molecules can also act as signal transducing molecules, by delivering intracellular signals in response to T cell contact. Evidence in support of this hypothesis has come from studies on the effects of antigen-specific, MHC-restricted T cells and anti-class II antibodies on the activation of B cells.[177-183] As discussed elsewhere in this volume[11] these studies have focused primarily on the idea that an MHC class II-associated signaling pathway is involved in the regulation of B cell proliferation and differentiation into antibody-secreting cells. Indeed, recent evidence indicates that signaling through the MHC class II molecule does lead to proliferation of appropriately primed B cells.[184,185] As will be discussed below, signal generation through the MHC class II molecule appears to play a role in antigen presentation as well.

Direct evidence that cross-linking of MHC-class II molecules leads to the generation of a second messenger in B cells was reported by Chen et al.[186] who showed that anti-MHC-induced cross-linking of surface MHC class II molecules on freshly isolated murine splenic B cells induced a rapid and transient increase of intracellular cAMP levels followed by the translocation of protein kinase C (PKC) from the cytoplasm to the nucleus. Since cAMP analogs were found to induce PKC translocation,[187] it was proposed that, during antigen presentation, ligation of MHC class II molecules by helper T cells leads to increased cAMP production which in turn causes PKC translocation to the nucleus. The MHC class II-mediated signal transduction pathway appears to be distinct from that used by the surface immunoglobulin (sIg) receptor, in that no modulation of intracellular free calcium or of inositol lipid hydrolysis has been demonstrated following antibody-mediated MHC class II signaling in either human or murine B cells.[188-190] The consequences of PKC translocation to the nucleus as a result of MHC class II cross-linking are as yet unknown. There are a variety of subtypes of PKC that possess different biological roles depending on the tissue distribution and PKC isotype involved (for review, see Reference 191). Although it has been shown that both the sIg and the MHC class II signaling pathways compete for the same pool of intracellular PKC,[186] the fact that both pathways show a transient and rapid translocation of PKC suggests that it is possible for both pathways to act sequentially during B cell activation.

Both the transmembrane and cytoplasmic domains of the MHC class II molecule are remarkably conserved among MHC class II alleles and species,[192] consistent with the idea that these domains play a role in transmembrane signaling. Recently, two groups have constructed B cell transfectants expressing genetically engineered MHC class II molecules with truncated cytoplasmic domains in order to study the potential role of the cytoplasmic domains in transmembrane signaling and antigen-presentation.[193-195] It was found that anti-MHC-induced translocation of PKC from the cytoplasm to the nucleus was severely impaired in B cells expressing truncated MHC class II molecules as compared to B cells expressing wild-type I-A^k.[194-195] A dominant role for the β chain was suggested by Wade et al.,[195] since cells expressing truncated Aβ in the presence of wild-type Aα showed a greater signaling defect than cells expressing truncated Aα coupled to wild-type Aβ.

In most cases, the truncated I-A^k-expressing transfectants were fully functional in inducing IL-2 production by peptide-specific I-A^k-restricted T cell hybridomas.[193-195] However, truncation of the MHC class II domains completely abrogated antigen presentation to several autoreactive T cell hybrids.[194] The possibility that these cells failed to present self-antigen derived from the cytoplasmic domains of the MHC class II proteins was ruled out since the presence of the same sequence in other MHC α or β chains expressed in the same transfectants did not restore the capacity of these cells to activate autoreactive T hybrids. Moreover, some peptide-specific T cell hybrids also showed a reduction in activation by the truncated-MHC class II expressing B cells.

In collaboration with Glimcher's laboratory,[196] we found that the autoreactive T cell hybrids which failed to respond to truncated-MHC class II molecules on live B cells also failed to respond to purified wild-type Ia in planar membranes or to glutaraldehyde-fixed B cells expressing wild-type or truncated Ia. This led us to speculate that the antigen presenting defect in the truncated-MHC class II expressing B cells was directly related to the inability of these cells to generate signals through the MHC class II molecule. To test this hypothesis, B cells expressing truncated MHC class II molecules were treated with a cAMP analog (dibutyryl-cAMP, db-cAMP) and then tested for their ability to stimulate IL-2 production by a panel of T cell hybrids. We found that treatment of db-cAMP or the adenylate cyclase activator, forskolin, restored the ability of the truncated Ia transfectants to stimulate IL-2 release by otherwise nonresponsive T hybrids.[196] Activation was not due to a nonspecific mitogenic effect of the cAMP since anti-T cell receptor, or anti-MHC still blocked the T cell response to cAMP-treated cells. These results provided functional evidence that signaling via the cytoplasmic domains of MHC class II molecules leads to the generation of accessory signals in the B cell that are important in the activation of some T cells. Further characterization of the db-cAMP-induced signal showed that it is not transferable between cells and is destroyed by glutaraldehyde fixation but not by γ-irradiation (3000 rads) of the db-cAMP-treated cells.[196] The generation of the cAMP-induced signal requires protein synthesis. Dibutyryl-cyclic-AMP-treatment of the truncated-MHC-expressing B cells in the presence of cycloheximide prevented activation of the truncation sensitive hybrids but not of truncation insensitive hybrids (Y.S.P. and T.H.W., unpublished).

For macrophage IL-1 expression, it has been suggested that in addition to lymphokine or microbial-induced IL-1 induction, binding of the T cell to the APC can lead to generation of IL-1, perhaps by signaling through the MHC class II molecule.[169,197,198] However, a number of observations suggest that IL-1 is unlikely to be the missing costimulator induced by cAMP-treatment of the truncated-MHC expressing B cells. First, at least in macrophages, cAMP analogs have been shown to block IL-1 induction.[199] Second, treatment of allogeneic B lymphomas with db-cAMP followed by co-culture with the truncated MHC-expressing B cell did not restore function, while IL-1 should be transferable between cells. Finally, preliminary results show that addition of recombinant IL-1 fails to mimic the delivery of the db-cAMP-induced signal in the truncated MHC-expressing transfectants (Naveen Bangia, Y.S.P., and T.H.W., unpublished observation).

At least two distinct classes of models can be proposed to explain the augmentation of antigen-presenting ability induced via the MHC class II signaling pathway. First, since protein synthesis inhibitors prevent the delivery of the second signal, it is possible that a new cell surface protein or a locally acting lymphokine-like molecule interacts directly with the T cells to deliver the costimulatory signal for autoreactive T cell activation. Candidates for such a protein include molecules whose expression on the B cell surface is induced via MHC-restricted T cell-B cell contact. Further investigation of the role of MHC class II-specific signaling in the induction of B cell surface proteins and their potential role in antigen presentation is needed to address this possibility.

An alternate model is that cAMP induces intracellular mediators involved in the redistribution of MHC class II molecules and/or adhesion molecules to the T cell-APC interface during T cell activation. As described above, the redistribution of interacting ligands to the T cell-APC interface is an integral part of antigen presentation[200] and recent evidence suggests that such clustering might be augmented by T cell activation.[137,156] It is possible that in B cells, signaling through the MHC class II molecule leads to the expression of proteins which stabilize or promote the clustering of the MHC molecules to the T cell-APC interface, thereby increasing the avidity of the interaction. T cells which are of low avidity would require extensive aggregation of the target MHC molecules in order to gain the requisite number of complexes for activation, while higher avidity hybrids would require less extensive clustering and would respond to antigen-MHC in the absence of signaling through the MHC class II molecule. In support of such a model, Kammer et al.[201] have demonstrated a role for cAMP in receptor clustering on T cells. Clearly, further work on the relationship between the cell surface reorganization of MHC class II molecules and the cAMP pathway are needed to determine the role of MHC class II clustering in antigen presentation by B cells.

Although the nature of the accessory signal delivered as a consequence of aggregation of the MHC class II molecules during antigen presentation is as yet unknown, the results described above suggest that during antigen presentation by B cells, binding of the T cell receptor and/or the CD4 molecule to the antigen-MHC complex leads to the activation of the adenylate cyclase/cAMP pathway which in turn leads to secondary events among which is a signal important for the activation of autoreactive T cells. Our interpretation of this data is described in Figure 2. Specific T cell-B cell interaction leads to the generation of initial signals in both cells. In the B cell the production of cAMP leads to secondary events, among which is a signal important for the activation of some T cell hybrids. The combination of this signal and the signal induced via MHC/antigen binding to the T cell receptor/CD4 co-receptor complex leads to the induction of interleukin 2 production by these hybrids.

Intracellular regulators such as cAMP and inositol phosphates have been found to have multiple effects depending on the cell type and differentiation state.[202,203] It remains to be seen whether similar signaling events are mediated through MHC class II molecules in other antigen-presenting cells and whether such signals lead to distinct accessory signals. At the present time, the data suggest that the cAMP-mediated signal important for autoreactive T

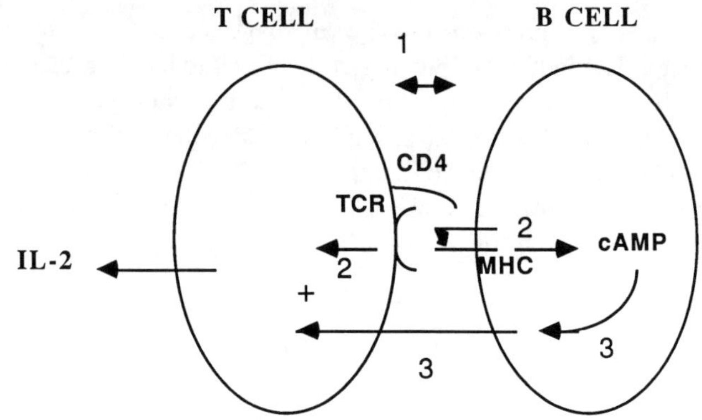

FIGURE 2. The role of signal transduction through the MHC class II molecule in antigen presentation by B cells. 1. T cell receptors and/or CD4 molecules interact with antigen/MHC complexes on the B cell surface. 2. Cross-linking of cell surface molecules leads to signal transduction in both cells. 3. cAMP produced in the B cell leads to secondary events among which is a signal important for T cell activation.

cell hybrids is distinct from the costimulator required for T_h1 clones, since the latter signal can be provided by a variety of allogeneic accessory cells, while the cAMP-induced signal must be on the same cell as the restricting element. Taken together, these data suggest that there are at least three types of costimulatory signal that have obligatory or enhancing effects during antigen recognition by T cell clones and hybrids (IL-1, the T_h1 costimulatory ligand, and the cAMP-induced signal described above). All of these signals can apparently be provided by B cells. However, much work needs to be done to clarify the nature of the costimulators, the regulation of their expression in resting and activated APC and their role in antigen presentation to naive and secondary T lymphocytes.

VII. THE IMPORTANCE OF B CELL ACTIVATION FOR ANTIGEN PRESENTATION *IN VITRO*

Chesnut et al.[8] first reported that resting B cells are incapable of presenting antigen, but acquire the capacity to activate T cells following stimulation by LPS. As recently reviewed by Pierce and collaborators,[53] it is now clear that small resting B cells can present antigen,[1,10,204,205] but that B cell antigen presenting ability can be significantly improved by treatment with bacterial LPS or IL-4.[9,206-208] Since resting B cells express only 5 to 10% the Ia of activated B cells and since IL-4 induces increased Ia expression in mouse B cells, the simplest interpretation of these data is that the improved antigen presenting ability of B cells upon activation is due to increased Ia expression.[209-210]

Kreiger et al.[211] have addressed the possibility that resting B cells are less efficient as APC than activated B cells due to changes in the MHC molecule themselves. Earlier studies had shown that there are glycosylation differences between the Ia molecules of macrophages and resting B cells[206] and that neuraminidase treatment can augment the antigen presenting capacity of normal B cells.[207] In the recent study, purified MHC class II molecules isolated from resting B cells, from B cells stimulated with LPS and dextran sulfate, from IL-4-stimulated B cells, or from B lymphomas, were equally capable of stimulating T cells when incorporated into planar membranes.[211] This rules out the possibility that changes in the MHC molecules themselves are responsible for the improvement of antigen presentation by activation of B cells. Furthermore, in the study of Kreiger et al.,[211] neuraminidase treatment of the purified Ia had no effect on its antigen presenting ability, although removal of sialic acid from either the B cells or the T cells did improve antigen presenting ability, presumably by allowing improved cell-cell adhesion.

Ashwell et al.[1] have suggested that the reason small resting B cells had initially appeared to be less capable of presenting antigen is that they were sensitive to the γ-irradiation treatments used in some studies. They showed that lightly irradiated (1000 rads), purified small resting B cells can in most cases, present antigen to T cell lines and hybridomas. However, irradiation at 3300 rads led to a greatly diminished antigen presentation ability. This radiation-sensitivity distinguishes B cells from macrophages and dendritic cells which are quite radioresistant.[212] These observations were extended by Jelachich et al.[10] Jelachich et al. separated unstimulated splenic B cells into groups on the basis of size (small B cells are thought to represent resting mature B cells, while the larger cell populations are thought to contain activated or developing cells). They found that all size groups were roughly equal in their ability to induce IL-2 production by a pigeon cytochrome *c*-specific T hybridoma. Furthermore, the antigen presentation ability in each B cell fraction could be improved by activation of the B cells with anti-immunoglobulin $(Fab')_2$s. In agreement with the studies of Ashwell et al.,[1] Jelachich et al.[10] found that all the B cell fractions were markedly radiation sensitive.

Recently, Ashwell et al.[213] have obtained functional evidence as to the nature of the radiation sensitivity of resting B cells. Using a T cell clone specific for pigeon cytochrome *c* 81-104 and I-E^k, they found that the antigen presenting defect in irradiated resting B cells could not be overcome by addition of peptide or by pre-pulsing with antigen prior to irradiation. When they used phosphatidylinositol (PI) hydrolysis in the T cell as a measure of T cell receptor occupancy with antigen/MHC they found that irradiation had a modest effect on PI turnover, suggesting some reduction in signaling through the T cell receptor. More striking was the observation that heavy irradiation of the stimulating B cells caused them to induce a hyporesponsive state in the responding T cells, similar to the effects described above for fixed APC. Since the presence of unirradiated allogeneic B cells prevented the induction of the hyporesponsive state by irradiated B cells, Ashwell and collaborators concluded[213] that a major effect of irradiation on B cells is the loss of costimulatory signals.

Taken together, these findings show that *in vitro*, resting B cells can present antigen to a variety of antigen primed T cell lines and hybridomas but that their antigen presenting ability is significantly improved upon activation.

VIII. HOW IMPORTANT ARE B CELLS AS APC *IN VIVO*?

Given the highly efficient mechanism that B cells possess for delivering antigen to the processing compartment (Section III) it seems likely that, at least in the secondary response to antigen when there are a significant number of specific B cells available, that B cells should be important as antigen presenting cells *in vivo*. However, the relative importance of B cells vs. macrophages or dendritic cells in antigen presentation in primary and secondary T cell responses *in vivo,* as well as the role of secreted immunoglobulin in this process remains a subject of much discussion.

Tew and collaborators[214,215] have described a role for follicular dendritic cells in trapping immune complexes produced during the secondary response to antigen. They have described beaded structures on the dendrites of follicular dendritic cells known as "iccosomes" for "immune complex coated bodies." Using the antigen horseradish peroxidase to facilitate visualization of this process, ultrastructural evidence was obtained that suggested that the iccosomes are released by the follicular dendritic cells and taken up by germinal center (GC) B cells.[216] Further work by this group[217] showed that GC B cells, isolated from antigen-primed animals one week after secondary immunization with ovalbumin, could activate an ovalbumin-specific T cell hybridoma to produce interleukin 2 in the absence of exogenous ovalbumin. These results strongly suggest that B cells can present antigen during the secondary response *in vivo*.

One of the more frequently used approaches to evaluate the role of B cells in antigen presentation *in vivo* has been the analysis of T cell responses in B cell deficient, anti-μ-treated mice.[218] These mice are generated by intraperitoneal injection of between 50 and 100 μg of anti-IgM antibody beginning at birth and continuing on alternate days throughout the life of the experimental animal. This regimen reduces IgM^+ cells in the spleen to 2% of total cells (from 50% in normal mice), and there is a corresponding disappearance of IgM in the sera.[219,220] Anti-μ-treated mice have a near normal number of T cells and retain the T cell functions of tumor allograft rejection and contact hypersensitivity.[219] Macrophage function, as assayed by the ability of spleen or peritoneal cells to stimulate primed T cells, likewise appears normal.[221] On the other hand, anti-μ-treated mice are deficient in certain idiotype-specific T helper cell populations.[222,223] Furthermore, Kim et al.[220] reported, in contradiction to other authors, that concanavalin A-induced proliferation of T cells in anti-μ-treated mice is 50% of normal, suggesting some perturbation of the T cell compartment.

As recently reviewed by Ashwell[224] experiments using the anti-μ-treated mouse model have supported an important role for B cells in the priming of lymph node T cells *in vivo*. Anti-μ-treated mice were found to be generally deficient in the priming of lymph node T cells to protein antigens and the priming defect could be compensated for by reinjection of B cells into the μ-depleted animals.[225,226] Ron and Sprent[225] showed that prior irradiation of the B cells used in the reconstitution prevented the restoration of T cell priming and that B-depleted lymph node cells also failed to restore priming, suggesting that the reconstitution of T cell priming was due to B cells and not contaminating macrophages or dendritic cells. Further evidence that B cells were the important cells in the reconstitution of priming in the anti-μ-treated mouse was provided by Kurt-Jones et al.[226] using hapten-specific B lymphocytes. Prior reconstitution of anti-μ-treated mice with hapten-specific B lymphocytes followed by challenge with haptenated-ovalbumin, restored priming of T cells to ovalbumin to a much greater extent when the B cells were specific for the same hapten as that used on the immunogen.

The possibility that the restoration of APC function was due to the binding of secreted immunoglobulin to Fc receptors on other APC, was tested by Kurt-Jones et al. using F1 anti-μ-treated mice.[226] They found that when donor B cells of one parental haplotype (A) were used to reconstitute F1 (A × B) anti-μ-treated mice, primed T cells isolated 2 weeks after antigenic challenge were responsive to MHC of either parental haplotype, suggesting that in this case circulating antibody might facilitate initial priming events. However, T cells isolated 4 months after immunization were restricted to the MHC of the donor haplotype, indicating an important role for B cells in antigen presentation in the later stages of the primary response. Furthermore, Ron et al.[227] showed that circulating Ig is not sufficient for restoration of *in vivo* priming, since injection of anti-KLH antibodies into anti-μ-treated mice challenged with KLH failed to elicit significant T cell priming to KLH.

Although the above data, suggest an important role for B cells in some aspect of priming of lymph node T cells *in vivo,* several studies have shown that splenic T helper cell functions of anti-μ-treated mice remain normal[222,223] or are not as profoundly disrupted as the priming of lymph node T cells.[228] This suggests that in the spleen other APC populations are sufficient for T cell priming. It should also be remembered that interpretations of anti-μ-treated mouse data are contingent on the assumption that the prolonged anti-IgM treatment adversely affects B cells, but not other antigen presenting cells in the lymph node. As pointed out by Ashwell in his recent review,[224] the reconstitution of T cell priming in the anti-μ-treated mouse has not been tested using purified non-B cell APC.

B cell antigen presentation *in vivo* has also been studied in other murine systems, though not as extensively. Since B cells but not dendritic cells or macrophages are radiation sensitive,[212] the inability of lymph node cells from irradiated animals to prime T cells is thought to reflect the importance of B cells in T cell priming. Ron and Sprent[225] reconstituted irradiated mice with purified normal T cells and showed that these animals were unable to respond to subsequent antigenic challenge with KLH unless purified B cells were also injected before the antigenic challenge. Janeway et al.[228] also reported that lymph node cells that have been depleted of Ig^+ cells are incapable of inducing primary T cell proliferation *in vitro,* again suggestive of a prominent role for B cells in T cell priming. These results are consistent with the anti-μ-treated mouse experiments, indicating an important role for B cells in some aspect of lymph node T cell priming. Again, the interpretation of these results are contingent on the assumption that irradiation only effects the B cells and not other APC and that the B cells used in the reconstitution are not contaminated by other APC.

A recent paper by Lasilla et al.[229] suggests that B cells are inactive in the priming of naive T cells *in vivo.* In chicken, B cell development can be prevented by the neonatal injection of cyclophosphamide and a mature B cell compartment can be subsequently restored by the injection of B progenitors derived from normal chickens. It is additionally possible to make a chimeric chicken where the donor B cell population and the host are of differing MHC types. Lassilia et al.[229] showed that such a chimeric chicken was capable of producing high titers of antibody against the T-independent antigen *Brucella abortus,* but was incapable of generating antibody against a T-dependent antigen, sheep red blood cells (SRBC). The lack of anti-SRBC antibody was interpreted to mean that T-B cooperation did not occur in the chimeras, ostensibly because the donor B cells were incapable of priming host T cells. Since the B and T cells were of disparate MHC types, it might be argued that T-B cooperation did not occur because there were no T cells specific for the donor B cell MHC. However, the addition of B cells and macrophages which were of one MHC type to a B cell-deficient chicken of another MHC type resulted in restoration of the anti-SRBC response. The absolute dependence of T-B cooperation on the macrophage population led Lassilla et al. to conclude that B cells alone are incapable of priming naive T cells in chicken. Whether this reflects an inability of the B cells to provide an essential costimulatory signal to the naive T cells or a qualitative or quantitative difference in SRBC processing and subsequent presentation by B cells vs. macrophages was not clear. Unfortunately, Lassila et al. did not look at other T-dependent antigens, so it is unclear whether this effect is peculiar to sheep red blood cells.

As pointed out by Lassilla et al.,[229] one would not expect B cells to play an important role in the priming of T cells. B cells are capable of processing and presenting their own surface immunoglobulin,[230] and if B cells could prime virgin T cells, then a vast repertoire of T cells specific for B cell idiotypic determinants would emerge. The simplest way to avoid generating B cell-specific T cells is to prevent B cells from priming T cells.

It is possible that the two experimental systems described above may reflect differences between the murine and avian immune systems or differences in the nature of the immunizing antigen. The anti-μ-treated mouse data do suggest that B cells are important at some stage during antigen presentation in the primary immune response in lymph nodes. However, at least in the case of sheep red blood cell-specific responses, B cells appear incapable of priming naive T cells in the absence of other APC. As suggested by Ron and Sprent,[225] a plausible scenario is that antigen presentation by macrophages or dendritic cells initiates the primary T cell response, and that B cells are involved in later stages of the response, such as clonal expansion. During the secondary response one expects B cells to play a greater role in antigen presentation, because of the larger pool of specific B cells due to clonal expansion as well as the higher affinity of the sIg receptors for antigen due to affinity maturation.

IX. SUMMARY

Antigen presentation by specific B cells begins with the efficient internalization of foreign antigen via surface immunoglobulin, followed by proteolysis of the antigen in an acidic, probably endosomal, compartment to generate peptide fragments. The peptide fragments ultimately reach a compartment where a subfraction of them bind to MHC class II molecules. The intracellular routes followed by antigen, the cellular location of the MHC class II-peptide association or the relative importance of newly synthesized vs. recycling MHC class II in this process are still not well defined. Newly synthesized MHC class II proteins are thought to encounter endocytosed antigen in the trans-Golgi or a post-Golgi vesicle. Subsequently, the MHC class II-peptide complex reaches the cell surface where it can be recognized by T cells.

The nature of the interaction between MHC and peptide is still poorly understood at the molecular level. The complex is unusually stable in a cell free system and the molecular basis for this stability is still unclear. Although MHC class II and peptide can associate in a cell free system and the resulting complexes, when reconstituted into supported lipid bilayers, can stimulate T cells, there remains much speculation as to the possibility that active cellular processes are involved in the intracellular transport of peptides and their association with or dissociation from MHC class II proteins. It is also becoming increasingly evident that the MHC class II-associated invariant chain plays a role in some aspect of this process.

T cell recognition of antigen/MHC on the surface of B cells is a complex, coordinated event, involving both the antigen-specific T cell receptor as well as a number of nonpolymorphic accessory molecules which are thought to redistribute to the T cell-B cell interface during recognition. In addition, poorly defined costimulatory signals are also required for effective antigen presentation to some T cells. T cell-B cell contact during antigen recognition appears to be modulated by signals in both cells leading to enhanced adhesion or presentation. While B cells are clearly competent antigen presenting cells *in vitro,* particularly after activation, their importance relative to macrophages and dendritic cells in stimulatory primary and secondary T cell responses *in vivo* is still controversial.

ACKNOWLEDGMENT

We thank Dr. Dana Skea for critical reading of the manuscript and Dr. Gillian Wu for helpful discussions. T.H.W. is indebted to Professor Harden McConnell, in whose laboratory my interest in antigen presentation was initiated. T.H.W. is a scholar of the Medical Research Council (M.R.C.) of Canada and Y.S.P. is funded by an M.R.C. studentship. Our research in this area is funded by the National Cancer Institute of Canada and the M.R.C. (Canada).

REFERENCES*

1. **Ashwell, J. D., Defranco, A. L., Paul, W. E., and Schwartz, R. H.,** Antigen presentation by resting B cells, *J. Exp. Med.,* 159, 881, 1984.

* The literature search for this chapter was completed November 1989.

2. **McKean, D. J., Infante, A. J., Nilson, A., Kimoto, M., Fathman, G. C., Walker, E., and Warner, N.,** Major histocompatibility-restricted antigen presentation to antigen-reactive T cells by B lymphocyte tumour cell lines, *J. Exp. Med.*, 154, 1419, 1981.
3. **Issekutz, T., Chu, E., and Geha, R. S.,** Antigen presentation by human B cells. T cell proliferation induced by Epstein-Barr virus B lymphoblastoid cells, *J. Immunol.*, 129, 1446, 1981.
4. **Glimcher, L. H., Kim, K.-J., Green, I., and Paul, W. E.,** Ia-antigen bearing B cell tumour lines can present protein antigen and alloantigen in a major histocompatibility-restricted fashion to antigen-reactive T cells, *J. Exp. Med.*, 155, 455, 1982.
5. **Walker, E., Warner, N. L., Chesnut, R., Kappler, J., and Marrack, P.,** Antigen-specific I region-restricted interactions *in vitro* between tumour cell lines and T cell hybridomas, *J. Immunol.*, 128, 2164, 1982.
6. **Kappler, J., White, J., Wegmann, D., Mustain, E., and Marrack, P.,** Antigen presentation by Ia^+ B cell hybridomas to H-2-restricted T cell hybridomas, *Proc. Natl. Acad. Sci. U.S.A.*, 79, 3604, 1982.
7. **Chesnut, R. W. and Grey, H. M.,** Studies on the capacity of B cells to serve as antigen-presenting cells, *J. Immunol.*, 126, 1075, 1981.
8. **Chesnut, R. W., Colon, S. M., and Grey, H. M.,** Antigen presentation by normal B cells, B cell tumours, and macrophages: functional and biochemical comparison, *J. Immunol.*, 128, 1764, 1982.
9. **Kakiuchi, T., Chesnut, R. W., and Grey, H. M.,** B cells as antigen-presenting cells: the requirement for B cell activation, *J. Immunol.*, 128, 2164, 1982.
10. **Jelachich, M. L., Lakey, E. K., Casten, L., and Pierce, S. K.,** Antigen presentation is a function of all B cell subpopulations separated on the basis of size, *Eur. J. Immunol.*, 16, 411, 1986.
11. **Bartlett, W. C. and Noelle, R.,** T cell-dependent B-cell activation, in *T-Cell Dependent and Independent B-Cell Activation,* Snow, E. C., Ed., CRC Press, Boca Raton, FL, 1990.
12. **Anderson, J.,** Thymus-dependent B cell responses, in *T-Cell Dependent and Independent B-Cell Activation,* Snow, E. C., Ed., CRC Press, Boca Raton, FL, 1990.
13. **Bottomly, K.,** T cell subsets and B cell activation, in *T-Cell Dependent and Independent B-Cell Activation,* Snow, E. C., Ed., CRC Press, Boca Raton, FL, 1990.
14. **Callard, R.,** The role of T cells in activation of human B cells, in *T-Cell Dependent and Independent B-Cell Activation,* Snow, E. C., Ed., CRC Press, Boca Raton, FL, 1990.
15. **Bjorkman, P. J., Saper, M. A., Samraoui, B., Bennet, W. S., Strominger, J. L., and Wiley, D. C.,** Structure of the human class I histocompatibility antigen HLA-A2, *Nature (London),* 329, 506, 1987.
16. **Bjorkman, P. J., Saper, M. A., Samraoui, B., Bennett, W. S., Strominger, J. L., and Wiley, D. C.,** The foreign antigen binding site and T cell recognition regions of class I histocompatibility antigens, *Nature (London),* 329, 512, 1987.
17. **Brown, J. H., Jardetsky, T., Saper, M. A., Samraoui, B., Bjorkman, P. J., and Wiley, D. C.,** A hypothetical model of the antigen binding site of Class II histocompatible molecules, *Nature (London),* 332, 845, 1988.
18. **Sung, E. and Jones, P. P.,** The invariant chain of murine Ia antigens: its glycosylation, abundance, and subcellular localization, *Mol. Immunol.*, 18, 899, 1981.
19. **Kvist, S., Wiman, K., Claesson, L., Peterson, P., and Dobberstein, B.,** Membrane insertion and oligometric assembly of HLA-DR histocompatibility antigens, *Cell,* 29, 61, 1982.
20. **Long, E. O.,** Intracellular traffic and antigen processing, *Immunol. Today,* 10, 232, 1989.
21. **Swain, S.,** T cell subsets and the recognition of MHC class, *Immunol. Rev.*, 74, 129, 1983.
22. **Rosenthal, A. S. and Shevach, E. M.,** Function of macrophages in antigen recognition by guinea pig T lymphocytes. I. Requirement for histocompatible macrophages and lymphocytes, *J. Exp. Med.*, 138, 1194, 1973.
23. **Benacerraf, B.,** A hypothesis to relate the specificity of T lymphocytes and the activity of I region-specific Ir genes in macrophages and B lymphocytes, *J. Immunol.*, 120, 1809, 1978.
24. **Unanue, E.,** Antigen presenting function of the macrophage, *Annu. Rev. Immunol.*, 2, 395, 1984.
25. **Gell, P. G. H. and Benacerraf, B.,** Studies on hypersensitivity. II. Delayed hypersensitivity to denatured proteins in guinea pigs, *Immunology,* 2, 64, 1959.
26. **Livingstone, A. M. and Fathman, C. G.,** The structure of T cell epitopes, *Annu. Rev. Immunol.*, 5, 477, 1987.
27. **Ziegler, K. and Unanue, E.,** Identification of a macrophage antigen-processing event for I-region-restricted antigen presentation to T lymphocytes, *J. Immunol.*, 127, 1869, 1981.
28. **Ziegler, K. and Unanue, E.,** Decrease in macrophage antigen catabolism caused by ammonia and chloroquine is associated with inhibition of antigen presentation to T cells, *Proc. Natl. Acad. Sci. U.S.A.*, 79, 175, 1982.
29. **Mellman, I., Fuchs, R., and Helenius, A.,** Acidification of the endocytic and exocytic pathways, *Annu. Rev. Biochem.*, 55, 663, 1986.
30. **Shimonkevitz, R., Kappler, J., Marrack, P., and Grey, H.,** Antigen recognition by H-2-restricted T cells. I. Cell free antigen processing, *J. Exp. Med.*, 158, 303, 1983.

31. **Watts, T. H., Brian, A. A., Kappler, J., Marrack, P., and McConnell, H. M.,** Antigen presentation by supported planar membranes containing affinity purified I-A^d, *Proc. Natl. Acad. Sci. U.S.A.*, 81, 7564, 1984.
32. **Streicher, H. Z., Berkower, I. J., Busch, M., Gurd, F. R. N., and Bersofsky, J. A.,** Antigen conformation determines processing requirements for T cell activation, *Proc. Natl. Acad. Sci. U.S.A.*, 81, 6831, 1984.
33. **Puri, J., Lonai, P., and Friedman, V.,** Antigen-Ia interaction and the proteolytic processing of antigen: the structure of the antigen determines its restriction to the A or E molecule of the major histocompatibility complex, *Eur. J. Immunol.*, 16, 1093, 1986.
34. **Buus, S. and Werdelin, O.,** A group specific inhibitor of both proteolytic degradation and presentation of dinitrophenyl-poly-L lysine by guinea pig accessory cells to T cells, *J. Immunol.*, 136, 2382, 1986.
35. **Blum, J. S. and Cresswell, P.,** Role for intracellular proteases in the processing and transport of class II HLA antigens, *Proc. Natl. Acad. Sci. U.S.A.*, 85, 3975, 1988.
36. **Allen, P. M. and Unanue, E. R.,** Differential requirements for antigen presentation by macrophages for lysozyme-specific T cell hybridomas, *J. Immunol.*, 132, 1077, 1984.
37. **Allen, P. M., Strydom, D. J., and Unanue, E. R.,** Processing of lysozyme by macrophages: identification of the determinant recognized by two T cell hybridomas, *Proc. Natl. Acad. Sci. U.S.A.*, 81, 2489, 1984.
38. **Allen, P. M., Matsueda, G. R., Haber, E., and Unanue, E. R.,** Specificity of the T cell receptor: two different determinants are generated by the same peptide and the I-A^k molecule, *J. Immunol.*, 135, 368, 1985.
39. **Sette, A., Adorini, L., Colon, S. M., Buus, S., and Grey, H. M.,** Capacity of intact proteins to bind to MHC class II molecules, *J. Immunol.*, 143, 1265, 1989.
40. **Donermeyer, D. L. and Allen, P. M.,** Binding to Ia protects an immunogenic peptide from proteolytic degradation, *J. Immunol.*, 142, 1063, 1989.
41. **Gammon, G., Shastri, N., Cogswell, J., Wilbur, S., Sadegh-Nasseri, S., Kryzych, U., Miller, A., and Sercarz, E.,** The choice of T-cell epitopes utilized on a protein antigen depends on multiple factors distant from, as well as at the determinant site, *Immunol. Rev.*, 98, 53, 1987.
42. **Shastri, N., Malissen, B., and Hood, L.,** Ia transfected L-cell fibroblasts, present a lysozyme peptide but not the native protein to lysozyme-specific T cells, *Proc. Natl. Acad. Sci. U.S.A.*, 82, 5885, 1985.
43. **Stockinger, B., Pessara, U., Lin, R. H., Habich, J., Grez, M., and Koch, N.,** A role of Ia-associated invariant chains in antigen processing and presentation, *Cell*, 56, 683, 1989.
44. **Michalek, M. T., Benacerraf, B., and Rock, K.,** Two genetically identical antigen presenting cell clones display heterogeneity in antigen processing, *Proc. Natl. Acad. Sci. U.S.A.*, 86, 3316, 1989.
45. **Shimonkevitz, R. S., Colon, S., Kappler, J., Marrack, P., and Grey, H. M.,** Antigen recognition by H-2 restricted T cells. II. A tryptic ovalbumin peptide that substitutes for processed antigen, *J. Immunol.*, 133, 2067, 1984.
46. **Zlotnik, A., Shimonkevitz, R. P., Gefter, M. L., Kappler, J., and Marrack, P.,** Characterization of the γ-interferon-mediated induction of antigen-presenting ability in P388D1 cells, *J. Immunol.*, 131, 2814, 1983.
47. **Demotz, S., Grey, H. M., Appella, E., and Sette, A.,** Characterization of a naturally processed MHC class II-restricted T cell determinant of hen egg lysozyme, *Nature (London)*, 342, 682, 1989.
48. **Falo, L. D., Benacerraf, B., and Rock, K. L.,** Phospholipase treatment of accessory cells that have been exposed to antigen selectively inhibits antigen-specific Ia-restricted, but not allospecific, stimulation of T lymphocytes, *Proc. Natl. Acad. Sci. U.S.A.*, 83, 6994, 1986.
49. **Matzinger, P. and Bevan, M. J.,** Why do so many lymphocytes respond to major histocompatibility antigens?, *Cell. Immunol.*, 29, 1, 1977.
50. **Kappler, J. W. and Marrack, P.,** T cells can distinguish between allogeneic major histocompatibility complex products on different cell types, *Nature (London)*, 332, 84, 1988.
51. **Falo, L. D., Benacerraf, B., Rothstein, L., and Rock, K. L.,** Cerulenin is a potent inhibitor of antigen processing by antigen-presenting cells, *J. Immunol.*, 139, 3918, 1987.
52. **Lakey, E. K., Margoliash, E., and Pierce, S. K.,** Identification of a peptide-binding protein that plays a role in antigen presentation, *Proc. Natl. Acad. Sci. U.S.A.*, 84, 1659, 1987.
53. **Pierce, S. K., Morris, J. F., Grusby, M. J., Kaumaya, P., van Buskirk, A., Srinivasan, M., Crump, B., and Smolenski, L.,** Antigen presenting function of B lymphocytes, *Immunol. Rev.*, 106, 149, 1988.
54. **Rothman, J. E.,** Polypeptide chain binding proteins: catalysts of protein folding and related processes in cells, *Cell*, 59, 591, 1989.
55. **Chiang, H.-L., Terlecky, S. R., Plant, C. P., and Dice, J. F.,** A role for a 70-kD heat shock protein in lysosomal degradation of intracellular proteins, *Science*, 246, 382, 1989.
56. **Chain, B. M., Kaye, P. M., and Shaw, M.-A.,** The biochemistry and cell biology of antigen processing, *Immunol. Rev.*, 106, 33, 1988.
57. **Diment, S. and Stahl, P.,** Macrophage endosomes contain proteases which degrade endocytosed protein ligands, *J. Biol. Chem.*, 260, 15311, 1985.

58. **Diment, S., Leech, M. S., and Stahl, P. D.,** Cathepsin D is membrane-associated in macrophage endosomes, *J. Biol. Chem.*, 263, 6901, 1988.
59. **McCoy, K. L., Miller, J., Jenkins, M., Ronchese, F., Germain, R. N., and Schwartz, R. H.,** Diminished antigen processing by endosomal acidification mutant antigen-presenting cells, *J. Immunol.*, 143, 29, 1989.
60. **Roff, C. F., Fuchs, R., Mellman, I., and Robbins, A. R.,** Chinese hamster ovary cell mutants with temperature-sensitive defects in endocytosis. I. Loss of function on shifting to the nonpermissive temperature, *J. Cell. Biol.*, 103, 2284, 1986.
61. **Roosnek, E., Demotz, S., Corradin, G., and Lanzavecchia, A.,** Kinetics of MHC-antigen complex formation on antigen-presenting cells, *J. Immunol.*, 140, 4079, 1988.
62. **Tsien, R. Y., Pozzan, T., and Rink, T. J.,** T cell mitogens cause early changes in cytoplasmic free Ca^{2+} and membrane potential in lymphocytes, *Nature (London)*, 295, 68, 1982.
63. **Watts, C.,** Rapid endocytosis of the transferrin receptor in the absence of bound transferrin, *J. Cell Biol.*, 100, 633, 1985.
64. **Adams, D. O. and Hamilton, T. A.,** The cell biology of macrophage activation, *Annu. Rev. Immunol.*, 2, 283, 1984.
65. **Lanzavecchia, A.,** Antigen uptake and accumulation in antigen-specific B cells, *Immunol. Rev.*, 99, 39, 1987.
66. **Malynn, B. A. and Wortis, H. H.,** Role of antigen specific B cells in the induction of SRBC-specific T cell proliferation, *J. Immunol.*, 132, 2253, 1984.
67. **Rock, K. L., Benacerraf, B., and Abbas, A. K.,** Antigen presentation by hapten-specific B lymphocytes, *J. Exp. Med.*, 160, 1102, 1984.
68. **Lanzavecchia, A.,** Antigen-specific interaction between T and B cells, *Nature (London)*, 314, 537, 1985.
69. **Tony, H.-P. and Parker, D. C.,** Major histocompatibility complex-restricted, polyclonal B response from helper T cell recognition of anti-immunoglobulin presented by small B lymphocytes, *J. Exp. Med.*, 161, 223, 1985.
70. **Howard, J. C.,** Immunological help at last, *Nature (London)*, 314, 494, 1985.
71. **Lanzavecchia, A., Abrignani, S., Sheidegger, D., Obrist, R., Dorken, B., and Moldenhauer, G.,** Antibodies as antigens. The use of mouse monoclonal antibodies to focus human T cells against selected targets, *J. Exp. Med.*, 167, 345, 1988.
72. **Snider, D. P. and Segal, D. M.,** Efficiency of antigen presentation after antigen targetting to surface IgD, IgM, MHC, FcγRII and B220 on murine splenic B cells, *J. Immunol.*, 143, 59, 1989.
73. **Casten, L. A., Kaumaya, P., and Pierce, S. K.,** Enhanced T cell responses to antigenic peptides targeted to B cell surface Ig, Ia or class I molecules, *J. Exp. Med.*, 168, 171, 1988.
74. **Schwartz, R. H.,** T lymphocyte recognition in association with gene products of the major histocompatibility complex, *Annu. Rev. Immunol.*, 3, 237, 1985.
75. **Kappler, J. W., Skidmore, B., White, J., and Marrack, P.,** Antigen-inducible, H-2-restricted, interleukin 2 producing T cell hybridomas. Lack of independent antigen and H-2 recognition, *J. Exp. Med.*, 153, 1198, 1981.
76. **Werdelin, O.,** Chemically related antigens compete for presentation by accessory cells to T cells, *J. Immunol.*, 129, 1883, 1982.
77. **Rock, K. L. and Benacerraf, B.,** Inhibition of antigen-specific T lymphocyte activation by structurally related Ir gene-controlled polymers. Evidence for specific competition for accessory cell presentation, *J. Exp. Med.*, 157, 1618, 1983.
78. **Babbitt, B. P., Allen, P. M., Matsueda, G., Haber, E., and Unanue, E. R.,** Binding of immunogenic peptides to Ia histocompatibility molecules, *Nature (London)*, 317, 359, 1985.
79. **Buus, S., Colon, S., Smith, C., Freed, J., Miles, C., and Grey, H. M.,** Interaction between processed ovalbumin peptide and Ia molecules, *Proc. Natl. Acad. Sci. U.S.A.*, 83, 3968, 1986.
80. **Buus, S., Sette, A., Colon, S., Miles, C., and Grey, H. M.,** The relation between Major Histocompatibility Complex (MHC) restriction and the capacity of Ia to bind immunogenic peptides, *Science*, 235, 1353, 1987.
81. **Lai, M.-Z., Ross, D. T., Guillet, J.-G., Briner, T. J., Gefter, M. L., and Smith, J. A.,** T lymphocyte response to bacteriophage λ. Recognition of the same peptide by Ia molecules of different haplotypes, *J. Immunol.*, 139, 3973, 1987.
82. **Sette, A., Buus, S., Colon, S., Miles, C., and Grey, H. M.,** Structural analysis of peptides capable of binding to more than one Ia antigen, *J. Immunol.*, 142, 35, 1989.
83. **Schaeffer, E. B., Sette, A., Johnson, D. L., Bekoff, M. C., Smith, J. A., Grey, H. M., and Buus, S.,** Relative contribution of "determinant selection" and "holes in the T cell repertoire" to T cell responses, *Proc. Natl. Acad. Sci. U.S.A.*, 86, 4649, 1989.
84. **Buus, S., Sette, A., Colon, S., Jenis, D. M., and Grey, H. M.,** Isolation and characterization of antigen-Ia complexes involved in T cell recognition, *Cell*, 47, 1071, 1986.

85. **Watts, T. H. and McConnell, H. M.,** High affinity fluorescent peptide binding to I-A^d in lipid membranes, *Proc. Natl. Acad. Sci. U.S.A.,* 83, 9660, 1986.
86. **Sadegh-Nasseri, S., and McConnell, H. M.,** A kinetic intermediate in the reaction of an antigenic peptide and I-E^k, *Nature (London),* 337, 274, 1989.
87. **Phillips, M. L., Yip, C. C., Shevach, E. M., and Delovitch, T. L.,** Photoaffinity labeling demonstrates binding between Ia molecules and nominal antigen on antigen-presenting cells, *Proc. Natl. Acad. Sci. U.S.A.,* 83, 5634, 1986.
88. **Luescher, I. F., Allen, P. M., and Unanue, E. R.,** Binding of photoreactive lysozyme peptides to murine histocompatibility class II molecules, *Proc. Natl. Acad. Sci. U.S.A.,* 85, 872, 1988.
89. **Wraith, D. C., Smilek, D. E., Mitchell, D. J., Steinman, L., and McDevitt, H. O.,** Antigen recognition in autoimmune encephalomyelitis and the potential for peptide-mediated immunotherapy, *Cell,* 59, 247, 1989.
90. **Buus, S. and Werdelin, O.,** Oligopeptide antigens of the angiotensin lineage compete for presentation by paraformaldehyde-treated accessory cells to T cells, *J. Immunol.,* 136, 459, 1986.
91. **Babbitt, B. P., Matsueda, G., Haber, E., Unanue, E. R., and Allen, P. M.,** Antigenic competition at the level of peptide-Ia binding, *Proc. Natl. Acad. Sci. U.S.A.,* 83, 4509, 1986.
92. **Guillet, J.-G., Lai, M.-Z., Briner, T. J., Smith, J. A., and Gefter, M. L.,** Interaction of peptide antigens and class II major histocompatibility complex antigens, *Nature (London),* 324, 260, 1986.
93. **Urban, J. L., Horvath, S. J., and Hood, L.,** Autoimmune T cells: immune recognition of normal and variant peptide epitopes and peptide based therapy, *Cell,* 59, 257, 1989.
94. **Ceppellini, R., Frumento, G., Ferrara, G. B., Tosi, R., Chersi, A., and Pernis, B.,** Binding of labelled influenza matrix peptide to HLA DR in living B lymphoid cells, *Nature (London),* 339, 392, 1989.
95. **Watts, T. H., Gaub, H. E., and McConnell, H. M.,** T cell mediated association of peptide antigen and MHC protein detected by resonance energy transfer in an evanescent wave-field, *Nature (London),* 320, 176, 1986.
96. **Watts, T. H.,** T cell activation by pre-formed, long-lived Ia-peptide complexes. Quantitative aspects, *J. Immunol.,* 141, 3708, 1988.
97. **Jensen, P. E.,** Stable association of processed antigen with antigen-presenting cell membranes, *J. Immunol.,* 143, 420, 1989.
98. **Lee, J. M. and Watts, T. H.,** On the dissociation and reassociation of Ia-peptide complexes: evidence that brief transit through an acidic compartment is not sufficient for binding site regeneration, *J. Immunol.,* 144, 1829, 1990.
99. **Lorenz, R. G. and Allen, P. M.,** Processing and presentation of self-peptides, *Immunol. Rev.,* 106, 115, 1988.
100. **Lakey, E. K., Margoliash, E., Fluoret, G., and Pierce, S. K.,** Peptides related to the antigenic determinant block T cell recognition of native protein as processed by antigen-presenting cells, *Eur. J. Immunol.,* 16, 721, 1986.
101. **Lorenz, R. G. and Allen, P. M.,** Direct evidence for functional self-protein/Ia-molecule complexes *in vivo, Proc. Natl. Acad. Sci. U.S.A.,* 85, 5220, 1988.
102. **Buus, S., Sette, A., Colon, S., and Grey, H. M.,** Autologous peptides constitutively occupy the antigen binding site on Ia, *Science,* 242, 1045, 1988.
103. **Buus, S., Sette, A., and Grey, H. M.,** The interaction between protein-derived immunogenic peptides and Ia, *Immunol. Rev.,* 98, 116, 1987.
104. **McMichael, A. J., Gotch, F. M., Santos-Aguado, J., and Strominger, J. L.,** Effect of mutations and variations of HLA-A2 on recognition of a virus peptide epitope by cytotoxic T lymphocytes, *Proc. Natl. Acad. Sci. U.S.A.,* 85, 9194, 1988.
105. **Davis, C. B., Buerstedde, J.-M., McKean, D. J., Jones, P. P., McDevitt, H. O., and Wraith, D. C.,** The role of polymorphic I-A^k β chain residues in presentation of a peptide from myelin basic protein, *J. Exp. Med.,* 169, 2239, 1989.
106. **Rosloniec, E. F., Vitez, L. J., Beck, B. N., Buerstedde, J.-M., McKean, D. J., Benoit, C., Mathis, D., and Freed, J. H.,** I-A^k polymorphisms define a functionally dominant region for the presentation of hen egg lysozyme peptides, *J. Immunol.,* 143, 50, 1989.
107. **Brett, S. J., McKean, D., York-Jolley, J., and Bersofsky, J. A.,** Antigen presentation to specific T cells by Ia molecules selectively altered by site directed mutagenesis, *Int. Immunol.,* 1, 130, 1989.
108. **Davis, C. B., Mitchell, D. J., Wraith, D. C., Todd, J. A., Zamvil, S. S., McDevitt, H. O., Steinman, L., and Jones, P. P.,** Polymorphic residues on the I-A$_\beta$ chain modulate the stimulation of T cell clones specific for the N-terminal peptide of the autoantigen myelin basic protein, *J. Immunol.,* 143, 2083, 1989.
109. **Allen, P. M., Matsueda, G. R., Evans, R. J., Dunbar, J. B., Marshall, G. R., and Unanue, E. R.,** Identification of the T-cell and Ia contact residues of a T-cell antigenic epitope, *Nature (London),* 327, 713, 1987.
110. **Allen, P. M., Babbitt, B. P., and Unanue, E. R.,** T-cell recognition of lysozyme: the biochemical basis of presentation, *Immunol. Rev.,* 98, 171, 1987.

111. **Sette, A., Buus, S., Colon, S., Smith, J. A., Miles, C., and Grey, H. M.,** Structural characteristics of an antigen required for its interaction with Ia and recognition by T cells, *Nature (London),* 328, 325, 1987.
112. **Watts, T. H., Gariépy, J., Schoolnik, G. K., and McConnell, H. M.,** T cell activation by peptide antigen: effect of peptide sequence and method of antigen presentation, *Proc. Natl. Acad. Sci. U.S.A.*, 82, 5480, 1985.
113. **Lorenz, R. G., Tyler, A. N., and Allen, P. M.,** Reconstruction of the immunogenic peptide RNase(43-56) by identification and transfer of critical residues into an unrelated peptide backbone, *J. Exp. Med.*, 170, 203.
114. **Delisi, C. and Bersofsky, J. A.,** T-cell antigenic sites tend to be amphipathic structures, *Proc. Natl. Acad. Sci. U.S.A.*, 82, 7048, 1985.
115. **Spouge, J. L., Guy, H. R., Cornette, J. L., Margalit, H., Cease, K., Bersofsky, J. A., and Delisi, C.,** Strong conformational propensities enhance T cell antigenicity, *J. Immunol.*, 138, 204, 1987.
116. **Margalit, H., Spouge, J. L., Cornette, J. L., Cease, K., Delisi, C., and Bersofsky, J. A.,** Prediction of immunodominant helper T cell antigenic sites from the primary sequence, *J. Immunol.*, 138, 2213, 1987.
117. **Rothbard, J. B. and Taylor, W. R.,** A sequence pattern common to T cell epitopes, *EMBO J.*, 7, 93, 1988.
118. **Sette, A., Buus, S., Appella, E., Smith, J. A., Chesnut, R., Miles, C., Colon, S. M., and Grey, H. M.,** Prediction of major histocompatibility complex binding regions of protein antigens by sequence pattern analysis, *Proc. Natl. Acad. Sci. U.S.A.*, 86, 3296, 1989.
119. **Pincus, M. R., Gerewitz, F., Schwartz, R. H., and Scheraga, H. A.,** Correlation between the conformation of cytochrome c peptides and their stimulatory activity in a T lymphocyte proliferation assay, *Proc. Natl. Acad. Sci. U.S.A.*, 80, 3297, 1983.
120. **Schwartz, R. H., Fox, B. S., Fraga, E., Chen, C., and Singh, B.,** The T lymphocyte response to cytochrome c. V. Determination of the peptide size required for stimulation of T cell clones and assessment of the contribution of each residue beyond this size to antigenic potency, *J. Immunol.*, 135, 2598, 1985.
121. **Allen, P. M., Matsueda, G. R., Adams, S., Freeman, J., Roof, R. W., Lambert, L., and Unanue, E. R.,** Enhanced immunogenicity of a T cell immunogenic peptide by modifications of its N and C termini, *Int. Immunol.*, 1, 141, 1989.
122. **Carbone, F. R., Fox, B. S., Schwartz, R. H., and Patterson, Y.,** The use of hydrophobic, α-helix-defined peptides in delineating the T cell determinant for pigeon cytochrome c, *J. Immunol.*, 138, 1838, 1987.
123. **Sette, A., Buus, S., Colon, S., Miles, C., and Grey, H. M.,** I-A^d-binding peptides derived from unrelated protein antigens share a common structural motif, *J. Immunol.*, 141, 45, 1988.
124. **Sette, A., Lamont, A., Buus, S., Colon, S. M., Miles, C., and Grey, H. M.,** Effect of conformational propensity of peptide antigens in their interaction with MHC class II molecules. Failure to document the importance of regular secondary structures, *J. Immunol.*, 143, 1268, 1989.
125. **Harding, C. V. and Unanue, E. R.,** Antigen processing and intracellular Ia. Possible roles of endocytosis and protein synthesis in Ia function, *J. Immunol.*, 142, 12, 1989.
126. **Davis, J. E. and Cresswell, P.,** MHC class II molecules do not undergo significant endocytosis, in Abstracts of the American Society for Histocompatibility and Immunogenicity, *Hum. Immunol.*, September 1989, 19.
127. **Jensen, P. E.,** Protein synthesis in antigen processing, *J. Immunol.*, 141, 2545, 1988.
128. **Cresswell, P.,** Intracellular class II HLA antigens are accessible to transferrin-neuraminidase conjugates internalized by receptor mediated endocytosis, *Proc. Natl. Acad. Sci. U.S.A.*, 82, 8188, 1985.
129. **Koch, N., Lauer, W., Habicht, J., and Dobberstein, B.,** Primary structure of the gene coding for the murine Ia antigen-associated invariant chains (Ii). An alternatively spliced exon encodes a cysteine-rich domain highly homologous to a repetitive sequence of thyroglobulin, *EMBO J.*, 6, 1677, 1987.
130. **Holt, G. D., Sweidler, S. J., Freed, J. H., and Hart, G. W.,** Murine Ia-associated invariant chain's processing to the complex oligosaccharide forms and its dissociation from the I-A^k complex, *J. Immunol.*, 153, 399.
131. **McConnell, H. M., Watts, T. H., Weis, R. M., and Brian, A. A.,** Supported planar membranes in studies of cell-cell recognition in the immune system, *Biochim. Biophys. Acta,* 864, 95, 1986.
132. **Brian, A. A. and McConnell, H. M.,** Allogeneic stimulation of cytotoxic T cells by supported planar membranes, *Proc. Natl. Acad. Sci. U.S.A.*, 81, 6159, 1984.
133. **Walden, P., Nagy, Z. A., and Klein, J.,** Induction of T lymphocyte responses by liposomes carrying major histocompatibility complex molecules and foreign antigen, *Nature (London),* 315, 327, 1985.
134. **Walden, P., Nagy, Z. A., and Klein, J.,** Antigen presentation by liposomes: inhibition with antibodies, *Eur. J. Immunol.*, 16, 717, 1986.
135. **Coeshott, C. M., Chesnut, R. W., Kubo, R. T., Grammer, S. F., Jenis, D. M., and Grey, H. M.,** Ia-specific mixed leukocyte-reactive T cell hybridomas: analysis of their specificity by using purified class II MHC molecules in a synthetic membrane system, *J. Immunol.*, 136, 2832, 1986.

136. **Fox, B. S., Quill, H., Carlson, L., and Schwartz, R. H.,** Quantitative analysis of the T cell response to antigen and planar membranes containing purified MHC molecules, *J. Immunol.*, 138, 3367, 1987.
137. **Kupfer, A. and Singer, S. J.,** Cell biology of cytotoxic and helper T-cell functions, *Annu. Rev. Immunol.*, 7, 309, 1989.
138. **Davis, M. M. and Bjorkman, P. J.,** T-cell antigen receptor genes and T-cell recognition, *Nature (London)*, 334, 395, 1988.
139. **Dialynas, D. P., Wilde, D. B., Marrack, P., Pierres, A., Wall, K. A., Havran, W., Otten, G., Loken, M. R., Pierres, M., Kappler, J., and Fitch, F.,** Characterization of the murine antigenic determinant, designated L3T4a recognized by monoclonal antibody GK1.5: expression of L3T4a by functional T cell clones appears to correlate primarily with class II MHC antigen reactivity, *Immunol. Rev.*, 74, 29, 1983.
140. **Gay, D., Maddon, P., Sekaly, R., Talle, M. A., Godfrey, M., Long, E., Goldstein, G., Chess, L., Axel, R., Kappler, J., and Marrack, P.,** Functional interaction between human T cell protein CD4 and the major histocompatibility complex HLA-DR antigen, *Nature (London)*, 328, 626, 1987.
141. **Doyle, C. and Strominger, J. L.,** Interaction between CD4 and class II MHC molecules mediates cell adhesion, *Nature (London)*, 330, 256, 1987.
142. **Kupfer, A., Singer, S. J., Janeway, C. A., Jr., and Swain, S. L.,** Coclustering of CD4 (L3T4) molecule with the T-cell receptor is induced by specific direct interaction of helper T cells and antigen presenting cells, *Proc. Natl. Acad. Sci. U.S.A.*, 84, 5888, 1987.
143. **Owens, T., Fazekas de St Groth, B., and Miller, J. F. A. P.,** Aggregation of the T cell receptor with CD4 and other T-cell surface molecules enhances T-cell activation, *Proc. Natl. Acad. Sci. U.S.A.*, 84, 9209, 1987.
144. **Emmrich, F., Rieber, P., Kurrle, R., and Eichmann, K.,** Selective stimulation of human T lymphocyte subsets by heteroconjugates of antibodies to the T cell receptor and to subset-specific differentiation antigens, *Eur. J. Immunol.*, 18, 645, 1988.
145. **Rivas, A., Takada, S., Koide, J., Sondersstrup-McDevitt, G., and Engleman, E.,** CD4 molecules are associated with the antigen receptor complex on activated but not resting cells, *J. Immunol.*, 140, 2912, 1988.
146. **Rojo, J. M., Saizawa, K., and Janeway, C. A., Jr.,** Physical association of CD4 and the T cell receptor can be induced by anti-T cell receptor antibodies, *Proc. Natl. Acad. Sci. U.S.A.*, 8, 3311, 1989.
147. **Mittler, R. S., Goldman, S. J., Spitalny, G. L., and Burakoff, S. J.,** T-cell receptor-CD4 physical association in a murine T-cell hybridoma: induction by antigen receptor ligation, *Proc. Natl. Acad. Sci. U.S.A.*, 86, 8531, 1989.
148. **Rudd, C. E., Trevillyan, J. M., Dasgupta, J. D., Wong, L. L., and Schlossman, S. F.,** The CD4 receptor is complexed in detergent lysates to a protein-tyrosine kinase (pp58) from human T lymphocytes, *Proc. Natl. Acad. Sci. U.S.A.*, 85, 5190, 1988.
149. **Veillette, A., Bookman, M. A., Horak, E. M., and Bolen, J. B.,** The CD4 and CD8 T cell surface antigens are associated with the internal membrane tyrosine-protein kinase $p56^{lck}$, *Cell*, 55, 301, 1988.
150. **Veillette, A., Bookman, M. A., Horak, E. M., Samelson, S. E., and Bolen, J. B.,** Signal transduction through the CD4 receptor involves the activation of the internal membrane tyrosine-protein kinase $p56^{lck}$, *Nature (London)*, 338, 257, 1989.
151. **Janeway, C. A., Jr.,** The role of CD4 in T cell activation: accessory molecule or co-receptor?, *Immunol. Today*, 7, 234, 1989.
152. **Bierer, B. E. and Burakoff, S. J.,** T cell adhesion molecules, *FASEB J.*, 2, 2584, 1988.
153. **Springer, T. A., Dustin, M. L., Kishimoto, T. K., and Marlin, S. D.,** The lymphocyte function-associated LFA-1, CD2 and LFA-3 molecules: cell adhesion receptors of the immune system, *Annu. Rev. Immunol.*, 5, 223, 1987.
154. **Makgoba, M. W., Sanders, M. E., and Shaw, S.,** The CD-2-LFA-3 and LFA-1-ICAM pathways: relevance to T-cell recognition, *Immunol. Today*, 10, 417, 1989.
155. **Altmann, D. M., Hogg, N., Trowsdale, J., and Wilkinson, D.,** Cotransfection of ICAM-1 and HLA-DR reconstitutes human antigen presenting function in mouse L cells, *Nature (London)*, 338, 512, 1989.
156. **Dustin, M. L. and Springer, T. A.,** T-cell receptor cross-linking transiently stimulates adhesiveness through LFA-1, *Nature (London)*, 341, 619, 1989.
157. **Marlin, S. D. and Springer, T. A.,** Purified intercellular adhesion molecule-1 (ICAM-1) is a ligand for lymphocyte function-associated antigen 1 (LFA-1), *Cell*, 51, 813, 1987.
158. **Weiss, A., Imboden, J., Hardy, K., Manger, B., Terhorst, C., and Stobo, J.,** The role of the T3/antigen receptor complex in T cell activation, *Annu. Rev. Immunol.*, 4, 593, 1986.
159. **Mueller, D. L., Jenkins, M. K., and Schwartz, R. H.,** Clonal expansion versus clonal inactivation: a costimulatory signalling pathway determines the outcome of T cell antigen receptor occupancy, *Annu. Rev. Immunol.*, 7, 445, 1989.
160. **Nisbet-Brown, E. R., Lee, J. W., Cheung, R. K., and Gelfand, E. W.,** Antigen-specific and nonspecific mitogenic signals in the activation of human T cell clones, *J. Immunol.*, 138, 3713, 1987.

161. **Jenkins, M. K. and Schwartz, R. H.,** Antigen presentation by chemically modified splenocytes induces antigen-specific T cell unresponsiveness *in vitro* and *in vivo, J. Exp. Med.,* 165, 302, 1987.
162. **Quill, H. and Schwartz, R. H.,** Stimulation of normal inducer T cell clones with antigen presented by purified Ia molecules in planar lipid membranes: specific induction of a long-lived state of proliferative nonresponsiveness, *J. Immunol.,* 138, 3704, 1987.
163. **Ashwell, J. D., Jenkins, M. J., and Schwartz, R. H.,** Allogeneic non-T spleen cells restore the responsiveness of normal T cell clones stimulated with antigen and chemically modified antigen-presenting cells, *J. Immunol.,* 140, 3324, 1988.
164. **Lafferty, K. and Woolnough, J.,** The origin and mechanism of allograft rejection, *Immunol. Rev.,* 35, 231, 1977.
165. **Mueller, D. L., Jenkins, M. K., and Schwartz, R. H.,** An accessory cell-derived costimulatory signal acts independently of protein kinase C activation to allow T cell proliferation and prevent the induction of unresponsiveness, *J. Immunol.,* 142, 2617, 1989.
166. **Durham, S. K., Schmidt, J. A., and Oppenheim, J. J.,** Interleukin 1: an immunological perspective, *Annu. Rev. Immunol.,* 3, 263, 1985.
167. **Mizel, B.,** Interleukin 1 and T-cell activation, *Immunol. Today,* 8, 330, 1987.
168. **Kurt-Jones, E. A., Beller, D. I., Mizel, S. B., and Unanue, E. R.,** Identification of a membrane-associated interleukin 1 in macrophages, *Proc. Natl. Acad. Sci. U.S.A.,* 82, 1205.
169. **Weaver, C. T. and Unanue, E. R.,** T cell induction of membrane IL-1 on macrophages, *J. Immunol.,* 137, 3868, 1986.
170. **Muegge, K., Williams, T. M., Kant, J., Karin, M., Chiu, R., Schmidt, A., Siebenlist, U., Young, H. A., and Durum, S. K.,** Interleukin-1 costimulatory activity on the Interleukin-2 promotor via AP-1, *Science,* 246, 249, 1989.
171. **Mosmann, T. R. and Coffman, R. L.,** TH1 and TH2 cells: different patterns of lymphokine secretion lead to different functional properties, *Annu. Rev. Immunol.,* 7, 145, 1989.
172. **Kurt-Jones, E. A., Hamberg, S., O'Hara, J., Paul, W. E., and Abbas, A. K.,** Heterogeneity of helper/inducer T lymphocytes. I. Lymphokine production and lymphokine responsiveness, *J. Exp. Med.,* 166, 1774, 1987.
173. **Greenbaum, L. A., Horowitz, J. B., Woods, A., Pasqualine, T., Reich, E.-P., and Bottomly, K.,** Autocrine growth of $CD4^+$ T cells. Differential effects of IL-1 on helper and inflammatory T cells, *J. Immunol.,* 140, 1555, 1988.
174. **Weaver, C. T., Hawrylowicz, C. M., and Unanue, E. R.,** T helper cell subsets require the expression of distinct costimulatory signals by antigen presenting cells, *Proc. Natl. Acad. Sci. U.S.A.,* 85, 8181, 1988.
175. **Lichtman, A. H., Chin, J., Schmidt, J. A., and Abbas, A. K.,** Role of interleukin 1 in the activation of T lymphocytes, *Proc. Natl. Acad. Sci. U.S.A.,* 85, 9699, 1988.
176. **Hawrylowicz, C. M. and Unanue, E. R.,** Regulation of antigen-presentation-I. IFN-γ induces antigen-presenting properties on B cells, *J. Immunol.,* 141, 4083, 1988.
177. **Frosgren, S., Pobor, G., Coutinho, A., and Pierres, M.,** The role of I-A/E molecules in B lymphocyte activation. I. Inhibition of lipopolysaccharide-induced responses by monoclonal antibodies, *J. Immunol.,* 133, 2104, 1984.
178. **Corley, R. B., Locascio, N. J., Ovnic, M., and Haughton, G.,** Two separate functions of class II (Ia) molecules: T-cell stimulation and B cell excitation, *Proc. Natl. Acad. Sci. U.S.A.,* 82, 516, 1985.
179. **Bishop, G. A. and Haughton, G.,** Induced differentiation of a transformed clone of $Ly\text{-}1^+$ B cells by clonal T cells and antigen, *Proc. Natl. Acad. Sci. U.S.A.,* 83, 7410, 1986.
180. **Bishop, G. A. and Frelinger, J. A.,** Haplotype-specific differences in signaling by transfected class II molecules to a $Ly\text{-}1^+$ B cell clone, *Proc. Natl. Acad. Sci. U.S.A.,* 86, 5933, 1989.
181. **Clement, L. T., Tedder, T. F., and Gartland, G. L.,** Antibodies reactive with class II antigens encoded for by the major histocompatibility complex inhibit human B cell activation, *J. Immunol.,* 136, 2375, 1986.
182. **Krusemeier, M. and Snow, E. C.,** Induction of lymphokine responsiveness of hapten-specific B lymphocytes promoted through an antigen-mediated T helper lymphocyte interaction, *J. Immunol.,* 140, 367, 1988.
183. **Noelle, R. J., McCann, J., Marshall, L., and Bartlett, W. C.,** Cognate interactions between helper T cells and B cells. III. Contact-dependent, lymphokine-independent induction of B cell cycle entry by activated helper T cells, *J. Immunol.,* 143, 1807, 1989.
184. **Baluyut, A. R. and Subbarao, B.,** The synergistic effects of anti-IgM and monoclonal anti-Ia antibodies in induction of murine B lymphocyte activation, *J. Mol. Cell. Immunol.,* 4, 45, 1988.
185. **Cambier, J. C. and Lehmann, K. R.,** Ia-mediated signal transduction leads to proliferation of primed B lymphocytes, *J. Exp. Med.,* 170, 877, 1989.
186. **Chen, Z. Z., McGuire, J. C., Leach, K. L., and Cambier, J. C.,** Transmembrane signaling through B cell MHC class II molecules: anti-Ia antibodies induce protein kinase C translocation to the nuclear fraction, *J. Immunol.,* 138, 2345, 1987.

187. **Cambier, J. C., Newell, M. K., Justement, L. B., McGuire, J. C., Leach, K. L., and Chen, Z. Z.,** Ia binding ligands and cAMP stimulate nuclear translocation of PKC in B lymphocytes, *Nature (London),* 327, 629, 1987.
188. **Harris, L. K. and Cambier, J. C.,** B lymphocyte activation. Transmembrane signal transduction by membrane immunoglobulin in isolated cell membranes, *J. Immunol.,* 139, 963, 1987.
189. **Chartash, E. K., Imai, A., Gershengron, M. C., Crow, M. K., and Friedman, S. M.,** Direct human T helper cell-induced B cell activation is not mediated by inositol lipid hydrolysis, *J. Immunol.,* 140, 1974, 1988.
190. **Holte, H. Blomhoff, H. K., Beiske, K., Funderud, S., Torjesen, P., Gaudernack, G., Stokke, T., and Smeland, E. B.,** Intracellular events associated with inhibition of B cell activation by monoclonal antibodies to HLA class II antigens, *Eur. J. Immunol.,* 19, 1221, 1989.
191. **Nishizuka, Y.,** The molecular heterogeneity of protein kinase C and its implications for cellular regulation, *Nature (London),* 334, 661, 1988.
192. **Figueroa, F. and Klein, J.,** The evolution of MHC class II genes, *Immunol. Today,* 7, 75, 1986.
193. **Griffith, I. J., Ghogawala, Z., Nabavi, N., Myer, A., Golan, D. E., McKean, D. J., and Glimcher, L. H.,** The cytoplasmic domain affects membrane expression and function of an Ia molecule, *Proc. Natl. Acad. Sci. U.S.A.,* 85, 4847, 1988.
194. **Nabavi, N., Ghogawala, Z., Myer, A., Griffith, I., Wade, W. F., Chen, Z. Z., McKean, D. J., and Glimcher, L. H.,** Antigen presentation abrogated in cells expressing truncated Ia molecules, *J. Immunol.,* 142, 1444, 1989.
195. **Wade, W. F., Chen, Z. Z., Maki, R., McKercher, S., Palmer, E., Cambier, J. C., and Freed, J. H.,** Altered I-A protein-mediated transmembrane signaling in B cells that express truncated I-A^k protein, *Proc. Natl. Acad. Sci. U.S.A.,* 86, 6297, 1989.
196. **St-Pierre, Y., Nabavi, N., Ghogawala, Z., Glimcher, L. H., and Watts, T. H.,** A functional role for signal transduction via the cytoplasmic domains of MHC class II proteins, *J. Immunol.,* 143, 808, 1989.
197. **Palacios, R.,** Monoclonal antibodies against human Ia antigens stimulate monocytes to secrete interleukin 1, *Proc. Natl. Acad. Sci. U.S.A.,* 82, 6652, 1985.
198. **Wasik, M. A., Donnelly, R. P., and Beller, D. I.,** Lymphokine-independent induction of macrophage membrane IL-1 by autoreactive T cells recognizing either class I or class II MHC determinants, *J. Immunol.,* 141, 3456, 1988.
199. **Brandwein, S. R.,** Regulation of interleukin 1 production by mouse peritoneal macrophages: effects of arachidonic acid metabolites, cyclic nucleotides, and interferons, *J. Biol. Chem.,* 261, 8624, 1986.
200. **Kupfer, A. and Singer, S. J.,** The specific interaction of helper T cells and antigen-presenting B cells. IV. Membrane and cytoskeletal reorganizations in the bound T cell as a function of antigen dose, *J. Exp. Med.,* 170, 1697, 1989.
201. **Kammer, G. M., Boehm, C. A., Rudolph, S. A., and Schultz, L. A.,** Mobility of the human T lymphocyte surface molecules CD3, CD4 and CD8: regulation by a cAMP-dependent pathway, *Proc. Natl. Acad. Sci. U.S.A.,* 85, 792, 1988.
202. **Berridge, M. J. and Irvine, R. F.,** Inositol phosphates and cell signalling, *Nature (London),* 341, 197, 1989.
203. **Kammer, G. M.,** The adenylate cyclase-cAMP-protein kinase A pathway and regulation of the immune response, *Immunol. Today,* 9, 222, 1988.
204. **Tony, H.-P. and Parker, D. C.,** Major histocompatibility complex-restricted, polyclonal B responses from helper T cell recognition of anti-immunoglobulin presented by small B lymphocytes, *J. Exp. Med.,* 161, 223, 1985.
205. **Grosselin, E. J., Tony, H.-P., and Parker, D. C.,** Characterization of antigen processing and presentation by resting B lymphocytes, *J. Immunol.,* 140, 1408, 1988.
206. **Cowing, C. and Chapdelaine, J. M.,** T cells discriminate between Ia antigens on allogeneic accessory cells and B cells: a potential function for carbohydrate side chains on Ia molecules, *Proc. Natl. Acad. Sci. U.S.A.,* 80, 6000, 1983.
207. **Frohman, M. and Cowing, C.,** Presentation of antigen by B cells: functional dependence on radiation dose, interleukins, cellular activation, and differential glycosylation, *J. Immunol.,* 134, 2269, 1985.
208. **Kreiger, J., Chesnut, R. W., and Grey, H. M.,** Capacity of B cells to function as stimulators of a primary mixed leukocyte reaction, *J. Immunol.,* 137, 3117, 1986.
209. **Roehm, N. W., Leibson, J. L., Zlotnik, A., Kappler, J., Marrack, P., and Cambier, J. C.,** Interleukin-induced increase in Ia expression by normal mouse B-cells, *J. Exp. Med.,* 160, 679, 1984.
210. **Cambier, J. C., Justement, L. B., Newall, M. K., Chen, Z. Z., Harris, L. K., Sandoval, V. M., Klemsz, M. J., and Ransom, J. T.,** Transmembrane signals and intracellular "Second messengers" in the regulation of quiescent B-lymphocyte activation, *Immunol. Rev.,* 95, 37, 1987.
211. **Kreiger, J., Jenis, D. M., Chesnut, R. W., and Grey, H. M.,** Studies on the capacity of intact cells and purified Ia from different B cell sources to function in antigen presentation to cells, *J. Immunol.,* 140, 388, 1988.

212. **Anderson, R. E., Sprent, J., and Miller, J. F. A. P.,** Radiosensitivity of T and B lymphocytes, *Eur. J. Immunol.*, 4, 199, 1974.

213. **Ashwell, J. D., Jenkins, M. K., and Schwartz, R. H.,** Effect of gamma radiation on resting B lymphocytes. II. Functional characterization of the antigen-presentation defect, *J. Immunol.*, 141, 2536, 1988.

214. **Tew, J. G., Kosco, M. H., and Szakal, A. K.,** The alternative antigen pathway, *Immunol. Today*, 10, 229, 1989.

215. **Szakal, A. K., Kosco, M. H., and Tew, J. G.,** Microanatomy of lymphoid tissue during humoral immune responses: structural functional relationships, *Annu. Rev. Immunol.*, 7, 91, 1989.

216. **Szakal, A. K., Kosco, M. H., and Tew, J. G.,** A novel *in vivo* follicular dendritic cell-dependent iccosome-mediated mechanism for delivery of antigen to antigen-processing cells, *J. Immunol.*, 140, 341, 1988.

217. **Kosco, M. H., Szakal, A. K., and Tew, J. G.,** *In vivo* obtained antigen presented by germinal center B cells to T cells *in vitro, J. Immunol.*, 140, 354, 1988.

218. **Cooper, M. D., Kearney, J. F., Gathings, W. E., and Lawton, A. R.,** Effects of anti-Ig antibodies on the development and differentiation of B cells, *Immunol. Rev.*, 52, 29, 1980.

219. **Gordon, J.,** The B lymphocyte depleted mouse as a tool in immunobiology, *J. Immunol. Meth.*, 25, 227, 1979.

220. **Kim, K. J., Rollwagen, F., Asofsky, R., and Lefkovits, I.,** The abnormal function of T cells in chronically anti-μ-treated mice with no mature B lymphocytes, *Eur. J. Immunol.*, 14, 476, 1984.

221. **Ron, Y., De Baetselier, P., Gordon, J. Feldman, M., and Segal, S.,** Defective induction of antigen-reactive proliferating T cells in B cell-deprived mice, *Eur. J. Immunol.*, 11, 964, 1981.

222. **Bottomly, K., Janeway, C. A., Mathieson, B. J., and Mosier, D. F.,** Absence of an antigen-specific helper T cell required for the expression of the T 15 idiotype in mice treated with anti-μ antibody, *Eur. J. Immunol.*, 10, 159, 1980.

223. **Janeway, C. A., Murigita, R. A., Weinbaum, F. I., Asofsky, R., and Wigzell, H.,** Evidence for an immunoglobulin dependent antigen-specific helper cell, *Proc. Natl. Acad. Sci. U.S.A.*, 74, 4582, 1977.

224. **Ashwell, J. D.,** Are B lymphocytes the principal antigen-presenting cells *in vivo?, J. Immunol.*, 140, 1697, 1988.

225. **Ron, Y. and Sprent, J.,** T cell priming *in vivo;* a major role for B cells in presenting antigen to T cells in lymph nodes, *J. Immunol.*, 138, 2848, 1987.

226. **Kurt-Jones, E. A., Liano, D., HayGlass, K. A., Benacerraf, B., Sy, M.-S., and Abbas, A. K.,** The role of antigen-presenting B cells in T cell priming *in vivo, J. Immunol.*, 140, 3773, 1988.

227. **Ron, Y., De Baetselier, P., Tzehoval, Gordon, J., Feldman, M., and Segal, S.,** Defective induction of antigen-reactive proliferating T cells in B cell-deprived mice, *Eur. J. Immunol.*, 13, 167, 1983.

228. **Janeway, C. A., Ron, Y., and Katz, M. E.,** The B cell is the initiating antigen-presenting cell in peripheral lymph nodes, *J. Immunol.*, 138, 1051, 1987.

229. **Lassila, O., Vainio, O., and Matzinger, P.,** Can B cells turn on virgin T cells?, *Nature (London)*, 334, 253, 1988.

230. **Weiss, S. and Bogen, B.,** B-lymphoma cells process and present their endogenous immunoglobulin to major histocompatibility restricted T cells, *Proc. Natl. Acad. Sci. U.S.A.*, 86, 282, 1989.

Chapter 6

T CELL-DEPENDENT B-CELL ACTIVATION

William C. Bartlett and Randolph J. Noelle

TABLE OF CONTENTS

I. EARLY STUDIES OF LYMPHOCYTE INTERACTIONS IN THE DEVELOPMENT OF HUMORAL IMMUNE RESPONSES

A. DISCOVERY OF THYMUS-DEPENDENT IMMUNE RESPONSES

The work of Claman, Davies, Miller and Mitchell and Katz and co-workers elucidated the requirements for helper T cell (T_h) for the generation of antibody responses.[1-5] The first direct evidence that interactions between T and B cells were required for the humoral immune response to SRBC was provided by Claman in 1966.[6] In these experiments, lethally irradiated mice were reconstituted with spleen, thymus, bone marrow, or thymus plus bone marrow from normal or immune syngeneic mice. Following administration of antigen (SRBC), spleens were removed and a splenic focus assay was performed to quantitate the antibody response to SRBC. Only when thymus *and* bone marrow (or spleen) from immune donors were transferred to recipients was an anti-SRBC response observed. These authors concluded that ''auxillary cells'' in the thymus and ''effector cells'' in the bone marrow cooperated in the development of antibody response. Later studies by Davies[7] and Miller and Mitchell[8] established bone marrow cells as the cells responsible for antibody production and the cells from the thymus as the ''auxillary cells''.

B. THE HAPTEN-CARRIER EFFECT AND COOPERATIVE INTERACTIONS BETWEEN T AND B LYMPHOCYTES

Optimal immune responses to small haptenic determinants required the covalent coupling of the hapten onto an immunogenic carrier. It had been realized that previous immunization with carrier significantly improved the response to the hapten, the so-called ''carrier effect''. Since it was known that the anti-hapten response was highly specific for the hapten and that the specificity of the antibody accurately represented the specificity of the antibody producing cell, it appeared that the other regulatory mechanisms were operative with regard to the carrier. It was suggested that distinct cellular populations recognized haptenic determinants and carrier determinants, respectively.

Direct evidence for the *in vivo* cooperative interactions of lymphocytes for the anti-hapten response was provided by Mitchison.[9] The system used to demonstrate hapten carrier effect was as follows: spleen cells from 4-hydroxy-5-iodo-3-nitrophenylacetyl (NIP)-OVA immune mice were transferred into syngeneic recipients and produced anti-NIP upon challenge with homologous antigen (NIP-OVA), but not NIP-BSA. When BSA-immune spleen cells were co-transferred with NIP-OVA-immune spleen cells and challenged with NIP-BSA, a secondary response was also observed. The authors concluded that the transfer of carrier-primed cells permitted or ''helped'' the antibody response to the hapten *only* if the hapten-carrier was provided. Raff[10] extended these studies to show that the cells that recognized the carrier were T cells and the cells which produced antibodies to the hapten were B cells. These results were confirmed in the laboratories of Rajewsky[11] and Katz.[12,13]

The first studies to report cooperative interactions *in vitro* were reported by Kettman

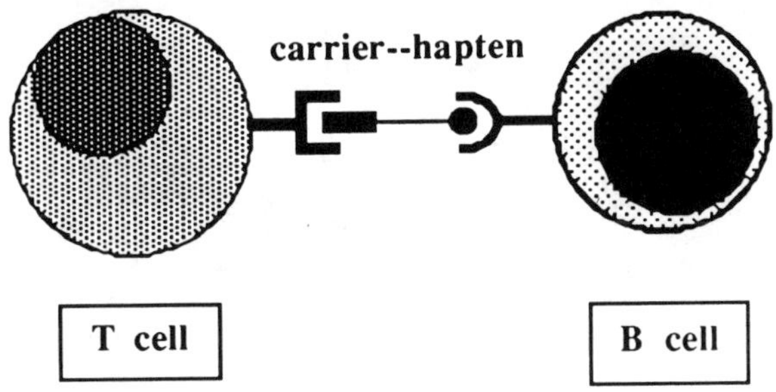

FIGURE 1. Cognate interactions are mediated by the respective recognition of haptenic and carrier determinants by the B and T cell.

and Dutton.[14] These investigators were the first to demonstrate secondary anti-hapten responses to hapten-coupled sheep erythrocytes. *In vitro* responses to protein-hapten conjugates were first reported by Bullock and Rittenberg[15,16] in which splenocytes from KLH-TNP immunized mice produced IgM and IgG anti-TNP when cultured in the presence of TNP-KLH. Later studies by Cheers et al.,[17] and Feldmann and Basten[18-21] delineated the respective roles of T and B cells in such *in vitro* systems. Mosier and co-workers[22-24] demonstrated that cells that were adherent to plastic (macrophages) were also required for the development of antibody responses *in vitro*. The original associative recognition model of Mitchison proposed that receptors on T and B cells recognized carrier determinants and haptenic determinants, respectively. As a result of this bridging, it was suggested that the T cell delivered an activation signal to the B cells. These and other studies[25-27] were responsible for initial models of T-B cell interactions which suggested that membrane (m)Ig on B cells served to focus antigen via binding of haptenic determinants. As a result hapten-linked carrier determinants were focused to the T cell (see Figure 1).

C. GENETICALLY RESTRICTED INTERACTIONS

Products of the MHC complex were implicated in the regulation of T_h-B cell interactions and the development of humoral immune responses.[27-32] Reports from Janeway[33] and Katz[34] using H-2 congenic strains of mice indicated that genes localized in the I-A subregion of H-2 regulated the interaction of T_h and B cells. At the same time,[35] it was reported that the response of antigen-primed T cells and macrophages was also MHC-restricted and this response was blocked by anti-MHC antisera. This observation and the results of others strongly suggested that the MHC restriction observed between T cells and accessory cells (or B cells) was due to serologically defined antigens on the surface of the interacting cells. Results from von Boehmer[36] were apparently contradictory and indicated that MHC incompatible T_h could productively interact with B cells. These experiments utilized double radiation bone marrow chimeras (P1/P2) and showed that P1 T cells could interact with P2 B cells. It is now well understood that the P1 and P2 precursor T cells in the P2 environment learned their genetic restriction as P2. The term ''adaptive differentiation'' was coined to describe the plasticity of the developing immune response. From these studies it was recognized that cognate interactions were genetically restricted by products of the I-subregion of the MHC complex. With recent studies[37,38] it has become evident that the T cell receptor recognizes the ''processed'' antigen and class II MHC products on the surface of the B cell or APC. With this understanding a modified version of MHC-restricted T_h-B cell interaction was devised. In this model mIg on the B cell surface served to focus carrier determinants for the T_h. In addition, cell surface molecules on the T_h recognized class II MHC molecules and associated carrier determinants on the B cell (see Figure 2). The model presented assumed that the TCR recognized the processed antigen and MHC class II molecules independently.

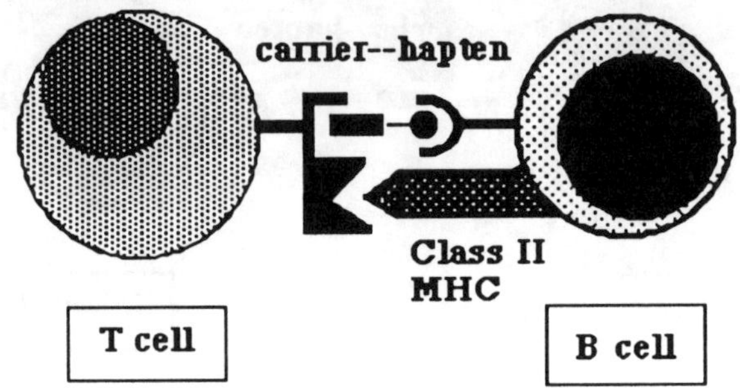

FIGURE 2. Cognate interactions are antigen specific and genetically restricted.

This model did not adequately resolve how a B cell that recognizes native antigen could assist in the presentation of ''processed'' antigen to the T cell.

D. ROLE OF THE MACROPHAGE IN HUMORAL IMMUNE RESPONSES

Although the studies cited above clearly established a requirement for class II-restricted recognition by the T_h, whether this restriction was between the T_h and the responding B cells or between the T_h and accessory cells was still unresolved. Studies by Katz,[39] Sprent,[40] and Swierkosz et al.[41] all supported the notion of direct T_h-B cell interactions in the develop of antibody responses. However, studies by Erb and Feldmann[42] and Singer et al.[43-45] indicated that T_h-accessory cell interactions were genetically-restricted and T_h-B cell interactions were not. A complex model of T_h-B-accessory cell interactions (Figure 3) was developed to explain the role of APC, T cell recognition of ''processed'' antigen and interactions with B cells.

E. ANTIGEN CAPTURE AND PROCESSING BY B CELLS

At this time it was known that the T cell receptor for antigen recognized ''processed'' carrier and the mIg receptor on B cells recognized native haptenic determinants. Definitive experiments establishing that B cells could selectively process and present antigen to T cells were first reported by Chesnut and Grey.[46] Using rabbit anti-mouse Ig (RaMIg), they showed that B cells could process and present RaMIg and activate RIg-specific T cells. These studies clearly established that B cells could present TD antigens at extraordinarily low concentrations of antigen. Later studies by Abbas and co-workers[47-49] used purified antigen-specific B cells and verified that TD antigens were processed and presented to T cells at low concentrations. Lanzavecchia[50] using tetanus toxoid specific, EBV-transformed human B cell clones and Abbas et al.[47] using purified antigen-specific B cells established that the major function of mIg was to capture antigen at low concentrations to facilitate antigen specific processing (Figure 4). These studies went on to show that mIg played no direct role in antigen presentation per se. That is, subsequent to capture and processing, the processed antigen was presented on the B cell surface in the context of class II MHC molecules.

Numerous ancillary issues have been raised since the discovery that B cells present antigen directly to T cells. Studies have indicated that the level of class II MHC antigen expression correlates with the antigen-presenting activity of the B cell. It is evident that cells not expressing class II MHC do not present antigen, but it is not clear how the level of class II MHC antigen effects the antigen-presenting capacity of B cell populations. When resting B cells are treated with IL4, a log increase in the number of class II MHC molecules is observed.[51] Sanders et al.[52] have noted that this increase in class II MHC molecules increases the frequency of class II-restricted T_h-B cell conjugates two- to threefold. Studies from our laboratory do not find that pretreatment of B cells with IL4 to have a dramatic

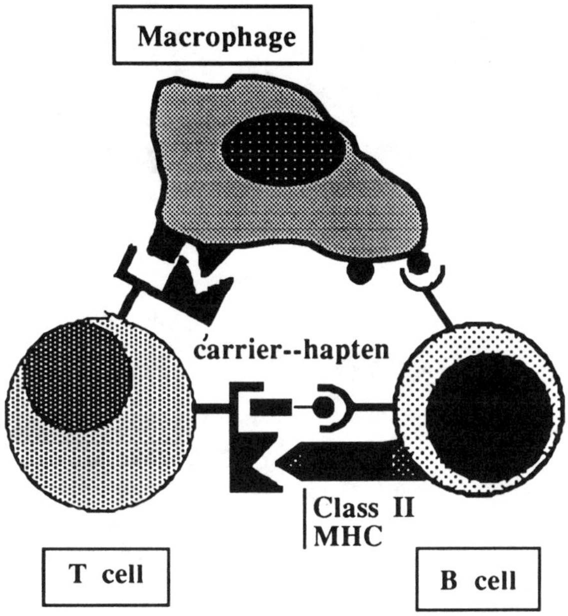

FIGURE 3. Cognate interactions and the involvement of accessory cells.

effect on the frequency of T-B cell conjugates (unpublished observations). This finding is in accord with the similar capabilities of resting or IL4-treated B cells to serve as allostimulators, i.e., although a huge increase in class II MHC molecules is observed with IL4 treatment, no coordinate increase in allopresentation was observed.[53] There are conditions that do result in increased B cell antigen-presentation. A number of groups have shown that activated B cells have a greatly enhanced presentation capacity.[54,55] Minimally, stimulation with LPS or stimulation with IL4 and PMA results in enhanced B cell antigen-presentation. Studies by Frohman and Cowing[54] indicated that glycosylation differences in the class II MHC molecules on resting and activated B cells may account for the relative antigen-presentation capacity of the two populations. These authors treated resting B cells with neuraminidase and showed in increased antigen-presentation capabilities, suggesting that the removal of terminal sialic acid residues from class II molecules or other glycosylated cell surface structures increased antigen-presentation. These results were questioned by experiments showing that purified class II MHC molecules isolated from B cells untreated or treated with LPS, or IL4 and incorporated into lipid bilayers were equally effective on a molar basis.[53] This indicated that class II MHC molecules derived from resting or activated B cells did not *intrinsically* have different antigen-presentation capabilities. In addition, treatment of isolated class II MHC molecules with neuraminidase, to remove terminal sialic acid residues did not increase the antigen presentation capabilities of the class II molecules. Therefore, it is likely that the desialylation of non-class II MHC molecules was crucial to increased T_h-B cell interaction. Studies by Kease et al.[56] showed that neuraminidase-treatment of resting, antigen-specific B cells greatly increased the frequency with which these B cells could form conjugates with antigen-specific T cell clones. Therefore, alterations in the carbohydrate content of membrane components, or changes in the overall charge of the B cell is likely the basis for increased antigen-presentation by neuraminidase-treated B cells. The same effect, that is, increased T_h-B cell conjugate formation could be achieved by the crosslinking of mIg. This suggests that the engagement of the antigen receptor on the B cell surface may predispose the B cell to forming a conjugate with an antigen-specific T cell.

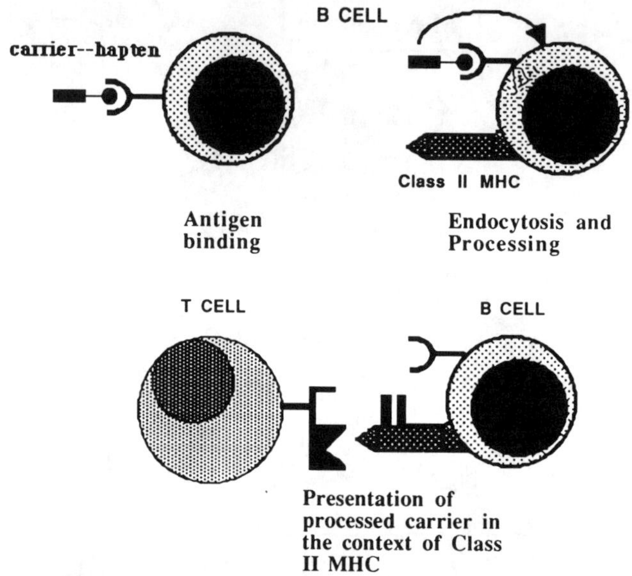

FIGURE 4. Cognate interactions. Antigen-specific processing by B cells.

Similarly, Krieger et al.[53] indicated that PMA treatment of B cells also increased their ability to serve as stimulators of 1° MLR. Therefore, it appears that activation of a PKC cascade, by PMA or mIg crosslinking may increase the ability of B cells to form antigen-specific conjugates with T_h. Whether this is due to a generalized alteration in the carbohydrate content of the cell surface, the alteration in the carbohydrate content of a specific cell surface protein or the expression of new adhesion molecules has not been resolved.

II. MODEL *IN VITRO* SYSTEMS FOR THE STUDY OF INTERACTIONS BETWEEN HELPER T CELLS AND B CELLS

A. ANTIGEN-SPECIFIC, CLASS II-RESTRICTED T_h-B CELL INTERACTIONS USING PURIFIED ANTIGEN-SPECIFIC B CELLS

All of the classical studies of cognate interactions relied on the specificity of the antibody produced by the B cell to measure B cell activation. That is, the success of T_h-B *in vitro* interactions was measured by an increase in the number of antigen-specific plaque forming cells. This assay allowed the detection of activated B cells that were present in the population at very low frequencies. Although this was a sensitive and reliable measure to detect T_h-dependent B cell differentiation, little was known concerning the requirements for T_h-dependent B cell interactions, B cell activation or B cell growth. To measure either T_h-dependent B cell activation or growth, a much higher frequency of responsive B cells was required. One approach that allowed the direct measurement of antigen-specific B cell activation by T_h was through the purification of antigen-specific B cells (so-called antigen-binding cells; ABC). Snow et al.[57] reported a rosetting procedure that afforded high numbers of purified, naive, trintrophenol (TNP)-specific B cells that were responsive to TI and TD activation. Briefly, splenocytes from unprimed mice were incubated with trinitrophenylated (TNP) horse erythrocytes. After a series of velocity and density gradient sedimentations, a population which was >90% TNP-specific was obtained. Using these TNP-ABC B cells, the requirements of T_h-dependent B cell growth was investigated. To induce B cell proliferation to a hapten-carrier, there was a requirement for the appropriate hapten-carrier, carrier-primed T cells and class II identity between the T_h and the B cells.[58,59] In agreement with studies reported by Cambier and co-workers,[60] it was shown that TD antigen alone or in the presence of lymphokines *did not induce B cell growth or entry into the cell cycle.* This was unexpected since at this time, studies showed that anti-Ig and lymphokines induced B cell growth.[61] Therefore anti-Ig was not a model system for the effects of TD antigens on B cells, since TD antigens did not induce B cell cycle entry. Further, studies using TNP-ABC and carrier-primed T cells indicated that T_h-B cell contact was required initially to induce B cell cycle entry, after which time, lymphokines could supplant the function of the T_h.[58] These studies were the first to show that clonal B cell expansion required hapten-carrier linkage, class II identity and T_h-B cell interaction. Later studies by Marrack et al.[62] and Sanders et al.[63] were important in that they established that the restrictions of cognate interactions (genetic restrictions and antigen-specificity) was at the level of conjugate formation between the antigen-presenting B cell and the T_h.

B. SMALL B CELL AND LARGE B CELL RESPONSES TO COGNATE HELP

The requirements for B cell growth and differentiation are dependent upon whether the B cells are resting or activated.[64,65] *In vivo* activated B cells will grow and differentiate when placed into culture with T cell factors. Hence, the continued growth and differentiation of activated B cells can be antigen-nonspecific, class II-*unrestricted* fashion and supported by soluble factors. In contrast, the induction of resting B cells to growth and antibody production requires class II-restricted, antigen-specific interactions with T_h and lymphokines. The finding that the activation requirements of resting vs. activated B cells were different was a major advance in our being able to critically dissect the activation requirements of resting B cells.

C. ARTIFICIAL BRIDGES FOR T_h-B CELL INTERACTIONS

A number of investigators have examined the regulation of T_h-B cell interactions, using artificial "bridges" to facilitate T_h-B cell interaction. The use of "artificial" focusing of T_h on B cells circumvented the low frequency of T_h-B interactions observed in antigen-specific T_h-B systems. This approach also permitted insights into the respective role of antigen, genetic restrictions and mIg in cognate interactions.

1. B Cell Surface Antigens as a Bridge

Studies by Katz[66] emphasized that although antigen-specific interactions required MHC identity between the T_h and B cells, allogeneic T cells were able to activate histo-incompatible B cells non-specifically through an "allogeneic" effect. Cognizant of these studies, Augustin and Coutinho[67,68] developed enriched populations of T_h that recognized B cells by virtue of their expression of MHC products. This enabled polyclonal stimulation of B cells on a population-wide scale as seen with mitogens such as LPS, but induced by the physiologic mediators of B cell activation, T cells. These studies demonstrated that B cell activation required that T_h cells directly recognize antigen on the responding cell. This suggested that T cell-dependent B cell activation was not initiated by soluble products generated by activated T_h cells, but by some form of physical contact between the T and B lymphocytes. Julius et al.[64] developed male antigen-specific, M locus specific and H-2-specific T cells and showed that these T cell could also support polyclonal B cell activation as long as B cells expressed the relevant membrane protein. Unlike most other polyspecific T_h-dependent B cell systems, anti-Ig was *required* as a second signal in the stimulation of resting B cells. Resolution of this contradiction is provided by a recent publication by Pereira et al.[69] that shows that mIg crosslinking greatly enhances TD responses if T_h are limiting. Specifically, using alloreactive T cell clones, these investigators showed that anti-μ or anti-δ increased effective helper activity by 10- to 20-fold. Therefore it appears that crosslinking of mIg by TD antigen may render the B cell more receptive to interactions with T_h or more responsive to T_h-derived signals.

Autoreactive T_h, which react to syngeneic Ia molecules in the absence of antigen, were shown to polyclonally activate populations of B cells in the absence of antigen. Studies by Nagarkatti et al.[70] suggested that direct binding of T cells to B cells, via B cell expressed Ia, was sufficient to initiate growth and differentiation in resting B cells. The requirement for cell contact was strongly implied, as SN from activated autoreactive T cells failed to activate cultures of resting, antigen-specific B cells. Using antigen-specific B cells and autoreactive T_h, Nagarkatti et al. showed that the addition of specific antigen did not alter the growth characteristics of antigen-specific B cells but significantly augmented the antibody response. Saito et al.[71] independently showed that an Ia-reactive T_h clone can stimulate approximately every hundredth B cell to proliferate and produce antibody. In addition, when this T_h clone was transferred *in vivo* along with B cells and antigen into irradiated recipients, the T_h cells were able to induce heightened antigen-specific responses of B cells in the presence of antigen, as compared with mice reconstituted with B cells and antigen alone. Taken together, T_h that recognize integral B cell membrane proteins, like class II or class I MHC can induce B cell growth and differentiation in the absence of mIg crosslinking of antigen. However, it appears that under conditions of limited T cell help, mIg crosslinking can augment B cell responsiveness.

2. Haptens and Soluble Antigens as Bridges

In 1980, Martinez-Alonzo et al.[72] described a series of experiments using murine T_h lymph node cells (LNC) which had been primed to one of three haptens, and tested for helper function by co-culture with directly haptenated spleen or B cells. They found that while the LNC did not induce PFC formation in cultures of normal spleen cells, they induced a significant number of both IgM and IgG PFC in hapten-modified, syngeneic spleen cell cultures. As with auto- and alloreactive T_h, these results suggested that Ig-mediated antigen recognition by responding B cells *was not essential* for growth and differentiation of that responding B cell. A companion paper[73] showed that the hapten-specific T_h activity described above was MHC-restricted. This suggested that T_h helper activity required T_h recognition of both specific antigen, in this case, hapten, and MHC on the target B cell. They also reported that non-haptenated resting B cells that were cultured with haptenated, irradiated, spleen cells and hapten-specific T_h, were not activated. That is, ''bystander'' B cells were not activated in this system. However, when LPS-treated nonirradiated B cell blasts (bystander) were co-cultured with irradiated haptenated B cells and hapten-specific T_h, bystander B cell blasts grew and differentiated. The mechanism of this effect was presumably via T_h-derived, soluble factors. This suggests that the initial specific interaction between T_h and B cells required physical contact and causes the induction of reactivity to soluble factors by the interacting B cell. Subsequently, studies in the human system[74,75] showed that a TNP-specific T_h clone activated B cell subsets by two distinct mechanisms. Large B cells were polyclonally activated nonspecifically and in a contact-independent and unrestricted manner (factors); small B cell activation required antigen (specific haptenation of the responding B cell) and was genetically-restricted. Using this system, insights into the requirement for the expression of a B cell-restricted cell surface activation antigen, BLAST-2, were afforded. Using TNP-coupled B cells and a TNP-specific T_h clone, Chartash et al.[75] found that direct T_h-B cell contact, rather than T_h-derived lymphokines, induced BLAST-2 (now known as CD23) antigen expression most efficiently. Finally, recent work has shown that the *reverse* system also effectively activates B cells. That is, anti-TNP B cells were induced to respond when TNP was directly coupled to the T_h surface.[76] The stimulation of B cells seen in this system was H-2 restricted, in a manner similar to conventional interactions. Restriction in this system suggested that the B cell activation occurred as a direct result of physical interactions between molecules present on the B and T_h cell surface, rather than from high local concentrations of T_h-derived lymphokines. These studies emphasized that mIg engagement was not required for T_h-dependent B cell signaling and supported the concept that resting B cells and activated B cells had distinct activation requirements.

Jones and Janeway[33] established that antigen-specific T cells could induce the polyclonal activation of B cells given high concentrations of the antigen to which the T_h were specific. In these studies, ovalbumin (OVA)-specific T_h (mitomycin-treated) were co-cultured with B cells. Vigorous B cell proliferation was observed in the presence of OVA. Similar findings were reported by Ratcliffe and Julius[77] and Defranco et al.[78] using *resting* B cells. It was proposed that at high antigen concentrations resting B cells nonspecifically processed and presented antigen to antigen-specific T cells. As a result, cognate interactions and B cell activation resulted.

3. Antibodies as Bridges

An alternative to directly focusing T_h on the B cell via recognition of directly coupled haptens or integral B cell membrane proteins was to use antibodies to bridge the two cells. Focusing T cell help via anti-B cell antibodies was exploited by Parker and co-workers[79-81] in a series of manuscripts. These investigators used rabbit gamma globulin (RGG)-specific T_h clones linked to B cells via an arsenated, anti-mouse IgD plus rabbit Fab′ anti-arsenate antibody. B cells cultured under these conditions proliferated, suggesting that H-2-restricted antigen-specific contact between the two lymphocytes was sufficient to activate the interacting resting B cell. This system allowed for antigen-specific T_h activation of a large population of B cells without requiring the isolation of antigen-specific B cells. Similar studies were performed using RGG-specific T_h clones and Rab anti-mouse Ig. In this system, submitogenic concentrations of monovalent or divalent rabbit anti-mouse Ig were sufficient to bridge T_h and B cells. This system retained the advantage described above without the added complication of the arsenate link. It has been shown that small B cells were quite efficient at presenting rabbit anti-IgM or IgD to RGG-specific T_h cells, establishing that mIgM and mIgD as functional mIg receptors for antigen processing. Subsequent work, using Fab′s of anti-Ig showed that crosslinking of membrane Ig was not essential for efficient presentation of RGG to T_h cells. Therefore, in accord with other T_h-B cell systems, signals transduced through mIg did not appear to be necessary for T cell help. MIg did appear to be specialized for antigen-presentation as rabbit anti-Ig was presented more efficiently than rabbit antibody directed against other B cell surface antigens. In addition, it appeared that cell-cell contact rather than T_h-derived soluble factors were responsible for B cell activation, as syngeneic B cells which were partitioned across a semi-permeable membrane from T_h and APC were not activated.

With the advent of transgenic mice, mice expressing single heavy chain and light chain specificities have been afforded. Recent studies from Julius et al.[82] used mice made transgenic for an anti-TNP antibody, so that effectively all B cells obtained from these mice were TNP-specific. In this system, histoincompatible $V_\beta 8$-bearing T_h clones stimulated TNP-transgenic B cells when cultured with TNP-conjugated, $V_\beta 8$-specific anti-T cell receptor (TCR) antibody (F23.1). B cells cultured with $V_\beta 8$-bearing T_h clones, but not $V_\beta 8$-negative T_h clones expressed a significant PFC response. Importantly, they also found that B cells cultured with a $V_\beta 8$-positive T_h clone and *unconjugated* divalent F23.1 responded with a vigorous PFC response. In contrast to all previous studies with antibody bridges, these data suggested that "bridging" between T_h and B cells was *not essential* for B cell activation if the antibody independently activated the T_h. That is to say that TCR cross-linking of the co-cultured T_h was sufficient to induce significant helper activity in the presence or absence of an antibody-bridge. Results of systems utilizing anti-TCR or anti-CD3 mAbs in the study of T_h-B cell interactions has elucidated the effector phase of T cell helper and will be extensively discussed below.

4. Anti-TCR/CD3 Antibodies and Cognate Interactions

As cited above, T_h which have been activated with anti-TCR complex antibodies activated resting B cells in the absence of an antigen-based linkage. Numerous reports exist documenting this phenomenon.[83-88] A common conclusion of these studies was that T_h which have been stimulated by anti-CD3 antibodies express the capacity to activate B cells in the absence of specific antigen, and in a contact-dependent manner. Further, it was also shown that this activation does not require cognate cell interactions, as histoincompatible activated T_h are still able to activate resting B cells. The results of these studies were contrary to the accepted dogma that resting B cells were activated exclusively as a result of "bridged", genetically restricted contact between T_h and B cells. To explain the lack of genetic restrictions, production of soluble factors by T_h were proposed to account for the induction of B cell cycle entry. Precedence for such factors had been established, in that activated T_h were shown to produce factors capable of inducing B cell cycle entry.[89,90] The fact that the anti-TCR effect was factor mediated was discounted since it was clearly shown that physical separation of the B cell from the activated T_h eliminated B cell activation.[83,86] Therefore, it appeared that once a T_h is activated, the mechanisms through which B cell activation are effected are not genetically-restricted, non-antigen-specific, yet contact-dependent.

The *bystander effect* is the antigen-induced equivalent of responses observed with T_h activation induced by anti-TCR mAbs. The activation of allogeneic B cells as a result of coculture with responding, syngeneic T_h and B cells has been referred to as "bystander" B cell activation.[91] At first, "bystander" B cell activation was largely recognized as the response of large (activated) B cells responding to the elaboration of soluble factors by T_h.[91] However, even when resting (bystander) B cells were exclusively used, B cell responses were still observed. Evidence shows that T_h, activated by syngeneic B cells and antigen, induce the proliferation of allogeneic, resting B cells via a non-class II restricted, antigen nonspecific, contact-dependent mechanism. The mechanism is likely identical to that employed by anti-TCR activated T_h. We have shown that anti-CD3 activated T_h form physical conjugates with resting B cells in a genetically unrestricted, antigen-nonspecific fashion.[92] We have proposed that this phenomenon represents the *effector phase* of T cell help. That is, T_h that are activated by antigen-presenting B cells bind to B cells in a class II-restricted and antigen-specific fashion (*recognition phase*). As a result of this interaction, T_h are activated and reciprocally activate B cells in a non-class II-restricted, antigen-nonspecific specific manner (*effector phase*). Anti-TCR/CD3 stimulation of T_h bypasses the recognition phase of T cell help and induce the expression of effector phase activities. A complete description of the distinct phases of cognate activation is discussed in later sections.

A number of the model *in vitro* systems described above have allowed the investigation of TD B cell responses at a polyclonal level. By developing T_h lines or clones that recognize B cell surface molecules, the role of mIg crosslinking has been elucidated. It is evident that mIg crosslinking is not necessary for TD B cell activation. It may be that mIg cross-linking enhances the ability of B cells to interact with T_h, as reported by Kearse,[56] or enhances the reception of T_h-derived signals. It has become apparent that the restrictions that govern responsiveness of B cells to TD antigens are those that govern the initial physical interaction of T_h with B cells. Therefore, physiological B cell activation, growth and differentiation requires associative recognition of antigen and class II MHC by the T_h. Studies by Parker and co-workers[79,80] using rabbit anti-mouse Ig and RGG-specific T_h clones emphasized the difference between mitogenic and non-mitogenic anti-Ig activities. Low concentrations of monovalent anti-Ig were not mitogenic, although they did function efficiently as antigen. These studies showed that cross-linking of Ig was not required for efficient antigen processing and presentation. Finally, results with anti-TCR complex-induced T_h-B cell interactions shed light on the effector phase of T cell help. That is, it appears that once a T_h becomes activated, it has the capacity to induce B cell cycle entry via a contact-dependent, unrestricted mechanism.

III. CELL SURFACE MOLECULES INVOLVED IN THE PHYSICAL INTERACTIONS OF T_h AND B CELLS

In general, T_h perform two functions in the induction of B cell growth and differentiation. First, T_h interact with B cells and likely deliver activation signals. Second, T_h elaborate soluble factors that sustain and promote the activation of activated B cells. A number of different approaches have been developed to address the function of specific T and B lymphocyte surface molecules in mediating the first component of cognate activation, namely, T_h-B cell conjugation formation and intercellular signaling. Monoclonal antibodies specific for T and/or B cell surface molecules have been used to inhibit the functional responses of B cells to help, that being proliferation or differentiation. More incisive evidence as to the function of specific cell surface molecules has been provided by direct measurement of the ability of mAbs to interfere with conjugate formation between T and B cells. Finally, mAbs have been used independent of T cells to ask whether engagement or crosslinking of these cell surface molecules results in B cell triggering. Prominent T and B cell membrane proteins that have been shown to be involved in cognate interactions are detailed below.

A. T CELL ANTIGEN RECEPTOR (TCR)/PROCESSED ANTIGEN-POLYMORPHIC CLASS II

It is well-established that T cell recognition of antigen in association with MHC is mediated by a single T cell receptor.[93,94] Structural studies on the TCR have established that the heterodimeric TCR is composed of a single acidic α chain (43 to 49 kDa) disulfide-linked to a single, more basic β chain (38 to 44 kDa). It is the recognition by the TCR of processed Ag/MHC on the B cell surface that is responsible for the initial class II-restricted, antigen-specific physical interactions between T and B cells. Antigen-specific T cells clones will not form physical conjugates with allogeneic B cells in the presence of the appropriate antigen. Therefore, those restrictions that limit the interactions with antigen-presenting B cells limit the humoral immune response to T-dependent antigens. Additional evidence showing the essential function of class II MHC molecules in the physical interactions between antigen-specific T cell clones and antigen-specific, resting B cells comes from inhibition studies with anti-class II mAbs. Sanders et al.[63,95] showed that anti-class II antibodies inhibit the formation of antigen-specific interactions between cloned T_h and antigen-specific B cells. These authors also showed that increased expression of class II MHC (by IL4) modestly increased the frequency of T_h-B cell interactions,[52,95] lending support to the notion that the level of class II MHC antigen expression might modestly influence T_h-B interactions.

B. CD4 MOLECULES/CLASS II MHC

CD4 is a membrane glycoprotein selectively expressed on the surface of helper T cells. *In vitro* CD4 plays an essential role in T_h activation and in the T_h-dependent triggering of B cells.[96-98] Because some studies have shown that anti-CD4 can *inhibit* T cell activities, it was difficult to determine if the primary effect of anti-CD4 was on inhibition of T_h function or on blocking T_h-B cell interactions. Studies by Sanders et al.[63] and Bartlett et al.[99] clearly showed that anti-CD4 blocked class II-restricted, antigen-specific conjugate formation between T_h and B cells. Early suggestive studies[96-98] indicated that CD4 recognized class II MHC molecules and played an important role in strengthening the interactions between APC and the T_h. Later studies from Doyle and Strominger[100] and Gay et al.[101] showed that the interaction of CD4 and the nonpolymorphic domains of class II MHC molecules mediated intercellular adhesion. Emerging from these studies was the concept that the additive adhesiveness of the TCR and CD4 for their respective ligands contributed to the sum total adhesiveness of the T_h for the APC. Recent studies by Kupfer and Singer[102] and Rojo et al.[103] suggest that occupancy of TCR promotes TCR-CD4 interactions and that the CD4-TCR complex acts as a single receptor unit for signal transduction and adhesion.

C. LFA-1/ICAM-1

LFA-1 is a monomorphic integral membrane protein that is a member of the integrin family of molecules. As such LFA-1 is composed of two transmembrane polypeptides chains, α and β, linked to the cytoskeleton on the cytoplasmic face of the plasma membrane.[104] Evidence has been presented that the ligand for LFA-1 is the intercellular adhesion molecule 1 (ICAM-1) another integral membrane protein.[105,106] In such a way, specific interactions between LFA-1 and ICAM-1 mediate cell adhesion. In general, anti-LFA-1 mAbs have been shown to inhibit cytotoxic-target cell interactions[107,109] and T_h-B cell interactions.[63] Numerous studies have reported that anti-LFA-1 inhibits thymus-dependent B cell activation.[109-112] Recent studies by Sanders et al.[63] showed that anti-LFA-1 inhibited conjugate formation between T_h and B cells. In the human system, it has been reported that anti-LFA-1 and anti-ICAM-1 inhibits TD B cell responses.[113] The functional role of LFA-1 has also been assessed using lymphocytes from patients with leukocyte adhesion deficiency, an inherited immunodeficiency characterized by a defective expression of LFA-1, MAC-1 and p150, 95.[114] It was shown that the antigen-specific helper function of LFA-1$^-$ T cells for histocompatible LFA-1$^+$ B cells was reduced. The authors further showed that antigen-nonspecific adherence of LFA-1$^-$ T cells to LFA-1$^+$ B cells was reduced but LFA-1$^-$ B cells bound normally to LFA-1$^+$ T cells. This suggested that the expression of LFA-1 on the T cell was more important than that on the B cell for interactions. Because the level of ICAM-1 increases upon B cell activation, one must consider the fact that adhesiveness between T_h and B cells may increase as B cell activation commences. Recent studies[115] indicate that the LFA-1 molecule goes through a high affinity transition upon T cell activation. This was shown by assessing the homotypic binding capacity of resting vs. activated T cells. T cells activated with anti-CD3 exhibited enhanced homotypic binding. At present it is not clear what biochemical changes in integrin molecules account for their shift to high affinity binding or how increased integrin density on the cell surface may contribute to intercellular adhesion or signaling. Although it is evident that LFA-1/ICAM-1 interactions are likely important in T_h-B cell adhesion, it is not clear if this interaction plays a role in intercellular signaling. Recent studies by Kupfer and Singer[116] indicate that LFA-1 and talin, an intracellular cytoskeletal protein, were concentrated at the site of T_h-B interaction. This indicates that there are immediate intracellular changes associated with T_h-B cell interaction.

D. LFA-3/CD2

Little information is currently available as to the role of LFA-3/CD2 interactions in T_h-B cell interactions. In one study,[117] it was reported that autologous RBC (which bind CD2) cultured with unfractionated human mononuclear cell cultures, enhanced B cell responses to antigens and mitogens. Under these conditions, anti-LFA-3 or anti-CD2 monoclonal antibody treatment abrogated this enhanced response, indicating that interaction between LFA-3 and CD2 may play some role in the physical interaction between T_h and B cells.

IV. THE RESPECTIVE CONTRIBUTION OF CELL-CELL CONTACT AND LYMPHOKINES IN THE ACTIVATION OF B CELLS TO GROWTH AND DIFFERENTIATION TO BY T_h AND ANTIGEN

A. INTRODUCTION

The immune system uses a basic program of specific activation followed by nonspecific amplification. Bretscher[118] proposed that lymphocyte growth and differentiation was regulated by two signals; an antigen-specific signal (signal 1) and a nonspecific signal (signal 2). Contemporary investigations into T lymphocyte activation have most clearly provided support for this hypothesis. That is, the recognition of class II-associated antigen by the T cell antigen receptor triggers the T lymphocyte into the early phases of the cell cycle (signal 1). A second "nonspecific" signal, interleukin-2 (IL2) supports continued growth of T lymphocytes.[119] Once activated, B cells similarly become responsive to the growth and differentiative effects of lymphokines;[64,65,120,121] however, there is a paucity of information as to the physiological signals (signal 1) that initiate T cell-dependent entry of B lymphocytes into the cell cycle and result in the acquisition of lymphokine responsiveness by B cells. Although thymus-dependent (TD) antigens bind to and cap mIgM and mIgD,[122] TD antigens *do not* induce antigen-specific B cells to enter the cell cycle.[60] Crosslinking of mIg by TD-antigen (in any physical form) in the presence or absence of growth factors did not induce the growth of antigen-specific B lymphocytes. Therefore, crosslinking of mIg by TD antigens is not signal 1 as defined by cell cycle entry or by acquisition of lymphokine responsiveness. The differences between anti-Ig and TD antigen in eliciting B cell growth have not been resolved. However, it may be suggested that anti-Ig is a polyclonal analog of a thymus-independent type II antigen.

We would propose that *T_h-B interaction* is signal 1 and is essential for efficient B cell cycle entry and the acquisition of B cell lymphokine responsiveness. The mechanism of this activation is still debated. Because T_h both physically interact with B cells and also produce lymphokines, it has been difficult to decipher which of these functions is critical for providing signal 1. Possible T_h effector mechanisms that could account for inducing B cell cycle entry and B cell lymphokine responsiveness (signal 1) can be summarized as follows:

1. One possible model for B cell activation is that the cell-cell contact of T_h and B cells per se does not activate the B cell, but the local and/or directional release of identified lymphokines (IL2, IL4, IFN-γ, IL5, etc.) results in B cell activation. Although directional release of lymphokines has been reported,[124] none of the molecularly cloned lymphokines causes resting B cells to transit from Go to S, in the absence of an activation signal (i.e., anti-Ig or T_h). Therefore, the release theory of local IL2, IL4, IL5, IL6, or IFN-γ cannot fully account for the efficient entry of B cells into cycle this is observed during cognate interactions with T_h. One recent observation that is an exception, is the recent observation that B cells stimulated with high levels of IL4 and IL5 subsequently become responsive to IL2.[125]
2. A second model suggests that T_h release short-range acting, labile factors (yet to be identified) or soluble antigen-specific T cell factors drive B cells into cycle. Although cloned lymphokines cannot account for B cell cycle progression, this model suggests that the local or directional release of unidentified factors may drive B cells into cycle. With noted exceptions,[89,90] the isolation and biochemical characterization of nonspecific T cell-derived factor(s) that can cause G_0 to G_1 transition of B cells has not been generally successful. Further, the argument that these factors are extremely labile is inconsistent with the biological stability of all other cloned lymphokines.

   Studies by Feldmann,[126] Taussig,[127] and more recently by Hodes and co-workers[128,129] have reported the activity of antigen-specific helper factors. Feldmann and colleagues were the first to develop an *in vitro* system to test for the activity of antigen-specific, genetically restricted factors. Antigen-specific T cell hybridomas were established that produced antigen-specific helper factors. Although numerous studies have detailed the antigen specificity and genetic restrictions of such factors, none have resolved the biochemical nature of the factors. Recent studies from Hodes and co-workers[128,129] have reported the production of antigen-specific helper factors from T_h clones. These investigators showed that found SN from T_h clones which had been stimulated with either antigen-pulsed APC or anti-CD3 antibody were able to induce specific antibody production by B cells in the presence of antigen. These SN were found to have the same specificities as the T_h clones they were derived from; that is, they were MHC-class II restricted, and carrier-specific. These T_h required the presence of specific antigen for induction of helper factor (HF) release into the SN, and the HF required the presence of specific antigen to activate the B cells.

3. A third model states that the interaction of T_h cell surface molecules and B cell surface molecules activates the B cells to a stage whereby B cells become responsive to the growth and differentiative effects of lymphokines. The salient feature of this model is that the physical interaction of T_h and B cells is itself an activation signal that is independent of the release of soluble factors by the T_h. It has been shown that physical contact of T_h and B cells triggers changes in lymphocyte physiology as measured by the dynamic changes in the cytoarchitecture of interacting cells.[102,116,130-132] Rapid alterations in the distribution of cytoskeletal proteins and membrane proteins have been reported when T_h and B cells interact. It has been shown that alterations in the orientation of the microtubule-organizing center (MTOC) and talin occur within the T_h upon interaction with class II-compatible, antigen-presenting B cells. Co-clustering of the CD4 and T_h antigen receptor has also been observed. These studies suggest that membrane protein-cytoskeletal interactions that result from cell-cell contact may play a fundamental role in the communication of T_h and B cells. Data from our lab that show the physical interaction of T_h and B cells induces B cell cycle entry. The acquisition of B cell lymphokine responsiveness is presented in the next section.

B. STUDIES INVOLVED IN DIRECTLY ASSESSING THE CONTRIBUTION OF CELL CONTACT TO B CELL ACTIVATION

With the multitude of activities of lymphokines on resting B cells, it was tempting to speculate that lymphokines may be responsible for the initial activation of B cells by T_h. Studies indicated that T_h can release lymphokines directionally and rapidly.[124] To more critically assess the respective contribution of lymphokines and cell contact in B cell activation we cloned T_h that were incapable of lymphokine production under certain conditions and asked if they could activate B cells.

1. A T_h Clone Incapable of Lymphokine Production Can Activate B Cells to Respond to Lymphokines

Th-3.1 is an ovalbumin-specific T_h clone that required antigen and IL2 to induce B cell growth.[59,99] IL2 was required for the synthesis of other lymphokines (IL4, IFN-γ, etc.) by Th-3.1. Therefore, in the absence of IL2, no lymphokine production was measurable and no B cell growth or differentiation was induced. Th-3.1 is different from all other T_h described in the literature in that IL2 was essential for the expression of helper activity and the production of other lymphokines. The addition of a lymphokine-rich conditioned medium, EL4 SN (in which the level of IL2 was substimulatory for lymphokine production by Th-3.1), replaced the need for synthesis of lymphokines by Th 3.1. In the presence of EL4 SN, antigen, Th-3.1, and B cells, class II-restricted, antigen-specific T cell help resulted in B cell growth and differentiation. Therefore, it appeared that Th-3.1 activated B cells in a class II-restricted and antigen-specific fashion that resulted in B cell responsiveness to exogenous lymphokines. Depletion of IL4 from the EL4 SN (EL4 SN-IL4) ablated the Th-dependent B cell response, yet purified IL4 alone was completely ineffective in stimulating B cell responses. This established that IL4 was essential for B cell responses, but not sufficient, for B cell proliferative responses. Pre-culture of B cells with IL4 supported the need for IL4 in culture, showing that the role of IL4 was exclusively at the level of the B cell. These data suggested that T cell help could be dissected into two components, one that activated B cells in a class II-restricted and antigen-specific manner and another that could be replaced by lymphokines. We proposed that the former component was dependent upon T_h-B cell contact.

2. T_h-B Cell Contact Induces B Cell Cycle Entry

Based on these results it was predicted that in the absence of IL2, Th-3.1 performed only contact-dependent functions. Lymphokine production (IFN-γ, IL4 and IL5) by Th-3.1 was shown to be dependent on antigen and IL2. Therefore, it appeared that the requirement for IL2 to support B cell proliferation was indirect in that it reflected the IL2 requirement for lymphokine production by Th-3.1. With the observation that contact vs. soluble T_h effector functions could be dissected by the availability of IL2, the contact-dependent functions performed by Th-3.1 were investigated.[99] In contrast to the absolute requirement of IL2 for Th-3.1-induced B cell *proliferation,* IL2 was not required for the formation of stable Th-3.1-B cell conjugates. Moreover, the B cells conjugated with Th-3.1 were induced to enter the G1 phase of the cell cycle. In the absence of IL2 (a condition that blocked lymphokine production by Th-3.1) and in the presence of antigen, approximately 12 to 13% of B cells formed stable physical conjugates with Th-3.1. Under conditions that promoted T_h-B cell interactions, Th-3.1 induced 10 to 20% of resting B cells to enter G_1. The inclusion of anti-IL2 receptor (R) mAb or anti-IL4, or exogenous IL2, did not influence the frequency of B cells entering cycle. These data suggested that production of lymphokines by Th-3.1 was not required for T_h-induced entry of B cells into G_1, but was required for progression of B cells into S phase.

The data suggested that the lymphokine-independent induction of B cells into G_1 by Th-3.1 was a cell contact-dependent event. Direct proof that Th-3.1-B cell contact was necessary for B cell cycle entry was provided by comparative *in situ* analysis of the RNA synthetic activity of B cells in physical contact with Th-3.1 or not in contact with Th-3.1. For this analysis, Th-3.1 and B cells were prestained with spectrally distinct lipophilic dyes, DiO and DiI. These dyes have been shown not to influence cell function (unpublished data). Prestained Th-3.1 and B cells were cultured in the presence of antigen and pulsed with ^{3}H-uridine. At the end of pulse period, T_h-B cell conjugates were gently transferred to glass slides and developed for autoradiography. The numbers of grains over those B cells interacting with Th-3.1 vs. the number of grains over those B cells not interacting with Th-3.1 were quantitated. Expression of high level RNA synthesis (a marker of cell cycle entry) was far more frequent in B cells in contact with Th-3.1 than in B cells not in contact with Th-3.1. These data indicated that contact rather than broadly diffusable factors induced the entry of B cells into G_1.

3. Fixed, Activated Antigen-Specific T_h Induce B Cell Cycle Entry in a Non-Class II-Restricted, Antigen-Nonspecific Manner

The previous studies suggested that contact of T_h and B cells resulted in the entry of B cells into G_1. These studies could not conclusively rule out the alternative explanation that the production of low levels of soluble factors by Th-3.1 was responsible for inducing B cell cycle entry. One way to conclusively dismiss the contribution of soluble factors was to metabolically inactivate Th-3.1 by fixing them with paraformaldehyde and demonstrate that fixed Th-3.1 could still drive B cells into the cell cycle.[92] Repeated attempts using fixed resting Th-3.1 to induce B cell cycle entry failed. This suggested that (1) soluble factors were involved in B cell cycle entry or (2) that the metabolic inhibition resulting from fixation prevented the induction of the effector phase of T cell help (defined as that phase that induces B cell to enter G_1). That is, we reasoned that although resting Th-3.1 formed conjugates with B cells, this interaction might not signal B cell cycle entry. Our hypothesis was that Th-3.1 had to be activated prior to fixation in order to reciprocally activate the B cell. To test this hypothesis, activated Th-3.1 were generated using insolubilized anti-T3. Prior to assessing the biological activities of fixed, anti-T3 activated Th-3.1 (Th-3.1^{T3}), we examined whether any of the biological properties or restrictions of viable Th-3.1^{T3} were altered as a consequence of anti-T3 activation. First, the restrictions for T_h-B interactions were investigated. Viable Th-3.1^{T3} formed stable conjugates with B cells in a class II-*unrestricted* and *antigen nonspecific* fashion. In contrast to Th-3.1,^{T3} the interaction of resting Th-3.1 and B cells was *restricted by antigen and class II MHC*. Therefore, after anti-T3 activation, the rules governing conjugate formation with B cells were changed. These nonrestricted interactions were productive in the sense that Th-3.1^{T3} induced both syngeneic and allogeneic B cells to DNA synthesis in an antigen-nonspecific manner.

To gain insights into the cell surface molecules that were responsible for stable conjugate formation between Th-3.1^{T3} and B cells, mAbs were used to block interactions. Unlike the results with resting Th-3.1 and antigen, inhibition of Th-3.1^{T3}-B cell interaction was inhibited to background levels only in the presence of both anti-IA and anti-IE mAbs or with anti-CD4.[133] This indicated that the interaction of Th-3.1^{T3} with B cells was via recognition of monomorphic domains of class II MHC molecules, since it has been reported that CD4 recognizes monomorphic domains of class II MHC molecules.[100,101]

We asked if Th-3.1^{T3} could stimulate B cells cycle progression by a cell contact-dependent, lymphokine-independent mechanism. Th-3.1^{T3} were fixed (FX) with paraformaldehyde and tested. When FX Th-3.1^{T3} were titered into culture with resting B cells, FX Th-3.1^{T3}, but not FX resting Th-3.1 induced enhanced B cell RNA synthesis (B cell entry into G_1).[92] Proof that FX Th-3.1 were metabolically dead was provided by the inability of FX Th-3.1 to synthesize RNA and protein and IL4. Trypsin-treatment of Th-3.1^{T3} prior to fixation ablated their ability to activate B cells, indicating that cell surface proteins on Th-3.1^{T3} were required for inducing enhanced B cell RNA synthesis. IL2, IL4 and IFN-γ were not involved in mediating the entry of B cells into cycle since anti-IL4, anti-IL2R or anti-IFN-γ were without effect on increased B cell RNA synthesis induced by FX Th-3.1^{T3}.

Although FX Th-3.1^{T3} induced B cells to enter G_1, these fixed cells did not induce the B cells to continue into S phase. FX Th-3.1^{T3} did not induce substantial levels of B cell DNA synthesis unless exogenous lymphokines were added. When stimulated with FX Th-3.1^{T3}, B cell DNA synthesis was enhanced four- to fivefold by EL4 SN. Depletion of IL4, but not IL2, from EL4 SN ablated its supportive capabilities. However, IL4, alone, was completely ineffective in supporting entry into S phase. Therefore, both IL4 and another activity(ies) in EL4 SN were necessary for B cell progression into S phase. Taken together, these data showed that after the T_h activation of B cells, B cells required IL4 and another lymphokine, ILX, to progress into S phase (unpublished data).

4. T_h Activation is Required to Induce B Cell Cycle Entry

We reasoned that if the effector phase of T cell help is class II-unrestricted and antigen-nonspecific then the polyclonal activation of B cells by activated T_h *in vivo* poses a significant dilemma. If T_h remain activated for extended periods of time, polyclonal B cell activation and the production of antibody with irrelevant specificities may result. However, if the duration of the effector phase induced as a result of activation was transient, only that B cell with which the T_h initially interacted via antigen recognition would be activated. The kinetics of anti-T3 induced effector phase activity were determined.[133] Th-3.1 was activated with anti-T3 for various time periods, FX and tested for effector activity by co-culture with B cells. Maximal effector function required 4 to 8 h of stimulation with anti-T3. The effector activity waned by 72 h of activation with anti-T3. The transient expression of effector cell activity may be important to limit the extent of polyclonal B cell activation under physiological conditions. That is, only the B cell with which the T_h initially interacted may be activated by effector phase activities due to the transient nature of the effector function.

To establish if synthesis of RNA and hence *de novo* expression of a T_h protein were required for effector phase activity, actinomycin D (0.1 μg/ml) was added to Th-3.1 activated with anti-T3. The AD-treated Th-3.1^{T3} did not form conjugates with B cells (unpublished data) or express effector phase activity.[133] The appearance of T_h effector phase activity was not explained by an increase in expression of LFA-1, CD4, Thy-1, or ICAM-1 since following anti-T3 activation the expression of these molecules was unaltered. Because *de novo* RNA synthesis (and presumably protein synthesis) was essential for effector function and we know that the effector function was a cell contact-dependent event, it is likely that the *de novo* expression of some T_h cell surface antigen is critical. On the other hand, it is also clear that CD4 clusters in response to T_h-B interaction, is involved in T_h-B cell interaction and perhaps B cell triggering. Therefore, it is reasonable to assume that molecules normally resident on the T_h surface as well as those induced by anti-T3 contribute to effector phase function.

Signaling via direct T_h-B cell contact has also been indicated by a number of other studies. Krusemeier and Snow[134] showed that cyclosporin-treated T_h, which were incapable of lymphokine production, activated B cells to a state of lymphokine responsiveness. In addition, fixed T_h[132] or purified plasma membranes[135] from mutant thymoma cells have been shown to activate B cells. These accounts together with the data from our lab establish intercellular contact between T_h and B cells as a viable mechanism for B cell triggering.

C. THE ROLE OF LYMPHOKINES AND CYTOKINES IN TD B CELL GROWTH AND DIFFERENTIATION

The role of antigen-nonspecific helper factors in the antibody response to sheep erythrocytes was vigorously pursued by a number of laboratories. Early studies indicated that a single 40 kDa T cell replacing factor (TRF, now known as IL5) was sufficient to replace the need for T cells in the antibody response to sheep erythrocytes.[136-139] Early studies by Marrack et al.[139] showed that IL1, IL2, and a T cell replacing factor (TRF) were required for supporting anti-erythrocyte responses and the concept that multiple lymphokines were required to achieve maximal B cell growth and differentiation was born. With the advent of hybridoma technology and recombinant DNA technology, as well as the development of assays to measure the activity of single immunological active molecules, lymphokines and monokines were molecularly cloned and assayed for function. Although many of these lymphokines can exert indirect effects on antibody production by effecting T_h or macrophage activities, we will deal exclusively the direct effects of interleukins on mature B cells. A recent comprehensive review[140] deals with the effect of these lymphokines on T_h, macrophages and B cells.

IL1—IL1 exists in two biochemically distinct, but related forms, IL1α and IL1β.[141] It appears that both forms of IL1 bind to the same receptor and elicit similar functional activities. Alone, IL1 has not been shown to exert any activities on mature B cells. In combination with other lymphokines, IL1 has been reported to enhance B cell growth and/or differentiation.[142,143] Using resting, affinity-purified, antigen-specific B cells Stein et al.[142] showed that IL1 approximately doubled the proliferative response to TNP-Ficoll and IL4. This is in accord with previous studies from Howard et al. indicating that IL1 potentiated the response to anti-Ig.[143] More recent work has reevaluated the role of IL1 and TRF in the development of anti-SRBC responses.[144] IL1 induced a significant increase in the anti-SRBC responses from normal B cells but had little influence on the response of antigen-primed B cells (the presumed target of TRF or IL5). IL1 did enhance the response of antigen-primed B cells to IL5, by acting early and IL5 acting late. These results suggest that IL1 acts primarily on resting B cells as a differentiation inducing factor in the presence of antigen and also acts as in concert with IL5.

IL2—Studies from the laboratories of Marrack and Kappler[139] and Swain and Dutton[145] were among the first to show a role of IL2 in supporting *in vitro* antibody responses to sheep erythrocytes. With the advent of biochemical methods to resolve the independent chains of the IL2 R, regulation of expression of p55 and p75 have been afforded. Recent evidence indicates that the activation of human B cells with SAC results in the heightened expression of p75, together or independent of the expression of p55. Those B cells expressing both p55 and p75 were induced to differentiation in the presence of low concentrations of IL2 (1 to 100 U/ml), suggesting that as B cells are induced to enter the cell cycle, the high affinity IL2 R is expressed.[146,147] The fact that activation of B cells results in their responsiveness to IL2 is consistent to reports by Zubler et al. showing that treatment of murine B cells with anti-Ig and LPS results in the expression of p55 and responsiveness to IL2.[148] Apparently, cell cycle entry is not essential for the heightened expression of IL2 R. In the murine system, it has been reported that the expression of p55 and p75 on resting B cells is regulated by IL5 and IL4, respectively.[132] That is, the culture of resting B cells with *high* concentrations of IL4 (1000 U/ml) induced the expression of the p75 and with IL5 the p55 components of the IL2 R. B cells treated with both IL4 and IL5 expressed high affinity IL2 R and were growth responsive to low concentrations of IL2.[149] As with the expression of class II MHC molecules, the expression of IL2 R appears to be regulated by the cell cycle status of the B cells, as well as by lymphokines. In that the expression of IL2 R appears to be an important element in the continued growth of activated B cells, it is important to determine if T_h-B cell contact results in increased IL2 R expression and the acquisition of IL2 responsiveness.

IL4—IL4 is a polypeptide of 21,000 MW.[150] It was originally described as a B cell growth factor (BCGF) because it promoted the transit through S of B cells triggered with anti-Ig.[151] Resting B cells express receptors for IL4[152] and as a consequence are responsive to IL4. IL4 induces the heightened expression of class II MHC molecules,[51,153] $Fc_\epsilon RII$[154,155] and the p75 chain of the IL2 R[149] on resting B cells. IL4 also exerts dramatic influence on the isotype of antibody produced by activated B cells, inducing the selective increase in IgG_1 and IgE production.[156-158] Concerning the role of IL4 in cognate activation and B cell cycle progression a number of points are worthy of mention. First, although IL4 induces a dramatic increase in class II MHC expression, it does not appear that a major effect of IL4 is to increase the frequency of T_h-B cell interactions. Studies by Sanders et al.[52] addressed this issue and showed that IL4 induced a 2- to 2.5-fold increase in frequency of T_h-B cell couples. In our own studies, we have not observed significant increases in antigen-specific T_h-B cell interactions with IL4-treated B cells. Second, it appears that under certain circumstances of T_h-B cell interaction, IL4 is essential, but not sufficient for B cell cycle progression. Reports from Dubois,[159] and others[59,160] have documented that IL4 is essential for B cell cycle progression using Th_2 or Th_2-like clones. The essentiality of IL4 for B cell growth has been examined indirectly through the use of T_h1. Numerous reports have shown that T_h1 can provide helper activity to B cells,[161-163] whereas others have shown little helper activity provided by T_h1.[160] In contrast, work by Owens et al.[86,87] have shown that T_h1 can induce a high frequency of B cells to grow and differentiate in the apparent absence of IL4. Developing a model for the mechanism of action of IL4 is difficult since receptors for IL4 are expressed on resting B cells and are modestly increased in number of affinity by B cell activation.[164] We have shown that antigen-pulsed B cells precultured with IL4 will enter S phase when they encounter antigen-specific T_h.[59] It seems that IL4 may change the reception of T_h-derived signals to the B cells. In other words, IL4 treatment predisposes the B cell to transit from G_1 to S when stimulated by a T_h. This function of IL4 is consistent with reports of its role in anti-Ig stimulation of B cells.

IL5—The first report of replacing T cell function with soluble factors was observed in the antibody response to heterologous erythrocytes.[138] The factors responsible for supporting anti-erythrocyte responses was referred to as T cell replacing factors (TRF). Two groups of investigators, Takatsu et al.[137] and Swain[145,165] reported the existence of similar factors that induced B cell differentiation, B cell growth to dextran sulfate and promoted the growth of BCL_1. This factor was discrete from other lymphokines. This activity was molecularly cloned.[166,167] and is referred to as IL5. Similar to IL4, IL5 can exert effects on resting B cells. As cited previously, it appears that IL5 induces the expression of the p55 subunit of the IL2 R on resting B cells and may be essential for B cells to acquire responsiveness to IL2.[149] Similar to IL4, IL5 has diverse activities on B cells in growth promotion,[168,169] inducing increased IgM production[170,171] and causing isotype switching[173,174] to IgA. IL5 has been shown to induce B cell proliferation of B cells stimulated with dextran sulfate, but not anti-Ig.[169] IL5 has also been shown to augment the IgM response of B cells from primed mice.[172]

IFN-γ—IFN-γ can synergize with other cytokines (IL1 and IL2) in the humoral immune response to heterologous erythrocytes and enhance Ig secretion by resting and activated B cells.[175,176] Although IFN-γ acts to promote B cell differentiation, it has been shown that IFN-γ has potent inhibitory effects on B cell activation, proliferation, and differentiation.[177-182] For example, IFN-γ antagonizes the actions of IL4 on B cells by inhibiting class II antigen induction,[181,182] CD23 induction[154] and immunoglobulin isotype switching to IgG_1 and IgE.[162,183] IFN-γ does, however, enhance the secretion of IgG_{2a} in LPS or antigen mediated T_h-stimulated B cells.[186] These data suggest that while IFN-γ is capable of acting as a B cell differentiation factor, particularly inducing IgG_{2a}, it also acts to inhibit the early IL4-mediated effects on B cells.

D. SUBPOPULATIONS OF T_h AND THE PRODUCTION OF LYMPHOKINES

Recently, Mossman and co-workers[187] have resolved that T_h can be functionally divided into at least two subpopulations of T_h depending on the array of lymphokines that are produced. A number of recent reviews,[140,187,189] as well as a chapter in this book, describes these subpopulations of T cell in detail and therefore we will only deal with them in a cursory fashion. By determining the pattern of lymphokine production by cloned T_h, it was shown that one type of T_h (T_h1) produce IL2, IFN-γ and IL3, and another type of T_h(T_h2) produce IL4, IL5, and IL3. Continued studies by Mossman and co-workers[187] and Bottomly and colleagues[188] have distinguished these two T_h into functionally distinct subpopulations of T_h. It has been proposed that T_h1 are inflammatory helper T cells and T_h2 are helper T cells for antibody responses. As stated previously, it is clear from a number of studies that *both* T_h1 and T_h2 can help antibody responses. Studies by Vitetta et al.[140] have shown that both T_h1 and T_h2 can form physical conjugates with B cells. Given the fact that T_h1 have been shown to induce B cell differentiation,[86] it appears that the interactions of either T_h1 or T_h2 with B cells is productive. Our own studies (unpublished data) indicate that this is the case in that anti-T3 activated T_h1 and T_h2 form physical conjugates with B cells and trigger B cell cycle entry.

E. A MODEL FOR T_h-DEPENDENT B CELL ACTIVATION (FIGURE 5)

A vital observation to our understanding of T cell help was that help is partitioned into two phases: the recognition and effector phases. Resting T_h recognize processed antigen on the surface of B cells and form physical stable conjugates. The formation of T_h-B cell conjugates is dependent upon the expression on the B cell of the appropriate class II restricting element together with the nominal processed antigen to which the T_h is specific. The stability of class II-restricted, antigen-specific conjugates is highly dependent upon the binding of ICAM-1 to LFA-1, CD4 to monomorphic domains of class II MHC molecules and TCR recognition of the class II MHC/Ag complex. We propose that this interaction primarily results in T_h activation and does not result in the reciprocal activation of the B cell. The observation that FX resting T_h bind to antigen-presenting B cells, but do not activate them indicated that this phase of conjugate formation does not signal the B cell to enter the cell cycle. As a result of class II-restricted interactions, clustering of the TCR and CD4, but not Ia, is observed. Within 4 to 8 h of activation, T_h express novel proteins on their cell surface and/or increase the expression of existing membrane proteins. As a consequence of the changes in the expression of T_h surface proteins, the restrictions that originally led to the

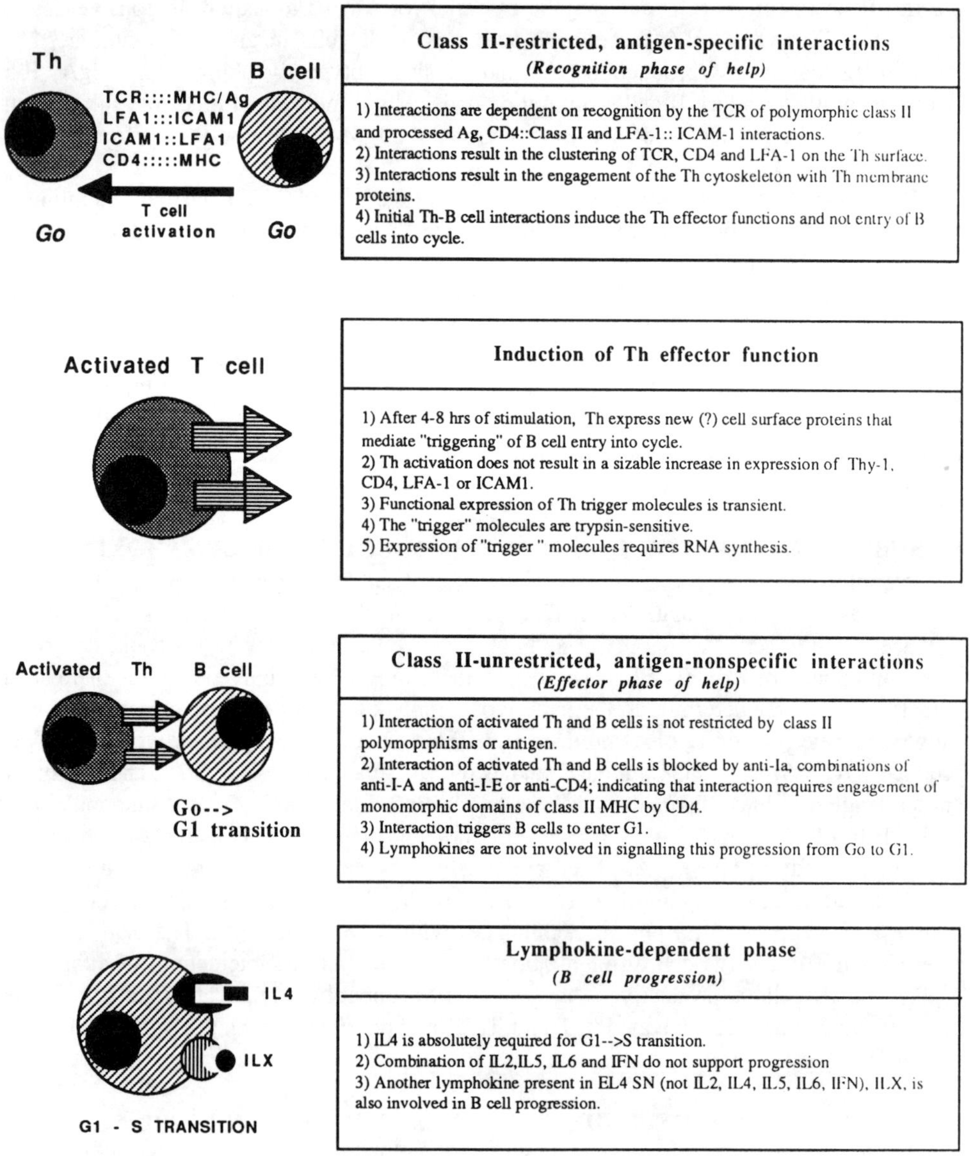

FIGURE 5. A model for cognate help.

formation of antigen-specific conjugates change. No longer does TCR recognition of antigen and polymorphic class II domains play a dominant role in stability of the T_h-B cell conjugate. As T_h activation proceeds, CD4 recognition of monomorphic domains of class II MHC becomes predominant. *De novo* expression of T_h surface molecules and altered topographical distribution of resident T_h surface proteins probably both contribute to effector function. As a result of the altered expression of T_h membrane proteins, T_h effector function is expressed. T_h effector function induces enhanced RNA synthesis in B cells that are in physical contact with effector phase T_h. Coincident with this entry into the cell cycle, B cells also become competent to respond to lymphokines that support their progression into S phase. Coordinate with the expression of the effector phase of T cell help, T_h begin to produce lymphokines that support the progression of B cells from G_1 to S. The lymphokines involved in B cell cycle progression are IL4 and another unidentified lymphokine, ILX.

REFERENCES

1. **Claman, H. N. and Chaperon, E. A.,** Immunologic complementation between thymus and marrow cells—a model for the two-cell model of immunocompetence, *Transplant. Rev.,* 1, 92, 1969.
2. **Claman, H. N. and Mosier, D. E.,** Cell-cell interactions in antibody production, *Prog. Allergy,* 16, 40, 1972.
3. **Cudkowicz, G. and Lotzova, E.,** Hemopoietic cell-derived components of the major histocompatibility complex of mice: identification of responsive and unresponsive recipients for bone marrow transplants, *Transplant. Proc.,* 5, 1399, 1973.
4. **Katz, D. H. and Benacerraf, B.,** The regulatory influence of activated T cells on B cell responses to antigen, *Adv. Immunol.,* 15, 1, 1972.
5. **Miller, J. F. A. P. and Mitchell, G. F.,** Thymus and antigen-reactive cells, *Transplant. Rev.,* 1, 3, 1969.
6. **Claman, H. N., Chaperon, E. A., and Triplett, R. F.,** Thymus-marrow cell combinations, synergism in antibody production, *Proc. Soc. Exp. Biol. Med.,* 122, 1167, 1966.
7. **Davies, A. J. S., Leuchers, E., Wallis, V., Marchant, R., and Elliott, E. V.,** The failure of thymus derived cells to produce antibody, *Transplantation,* 5, 222, 1967.
8. **Miller, J. F. A. P. and Mitchell, G. F.,** The thymus and the precursors of antigen reactive cells, *Nature (London),* 216, 659, 1967.
9. **Mitchison, N. A.,** Cell populations involved in immune responses, in *Immunological Tolerance,* Landy, M. and Braun, W., Eds., Academic Press, New York, 1969, 149.
10. **Raff, M. C.,** Role of thymus derived lymphocytes in the secondary humoral immune response in mice, *Nature (London),* 226, 1257, 1970.
11. **Rajewsky, K., Schirrmacher, V., Nase, S., and Jerne, N. K.,** The requirement of more than one antigenic determinant for immunogenicity, *J. Exp. Med.,* 129, 1131, 1969.
12. **Katz, D. H., Paul, W. E., Goidl, E. A., and Benacerraf, B.,** Carrier function in anti-hapten immune responses. I. Enhancement of primary and secondary anti-hapten antibody responses by carrier preimmunization, *J. Exp. Med.,* 132, 261, 1970.
13. **Paul, W. E., Katz, D. H., Goidl, E. A., and Benacerraf, B.,** Carrier function in anti-hapten immune responses. II. Specific properties of carrier cells capable of enhancing anti-hapten antibody responses, *J. Exp. Med.,* 132, 283, 1970.
14. **Kettman, J. and Dutton, R. W.,** An in vitro primary immune response to 2,4,6-trinitrophenyl substituted erythrocytes: response against carrier and hapten, *J. Immunol.,* 104, 1558, 1970.
15. **Bullock, W. W. and Rittenberg, M. B.,** In vitro-initiated secondary anti-hapten response. II. Increasing cell avidity for antigen, *J. Exp. Med.,* 132, 926, 1970.
16. **Bullock, W. W. and Rittenberg, M. B.,** Kinetics of in vitro initiated secondary anti-hapten response: induction of plaque-forming cells by soluble and particulate antigen, *Immunochemistry,* 7, 310, 1970.
17. **Cheers, C., Breitner, J. C. S., Little, M., and Miller, J. F. A. P.,** Cooperation between carrier-reactive and hapten-sensitive cells in vitro, *Nature (London) New Biol.,* 232, 248, 1971.
18. **Feldmann, M.,** Cell interactions in the immune response in vitro. V. Specific collaboration via complexes of antigen and thymus-derived cell immunoglobulin, *J. Exp. Med.,* 136, 737, 1972.
19. **Feldmann, M. and Basten, A.,** Specific collaboration between T and B lymphocytes across a cell impermeable membrane *in vitro, Nature (London) New Biol.,* 237, 13, 1972.
20. **Feldmann, M. and Basten, A.,** Cell interactions in the immune response in vitro. IV. Comparison of the effects of antigen-specific and allogeneic thymus-derived cell factors, *J. Exp. Med.,* 136, 722, 1972.
21. **Feldmann, M. and Basten, A.,** Cell interactions in the immune response in vitro. III. Specific collaboration across a cell impermeable membrane, *J. Exp. Med.,* 136, 49, 1972.
22. **Mosier, D. E. and Coppelson, L. W.,** A three-cell interaction required for the induction of the primary immune response in vitro, *Proc. Natl. Acad. Sci. U.S.A.,* 61, 542, 1968.
23. **Mosier, D. E.,** Cell interactions in the primary immune response in vitro: a requirement for specific cell clusters, *J. Exp. Med.,* 129, 351, 1969.
24. **Mosier, D. E., Fitch, F. W., Rowley, D. A., and Davies, A. J. S.,** Cellular defects in thymectomized mice, *Nature,* 225, 276, 1970.
25. **Kindred, B. and Shreffler, D. C.,** H-2 dependence of cooperation between T and B cells, *J. Immunol.,* 109, 940, 1972.
26. **Katz, D. H., Graves, M., Dorf, M. E., DiMuzio, H., and Benacerraf, B.,** Cell interactions between histoincompatible T and B lymphocytes. VII. Cooperative responses between lymphocytes are controlled by genes in the *I* region of the H-2 complex, *J. Exp. Med.,* 141, 263, 1975.
27. **Sprent, J.,** Restricted helper function of F_1 hybrid T cells positively selected to heterologous erythrocytes in irradiated parental strain mice. I. Failure to collaborate with B cells of the opposite parental strain not associated with active suppression, *J. Exp. Med.,* 147, 1142, 1978.
28. **Kindred, B. and Shreffler, D. C.,** H-2 dependence of cooperation between T and B cells, *J. Immunol.,* 109, 940, 1972.

29. **Sprent, J.,** Restricted helper function of F_1 hybrid T cells positively selected to heterologous erythrocytes in irradiated parental strain mice. II. Evidence for restrictions affecting helper cell induction and T-B collaboration, both mapping to the K-end of the H-2 complex, *J. Exp. Med.*, 147, 1159, 1978.
30. **Katz, D. H., Armerding, D., and Eshhar, Z.,** The role of products of the histocompatibility gene complex in immune responses, in *Lymphocyte Differentiation, Recognition, and Regulation,* Katz, D. H. and Benacerraf, B., Eds., Academic Press, New York, 541.
31. **Pierce, S. K., Klinman, N. R., Maurer, P. H., and Merryman, C. F.,** Role of the major histocompatibility gene products in regulating the antibody response to dinitrophenyl poly (L-Gly^{35}, L-Ala^{35}, L-Phe^{9}), *J. Exp. Med.*, 152, 336, 1980.
32. **Katz, D. H.,** *Lymphocyte Differentiation, Recognition, and Regulation,* Academic Press, New York, 1977, 550.
33. **Jones, B. and Janeway, C. A., Jr.,** Cooperative interaction of B lymphocytes with antigen-specific helper T lymphocytes is MHC restricted, *Nature,* 292, 547, 1981.
34. **Katz, D. H.,** *Lymphocyte Differentiation, Recognition and Regulation,* Academic Press, New York, 1977.
35. **Rosenthal, A. S. and Shevach, E. M.,** Function of macrophages in antigen recognition by guinea pig T-lymphocytes, *J. Exp. Med.*, 138, 1194, 1973.
36. **von Boehmer, H., Hudson, L., and Sprent, J.,** Collaboration of histocompatible T and B lymphocytes using cells from tetraparental bone marrow chimeras, *J. Exp. Med.*, 142, 989, 1975.
37. **Yague, J., White, J., Coleclough, C., Kappler, J., Palmer, E., and Marrack, P.,** The T cell receptor: The α and β chains define idiotype, and antigen and MHC specificity, *Cell,* 42, 81, 1985.
38. **Dembic, Z., Haas, W., Weiss, S., McCubrey, J., Kiefer, H., von Boehmer, H., and Steinmetz, M.,** Transfer of specificity by murine α and β T-cell receptor genes, *Nature,* 320, 232, 1986.
39. **Katz, D. H. and Bennacerraf, B.,** The function and interrelationship of T cell receptors, Ir genes, and other histocompatibility gene products, *Transplant. Rev.*, 22, 175, 1975.
40. **Sprent, J.,** Restricted helper function of F1 hybrid T cells previously selected to heterologous erythrocytes in irradiated parental strain mice. I. Failure to collaborate with B cells of the opposite strain not associated with active suppression, *J. Exp. Med.*, 147, 1142, 1978.
41. **Swierkosz, J. E., Rock, K., Marrack, P., and Kappler, J. W.,** The role of H-2-linked genes in helper T-cell function. II. Isolation on antigen-pulsed macrophages of two specific populations of F_1 helper T cells each specific for antigen and one set of parental H-2 products, *J. Exp. Med.*, 147, 554, 1978.
42. **Erb, P. and Feldmann, M.,** The role of macrophages in the generation of T helper cells. II. The genetic control of macrophage-T-cell interaction for helper cell induction with soluble antigen, *J. Exp. Med.*, 142, 460, 1975.
43. **Singer, A., Hathcock, K. S., and Hodes, R. J.,** Cellular and genetic control of antibody responses. V. Helper T cell recognition of H-2 determinants on accessory cells but not B cells, *J. Exp. Med.*, 149, 1208, 1979.
44. **Singer, A., Hathcock, K. S., and Hodes, R. J.,** Cellular and genetic control of antibody responses. VIII. MHC restricted recognition of accessory cells, not B cells, by parent-specific subpopulations of normal F_1 T helper cells, *J. Immunol.*, 125, 914, 1980.
45. **Singer, A., Hathcock, K. S., and Hodes, R. J.,** Self-recognition in allogeneic radiation bone marrow chimeras. A radiation-resistant host element dictates the self-specificity and immune response gene phenotype of T-helper cells, *J. Exp. Med.*, 153, 1286, 1981.
46. **Chesnut, R. and Grey, H.,** Studies on the capacity of B cells to serve as antigen-presenting cells, *J. Immunol.*, 126, 1075, 1981.
47. **Abbas, A. K., Haber, S., and Rock, K. L.,** Antigen presentation by hapten-specific B lymphocytes. II. Specificity and properties of antigen-presenting B lymphocytes, and function of immunoglobulin receptors, *J. Immunol.*, 135, 1661, 1985.
48. **Rock, K. L., Benacerraf, B., and Abbas, A. K.,** Antigen presentation by hapten-specific B lymphocytes. I. Role of surface immunoglobulin receptors, *J. Exp. Med.*, 160, 1102, 1984.
49. **Rock, K. L., Haber, S. I., Liano, D., Benacerraf, B., and Abbas, A. K.,** Antigen presentation by hapten-specific B lymphocytes. III. Analysis of the immunoglobulin-dependent pathway of antigen presentation to interleukin 1-dependent T lymphocytes, *Eur. J. Immunol.*, 16, 1407, 1986.
50. **Lanzavecchia, A.,** Antigen-specific interaction between T and B cells, *Nature,* 314, 537, 1985.
51. **Noelle, R. J., Krammer, P. H., Ohara, J., Uhr, J. W., and Vitetta, E. S.,** Increased expression of Ia antigens on resting B cells: an additional role for B-cell growth factor, *Proc. Natl. Acad. Sci. U.S.A.*, 81, 6149, 1984.
52. **Sanders, V. M., Fernandez Botran, R., Uhr, J. W., and Vitetta, E. S.,** Interleukin 4 enhances the ability of antigen-specific B cells to form conjugates with T cells, *J. Immunol.*, 139, 2349, 1987.
53. **Krieger, J. I., Grammer, S. F., Grey, H. M., and Chesnut, R. W.,** Antigen presentation by splenic B cells: Resting B cells are ineffective, whereas activated B cells are effective accessory cells for T cell responses, *J. Immunol.*, 135, 2937, 1985.
54. **Frohman, M. and Cowing, C.,** Presentation of antigen by B cells: functional dependence on radiation dose, interleukins, cellular activation, and differential glycosylation, *J. Immunol.*, 134, 2269, 1985.

55. **Chesnut, R. W., Colon, S. M., and Grey, H. M.,** Requirements for the processing of antigens by antigen-presenting B cells. I. Functional comparison of B cell tumors and macrophages, *J. Immunol.*, 129, 2382, 1982.
56. **Kearse, K. P., Cassatt, D. R., Kaplan, A. M., and Cohen, D. A.,** The requirement for surface Ig signalling as a prerequisite for T cell:B cell interactions. A possible role for desialylation, *J. Immunol.*, 140, 1770, 1988.
57. **Snow, E. C., Vitetta, E. S., and Uhr, J. W.,** Activation of antigen-enriched B cells. I. Purification and response to thymus-independent antigens, *J. Immunol.*, 130, 607, 1983.
58. **Noelle, R. J., Snow, E. C., Uhr, J. W., and Vitetta, E. S.,** Activation of antigen-specific B cells: role of T cells, cytokines, and antigen in induction of growth and differentiation, *Proc. Natl. Acad. Sci. U.S.A.*, 80, 6628, 1983.
59. **Noelle, R. J. and McCann, J.,** Cognate interactions between helper T cells and B cells. I. Cloning and helper activity of a lymphokine-dependent helper T cell clone (Th-3), *J. Mol. Cell. Immunol.*, 4, 161, 1989.
60. **Cambier, J. C., Monroe, J. G., and Neale, N. J.,** Definition of conditions that enable antigen-specific activation of the majority of isolated trimitrophenol-binding B cells, *J. Exp. Med.*, 156, 635, 1982.
61. **Howard, M. and Paul, W. E.,** Regulation of B-cell growth and differentiation by soluble factors, *Annu. Rev. Immunol.*, 1, 307, 1983.
62. **Marrack, P., Skidmore, B., and Kappler, J. W.,** Binding of antigen-specific, *H-2*-restricted T cell hybridomas to antigen-pulsed adherent cell monolayers, *J. Immunol.*, 130, 2088, 1983.
63. **Sanders, V. M., Snyder, J. M., Uhr, J. W., and Vitetta, E. S.,** Characterization of the physical interaction between antigen-specific B and T cells, *J. Immunol.*, 137, 2395, 1986.
64. **Julius, M. H., von Boehmer, H., and Sidman, C. L.,** Dissociation of two signals required for activation of resting B cells, *Proc. Natl. Acad. Sci. U.S.A.*, 79, 1989, 1982.
65. **Andersson, J., Schrier, M. H., and Melchers, F. T.,** T cell-dependent B cell stimulation is H-2 restricted and antigen-dependent only at the resting B cell level, *Proc. Natl. Acad. Sci. U.S.A.*, 77, 1612, 1980.
66. **Katz, D. H.,** *Lymphocyte Differentiation, Recognition, and Regulation,* Academic Press, New York, 1977.
67. **Augustin, A. A. and Coutinho, A.,** Specific T helper cells that activate B cells polyclonally. In vivo enrichment and cooperative function, *J. Exp. Med.*, 151, 587, 1980.
68. **Coutinho, A. and Augustin, A. A.,** Major histocompatibility complex-restricted and unrestricted T helper cells recognizing minor histocompatibility antigens of B cell surfaces, *Eur. J. Immunol.*, 10, 535, 1980.
69. **Pareira, P., Forsgen, S., Portnoi, D., Bandeira, A., Martinez-Alonso, C., and Coutinho, A.,** The role of immunoglobulin receptors in cognate T-B cell collaboration, *Eur. J. Immunol.*, 16:355, 1986.
70. **Nagarkatti, P. S., Snow, E. C., and Kaplan, A. M.,** Characterization and function of autoreactive T-lymphocyte clones isolated from normal, unprimed mice, *Cell. Immunol.*, 94, 32, 1985.
71. **Saito, T. and Rajewsky, K.,** A self-Ia reactive T cell clone directly stimulates every hundredth B cell and helps antigen-specific B cell responses, *Eur. J. Immunol.*, 15, 927, 1985.
72. **Martinez-Alonso, C., Coutinho, A., Bernabe, R. R., Augustin, A., Haas, W., and Pohlit, H.,** Hapten-specific helper T cells. I. Collaboration with B cells to which the hapten has been directly coupled, *Eur. J. Immunol.*, 10, 403, 1980.
73. **Martinez-Alonso, C., Coutinho, A., Bernabe, R. R., Haas, W., and Pohlit, H.,** Hapten-specific helper T cells. II. Genetic determination of functional recognition, *Eur. J. Immunol.*, 10, 411, 1980.
74. **Principato, M. A., Thompson, G. S., and Friedman, S. M.,** A cloned major histocompatibility complex-restricted trinitrophenyl-reactive human helper T cell line that activates B cell subsets via two distinct pathways, *J. Exp. Med.*, 158, 1444, 1983.
75. **Chartash, E. K., Imai, A., Gershengorn, M. C., Crow, M. K., and Friedman, S. M.,** Direct human T helper cell-induced B cell activation is not mediated by inositol lipid hydrolysis, *J. Immunol.*, 140, 1974, 1988.
76. **Martinez-Alonso, C., Bandeira, A., Toribio, M. L., Coutinho, A., Marcos, M. A., and Pareira, P.,** Back-stimulation of B lymphocytes binding to helper T cell surface antigens, *J. Immunol.*, 18, 1895, 1988.
77. **Ratcliffe, M. J. H. and Julius, M. H.,** Two classes of bystander B cell response: activation requirements reflect those of B cells in general, *J. Immunol.*, 131, 581, 1983.
78. **DeFranco, A. L., Ashwell, J. D., Schwartz, R. H., and Paul, W. E.,** Polyclonal stimulation of resting B lymphocytes by antigen-specific T lymphocytes, *J. Exp. Med.*, 159, 861, 1984.
79. **Tony, H.-P., Phillips, N. E., and Parker, D. C.,** Role of membrane immunoglobulin (Ig) crosslinking in membrane Ig-mediated, major histocompatibility-restricted T cell-B cell cooperation, *J. Exp. Med.*, 162, 1695, 1985.
80. **Tony, H.-P. and Parker, D. C.,** Major histocompatibility complex-restricted, polyclonal B cell responses resulting from helper T cell recognition of anti-immunoglobulin presented by small B lymphocytes, *J. Exp. Med.*, 161, 223, 1985.
81. **Gosselin, E. J., Tony, H.-P., and Parker, D. C.,** Characterization of antigen processing and presentation resting lymphocytes, *J. Immunol.*, 140, 1413, 1988.

82. **Julius, M. H., Rammensee, H.-G., Ratcliffe, M. J. H., Lamers, M. C., Lanshorne, J., and Kohler, G.,** The molecular interactions with helper T cells which limit antigen-specific B cell differentiation, *Eur. J. Immunol.,* 18, 381, 1988.
83. **Hirohata, S., Jelinek, D. F., and Lipsky, P. E.,** T cell-dependent activation of B cell proliferation and differentiation by immobilized monoclonal antibodies to CD3, *J. Immunol.,* 140, 3726, 1988.
84. **Stohl, W., Posnett, D. N., and Chiorazzi, N.,** Induction of T cell-dependent B cell differentiation by anti-CD3 monoclonal antibodies, *J. Immunol.,* 138, 1667, 1987.
85. **Julius, M. H. and Rammensee, H.-G.,** T helper cell-dependent induction of resting B cell differentiation need not require cognate cell interactions, *Eur. J. Immunol.,* 18, 375, 1988.
86. **Owens, T.,** A noncognate interaction with anti-receptor antibody-activated helper T cells induces small murine B cells to proliferate and to secrete antibody, *Eur. J. Immunol.,* 18, 395, 1988.
87. **Riedel, C., Owens, T., and Nossal, G. J. V.,** A significant proportion of normal resting B cells are induced to secrete immunoglobulin through contact with anti-receptor antibody-activated helper T cells in clonal cultures, *Eur. J. Immunol.,* 18, 403, 1988.
88. **Romagnani, S., del Prete, G., Maggi, E., and Ricci, M.,** Activation through CD3 molecule leads a number of human T cell clones to induce IgE synthesis in vitro by B cells from allergic and nonallergic individuals, *J. Immunol.,* 138, 1744, 1987.
89. **Leclercq, L., Cambier, J. C., Mishal, Z., Julius, M. H., and Theze, J.,** Supernatant from a cloned helper T cell stimulates most small resting B cells to undergo increased I-A expression, blastogenesis, and progression through cell cycle, *J. Immunol.,* 136, 539, 1987.
90. **Diu, A., Gougon, M.-L., Moreau, J.-L., Reinherz, E., and Theze, J.,** Activation of resting human B cells by helper T-cell clone supernatant: characterization of a human B-cell-activating factor, *Proc. Natl. Acad. Sci. U.S.A.,* 84, 9140, 1987.
91. **Julius, M. H.,** Reciprocity in lymphocyte interactions, *Immunol. Rev.,* 95, 177, 1987.
92. **Noelle, R. J., McCann, J., Marshall, L., and Bartlett, W. C.,** Cognate interactions between helper T cells and B cells. III. Contact-dependent, lymphokine-independent induction of B cell cycle entry by activated helper T cells, *J. Immunol.,* 143, 1807, 1989.
93. **Yague, J., White, J., Coleclough, C., Kappler, J., Palmer, E., and Marrack, P.,** The T cell receptor: the α and β chains define idiotype, and antigen and MHC specificity, *Cell,* 42, 81, 1985.
94. **Dembic, Z., Haas, W., Weiss, S., McCubrey, J., Kiefer, H., von Boehmer, H., and Steinmetz, M.,** Transfer of specificity by murine α and β T-cell receptor genes, *Nature,* 320, 232, 1986.
95. **Vitetta, E. S., Bossie, A., Fernandez-Botran, R., Myers, C. D., Oliver, K. G., Sanders, V. M., and Stevens, T. L.,** Interaction and activation of antigen-specific T and B cells, *Immunol. Rev.,* 99, 193, 1987.
96. **Dialynas, D. P., Quan, Z. S., Wall, K. A., Pierres, A., Quinta, J., Loken, M. R., Pierres, M., and Fitch, F. W.,** Identification of the murine T cell surface molecule, designated L3T4, identified by monoclonal antibody GK1.5: similarity of L3T4 to the human Leu-3/T4 molecule, *J. Immunol.,* 131, 2445, 1983.
97. **Dialynas, D. P., Wilde, D. B., Marrack, P., Pierres, A., Wall, K. A., Havran, W., Otten, G., Loken, M. R., Pierres, M., Kappler, J., and Fitch, F. W.,** Characterization of the murine antigenic determinant, designated L3T4a, recognized by monoclonal antibody GK1.5: expression of L3T4a by functional T cell clones appears to correlate primarily with class II MHC antigen-reactivity, *Immunol. Rev.,* 74, 29, 1983.
98. **Wilde, D. B., Marrack, P., Kappler, J., Dialynas, D. P., and Fitch, F. W.,** Evidence implicating L3T4 in class II-MHC reactivity: monoclonal antibody GK1.5 (anti-L3T4a) blocks class II MHC antigen-specific proliferation, release of lymphokines, and binding by cloned murine helper T lymphocyte lines, *J. Immunol.,* 131, 2178, 1983.
99. **Bartlett, W. C., Michael, A., McCann, J., Yuan, D., Claassen, E., and Noelle, R. J.,** Cognate interactions between helper T cells and B cells. II. Dissection of cognate help by using a class II-restricted, antigen-specific, IL-2-dependent helper T cell clone, *J. Immunol.,* 143, 1745, 1989.
100. **Doyle, C. and Strominger, J. L.,** Interaction between CD4 and class II MHC molecules mediates cell adhesion, *Nature,* 330, 256, 1987.
101. **Gay, D., Maddon, P., Sekaly, R., Talle, M. A., Godfrey, M., Long, E., Goldstein, G., Chess, L., Axel, R., Kappler, J., and Marrack, P.,** Functional interaction between human T-cell protein CD4 and the major histocompatibility complex HLA-DR antigen, *Nature,* 328, 626, 1987.
102. **Kupfer, A. and Singer, S. J.,** The specific interaction of helper T cells and antigen-presenting B cells. IV. Membrane and cytoskeletal reorganizations in the bound T cell as a function of antigen dose, *J. Exp. Med.,* 170, 1697, 1989.
103. **Saizawa, K., Rojo, J., and Janeway, C. A., Jr.,** Evidence for a physical association of CD4 and the CD3:alpha:beta T-cell receptor, *Nature,* 328, 260, 1987.
104. **Springer, T. A., Dustin, M. L., Kishimoto, T. K., and Marlin, S. D.,** The lymphocyte function-associated LFA-1, CD2, and LFA-3 molecules: cell adhesion receptors of the immune system, *Annu. Rev. Immunol.,* 5, 223, 1987.

105. **Rothlein, R., Dustin, M. L., Marlin, S. D., and Springer, T. A.,** A human intercellular adhesion molecule (ICAM-1) distinct from LFA-1, *J. Immunol.,* 137, 1270, 1986.
106. **Makgoba, M. W., Sanders, M. E., Ginther Luce, G. E., Dustin, M. L., Springer, T. A., Clark, E. A., Mannoni, P., and Shaw, S.,** ICAM-1 a ligand for LFA-1 dependent adhesion of B, T, and myeloid cells, *Nature,* 331, 86, 1988.
107. **Davignon, D., Martz, E., Reynolds, T., Kurzinger, K., and Springer, T. A.,** Lymphocyte function-associated antigen 1 (LFA-1): a surface antigen distinct from Lyt-2,3 that participates in T lymphocyte mediated killing, *Proc. Natl. Acad. Sci. U.S.A.,* 78, 4535, 1981.
108. **Pierres, M., Goridis, C., and Goldstein, P.,** Inhibition of murine T cell-mediated cytolysis and T cell proliferation by a rat monoclonal antibody immunoprecipitating two lymphoid cell surface polypeptides of 94,000 and 180,000 molecular weight, *Eur. J. Immunol.,* 12, 60, 1982.
109. **Davignon, D., Martz, E., Reynolds, T., Kurzinger, K., and Springer, T. A.,** Monoclonal antibody to a novel lymphocyte function-associated antigen (LFA-1): mechanism of blockade of T lymphocyte-mediated killing and effects on other T and B lymphocyte functions, *J. Immunol.,* 127, 590, 1981.
110. **Fischer, A., Durandy, A., Sterkers, G., and Griscelli, C.,** Role of the LFA-1 molecule in cellular interactions required for antibody production in humans, *J. Immunol.,* 136, 3198, 1986.
111. **Tedder, T. F., Schmidt, R. E., Rudd, C. E., Kornacki, M. M., Rizt, J., and Schlossman, S. F.,** Function of the LFA-1 and T4 molecules in the direct activation of resting human B lymphocytes by T lymphocytes, *Eur. J. Immunol.,* 16, 1539, 1986.
112. **Howard, D. R., Eaves, A. C., and Takei, F.,** Lymphocyte function-associated antigen (LFA-1) is involved in B cell activation, *J. Immunol.,* 136, 4013, 1986.
113. **Warwyk, S. O., Novotny, J. R., Wicks, I. P., Wilkinson, D., Maher, D., Salvaris, E., Welch, K., Fecondo, J., and Boyd, A. W.,** The role of the LFA-1/ICAM-1 interaction in human leukocyte homing and adhesion, *Immunol. Rev.,* 108, 135, 1989.
114. **Anderson, D. C. and Springer, T. A.,** Leukocyte adhesion deficiency: an inherited defect in the Mac-1, LFA-1, and p150,95 glycoproteins, *Annu. Rev. Med.,* 38, 175, 1987.
115. **van Kooyk, Y., van de Wiel-van Kemenade, P., Weder, P., Kuijpers, T. W., and Figdor, C. G.,** Enhancement of LFA-1-mediated cell adhesion by triggering through CD2 or CD3 on T lymphocytes, *Nature,* 342, 811, 1989.
116. **Kupfer, A., Swain, S. L., and Singer, S. J.,** The specific direct interaction of helper T cells and antigen-presenting B cells. II. Reorientation of the microtubule organizing center and reorganization of the membrane associated cytoskeleton inside the bound helper T cells, *J. Exp. Med.,* 165, 1565 1987.
117. **Virella, G., Rugeles, M. T., Hyman, B., La Via, M., Goust, J. M., Frankis, M., and Bierer, B. E.,** The interaction of CD2 with its LFA-3 ligand expressed by autologous erythrocytes results in enhancement of B cell responses, *Cell. Immunol.,* 116, 308, 1988.
118. **Bretscher, P. A.,** The two signal model for B cell induction, *Transplant Rev.,* 23, 37, 1975.
119. **Cantrell, D. A. and Smith, K. A.,** The interleukin-2 T-cell system: a new cell growth model, *Science,* 224, 1312, 1984.
120. **Howard, M. and Paul, W. E.,** Regulation of B-cell growth and differentiation by soluble factors, *Annu. Rev. Immunol.,* 1, 307, 1983.
121. **Snow, E. C. and Noelle, R. J.,** Thymus-dependent antigenic stimulation of hapten-specific B lymphocytes, *Immunol. Rev.,* 99, 173, 1987.
122. **Goding, J. W. and Layton, J. E.,** Antigen-induced co-capping of IgM and IgD-like receptors on murine B cells, *J. Exp. Med.,* 144, 852, 1976.
123. **Snow, E. C., Noelle, R. J., Uhr, J. W., and Vitetta, E. S.,** Activation of antigen-enriched B cells. II. Role of linked recognition in B cell proliferation to thymus-dependent antigens, *J. Immunol.,* 130, 614, 1983.
124. **Poo, W. J., Conrad, L., and Janeway, C. A., Jr.,** Receptor-directed focusing of lymphokine release by helper T cells, *Nature,* 332, 378, 1988.
125. **Loughnan, M. S., Takatsu, K., and Nossal, G. J.,** T-cell-replacing factor (interleukin 5) induces expression of interleukin 2 receptors on murine splenic B cells, *Proc. Natl. Acad. Sci. U.S.A.,* 84, 5399, 1987.
126. **Feldmann, M.,** Cell interactions in the immune response *in vitro*. Specific collaboration via complexes of antigen and thymus-derived cell immunoglobulin, *J. Exp. Med.,* 136, 737, 1972.
127. **Taussig, M. J., Mozes, E., and Isac, R.,** Antigen-specific thymus cell factors in the genetic control of the immune response, *J. Exp. Med.,* 140, 301, 1974.
128. **Guy, R. and Hodes, R. J.,** Antigen-specific, MHC-restricted B cell activation by cell-free Th2 cell products. Synergy between antigen-specific helper factors and IL-4, *J. Immunol.,* 143, 1433, 1989.
129. **Guy, R., Ullrich, S. J., Foo-Philips, M., Hathcock, K. S., Apella, E., and Hodes, R. J.,** Antigen-specific helper function of cell-free T cell products bearing TCR V beta 8 determinants, *Science,* 244, 1477, 1989.
130. **Kupfer, A., Swain, S. L., Janeway, C. A., Jr., and Singer, S. J.,** The specific direct interaction of helper T cells and antigen-presenting B cells, *Proc. Natl. Acad. Sci. U.S.A.,* 83, 6080, 1986.

131. **Kupfer, A., Singer, S. J., Janeway, C. A., Jr., and Swain, S. L.,** Coclustering of CD4 (L3T4) molecule with the T-cell receptor is induced by specific direct interaction of helper T cells and antigen-presenting cells, *Proc. Natl. Acad. Sci. U.S.A.,* 84, 5888, 1987.
132. **Swain, S. L. and Dutton, R. W.,** Consequences of the direct interaction of helper T cells with B cells presenting antigen, *Immunol. Rev.,* 99, 263, 1987.
133. **Bartlett, W. C., McCann, J., Shepherd, D. M., and Noelle, R. J.,** Cognate interactions between helper T cells and B cells. IV. Helper T cell activation is essential for B cells to trigger B cell cycle entry, submitted.
134. **Krusemeier, M. and Snow, E. C.,** Induction of lymphokine responsiveness of hapten-specific B lymphocytes promoted through an antigen-mediated T helper lymphocyte interaction, *J. Immunol.,* 140, 367, 1987.
135. **Sekita, K., Straub, C., Hoessli, D., and Zubler, R. H.,** B cell-stimulating activity of lymphoid cell membrane fractions, *Eur. J. Immunol.,* 18, 1405, 1988.
136. **Schimpl, A. and Wecker, E.,** Replacement of T cell function by a T cell producer, *Nature,* 237, 15, 1972.
137. **Takatsu, K., Tanaka, K., Tominaga, A., Kumahara, Y., and Hamaoka, T.,** Antigen-induced T cell-replacing factor (TRF). III. Establishment of a T cell hybrid clone continuously producing TRF and functional analysis of released TRF, *J. Immunol.,* 125, 2076, 1980.
138. **Watson, J., Gillis, S., Marbrook, J., Mochizuki, D., and Smith, K. A.,** Biochemical and biological characterization of lymphocyte regulatory molecules. I. Purification of a class of murine lymphokines, *J. Exp. Med.,* 150, 849, 1979.
139. **Marrack, P., Graham, S. D., Kushnir, E., Leibson, H. J., Roehm, N., and Kappler, J. W.,** Nonspecific factors in B cell responses, *Immunol. Rev.,* 63, 33, 1982.
140. **Vitetta, E. S., Fernandez-Botran, R., Myers, C. D., and Sanders, V. M.,** Cellular interactions in the humoral immune response, *Adv. Immunol.,* 45, 1, 1989.
141. **Dinarello, C. A.,** Biology of interleukin 1, *FASEB J.,* 2, 108, 1988.
142. **Stein, P., DuBois, P., Greenblatt, D., and Howard, M.,** Induction of antigen-specific proliferation in affinity-purified small B lymphocytes: requirement for BSF-1 by Type 2 but not Type 1 thymus independent antigens, *J. Immunol.,* 136, 2080, 1986.
143. **Howard, M., Mizel, S. B., Lachman, L., Ansel, J., Johnson, B., and Paul, W. E.,** *J. Exp. Med.,* 157, 1529, 1983.
144. **Koyuma, N., Harada, N., Takahashi, T., Mita, S., Okamura, H., Tominaga, A., and Takatsu, K.,** Role of recombinant interleukin 1 compare to recombinant interleukin-5 in B cell differentiation, *Immunology,* 63, 277, 1988.
145. **Swain, S. L., Wetzel, G. D., Soubiran, P., and Dutton, R. W.,** T cell replacing factors in the B cell response to antigen, *Immunol. Rev.,* 63, 111, 1982.
146. **Saiki, O., Tanaka, T., Doi, S., and Kishimoto, S.,** Expression and the functional role of a p70/75 interleukin 2-binding molecule in human B cell, *J. Immunol.,* 140, 853, 1988.
147. **Bich-Thuy, L. and Fauci, A. S.,** Direct effect of interleukin 2 on the differentiation of human B cells which have not been preactivated in vitro, *Eur. J. Immunol.,* 15, 1075, 1985.
148. **Zubler, R. H., Lowenthal, J. W., Erard, F., Hashimoto, N., Devos, R., and MacDonald, H. R.,** Activated B cells express receptors for, proliferate in response to, pure interleukin 2, *J. Exp. Med.,* 160, 1170, 1984.
149. **Loughnan, M. S. and Nossal, G. J.,** Interleukins 4 and 5 control expression of IL-2 receptor on murine B cells through independent induction of its two chains, *Nature,* 340, 76, 1989.
150. **Paul, W. E. and Ohara, J.,** B-cell stimulatory factor-1/Interleukin 4, *Annu. Rev. Immunol.,* 5, 429, 1987.
151. **Howard, M., Farrar, J., Hilfiker, M., Johnson, B., Takatsu, K., Hamaoka, T., and Paul, W. E.,** Identification of a T cell-derived B cell growth factor distinct from interleukin 2, *J. Exp. Med.,* 155, 914, 1982.
152. **Lowenthal, J. W., Castle, B. E., Christiansen, J., Schreurs, J., Rennick, D., Arai, N., Hoy, P., Takebe, Y., and Howard, M.,** Expression of high affinity receptors for murine interleukin 4 (BSF-1) on hemopoietic and nonhemopoietic cells, *J. Immunol.,* 140, 456, 1988.
153. **Roehm, N. W., Leibson, H. J., Zlotnick, A., Kappler, J., Marrack, P., and Cambier, J. C.,** Interleukin-induced increase in I-A expression by normal mouse B cells, *J. Exp. Med.,* 160, 679, 1984.
154. **Conrad, D. H., Waldschmidt, T. S., Lee, W. T., Rao, M., Keegan, A. D., Noelle, R. J., Lynch, R. G., and Kehry, M. R.,** Effect of B cell stimulatory factor-1 (interleukin 4) on Fcg and Fce receptor expression on murine B lymphocytes, *J. Immunol.,* 139, 2290, 1987.
155. **DeFrance, T., Aubry, J. P., Rousset, F., Vandervliet, B., Bonnefoy, J. Y., Arai, N., Takebe, Y., Yokota, T., Lee, F., Arai, K., DeVries, J., and Banchereau, J.,** Human recombinant interleukin 4 induces Fce receptors (CD23) on normal human B lymphocytes, *J. Exp. Med.,* 165, 1459, 1987.
156. **Isakson, P. C., Pure, E., Vitetta, E. S., and Krammer, P. H.,** T cell-derived B cell differentiation factor(s). Effect on the isotype switch of murine B cells, *J. Exp. Med.,* 155, 734, 1982.

157. **Vitetta, E. S., Ohara, J., Myers, C. D., Layton, J. E., Krammer, P. H., and Paul, W. E.,** Serological, biochemical, and functional identity of B cell-stimulatory factor 1 and B cell differentiation factor for IgG_1, *J. Exp. Med.,* 162, 1726, 1985.
158. **Coffman, R. L. and Carty, A.,** T cell activity that enhances polyclonal IgE production and its inhibition by interferon-γ, *J. Immunol.,* 136, 949, 1986.
159. **Dubois, P., Stein, P., Ennist, D., Greenblatt, D., Mosmann, T. R., and Howard, M.,** Requirement for BSF-1 in the induction of antigen-specific B cell proliferation by a thymus-dependent antigen and carrier reactive T cell line, *J. Immunol.,* 139, 1927, 1987.
160. **Boom, W. H., Liano, D., and Abbas, A. K.,** Heterogeneity of helper/inducer T lymphocytes. II. Effects of IL4- and IL2-producing T cell clones on resting B lymphocytes, *J. Exp. Med.,* 167, 1350, 1988.
161. **Mossman, T. R., Cherwinski, H., Bond, M. W., Giedlin, M. A., and Coffman, R. L.,** Two types of murine helper T cell clones. I. Definition according to profiles of lymphokine activities of secreted proteins, *J. Immunol.,* 136, 2348, 1986.
162. **Stevens, T. L., Bossie, A., Sanders, V. M., Fernandez-Botran, R., Coffman, R. L., Mosmann, T. R., and Vitetta, E. S.,** Regulation of antibody isotype secretion by subsets of antigen specific helper T cells, *Nature,* 334, 255, 1988.
163. **Coffman, R. L., Seymour, B. W., Lebman, D. A., Hiraki, D. D., Christiansen, J. A., Shrader, B., Cherwinski, H. M., Savelkoul, H. F., Finkelman, F. D., Bond, M. W., and Mosmann, T. R.,** The role of helper T cell products in mouse B cell differentiation and isotype regulation, *Immunol. Rev.,* 102, 5, 1988.
164. **Ohara, J. and Paul, W. E.,** Receptors for B-cell stimulatory factor-1 expressed on cells of haematopoietic lineage, *Nature,* 325, 537, 1987.
165. **Swain, S. L. and Dutton, R. W.,** Production of a B cell growth-promoting activity, (DL)BCG from a cloned T cell line and its assay on the BCL_1 B cell tumor, *J. Exp. Med.,* 156, 1821, 1982.
166. **Yokota, T., Coffman, R. L., Hagiwara, H., Rennick, D. M., Takebe, Y., Yokota, K., Gemmell, L., Shrader, B., Yang, G., Meyerson, P., Luh, J., Hoy, P., Pene, J., Briere, F., Banchereau, J., DeVries, J., Lee, F. D., Arai, N., and Arai, K.,** Isolation and characterization lymphokine cDNA clones encoding mouse and human IgA-enhancing factor and eosinophil colony-stimulation factor activities: relationship to interleukin 5, *Proc. Natl. Acad. Sci. U.S.A.,* 84, 7388, 1987.
167. **Kinashi, T., Harada, N., Severinson, E., Tanabe, T., Sideras, P., Kinishi, M., Azum, C., Tominaga, A., Bergstedt-Lindqvist, S., Takahashi, M., Matsuda, F., Yaoita, Y., Takatsu, K., and Honjo, T.,** Cloning of complementary DNA encoding T-cell replacing factor and identity with B-cell growth factor II, *Nature,* 324, 70, 1986.
168. **Nakanishi, K., Howard, M., Muraguchi, A., Takatsu, K., Hamaoka, T., Koshland, M. E., and Paul, W. E.,** Identification of two distinct T cell-replacing factor (TRF), *J. Immunol.,* 130, 2219, 1983.
169. **Swain, S. L., Howard, M., Kappler, J., Marrack, P., Watson, J., Booth, R., Wetzel, G. D., and Dutton, R. W.,** Evidence for two distinct classes of murine B cell growth factors with activities in different functional assays, *J. Exp. Med.,* 158, 822, 1983.
170. **Takatsu, K., Tanaka, K., Tominaga, A., Kumahara, Y., and Hamaoka, T.,** Antigen-induced T cell-replacing factor (TRF). III. Establishment of T cell hybrid clone continuously producing TRF and functional analysis of released TRF, *J. Immunol.,* 125, 2646, 1980.
171. **Pure, E., Isakson, P. C., Takatsu, K., Hamaoka, T., Swain, S. L., Dutton, R. W., Dennert, G., Uhr, J. W., and Vitetta, E. S.,** Induction of B cell differentiation by T cell factors. I. Stimulation of IgM secretion by products of a T cell hybridoma and a T cell line, *J. Immunol.,* 127, 1953, 1981.
172. **Takatsu, K., Tanaka, K., Tominaga, A., Kumahara, Y., and Hamaoka, T.,** Antigen-induced T cell-replacing factor (TRF). III. Establishment of T cell hybrid clone continuously producing TRF and functional analysis of released TRF, *J. Immunol.,* 125, 2646, 1980.
173. **Murray, P. D., McKenzie, D. T., Swain, S. L., and Kagnoff, M. F.,** IL-5 and IL-4 produced by Peyer's patch T cells selectively enhance IgA expression, *J. Immunol.,* 138, 2669, 1987.
174. **Tonkonogy, S. L., McKenzie, S. T., and Swain, S. L.,** Regulation of isotype production by IL4 and IL5. Effects of lymphokines on Ig production depend on the state of activation of the responding B cells, *J. Immunol.,* 142, 4351, 1989.
175. **Liebson, H. J., Gefter, M., Zlotnik, A., Marrack, P., and Kappler, J. W.,** Role of interferon in antibody-producing responses, *Nature,* 309, 799, 1984.
176. **Sidman, C. L., Marshall, J. D., Schultz, L. D., Gray, P. W., and Johnson, H. M.,** Interferon is one of several direct B cell-maturing lymphokines, *Nature,* 309, 801, 1984.
177. **Rasmussen, R., Takatsu, K., Harada, N., Takahashi, T., and Bottomly, K.,** T cell dependent hapten-specific and polyclonal B cell responses require release of interleukin 5, *J. Immunol.,* 14, 705, 1988.
178. **Mond, J. J., Finkelman, F. D., Sarma, C., Ohara, J., and Serrate, S.,** Recombinant interferon-γ inhibits the B cell proliferative response stimulated by soluble but not by sepharose-bound anti-immunoglobulin antibody, *J. Immunol.,* 135, 2513, 1986.

179. **Rabin, E. M., Mond, J. J., Ohara, J., and Paul, W. E.,** Interferon-γ inhibits the action of B cell stimulatory factor (BSF)-1 on resting B cells, *J. Immunol.*, 137, 1573, 1986.
180. **Coffman, R. L. and Carty, J.,** A T cell activity that enhances polyclonal IgE production and inhibition by interferon-γ, *J. Immunol.*, 136, 949, 1986.
181. **Mond, J. J., Carman, J., Sarma, C., Ohara, J., and Finkleman, F. D.,** Interferon suppresses B cell stimulation factor (BSF-1) induction of class II MHC determinants on B cells, *J. Immunol.*, 137, 3534, 1986.
182. **Oliver, K., Krammer, P. H., Tucker, P. W., and Vitetta, E. S.,** The effects of cytokines and adherent cells on the interleukin 4-mediated induction of Ia antigens on resting B cells, *Cell Immunol.*, 106, 42, 1987.

182a. **Snapper, C. M. and Paul, W. E.,** Interferon-γ and B cell stimulatory factor-1 reciprocally regulate isotype production, *Science*, 236, 944, 1987.

183. **Reynolds, D. S., Boom, W. H., and Abbas, A. K.,** Inhibition of B lymphocyte activation by interferon-γ, *J. Immunol.*, 139, 767, 1987.
184. **Stevens, T. L., Bossie, A., Sanders, V. M., Fernandez-Botran, R., Coffman, R. L., Mosmann, T. R., and Vitetta, E. S.,** *Nature*, 334, 255, 1988.
185. **Finkelman, F. D., Ohara, J., Goroff, D. K., Smith, J., Villacreses, N., Mond, J. J., and Paul, W. E.,** Production of BSF-1 during an *in vivo*, T-dependent immune response, *J. Immunol.*, 137, 2878, 1986.
186. **Snapper, C. V., Peschell, C., and Paul, W. E.,** IFN-γ stimulates $IgG2_a$ secretion by murine B cells stimulated with bacterial lipopolysaccharide, *J. Immunol.*, 140, 2121, 1988.
187. **Mosmann, T. R. and Coffman, R. L.,** TH1 and TH2 cells: different patterns of lymphokine secretion lead to different functional properties, *Annu. Rev. Immunol.*, 7, 145, 1989.
188. **Killar, L., MacDonald, G., West, J., Woods, A., and Bottomly, K.,** Cloned, Ia-restricted T cells that do not produce interleukin 4 (IL4)/B cell stimulatory factor 1 (BSF-1) fail to help antigen specific B cells, *J. Immunol.*, 138, 1674, 1987.
189. **Bottomly, K.,** A functional dichotomy in $CD4^+$ T lymphocytes, *Immunol. Today*, 9, 268, 1988.

Chapter 7

T CELL DEPENDENT ACTIVATION OF HUMAN B CELLS

Robin E. Callard

TABLE OF CONTENTS

I. INTRODUCTION

Antibody responses by human B lymphocytes are mostly T cell dependent. Many clinical studies of patients with T cell immunodeficiencies have shown that functional T cells are required for normal humoral immunity.[1] In some instances, however, the underlying defect is expressed in both T cells and B cells making it difficult to determine whether T cells are required for normal antibody responses. There is also a vast literature describing T cell dependent activation of human B cells *in vitro* where it is easier to dissect the cell interactions involved. For the most part, these studies have relied on the use of polyclonal activators such as pokeweed mitogen (PWM) which require T cells or T cell derived factors to stimulate immunoglobulin secretion by human B cells. Only Epstein-Barr virus (EBV) is able to induce polyclonal Ig secretion by human B cells in the complete absence of T cells.[2] More recently, specific antibody responses *in vitro* to a wide range of different antigens have also been used to investigate T cell interactions with B cells in man.

Although in broad terms the mechanisms of T_h collaboration with B cells appear similar in humans and mice, a number of important differences in detail have been reported. This is not surprising considering that the evolutionary lines giving rise to humans and mice diverged between 60 and 80 million years ago. The aim of this chapter is to describe what is currently known about collaboration between T helper cells (T_h) and B cells in man, and to highlight the features which appear to be unique to human antibody responses. By taking this approach, it is hoped to provide some insight into how the immune system has adapted to the needs of human evolution, and to help understand the defects which can give rise to immunological disease. For the most part, the discussion will concentrate on T cell activation of B cells in specific antibody responses, but it will also be useful to consider briefly some aspects of T_h-B cell interactions required for Ig production to polyclonal mitogens.

II. ANTIGEN SPECIFIC RESPONSES

A. T CELL DEPENDENT ANTIBODY RESPONSES BY HUMAN LYMPHOCYTES *IN VITRO*

Specific antibody responses by human lymphocytes have been obtained to a variety of different antigens including tetanus toxoid;[3,4] keyhole limpet hemocyanin (KLH);[4] several viruses such as influenza,[5] herpes,[6] polio,[7] and rabies;[8] parasites;[9] polysaccharides;[10,11] sheep red blood cells;[12] and haptenated (TNP) *Brucella abortus,*[13] KLH,[14] or polyacrylamide beads;[15] (Table 1). The majority of these antigens are T cell dependent. Only TNP-BA is clearly T cell independent,[13,16] whereas Streptococcal carbohydrate A is partially T cell dependent.[11]

In trying to understand the role of T cell help for antibody responses to these antigens, it is important to consider the nature of the antigen, and whether a primary or memory response is being elicited. Most, if not all, *in vitro* antibody responses by human lymphocytes are to recall antigens, in some cases as a result of multiple immunizations: for example to

TABLE 1
T-Cell Dependence of Antigens Used to Elicit Specific Antibody Response by Human Lymphocytes

Antigen	T-cell dependence	Isotypes produced[a]
Influenza virus	Yes	IgG(1,4 & 3) IgA IgM
Herpes varicella/zoster	Yes	IgG
Tetanus toxoid	Yes	IgG
KLH	Yes	IgM
Surface protein antigen from *Rickettsiae typhi*	Yes	IgG
Mannan	Yes	IgM IgG(2)
TNP-polyacrylamide beads	Yes	IgM
Streptococcal carbohydrate	Partial	IgM IgG
TNP-*Brucella abortus*	No	IgM

[a] Immunoglobulin isotypes in descending order of the amount produced *in vitro* to each of the different antigens listed.

influenza virus or tetanus toxoid. In our experience, it has not been possible to obtain *in vitro* antibody responses from naive donors. For example, PBMC from individuals who have never been exposed to herpes varicella/zoster, as determined by the absence of serum antibodies and specific T cell proliferation, and no clinical history of infection, do not make detectable antibody *in vitro*. KLH has been said to elicit a primary IgM response, but exposure to cross-reacting hemocyanins by eating some seafoods cannot be excluded, and serum antibodies to KLH are common (Prentice, personal communication). Nevertheless, the possibility that different T helper cell functions are required for primary and secondary responses should be kept in mind.

It is also important to consider the Ig isotype made in response to any particular antigen (Table 1). Antibody to influenza virus produced *in vitro* (as it is *in vivo*) is predominantly IgG1 and IgG4 with occasional IgG3 but no IgG2.[17] Some IgM and IgA but never IgE can also be detected, especially in responses by tonsillar mononuclear cells.[18] In contrast, responses *in vitro* to the polysaccharide mannan from *Candida albicans* are mostly IgM, IgG1 and IgG2 (Durandy, personal communication), whereas responses to the group A streptococcal carbohydrate (A-CHO) are mostly IgM although some IgG can be detected.[19,20] The signals which determine the Ig isotype and IgG subclass produced in response to a particular antigen are not known, but there is accumulating evidence in man as well as in mice that the set of cytokines produced by the T helper cells is an important if not deciding factor.

B. T-HELPER CELLS

The requirement for T helper cells in specific antibody responses has been amply demonstrated by experiments in which blood or tonsillar B cells were depleted of T cells, usually by rosetting with AET treated sheep red blood cells.[21] This technique has been shown to result in B cell preparations contaminated with between 1 and 10% of T cells depending on the precise procedure employed.[22] With adequate T cell depletion, antibody responses to T cell dependent antigens such as influenza virus or tetanus toxoid are reduced to background.[3,22,23] $CD3^+/CD4^+$ T helper cells are able to fully restore antibody production,[23] but no help has been reported with $CD8^+$ T cells, which in some systems will suppress the response.[3,23] Further subdivision of $CD4^+$ T cells has shown help to be confined to the

$CD4^+/CD45Ra^-$(2H4) $CD45Ro^+$(UCHL1) subset.[24,25] This phenotype was originally thought to define a subset responsible for induction of T cell help, and the reciprocal $2H4^+$ $UCHL1^-$ phenotype to define a suppressor inducer subset.[26,27] More recently, the loss of the CD45Ra marker and the expression of CD45Ro (UCHL1) has been shown to accompany T cell activation and the generation of memory T cells.[28-31] Help by memory T_h cells would of course be expected for responses to recall antigens such as influenza virus and tetanus toxoid.

Specific T helper cell clones have also been described; for example to influenza virus[32,33] and HSV.[6] In one study, antibody to hemagglutinin was obtained when B cells were stimulated with intact influenza virus in the presence of T cell clones specific for influenza matrix protein[32] suggesting associative recognition of two discrete antigens. T cell replacing factors (TRF) secreted by some T cell clones have been described,[32,33] but not properly characterized. In other studies, both IL-2 and low molecular weight BCGF ($BCGF_{low}$) have been shown to be potent TRFs,[34,35] and it is likely that one or other of these cytokines is responsible for the helper activity secreted by T cell clones. Secretion of IL-2 or $BCGF_{low}$ may also be the explanation for help by some cytotoxic T cell clones,[36] and for the lack of genetic restriction which has been observed in some instances.[6]

C. ABSENCE OF MHC CONTROLLED GENETIC RESTRICTION IN SPECIFIC ANTIBODY RESPONSES BY HUMAN LYMPHOCYTES

Allogeneic combinations of human B and T cells do not make antibody in response to stimulation with antigen.[3,22] In this respect, specific antibody responses differ from PWM stimulated Ig secretion in which T_h-B cell collaboration is entirely unrestricted.[37] In mapping experiments, combinations of allogeneic T cells and B cells which differed at either class I or class II regions of the MHC were unable to respond[38] suggesting that MHC (class II) controlled genetic restriction is not the only reason for the failure of incompatible Th and B cells to collaborate. It is now known that the inability to respond is due to the activation of suppressor T cells and not genetic restriction.[3,22,39] This somewhat surprising conclusion comes from experiments in which help by freshly prepared T cells was obtained in the absence of any class I or class II compatibility provided that T suppressor cells were removed either functionally by irradiation, or physically by depletion of the $CD8^+$ T suppressor/cytotoxic subset.[3,39-41] Under these conditions, allogeneic T cell help was shown to be antigen specific, and was not due to the release of nonspecific helper factors.[22,39] In the absence of allogeneic T suppressor cells, no evidence has been obtained for MHC controlled genetic restriction in specific antibody responses by human lymphocytes with any combination of allogeneic T cells and B cells.[22,38,39] Moreover, the requirement for antigen presenting cells in this response can be provided by allogeneic adherent cells (monocyte/macrophage) or allogeneic antigen pulsed adherent cells.[42]

The absence of MHC controlled genetic restriction in human antibody responses is not unique to one antigen as similar findings have been reported for both influenza virus and tetanus toxoid.[3,22] Rather, it seems to be a feature of human T_h-B cell interactions. The reasons for this are not clear, but may be related to the organization of the MHC in humans compared with the mouse. The potential number of unique class II structures on APC or B cells in individuals which are heterozygous at DP, DQ, and DR alpha and beta chain loci is clearly much greater than in inbred strains of mice. Accordingly, the potential for cross reactivity and therefore recognition of allogeneic class II/antigen complexes by the T cell receptor is also higher. Alternatively, as CD4 binding to class II is a feature of Th collaboration with B cells in specific antibody responses,[43] it seems possible that the absence of restriction might be due to interactions of this sort by-passing specific TCR recognition of antigen complexed with class II.

In contrast to the absence of restriction with normal unselected T cell populations, class II restriction has been observed with some but not all T helper cell clones.[6,44,45] For example, help provided by a hemagglutinin specific T cell clone in antibody responses to influenza virus was shown to be HLA-DR restricted and could be blocked with anti-DR but not anti-DC antibodies.[17] Interestingly, responses to the polysaccharide mannan from *Candida albicans* appear to be HLA-DQ restricted.[46] Whether this has anything to do with the different IgG subclasses made to the two antigens has yet to be determined. In those clones which are unrestricted, crossreactivity between allogeneic and autologous class II antigen complexes may occur. In this regard, TCR recognition of nonpolymorphic class II determinants has been put forward as an explanation for antigen specific, nonrestricted help exhibited by some clones.[47] Alternatively, it is possible that helper factors such as IL-2 or $BCGF_{low}$ released by allogeneic T cells or T cell clones may by-pass cognate T cell B cell interactions and induce antibody production in the absence of MHC compatibility.

III. POLYCLONAL RESPONSES

A. T CELL DEPENDENT RESPONSES TO POLYCLONAL ACTIVATORS

Helper T cells are required for optimal Ig production in response to most human B cell mitogens including PWM,[48] soluble *Staphylococcus aureus* protein A (SPA),[49] Bacto-streptolysin O (SLO),[2] and lipopolysaccharide (LPS).[50] Even the so called T cell independent mitogens such as water soluble extract from *Nocardia, Staphylococcus aureus* Cowan 1 (SAC), and PMA,[49,51] require T cells,[2,52,53] or T cell products (IL-4),[54] for optimal Ig secretion. Only EBV has been shown unequivocally to be a T cell independent mitogen for human B cells.[2,55,56]

B. ANTIBODY RESPONSES TO POKEWEED MITOGEN

PWM is the most commonly used T cell dependent mitogen for studies of human B cell differentiation. All the major immunoglobulin isotypes (IgM, IgG and IgA),[2] but not IgE,[57] are produced in cultures of blood or tonsillar mononuclear cells stimulated with PWM. The failure to make IgE may be due to an inability to activate appropriate (IL-4 secreting?) T helper cells, or to act on IgE specific B cell precursors. It cannot be explained by an absence of IgE specific B cells since significant IgE responses can be obtained from atopic and from normal donors by activation with HLA-DR specific T cell lines.[58,59] All four IgG subclasses are produced in response to PWM in approximately the same proportions as found in normal serum.[60-62] PWM seems therefore to be a crude activator of human B cells without the Ig isotype selectivity of either LPS or *Staph. aureus* which preferentially induce IgG2 and IgG3,[60] or specific antigens which give rise to restricted Ig class and IgG subclass responses (Callard and Hannam, in preparation). It is important however to appreciate that PWM activates only a small proportion of all B cells.[63] The majority of these are large low-density $sIgD^-$ (activated) B cells,[64,65] whereas those responding to LPS,[65] or specific antigen,[66] are mostly small high density (resting) B cells. It has also been shown that B cells secreting IgG or IgA in response to PWM are derived largely from $sIgM^-$ precursors, whereas $sIgM^+$ B cells mature almost exclusively into IgM secreting cells.[67-69] These results suggest that PWM activates only precommitted B cells and does not initiate switching.

C. T HELPER CELLS IN RESPONSES TO PWM

T cells responsible for help in PWM responses are $CD4^+$ whereas $CD8^+$ T cells are inhibitory.[70-72] Further subdivision of the $CD4^+$ subset with the monoclonal antibodies 2H4 (CD45Ra) and UCHL1 (CD45Ro) have shown that help for PWM responses is restricted to the $2H4^-$ $UCHL1^+$ subset,[24,26,31,73] now thought to be memory T cells.[28,30] Why help in polyclonal antibody responses to PWM should be confined to the $CD45Ra^-$ ($CD45Ro^+$) memory T cell subset is not clear although an explanation may be found in the different ability of these subsets to secrete IL-2. After activation, $CD45Ra^-$ cells produce maximal amounts of IL-2 within 24 h compared with CD45Ra$^+$ T cells which do not secrete IL-2 until much later.[74] Ig secretion by PWM stimulated B cell requires IL-2,[75,76] and it is possible that early secretion by the CD45Ra$^-$ subset may be essential for optimal help. The predominant importance of IL-2 for human B cell differentiation is a recurring theme in antibody responses to mitogens and antigens, and is discussed in more detail below.

T cell help for PWM induced antibody secretion is not MHC restricted and B cells will respond equally well in the presence of autologous or allogeneic T cells.[37] As a result, co-culture experiments with T cells and B cells from patients and normal donors can be performed in allogeneic combinations to define the underlying abnormality in certain immunological diseases.[2,77,78] T helper cells in PWM responses can be completely replaced with T cell conditioned medium, but the nature of the T cell replacing factor(s) responsible has not been defined.[37,66,79] The ease with which T cells can be replaced with TRF may help explain the absence of MHC controlled genetic restriction. Soluble mediators such as IL-2 secreted by T cells in response to PWM[75] may act on low density (activated) B cells to induce Ig secretion without the need for MHC restricted activation of T cells by antigen presenting cells or cognate interactions with B cells. This hypothesis is similar to the bystander effect in specific antibody responses,[80,81] and is consistent with the known activation of T cells by PWM and the 'activated' nature of B cells stimulated to Ig secretion by PWM.[64,65] Interestingly, evidence has been obtained recently for cognate T cell help in responses to PWM. In this study, it was found that T cells from an apparently normal sister of a patient who had died from severe combined immunodeficiency (a disease in which T cells are known to be defective) were unable to collaborate with MHC incompatible allogeneic B cells, but did collaborate with autologous B cells and with B cells from her histoidentical brother.[82] In this unusual case, it was suggested that secretion of B cell growth and differentiation factors (BCGF and BCDF) by T cells was defective, and that help depended on an alternative mechanism involving genetically restricted cognate interactions.

IV. T CELL ACTIVATION OF HUMAN B CELLS

A. MECHANISMS OF T_h-B CELL INTERACTIONS

It is well established that cell contact between specific T_h and B cells is required for optimal antibody production to T cell dependent antigens. In the first step of this process, antigen binding to B cell membrane Ig is internalized, processed and re-expressed on the surface in association with MHC class II.[83] T cell receptor (TCR) recognition of this complex together with binding of other (adhesion) molecules forms the basis of cognate interactions of T cell help. Intercellular signaling between T cells and B cells has also been shown to occur at a distance by the release of soluble factors (cytokines). Such a mechanism was first indicated by experiments in which conditioned medium from activated T cells was found to replace T_h in specific antibody responses.[84,85] Since then, a number of T cell derived B cell growth and differentiation factors have been described and their genes cloned. Those known to be important in human B cell responses are given in Table 2. The evidence for cognate vs. cytokine mediated help in human antibody responses is discussed in the following sections (IV.B and IV.C).

B. COGNATE INTERACTIONS BETWEEN T HELPER CELLS AND B CELLS

A requirement for cognate interactions between T_h and B cells has been demonstrated for antibody responses in man as it has in mice.[45,86-88] In most cases, however, this form of help has been shown to be required only by high density 'resting' B cells. For example, specific antibody production by high density B cells was found in one study to require MHC restricted antigen dependent cell contact with a (TNP) specific HLA-DR restricted T cell clone. In contrast, low density (activated) B cells were activated in a nonrestricted polyclonal

TABLE 2
T-Cell Derived Cytokines Active on Human Cells

Cytokine	Chromosome	Action on Human B Cells: Activation	Proliferation	Differentiation
IL-1	2	+?	+?	+?
IL-2	4	?	+	+
IL-4	5	+	+/−	+/−
IL-5	5	No	No	No
IL-6	7	?	+	+
IFN-gamma	12	−	+/−	+/−
TNF-alpha	6	?	+/−	+/−
TGF-beta	19	?	−	−
$BCGF_{low}$	1q21-32	+	+	+

Note: + = positive response; − = inhibition; No = no effect described; ? = not known.

manner by soluble factors.[45] Similarly, IL-2 has been shown to replace T cells in specific antibody responses by low density B cells, but in the same experiments T cells were required for antibody responses by high density B cells.[66,89] However, high density B cells cultured with T cells and antigen for 18 h and then depleted again of T cells were able to respond to IL-2 in the TRF assay.[90] These findings support the contention that small dense (resting) B cells have an absolute requirement for T cell activation before they are able to respond to B cell growth and differentiation factors, in this case IL-2.

1. Cell Surface Molecules Involved in Cognate Interactions

At least four membrane ligand receptor interactions are involved in cognate interactions between human T_h and B cells, each of which may deliver an activation signal to the B cell (Figure 1).

a. T Cell Receptor Recognition of Antigen Complexed with MHC Class II

The evidence for TCR recognition of antigen-class II complexes on B cells by T helper cells in human antibody responses is threefold. First, class II restricted help in specific antibody responses has been demonstrated for a number of antigen specific T cell clones.[32,33,45] Second, collaboration between T helper cell clones and B cells can be inhibited by antibody to the T cell receptor/CD3 complex or to MHC class II antigens.[43,91] Third, allospecific T cell clones are able to activate B cells to high rate Ig secretion.[58,59] The latter finding demonstrates the importance of T cell receptor binding to class II for B cell activation and differentiation.

b. CD4 Binding to Class II

Monoclonal CD4 antibodies inhibit specific antibody responses to T cell dependent antigens,[43] and IgE production by B cells stimulated with allospecific T cell clones.[91] In the former, synergistic inhibition was obtained with combinations of CD4 and MHC class II antibodies.[43] These results suggest that CD4 binding to class II is important for T_h cell interactions with B cells. Indeed, it has been suggested T_h activation of B cells by CD4 binding to class II and/or TCR recognition of non-polymorphic class II determinants may explain the unrestricted help observed with CD3 or mitogen activated T cells.[86,92,93]

c. LFA-1 ICAM-1/2 Interactions

Antibody to LFA-1 has been shown to block T cell induced proliferation of human B cells,[94] and T cell help in specific antibody responses.[95] Moreover, T cells from LFA-1

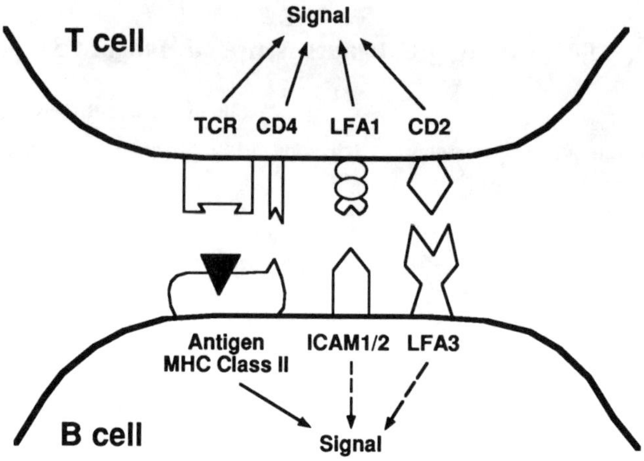

FIGURE 1. Receptor ligand interactions involved in cognate interactions between T_h and B cells.

deficient patients have an impaired ability to form conjugates with B cells, and are unable to function as T helper cells in specific antibody responses to influenza virus.[95,96] In these experiments, LFA-1$^+$ T cells were able to form conjugates with LFA-1$^-$ B cells, but LFA-1$^-$ T cells were unable to form conjugates, even with normal B cells. These findings suggest that binding of LFA-1 on T cells to ICAM1 or ICAM2 on B cells is required for cognate interactions. This conclusion is supported further by experiments in which preincubation of T cells but not B cells with anti-LFA-1 inhibited T cell dependent antibody responses.[95] In contrast, LFA-1 deficiency does not interfere with B cell responses to IL-4 or $BCGF_{low}$,[97] nor with antigen specific T cell proliferation, or with T helper cell function in polyclonal responses to PWM or specific antibody responses to mannan.[95] The importance of LFA-1 for antibody responses to protein but not polysaccharide antigens such as mannan may indicate a requirement for tighter T_h-B cell binding in the former. Alternatively, polysaccharide antigens may directly activate resting B cells and bypass an early T_h activation step.

d. LFA-3 Binding to CD2

Involvement of other adhesion molecules in T_h cell activation of human B cells is indicated by the inhibition of specific antibody responses following antibody blocking of CD2 on T_h cells, or LFA-3 on B cells.[96,98]

2. Signal Transduction Resulting from T_h Interactions with B Cells

An important question arising from the work with T_h-B cell interaction molecules is whether ligand receptor binding is a passive mechanism simply to bring the cells into close contact for responses to locally produced cytokines, or a signaling mechanism required for B cell activation. The weight of evidence available to date is consistent with the latter. For example, allospecific T cell binding to high density (resting) B cells is known to induce expression of B cell surface CD23.[99] Both live and fixed cells were able to elicit this response thus excluding indirect activation by the release of cytokines, particularly IL-4. No evidence for inositol phosphate breakdown or calcium release was obtained. In contrast, both anti-Ig and SAC activation of B cells resulted in hydrolysis of PIP2, but these mitogens were poor stimulators of CD23 expression.

Signal transduction by MHC class II on murine B cells has been reported by several groups. For example, antibody to MHC class II has been shown to increase intracellular cAMP, but not induce PIP2 hydrolysis or release of intracellular calcium.[100,101] Furthermore,

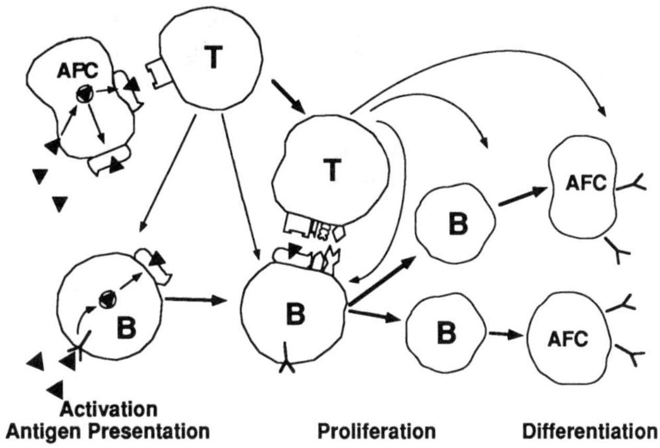

FIGURE 2. T cell regulation of B cell activation, antigen presentation, proliferation and differentiation into antibody forming cells (AFC). In this model it is proposed that antigen (▲) processing and re-expression on the surface of B cells precedes cognate interactions with T cells and response to B cell growth and differentiation factors (curved arrows). This early phase of B cell activation may also be regulated by T cell derived cytokines (thin straight arrows) released after activation of T cells by antigen presenting cells (APC).

both anti-class II, and cAMP analogs such as dibutyl-cAMP, or agents which elevate cAMP, have been found to induce protein kinase C (PKC) translocation from the cytosol to the nucleus.[101] Translocation of PKC away from the surface membrane may explain the inhibition of B cell responses to LPS by antibodies to class II antigens.[102,103] Direct evidence for signal transduction by class II on human B cells has not yet been reported, but antibodies to HLA class II antigens have been found to have both negative and positive effects on B cell function. In one study, antibodies to HLA-DR and HLA-DQ were found to inhibit the response of small resting $sIgM^+$ B cells to anti-IgM or SAC, but responses by low density 'activated' B cells were not affected.[104] As cognate T cell help is not required for B cell responses to these polyclonal activators, the results suggest that class II MHC molecules can deliver signals which inhibit early B cell activation events. Similar inhibition of B cell proliferation has been shown with a class II antibody which interfered with PIP2 hydrolysis and calcium flux induced with anti-Ig, but did not elevate cAMP.[105] In contrast, antibodies to HLA-DR have been shown to replace T_h cells in PWM induced polyclonal B cell proliferation and differentiation showing that class II antigens can also deliver a positive signal to B cells.[106] It is noteworthy that good responses were obtained in this study by prestimulation of B cells with PWM followed by treatment with anti-DR, but not in the reverse order. This is consistent with a recent study in which up to 95% of murine B cells were activated to enter the cell cycle by pretreatment with low doses of anti-IgM plus IL-4 followed by stimulation with immobilized anti-class II.[107] Subsequent addition of IL-4 and IL-5 in this system resulted in Ig secretion. Taken together, these studies show that signal transduction by B cell class II antigens after an initial exposure to antigen and a cytokine such as IL-4 is important (and may be essential) for activation and differentiation of resting B cells (Figure 2).

Signaling by other molecules involved in T_h-B cell interactions may also occur. There is evidence for transmembrane signaling by LFA-1,[108-110] but LFA-1 expression is required on T cells not B cells for conjugate formation and antibody responses.[96] Presumably, ICAM-1 or ICAM-2 on B cells binds to T cell LFA-1, but whether or not these molecules are signal transducers for B cells is not known. Experiments with blocking antibodies have also shown that binding of LFA-3 on B cells to CD2 on T_h is important for specific antibody responses.[96,98] LFA-3 is found in both transmembrane and PI linked forms and may be a signaling molecule for B cells as it is for T cells.[111]

C. ROLE OF CYTOKINES IN T_h ACTIVATION OF B CELLS

T cell derived cytokines are known to regulate B cell activation, proliferation (clonal expansion), differentiation into antibody secreting cells, and isotype switching (Table 2). It is also probable that some cytokines are important for antigen presentation by B cells, and for selection of diverging differentiation pathways into memory and antibody producing cells.

1. B Cell Activation

Under appropriate experimental conditions, ligand (antigen or anti-Ig) binding to B cell surface Ig triggers a cascade of biochemical events which primes the cell to respond to other signals but does not in itself induce cell division. Signaling by this pathway results in the hydrolysis of phosphatidylinositol 4,5-bisphosphate (PIP2) by polyphosphoinositide-specific phosphodiesterase (PPI-PDE) to yield inositol 1,4,5,-triphosphate (IP3) and 1,2-diacylglycerol (DAG). IP3 causes an increase in free intracellular calcium ($[Ca^{2+}]_i$), and DAG activates protein kinase C (PKC) (reviewed in References 112 and 113). Ultimately, these events result in specific gene activation, but the exact sequence of events from membrane to nucleus is not yet known. Recent experiments have shown that B cell responses to anti-Ig coupled to dextran can occur in the absence of a detectable increase in $[Ca^{2+}]_i$ or IP3 release from the breakdown of PIP2 suggesting that another signal transduction pathway may also be triggered by ligand binding to surface Ig.[114]

Except for IL-4, the intracellular signaling events triggered in B cells by T cell derived cytokines have not been determined.[115] In the mouse, IL-4 activation of B cells results in some protein phosphorylation, but does not appear to involve the breakdown of PIP2.[116] In contrast, activation of human B cells with IL-4 has been shown recently to stimulate a transient release of IP3 and elevation of $[Ca^{2+}]_i$ followed 10 to 15 min later by a rise in cAMP.[117] Both elements of this novel signaling pathway are required for the increase in expression of B cell surface CD23 in response to IL-4. Why the IL-4 signaling pathway in human B cells should be different to that in the mouse is not at all clear although it may be related to the different responses to IL-4 observed in the two species. Activation of resting human B cells with, for example, IL-4, anti-Ig, or PMA, results in the increased expression of cell surface activation antigens such as MHC class II, IgM, 4F2, CD23, IL-2 receptor (CD25) and IL-4 receptor. At the same time, other antigens such as IgD, CD5 and epitope c of CD24 are decreased or lost. The importance of these changes on the B cell surface are not always apparent although in the case of receptors for growth factors (IL-2 and IL-4) it is presumably in preparation for cell division and/or differentiation. Of interest is the dramatic increase (tenfold) in surface IgM on human B cells in response to IL-4.[118] Unlike CD23 which is increased in response to a number of different activation signals, the change in sIgM occurs only in response to IL-4. Moreover, it does not occur on mouse B cells. We now have evidence that the increases in CD23 and surface IgM on human B cells activated with IL-4 are regulated independently by two different biochemical pathways (Rigley et al.). There is evidence for IL-4 binding to low and high affinity receptors on human B cells,[119] but whether the independent regulation of CD23 and surface IgM can be explained by signaling through two separate IL-4 receptors has yet to be determined. If the function of IL-4 is to activate B cells to promote interactions with T_h, the increase in sIgM may be to enhance antigen uptake, processing, and presentation in association with MHC class II antigens which are also increased with IL-4. A role for IL-4 in B cell activation prior to interaction with T_h has also been indicated by experiments with murine cells.[120]

2. B Cell Proliferation

Co-stimulation of resting human B cells with anti-Ig and a T cell derived BCGF such as IL-2,[121,122] IL-4,[123] or $BCGF_{low}$[124] results in proliferation. TNF-alpha,[125,126] TNF-beta,[127] and IFN-gamma[128,129] are also growth factors in co-stimulation assays, but whether these act directly on B cells or indirectly through other cell types is not entirely clear. Other cytokines inhibit this response including IL-1[130] and TGF-beta.[131,132] IL-6 is a growth factor for some transformed B cells but does not induce proliferation in co-stimulation assays with anti-IgM, or SAC.[133-135] Importantly, IL-5 has no growth or differentiation factor activity on human B cells whatsoever,[136] although some indirect effects can occur in the presence of T cells.[137,138]

In contrast to its growth promoting action with anti-IgM, IL-4 is a potent inhibitor of B cell proliferation stimulated by IL-2.[139,140] The mechanism of this inhibition is not known although it does not seem to involve regulation of the IL-2 receptor since we have shown recently that IL-4 does not inhibit Tac expression induced with IL-2. The way in which B cell responses can be enhanced or inhibited by a single cytokine or other membrane signal depending on what other signals the cell has received is becoming a common finding. In addition to the example of IL-4, IFN-gamma is known to co-stimulate with anti-Ig, and inhibit B cell responses to IL-4. Similarly, we have recently found that CD19 monoclonal antibodies inhibit responses to anti-IgM plus IL-4 and enhance proliferation in response to EBV or mitogenic doses of PMA.

3. B Cell Differentiation

The best defined human B cell differentiation factors are IL-2 and IL-6. Both have been shown to induce Ig secretion by normal B cells stimulated with anti-Ig or SAC.[134,141] In co-stimulation assays with IL-2, B cell proliferation also occurs, and it is often difficult to distinguish between the two effects. With IL-6 however, increased Ig secretion occurs in the absence of any effect on cell division.[142,143]

Addition of IL-6 to PWM stimulated human PBMC has been shown to augment the production of all classes of immunoglobulin whereas the addition of a blocking IL-6 antibody diminished Ig secretion by as much as 90% without affecting proliferation.[144] These results suggest that IL-6 is an essential BCDF for T dependent (polyclonal) Ig secretion. On the other hand, our recent findings with specific antibody responses have led to an entirely different conclusion. In these experiments, addition of rIL-6 had no effect at all on specific antibody responses by tonsillar mononuclear cells to influenza virus, except at high doses (>1000 units/ml) when the response was partially inhibited (Callard, in press). Furthermore, a polyclonal blocking IL-6 antibody did not inhibit antibody production showing that IL-6 was not required for this response. Whether or not IL-6 is important for antibody production to other antigens, or in a primary rather than secondary response needs to be determined. A further discussion on cytokines in specific antibody responses is found in Section IV.C.5.

Several other cytokines have also been shown to increase Ig secretion by human B cells including IL-1, IFN-gamma and TNF.[121,125,126,145,146] None of these three cytokines has been unequivocally identified as a B cell differentiation factor in its own right, but each is able to augment differentiation in response to other factors particularly IL-2.[121,126,146] Although there is some evidence that IFN-gamma acts directly on B cells,[147] indirect action through non-B cells cannot be entirely ruled out. Interleukin 4 is not a BCDF in co-stimulation assays, but it has been shown to enhance Ig secretion by B cells prestimulated with SAC[54] or SAC plus IL-2.[148] IL-6 has recently been shown to be secreted by B cells in response to IL-4,[149] and may play a central role in IL-4 induced secretion of antibody. In all of these experiments, it is important to distinguish between signals which promote Ig secretion of differentiated cells and differentiation signals per se. Until the full sequence of events leading to antibody secretion from activation of B cells to interaction with T cells and differentiation into antibody secreting cells is properly defined, it will not be possible to determine the significance of the various 'differentiation' signals delivered by the different cytokines.

4. Cytokines and Switching

At least four different T cell derived cytokines have been shown to regulate immunoglobulin isotype production in the mouse. The best studied of these is IL-4 which selectively promotes the secretion of IgE and IgG1 by LPS stimulated spleen B cells.[150] TH2 cells also stimulate the production of these isotypes.[151] In similar studies, IL-5 and TGF-beta have been found to enhance IgA secretion, whereas IFN-gamma enhances IgG2a.[152-154] There are two independent mechanisms of isotype regulation which may be involved in selective isotype regulation. In the first, B cells already committed to the production of a particular isotype are selectively stimulated by the cytokine to proliferate and differentiate into antibody secreting cells. Alternatively, cytokines may signal B cells to switch by rearrangement of V_H to the required C_H region. Recent studies have shown IL-4 regulation of IgE and IgG1 to occur by the second mechanism. In contrast, IL-5 is thought to selectively stimulate B cells already committed to the production of IgA.[155-160] The mechanisms utilized by IFN-gamma and TGF-beta are less clear.

When considering isotype regulation in man, there are some similarities with the mouse, but there are also some important differences. A number of studies have shown that IL-4 selectively increases IgE secretion in man,[161,162] in some cases simply by incubating PBMC with IL-4.[162-164] This response is inhibited by IFN-gamma and by antibodies to CD23,[162-165] but is dependent on IL-6.[166] In all these experiments, IgE production is absolutely dependent on the presence of T cells,[161] and is inhibited by antibody to MHC class II, or to the T cell receptor.[91] Antibodies to CD4 and LFA-1 also inhibited IgE production implying a requirement for cognate interactions between autologous Th and B cells.[91]

Allospecific T cell clones, or antigen specific clones stimulated with antigen, mitogen (PHA) or CD3 antibody, may stimulate IgE production by normal human B cells.[59,93,167] In each case there is an absolute requirement for cognate interactions between the T cell clone and B cells,[91,167] and for subsequent activation by a soluble factor(s).[168] There is a high correlation between the ability to stimulate IgE production and secretion of IL-4 by T cell clones, and an inverse relationship with secretion of IFN-gamma.[161,169] Moreover, clones which do not secrete IL-4 do not support IgE responses.[169] The majority of human T cell clones, however, cannot be classified as TH1 or TH2 since IL-4, IFN-gamma and/or IL-2 are commonly produced by the same clone.[169,170]

The reason why T cells are required for IgE responses by human B cells in the experiments described above, but not in similar experiments with LPS stimulated mouse B cells, is not clear. Two possible explanations for this species difference should be considered. First, it is possible that there is a fundamental difference in the action of IL-4 on mouse and human B cells. The recent finding of a transmembrane signaling pathway stimulated by IL-4 in human B cells but not mouse B cells is consistent with this alternative.[117] Alternatively, an obligatory signal may be provided by T cells in the response by human B cells which LPS is able to provide in the mouse. In a recent study, IgE secretion was obtained from highly purified human B cells stimulated with EBV and IL-4,[171] suggesting that EBV may activate human B cells in a similar way to LPS in the mouse. Similar findings have been reported for PMA stimulated splenic B cells, but an indirect effect through the 3 to 7% T cells present in the B cell preparations could not be ruled out.[172] It is not yet known whether IL-4 stimulation of IgE production by human B cells is by selection of precommitted cells, or by switching. Resolution of this question is required before it can be unequivocally stated that IL-4 regulates IgE by the same mechanism in the two species.

Although the regulation of IgE production by IL-4 is similar in mouse and man, the same cannot be said for IgG subclass responses. In the studies undertaken so far, totally conflicting results have been obtained. For example, Flores-Romo et al. have shown that IL-4 increases IgG1-3 but not IgG4 secretion by B cells activated with PMA and ionomycin,[173] whereas Lundgren et al. found that IL-4 increased only IgG4 production by PMA stimulated human splenic B cells.[172] IgG4 is not equivalent to murine IgG1 either structurally or functionally, and the significance of this finding remains to be established. In another extensive study, the pattern of IgG subclass production obtained in response to different cytokines was shown to depend both on the tissue from which the B cells were prepared, and the means of B cell activation (SAC, PMA, etc.) (Calvert, submitted). As IgG subclasses in humans and mice have probably arisen independently at different times in evolution,[174] it is extremely unlikely that the same regulatory mechanisms would also have evolved in both species.[175] With this in mind, it is not surprising that IL-4 (and IFN-gamma) have not been shown to regulate IgG subclass responses in a comparable way in humans and mice.

5. Cytokines in Specific Antibody Responses

Most of what we know about cytokine action on B cells has come from studies with polyclonal activators such as anti-Ig, SAC, or PMA. As a result, very little is known about the function of cytokines in specific antibody responses except for some recent work on T cell replacing factors (TRF), and on requirements for IL-4 and IL-6.

Of all the recombinant factors known to regulate human B cell responses, only IL-2 and $BCGF_{low}$ have been shown to have TRF activity. Similar results have been obtained by two different groups. In one, IgM antibody production to the T cell dependent antigen TNP-PAA was obtained in cultures of purified B cells in the presence of IL-2.[98,176] Synergy between IL-2 and IFN-alpha was found in this system, but whether by direct action on B cells or indirectly through accessory cells was not determined.[176] IL-2 and $BCGF_{low}$ have also been identified as TRF in specific antibody responses to influenza virus.[34,35,90] This response is a secondary response to a recall antigen, with IgG1 and IgG4 being the predominant Ig isotypes produced.[17] In these experiments, IL-2 was shown to act directly on the B cells and not indirectly through T cells or other accessory cells.[34] Taken together, these results establish IL-2 as an important cytokine in specific antibody responses by human B cells.

In antibody responses to both TNP-PAA and influenza virus, IL-2 was found to be a TRF only for low/medium density B cells and was unable to restore antibody responses to high density 'resting' B cells.[66,89,90] Co-stimulation with other activation signals was unable to convert high density B cells to IL-2 responders, but overnight incubation with antigen and T cells was successful.[90] This observation is consistent with the commonly held dogma that resting B cells stimulated with T dependent antigens must receive a signal from T_h before they are able to respond to cytokines and differentiate into antibody secreting cells. In contrast, $BCGF_{low}$ was an extremely potent TRF for high density B cells in these experiments,[90] showing that T cells are not always obligatory for early B cell activation in responses to T dependent antigens.

The action of IL-4 in specific antibody responses has also been investigated. Addition of IL-4 to antigen (influenza virus) stimulated cultures of PBMC or TMC either had no effect, or significantly suppressed specific antibody production at doses as low as 10 units/ml (Optimal B cell proliferation in co-stimulation assays with anti-IgM was obtained with 100 to 200 units/ml of IL-4).[177] It was also found that a blocking IL-4 antibody did not inhibit specific antibody production showing that IL-4 was not required for responses to antigen. In contrast, inhibition was obtained with anti-Tac which is consistent with an important role for IL-2. IL-4 was also shown to inhibit antibody production by purified B cells when IL-2 was used as a TRF.[177,178] In both studies, the inhibitory effect of IL-4 was shown to be maximal during the first 3 days of culture and was lost by the end of day 4. No inhibition was obtained by exposure to IL-4 for the first day only suggesting that the

TABLE 3
Functional B Cell Surface Molecules

Antigen	Molecular ratio (M_r)	Signal	Possible function
CD19	95 K	Unknown	Inhibition/stimulation, homotypic adhesions
CD20	35 K	c-*myc*, G0-G1, no $[Ca^{2+}]_i$ phosphoprotein	Activation
CD21 (CR2 EBVR)	140 K	C3d receptor phosphoprotein	Activation
CD23	45 K	Unknown	Proliferation, IgE regulation
CD40	48 K	Not $[Ca^{2+}]_i$	Proliferation, rescue from apoptosis
Bp 95	95 K	$[Ca^{2+}]_i$	Activation, proliferation

effect occurred after antigenic stimulation.[178] Taken together, these results show that IL-4 inhibits early IL-2 dependent B cell activation or proliferation essential for specific responses to antigen. This conclusion is entirely consistent with other studies showing IL-4 inhibition of B cell proliferation but not differentiation in response to IL-2.[139,140] Similar studies have also been carried out with IL-6. Although blocking IL-6 antibodies inhibit PWM induced Ig secretion,[144] in our hands neither IL-6 nor a blocking IL-6 antibody had any effect at all on specific antibody responses to influenza virus (Callard, in press). These results underline the difficulties in drawing conclusions about the regulation of antibody responses in man from nonphysiological systems.

6. A Model for Cytokine Regulation of B Cell Growth and Differentiation

In most models of B cell responses to cytokines, activation by ligand (antigen or anti-Ig) binding to surface Ig primes the cell for proliferation and differentiation in response to growth and differentiation factors.[179] A moments reflection shows the inadequacy of this model. As illustrated in Figure 2, antigen binding to B cell membrane Ig is first internalized, processed, and then re-expressed on the surface in association with MHC class II.[83] Co-stimulation with antigen and IL-4 thus primes B cells for cognate interactions with T_h cells, and would not normally result in cell division. The increase in expression of surface IgM[118] and MHC class II[180] on B cells activated with IL-4 is consistent with this hypothesis. Note that the high doses of anti-IgM or crosslinked anti-IgM used to obtain proliferation of B cells with IL-4 may not accurately mimic stimulation of B cells with antigen.[181] After this activation step, cognate interactions with T cells followed by stimulation with growth and differentiation factors (e.g., IL-2) results in proliferation and differentiation. The appearance of IL-4 at this stage may be inhibitory.[177] This model more accurately reflects the events in a B cell response to T cell dependent antigens and is currently under investigation.

V. FUNCTIONAL B CELL SURFACE ANTIGENS

A number of B cell surface antigens defined with monoclonal antibodies are known to be important transmembrane signaling molecules (Table 3). Although these may not be directly involved in T_h interactions with B cells, they are undoubtedly important for controlling B cell function and are discussed briefly here. They will be referred to by their CD nomenclature as determined by the International Workshops on Human Leukocyte Differentiation Antigens. For a more comprehensive discussion of these molecules the reader is referred to the review by Clark and Ledbetter.[182]

A. CD19

CD19 is a heavily glycosylated M_r 95 K pan B cell marker. It has three extracellular

Ig-like domains, a single transmembrane region and a long serine/threonine-rich cytoplasmic tail.[183,184] CD19 monoclonal antibodies have been shown to either inhibit or enhance B cell responses depending on the nature of the other activating signal. For example, B cell proliferation in response to anti-IgM plus IL-4 or to low doses of PMA plus CD40 antibody, as well as specific antibody responses to influenza virus, are all inhibited by CD19 antibodies. In contrast, proliferation in response to EBV or high (mitogenic) concentrations of PMA, are significantly enhanced.[185,186] In the absence of cross linking, CD19 antibodies do not induce phosphatidylinositol metabolism or increase $[Ca^{2+}]_i$, and they do not interfere with signal transduction by either anti-IgM or IL-4 (Rigley and Callard, submitted). Taken together, these data show that CD19 signaling acts at a point distal to the second messenger cascades triggered by sIg and IL-4 receptor binding.

B. CD20

The CD20 antigen is a M_r 35 K non-glycosylated phosphoprotein found on most B cells. It has a small extracellular domain, three transmembrane regions, one of which spans the membrane twice without exiting from the membrane, and two (N and C terminal) intracellular tails.[182,187,188] One CD20 antibody (1F5) has been shown to increase expression of C3d receptors and MHC class II on resting B cells, and to stimulate B cell proliferation.[189-191] Stimulation of resting B cells with CD20 antibodies does not increase $[Ca^{2+}]_i$, and if added before anti-Ig does not modulate anti-Ig induced $[Ca^{2+}]_i$.[192,193] Some recent evidence suggests that CD20 may be an ion (Ca^{2+}) channel.[182] Activation of B cells with anti-Ig or PMA results in hyperphosphorylation and internalization of CD20 suggesting that it may function as a signal in a PKC dependent pathway.[182,194]

C. CD21

CD21 is the receptor for C3d and EBV. It is a nonphosphorylated M_r 140 K glycoprotein with a large extracellular domain consisting of 16 tandemly arranged repeating elements, a single transmembrane region, and a small cytoplasmic tail.[195] CD21 expression is restricted to B cells and is present in highest amounts on mature B cells but is lost on activation. CD21 monoclonal antibodies can stimulate proliferation of resting B cells,[196,197] but not much is known of signal transduction by this antigen. Antibody binding does not increase $[Ca^{2+}]_i$ but does enhance calcium release and proliferation in response to suboptimal doses of anti-IgM.[198]

D. CD23

This interesting B cell antigen is a low affinity receptor for IgE, an autocrine B cell growth factor and a signal transducing molecule. It is a M_r 45 K N-linked glycoprotein with a C-terminal extracellular region comprised of a reverse RGD triplet, a lectin like sequence and an IgE binding site, a transmembrane region and an N-terminal intracellular domain (reviewed in References 199 and 200). Two forms of CD23 have been described which differ only in their intracellular (N-terminal) domains derived by differential splicing.[201] One form is common to all cells expressing CD23 and the other is exclusive to B cells. It has been suggested that the former is important for IgE responses and the latter for immunoregulation. CD23 expression on B cells is increased on activation with, for example, IL-4, EBV and PMA. It is also increased on B cells as a consequence of cognate interactions with T_h cells which cannot be explained simply by the release of T cell derived cytokines such as IL-4.[47,99,202] Expression of CD23 by B cells and follicular dendritic cells in germinal centers has led Gordon et al. to postulate that release of s-CD23 as a consequence of B cell activation by T cells may be an autocrine mechanism for B cell clonal expansion.[199]

E. CD40

The CD40 antigen is a M_r 50 K glycoprotein found on B cells and carcinomas. It has significant homology with human nerve growth factor receptor suggesting that it may be a receptor for a cytokine.[203] The mature protein has an N-terminal extracellular region of 172 amino acids, a transmembrane region, and a cytoplasmic tail with six threonines and serines, but no tyrosines, as potential phosphorylation sites (reviewed in References 182 and 204). Co-stimulation of B cells with CD40 antibodies plus anti-IgM or PMA, but not IL-4 or $BCGF_{low}$ results in proliferation.[190,205] In experiments of this sort, CD40 antibodies act as a growth factor of similar potency to IL-4, but no known growth factor will stimulate proliferation of CD40 transformants.[182] CD40 monoclonal antibodies activate B cells and induce proliferation,[205] and it has recently been shown that CD40 signaling together with antigen rescues germinal center B cells undergoing affinity maturation (centrocytes) from apoptosis.[206]

F. Bp 95

This B cell antigen defined by the antibody G28-8 was classified as CD39 by the third international workshop, but has since been shown not to belong to this cluster. Activation of B cells with G28-8 antibody results in increased expression of MHC class II, and a G0 to G1 cell cycle transition. Intracellular Ca^{2+} concentrations are increased in resting, but not in activated (low density) B cells by G28-8.[182,207] In co-stimulation experiments, G28-8 plus IL-4 promotes strong B cell proliferation.[185,207]

G. WHAT ROLE DO THE B CELL SURFACE ANTIGENS PLAY IN T_h-B CELL INTERACTIONS

The above list of functionally important B cell surface molecules is not exhaustive, but it does serve to indicate the range and diversity of potential membrane interactions available to the B cell. To date, none of the natural ligands for these molecules has been identified. In principle, they may be soluble factors, or cell surface antigens on accessory cells including T cells.

Their functions in regulating B cell responses are likewise unknown although it is possible to speculate on their potential roles in determining the type of B cell response required. For example, CD40 appears to be important for rescuing antigen binding B cells from apoptosis during affinity maturation,[206] and others may be important for microenvironmental localization, activation of B cells for antigen presentation, stimulation of memory rather than antigen secreting cells, or immunoglobulin isotype regulation. Whatever their function, they must be taken into consideration when attempting to understand the various cell interactions involved in regulating antibody production by B cells.

VI. WHAT NEXT

In the past, it has been common to think of B cell differentiation as a linear transition from a homogeneous population of resting B cells to antibody secreting cells regulated by interactions with T cells and cytokines. This simple model has been useful, but is now inadequate to account for the complexity of B cell responses to the many different interacting signals which have recently been identified. When the possible combinations of different signals received by B cells from interactions with T cells, cytokines, and other ligands binding to B cell surface molecules, are taken into account, it is clear why earlier efforts to understand B cell regulation simply in terms of interactions with networks of T cell subsets have not been very successful. It is also essential to appreciate that B cell responses are more complex than considered hitherto. In addition to activation, proliferation, differentiation and isotype regulation, B cell membrane signaling must also be important for antigen presentation, antibody affinity maturation, generation of memory, tolerance, recirculation, and microenvironmental localization. Despite the complexity, the next decade should see major advances in our understanding of how interacting molecular signals at the cell surface precisely determine B cell function. At the same time, it will be essential to obtain a better understanding of B cell responses *in vivo*. It has become clear recently that the microenvironmental localization of B cell subpopulations in lymphoid tissues is fundamental to their function. For example, it is now known that virgin B cells which migrate to extrafollicular areas rich in interdigitating cells are replaced in a few days. In contrast, primed B cells found mostly in the follicles, in which antigen is present on dendritic cells, sometimes for many months.[208] In the near future, specific staining with monoclonal antibodies and/or *in situ* hybridization will permit the signals controlling B cell migration and their different responses in specific microenvironments to be defined. This integrated approach will provide a more complete picture of immunoregulation in man, and ultimately enable the mechanisms underlying various immunological diseases to be properly defined.

REFERENCES

1. **Ammann, A. J. and Hong, R.,** Disorders of the T cell system, in *Immunologic Disorders in Infants and Children,* Steihm, E. R., Ed., W. B. Saunders, Philadelphia, 1989, 257.
2. **Waldmann, T. A. and Broder, S.,** Polyclonal B cell activators in the study of the regulation of immunoglobulin synthesis in the human system, *Adv. Immunol.,* 32, 1, 1982.
3. **Brenner, M. K. and Munro, A. J.,** T-cell help in human in vitro antibody producing systems: role of inhibitory T-cells in masking allogeneic help, *Cell. Immunol.,* 57, 201, 1981.
4. **Lane, H. C., Volkman, D. J,. Whalen, G., and Fauci, A. S.,** In vitro antigen-induced, antigen-specific antibody production in man. Specific and polyclonal components, kinetics, and cellular requirements, *J. Exp. Med.,* 154, 1043, 1981.
5. **Callard, R. E.,** Specific in vitro antibody response to influenza virus by human blood lymphocytes, *Nature,* 282, 734, 1979.
6. **Williamson, S. A., Parish, N., Chambers, J. D., and Knight, R. A.,** Human immune responses to herpes simplex virus, varicella-zoster and cytomegalovirus in vitro, *Immunology,* 57, 437, 1986.
7. **Uytdehaag, F. G. C. M., Loggen, H. G., Logtenberg, T., Lichtveld, R. A., Van Steenis, B., Van Asten, J. A. A. M., and Osterhaus, A. D. M. E.,** Human peripheral blood lymphocytes from recently vaccinated individuals produce both type-specific and intertypic cross-reacting neutralizing antibody on in vitro stimulation with one type of poliovirus, *J. Immunol.,* 135, 3094, 1985.
8. **Uytdehaag, F. G. C. M., Osterhaus, A. D. M. E., Loggen, H. G., Bakker, R. H. L., Van Asten, J. A. A. M., Kreeftenberg, J. G., Van Der Marel, P., and Van Steenis, B.,** Induction of antigen-specific antibody response in human peripheral blood lymphocytes in vitro by a kidney cell vaccine against rabies virus (DKCV), *J. Immunol.,* 131, 1234, 1983.
9. **Nutman, T. B., Withers, A. S., and Ottesen, E. A.,** In vitro parasite antigen-induced antibody responses in human helminth infections, *J. Immunol.,* 135, 2794, 1985.
10. **Durandy, A., Fischer, A., and Griscelli, C.,** Specific in vitro anti-mannan-rich antigen of *Candida albicans* antibody production by sensitized human blood lymphocytes, *J. Clin. Invest.,* 71, 1602, 1983.
11. **Heilmann, C., Barington, T., and Sigsgaard, T.,** Subclass of individual IgA-secreting human lymphocytes. Investigation of in vivo pneumococcal polysaccharide-induced and in vitro mitogen-induced blood B cells by monolayer plaque-forming cell assays, *J. Immunol.,* 140, 1496, 1988.
12. **Hoffmann, M. K.,** Antigen-specific induction and regulation of antibody synthesis in cultures of human peripheral blood mononuclear cells, *Proc. Natl. Acad. Sci. U.S.A.,* 77, 1139, 1980.
13. **Golding, B., Chang, S. P., Golding, H., Jones, R. E., Pratt, K. L., Burger, D. R., and Rittenberg, M. B.,** Human lymphocytes can generate thymus-independent as well as thymus-dependent anti-hapten plaque forming cell responses in vitro, *J. Immunol.,* 127, 220, 1981.
14. **Morimoto, C., Reinherz, E. L., and Schlossman, S. F.,** Regulation of in vitro primary anti-DNP antibody production by functional subsets of T lymphocytes in man, *J. Immunol.,* 127, 69, 1981.
15. **Delfraissy, J.-F., Gelanaud, P., Dormont, J., and Wallon, C.,** Primary in vitro antibody response from human peripheral blood lymphocytes, *J. Immunol.,* 118, 630, 1977.
16. **Golding, B., Muchmore, A. V., and Blaese, R. M.,** Newborn and Wiskott-Aldrich patient B cells can be activated by TNP-brucella abortus: evidence that TNP-brucella abortus behaves as a T-independent type 1 antigen in humans, *J. Immunol.,* 133, 2966, 1984.

17. **Callard, R. E. and Hannam, S.,** IgG subclasses in specific antibody responses to influenza virus by human blood lymphocytes in vitro, submitted.
18. **McCaughan, G. W., Adams, E., and Basten, A.,** Human antigen-specific Iga responses in blood and secondary lymphoid tissue: an analysis of help and suppression, *J. Immunol.,* 132, 1190, 1984.
19. **Emmrich, F., Schilling, B., and Eichmann, K.,** Human immune response to group A streptococcal carbohydrate (A-CHO). I. Quantitative and qualitative analysis of the A-CHO-specific B cell population responding in vitro to polyclonal and specific activation, *J. Exp. Med.,* 161, 547, 1985.
20. **Bloem, A. C., Jurgens, R., Eichmann, K., and Emmrich, F.,** Human immune response to group A streptococcal carbohydrate (A-CHO). III. Comparison of the efficiencies of anti-idiotopic antibody and of nominal antigen in the induction of IgM anti-A-CHO-producing B cells, *Eur. J. Immunol.,* 18, 1959, 1988.
21. **Kaplan, M. E. and Clark, C.,** An improved rosetting assay for detection of human T lymphocytes, *J. Immunol. Methods,* 5, 131, 1974.
22. **Callard, R. E. and Smith, C. M.,** Histocompatibility requirements for T cell help in specific in vitro antibody responses to influenza virus by human blood lymphocytes, *Eur. J. Immunol.,* 11, 206, 1981.
23. **Callard, R. E., Smith, C. M., and Beverley, P. C. L.,** Phenotype of human T helper and suppressor cells in an in vitro specific antibody response, *Eur. J. Immunol.,* 12, 232, 1982.
24. **Morimoto, C., Letvin, N. L., Distaso, J. A., Brown, H. M., and Schlossman, S. F.,** The cellular basis for the induction of antigen-specific T8 suppressor cells, *Eur. J. Immunol.,* 16, 198, 1986.
25. **Beverley, P. C. L., Terry, L., and Pickford, A.,** T cell subsets and function, in *Progress in Immunology VI,* Cinader, B. and Miller, R. G., Eds., Academic Press, New York, 1986, 941.
26. **Morimoto, C., Letvin, N. L., Distaso, J. A., Aldrich, W. R., and Schlossman, S. F.,** The isolation and characterization of the human suppressor inducer T cell subset, *J. Immunol.,* 134, 1508, 1985.
27. **Calvo, C.-F., Bernard, A., Huet, S., Leroy, E., Boumsell, L., and Senik, A.,** Regulation of immunoglobulin synthesis by human T cell subsets as defined by anti-D44 monoclonal antibody within the CD4+ and CD8+ subpopulations, *J. Immunol.,* 136, 1144, 1986.
28. **Akbar, A. N., Terry, L., Timms, A., Beverley, P. C. L., and Janossy, G.,** Loss of CD45R and gain of UCHL1 reactivity is a feature of primed T cells, *J. Immunol.,* 140, 2171, 1988.
29. **Sanders, M. E., Makgoba, M. W., and Shaw, S.,** Human naive and memory cells: reinterpretation of helper-inducer and suppressor-inducer subsets, *Immunol. Today,* 9, 195, 1988.
30. **Merkenschlager, M., Terry, L., Edwards, R., and Beverley, P. C. L.,** Limiting dilution analysis of proliferative responses in human lymphocyte populations defined by the monoclonal antibody UCHL1: implications for differential CD45 expression in T cell memory formation, *Eur. J. Immunol.,* 18, 1653, 1988.
31. **Beverley, P. C. L.,** Human T cell subsets, *Immunology Lett.,* 14, 263, 1987.
32. **Lamb, J. R., Woody, J. N., Hartzman, R. J., and Eckels, D. D.,** In vitro influenza virus-specific antibody production in man: antigen-specific and HLA-restricted induction of helper activity mediated by cloned human T lymphocytes, *J. Immunol.,* 129, 1465, 1982.
33. **Fischer, A., Beverley, P. C. L., and Feldmann, M.,** Long-term human T-helper lines producing specific helper factor reactive to influenza virus, *Nature,* 294, 166, 1981.
34. **Callard, R. E., Smith, S. H., Shields, J., and Levinsky, R. J.,** T cell help in human antigen specific antibody responses can be replaced by interleukin 2 (IL-2), *Eur. J. Immunol.,* 16, 1037, 1986.
35. **Smith, S. H., Shields, J. G., and Callard, R. E.,** Human T cell replacing factors: a comparison of recombinant and purified human B cell growth and differentiation factors, *Eur. J. Immunol.,* 19, 2045, 1989.
36. **Yasukawa, M., Inatsuki, A., and Kobayashi, Y.,** Helper activity in antigen-specific antibody production mediated by CD4+ human cytotoxic T cell clones directed against herpes simplex virus, *J. Immunol.,* 140, 3419, 1988.
37. **Hirano, T., Kuritani, T., Kishimoto, T., and Yamamura, Y.,** In vitro immune response of human peripheral lymphocytes. I. The mechanism(s) involved in T cell helper functions in the pokeweed mitogen-induced differentiation and proliferation of B cells, *J. Immunol.,* 119, 1235, 1977.
38. **Callard, R. E., Winger, C. M., and Tiernan, S. L.,** Antigen specific suppression of human antibody responses by allogeneic T cells. III. Role of the major histocompatibility complex, *Hum. Immunol.,* 17, 69, 1986.
39. **Callard, R. E., Winger, C. M., and Tiernan, S. L.,** Antigen-specific suppression of human antibody responses by allogeneic T cells. I. Frequency and antigen specificity of allogeneic suppressor T cells and their role in major histocompatibility complex controlled genetic restriction, *J. Exp. Med.,* 159, 1225, 1984.
40. **Smith, C. M. and Callard, R. E.,** Antigen specificity and frequency of autologous and allogeneic helper T cells in the in vitro production of antibody against influenza virus by human blood lymphocytes, *Eur. J. Immunol.,* 12, 558, 1982.
41. **Callard, R. E., Tiernan, S. L., and Winger, C. M.,** Antigen specific suppression of human antibody responses by allogeneic T cells. II. Cell interactions involved in the generation of suppression, *Eur. J. Immunol.,* 16, 252, 1986.

42. **Mitchell, D. M., Beverley, P. C. L., Boyle, D. G., Winger, L. A., and Callard, R. E.,** Accessory cell and HLA compatibility requirements for the generation of specific in vitro antibody responses to influenza virus by human blood lymphocytes, *Immunology,* 50, 239, 1983.
43. **Fischer, A., Sterkers, G., Charron, D., and Durandy, A.,** Possible T4-HLA class II interaction as an essential event in antigen-specific helper T lymphocyte-dependent B cell activation, *Eur. J. Immunol.,* 16, 1111, 1986.
44. **Fischer, A., Sterkers, G., Charron, D., and Durandy, A.,** HLA class II restriction governing cell cooperation between antigen-specific helper T lymphocytes, B lymphocytes and monocytes for in vitro antibody production to influenza virus, *Eur. J. Immunol.,* 15, 620, 1985.
45. **Principato, M. A., Thompson, G. S., and Friedman, S. M.,** A cloned major histocompatibility complex-restricted trinitrophenyl-reactive human helper T cell line that activates B cell subsets via two distinct pathways, *J. Exp. Med.,* 158, 1444, 1983.
46. **Durandy, A., Fischer, A., Charron, D. J., and Griscelli, C.,** Restriction of the in vitro anti-mannan antibody response by HLA-DQ molecules, *Hum. Immunol.,* 16, 114, 1986.
47. **Friedman, S. M., Jover, J. A., Chartash, E. K., and Crow, M. K.,** Antigen-specific, MHC nonrestricted T helper cell-induced B cell activation, *J. Exp. Med.,* 164, 1773, 1986.
48. **Keightley, R. G., Cooper, M. D., and Lawton, A. R.,** The T cell dependence of B cell differentiation induced by pokeweed mitogen, *J. Immunol.,* 117, 1538, 1976.
49. **Romagnani, S., Amadori, A., Giudizi, M. G., Biagiotti, R., Maggi, E., and Ricci, M.,** Different mitogenic activity of soluble and insoluble staphylococcal protein A (SPA), *Immunology,* 35, 471, 1978.
50. **Fauci, A. S. and Pratt, K. R.,** Polyclonal activation of bone marrow-derived lymphocytes from human peripheral blood measured by a direct plaque-forming cell assay, *Proc. Natl. Acad. Sci. U.S.A.,* 73, 3676, 1976.
51. **Pryjma, J., Munoz, J., Galbraith, J. M., Fudenberg, H. H., and Virella, G.,** Induction and suppression of immunoglobulin synthesis in cultures of human lymphocytes: effects of pokeweed mitogen and *Staphylococcus aureus* cowan I, *J. Immunol.,* 124, 656, 1980.
52. **Ralph, P. and Kishimoto, T.,** Tumour promotor phorbol myristic acetate is a T cell dependent inducer of immunoglobulin secretion in human lymphocytes, *Clin. Immunol. Immunopathol.,* 22, 340, 1982.
53. **Ringden, O., Rynnel-dagoo, B., Waterfield, E. M., Moller, E., and Moller, G.,** Polyclonal antibody secretion in human lymphocytes induced by killed staphylococcal bacteria and by lipopolysaccharide, *Scand. J. Immunol.,* 6, 1159, 1977.
54. **Defrance, T., Vanbervliet, B., Pene, J., and Banchereau, J.,** Human recombinant IL-4 induces activated B lymphocytes to produce IgG and IgM, *J. Immunol.,* 141, 2000, 1988.
55. **Kirchner, H., Tosato, G., Blaese, R. M., Broder, S., and Magrath, I. T.,** Polyclonal immunoglobulin secretion by human B lymphocytes exposed to Epstein-Barr virus in vitro, *J. Immunol.,* 122, 1310, 1979.
56. **Bird, A. G. and Britton, S.,** A new approach to the study of human B lymphocyte function using an indirect plaque assay and a direct B cell activator, *Immunol. Rev.,* 45, 41, 1979.
57. **Sampson, H. A. and Buckley, R. H.,** Human IgE synthesis in vitro: a reassessment, *J. Immunol.,* 127, 829, 1981.
58. **Lanzavecchia, A., Parodi, B., and Celada, F.,** Activation of human B lymphocytes: frequency of antigen-specific B cells triggered by alloreactive or by antigen-specific T cell clones, *Eur. J. Immunol.,* 13, 733, 1983.
59. **Lanzavecchia, A.,** One out of five peripheral blood B lymphocytes is activated to high-rate Ig production by human alloreactive T cell clones, *Eur. J. Immunol.,* 13, 820, 1983.
60. **Walker, L., Johnson, G. D., and MacLennan, I. C. M.,** The IgG subclass responses of human lymphocytes to B-cell activators, *Immunology,* 50, 269, 1983.
61. **Scott, M. G. and Nahm, M. H.,** Mitogen-induced human IgG subclass expression, *J. Immunol.,* 133, 2454, 1984.
62. **Morell, A., Van Loghem, E., Nef, M., Theilkaes, L., and Skvaril, F.,** Determination of IgG subclasses and Gm allotypes in culture supernatants of pokeweed mitogen-stimulated human blood lymphocytes, *J. Immunol.,* 127, 1099, 1981.
63. **Stevens, R. H., Macy, E., and Thiele, C. J.,** Evidence that pokeweed-mitogen reactive B cells are precommitted in vivo to the high-rate secretion of a single immunoglobulin isotype in vitro, *Scand. J. Immunol.,* 14, 449, 1981.
64. **Kuritani, T. and Cooper, M. D.,** Human B cell differentiation. II. Pokeweed mitogen responsive B cells belong to a surface immunoglobulin D-negative subpopulation, *J. Exp. Med.,* 155, 1561, 1982.
65. **Dagg, M. K. and Levitt, D.,** Human B lymphocyte subpopulations. I. Differentiation of density separated B lymphocytes, *Clin. Immunol. Immunopathol.,* 21, 39, 1981.
66. **Callard, R. E. and Tiernan, S. L.,** Specific antibody responses by high and low density human peripheral blood B cells: T helper cells and T cell replacing factor (TRF) act on different B cell subpopulations, *Immunology,* 62, 451, 1987.

67. **Thiele, C. J. and Stevens, R. H.,** Antibody potential of human peripheral blood lymphocytes differentially expressing surface membrane IgM, *J. Immunol.*, 124, 1898, 1980.
68. **Saiki, O. and Ralph, P.,** IgM- and IgG-bearing peripheral blood lymphocytes differentiate to IgM but not IgG or IgA immunoglobulin-secreting cells, *Eur. J. Immunol.*, 12, 506, 1982.
69. **Levitt, D. and Dagg, M. K.,** Human B-lymphocyte subpopulations. II. Plasma cell differentiation of isotype-specific B-lymphocytes from peripheral blood, *Clin. Immunol. Immunopathol.*, 21, 50, 1981.
70. **Reinherz, E. L., Kung, P. C., Goldstein, G., and Schlossman, S. F.,** Further characterisation of the human inducer T cell subset defined by monoclonal antibody, *J. Immunol.*, 123, 2894, 1979.
71. **Reinherz, E. L., Morimoto, C., Penta, A. C., and Schlossman, S. F.,** Regulation of B cell immunoglobulin secretion by functional subsets of T lymphocytes in man, *Eur. J. Immunol.*, 10, 570, 1980.
72. **Reinherz, E. L. and Schlossman, S. F.,** The characterisation and function of human immunoregulatory T lymphocyte subsets, *Immunol. Today*, 10, 69, 1981.
73. **Smith, S. H., Brown, M. H., Rowe, D., Callard, R. E., and Beverley, P. C. L.,** Functional subsets of human helper-inducer cells defined by a new monoclonal antibody, UCHL1, *Immunology*, 58, 63, 1986.
74. **Dohlsten, M., Hedlund, G., Sjogren, H.-O., and Carlsson, R.,** Two subsets of human $CD4^+$ T helper cells differing in kinetics and capacities to produce interleukin 2 and interferon-gamma can be defined by the Leu18 and UCHL1 monoclonal antibodies, *Eur. J. Immunol.*, 18, 1173, 1988.
75. **Nakagawa, N., Nakagawa, T., Volkman, D. J., Ambrus, J. L., and Fauci, A. S.,** The role of interleukin 2 in inducing Ig production in a pokeweed mitogen-stimulated mononuclear cell system, *J. Immunol.*, 138, 795, 1987.
76. **Miedema, F., Van Oostveen, J. W., Sauerwein, R. W., Terpstra, F. G., Aarden, L. A., and Melief, C. J. M.,** Induction of immunoglobulin synthesis by interleukin 2 is T4+/T8− cell dependent. A role for interleukin 2 in the pokeweed mitogen-driven system, *Eur. J. Immunol.*, 15, 107, 1985.
77. **Platts-Mills, T. A. E., De Gast, G. C., Webster, A. D. B., Asherson, G. L., and Wilkins, S. R.,** Two immunologically distinct forms of late-onset hypogammaglobulinaemia, *Clin. Exp. Immunol.*, 44, 383, 1981.
78. **Waldmann, T. A., Broder, S., Krakauer, R., MacDermot, R. P., Durm, M., Goldman, C., and Meade, B.,** The role of suppressor cells in the pathogenesis of common variable hypogammaglobulinaemia and the immunodeficiency associated with myeloma, *Fed. Proc.*, 35, 2067, 1976.
79. **Janossy, G. and Greaves, M. F.,** Functional analysis of murine and human B lymphocyte subsets, *Transplant. Rev.*, 24, 177, 1975.
80. **Julius, M. H., Chiller, J. M., and Sidman, C. L.,** Major histocompatibility complex-restricted cellular interactions determining B cell activation, *Eur. J. Immunol.*, 12, 627, 1982.
81. **Whalen, B. J., Tony, H.-P., and Parker, D. C.,** Characterization of the effector mechanism of help for antigen-presenting and bystander resting B cell growth mediated by Ia-restricted Th2 helper T cell lines, *J. Immunol.*, 141, 2230, 1988.
82. **Yata, J., Kobayashi, N., Shimbo, T., Kakihara, U., Sugahara, M., and Yamakura, T.,** Requirement of HLA compatibility between T and B cells in mitogen induced in vitro immunoglobulin production, *J. Clin. Lab. Immunol.*, 1, 245, 1978.
83. **Lanzavecchia, A.,** Clonal sketches of the immune response, *EMBO J.*, 7, 2945, 1988.
84. **Schimpl, A. and Wecker, E.,** Replacement of T-cell function by a T-cell product, *Nature New Biol.*, 237, 15, 1972.
85. **Dutton, R. W.,** Separate signals for the initiation of proliferation and differentiation in the B cell response to antigen, *Transplant. Rev.*, 23, 66, 1975.
86. **Clement, L. T., Dagg, M. K., and Gartland, G. L.,** Small, resting B cells can be induced to proliferate by direct signals from activated helper T cells, *J. Immunol.*, 132, 740, 1984.
87. **Snow, E. C. and Noelle, R. J.,** Thymus-dependent antigenic stimulation of hapten specific B lymphocytes, *Immunol. Rev.*, 99, 173, 1987.
88. **Krusemeier, M. and Snow, E. C.,** Induction of lymphokine responsiveness of hapten-specific B lymphocytes promoted through an antigen-mediated T helper lymphocyte interaction, *J. Immunol.*, 140, 367, 1988.
89. **Richard, Y., Leprince, C., Treton, D., and Galanaud, P.,** Functional heterogeneity of nonresting B cells in human blood, *Eur. J. Immunol.*, 16, 1303, 1986.
90. **Callard, R. E. and Smith, S. H.,** Interleukin 2 (IL-2) and low molecular weight B cell growth factor ($BCGF_{low}$) are T cell replacing factors (TRF) for different subpopulations of human B cells, *Eur. J. Immunol.*, 18, 1635, 1988.
91. **Vercelli, D., Jabara, H. H., Arai, K.-I., and Geha, R. S.,** Induction of human IgE synthesis requires interleukin 4 and T/B cell interactions involving the T cell receptor/CD3 complex and MHC class II antigens, *J. Exp. Med.*, 169, 1295, 1989.
92. **Noelle, R. J., McCann, J., Marshall, L., and Bartlett, W. C.,** Cognate interactions between helper T cells and B cells. III. Contact-dependent, lymphokine-independent induction of B cell cycle entry by activated helper T cells, *J. Immunol.*, 143, 1807, 1989.

93. **Del Prete, G., Maggi, E., Macchia, D., Tiri, A., Parronchi, P., Ricci, M., and Romagnani, S.,** Human T cell clones can induce in vitro IgE synthesis in normal B cells regardless of alloantigen recognition or specificity for peculiar antigens, *Eur. J. Immunol.,* 16, 1509, 1986.
94. **Tedder, T. F., Schmidt, R. E., Rudd, C. E., Kornacki, M. M., Ritz, J., and Schlossman, S. F.,** Function of the LFA-1 and T4 molecules in the direct activation of resting human B lymphocytes by T lymphocytes, *Eur. J. Immunol.,* 16, 1539, 1986.
95. **Fischer, A., Durandy, A., Sterkers, G., and Griscelli, C.,** Role of the LFA-1 molecule in cellular interactions required for antibody production in humans, *J. Immunol.,* 136, 3198, 1986.
96. **Mazerolles, F., Lumbroso, C., Lecomte, O., Le Deist, F., and Fischer, A.,** The role of lymphocyte function-associated antigen 1 (LFA-1) in the adherence of T lymphocytes to B lymphocytes, *Eur. J. Immunol.,* 18, 1229, 1988.
97. **Shields, J. G., Smith, S. H., Strobel, S., Levinsky, R. J., Defrance, T., De Vries, J., Banchereau, J., and Callard, R. E.,** Response of LFA-1 deficient B cells to interleukin 4 (BSF-1) and low molecular weight B cell growth factor ($BCGF_{low}$), *Eur. J. Immunol.,* 18, 255, 1987.
98. **Emilie, D., Wallon, C., Galanaud, P., Fischer, A., Olive, D., and Delfraissy, J.-F.,** Role of the LFA3-CD2 interaction in human specific B cell differentiation, *J. Immunol.,* 141, 1912, 1988.
99. **Chartash, E. K., Imai, A., Gershengorn, M. C., Crow, M. K., and Friedman, S. M.,** Direct human T helper cell-induced B cell activation is not mediated by inositol lipid hydrolysis, *J. Immunol.,* 140, 1974, 1988.
100. **Chen, Z. Z., McGuire, J. C., Leach, K. L., and Cambier, J. C.,** Transmembrane signaling through B cell MHC Class II molecules: anti-Ia antibodies induce protein kinase C translocation to the nuclear fraction, *J. Immunol.,* 138, 2345, 1987.
101. **Cambier, J. C., Newell, M. K., Justement, L. B., McGuire, J. C., Leach, K. L., and Chen, Z. Z.,** Ia binding ligands and cAMP stimulate nuclear translocation of PKC in B lymphocytes, *Nature,* 327, 629, 1987.
102. **Forsgren, S., Pobor, G., Coutinho, A., and Pierres, M.,** The role of I-A/E molecules in B lymphocyte activation. I. Inhibition of lipopolysaccharide-induced responses by monoclonal antibodies, *J. Immunol.,* 133, 2104, 1984.
103. **Neiderhuber, J. E., Frelinger, J. A., Dugan, E., Coutinho, A., and Shreffler, D. C.,** Effects of anti-Ia serum on mitogenic responses. I. Inhibition of the proliferative response to B cell mitogen, LPS, by specific anti-Ia sera, *J. Immunol.,* 115, 1672, 1975.
104. **Clement, L. T., Tedder, T. F., and Gartland, G. L.,** Antibodies reactive with class II antigens encoded for by the major histocompatibility complex inhibit human B cell activation, *J. Immunol.,* 136, 2375, 1986.
105. **Holte, H., Blomhoff, H. K., Beiske, K., Funderud, S., Torjesen, P., Gaudernack, G., Stokke, T., and Smeland, E. B.,** Intracellular events associated with inhibition of B cell activation by monoclonal antibodies to HLA class II antigens, *Eur. J. Immunol.,* 19, 1221, 1989.
106. **Palacios, R., Martinez-Maza, O., and Guy, K.,** Monoclonal antibodies against HLA-DR antigens replace T helper cells in activation of B lymphocytes, *Proc. Natl. Acad. Sci. U.S.A.,* 80, 3456, 1983.
107. **Bartlett, W. C., Michael, A., McCann, J., Yuan, D., Claassen, E., and Noelle, R. J.,** Cognate interactions between helper T cells and B cells. II. Dissection of cognate help by using a class II-restricted, antigen-specific, IL-2-dependent helper T cell clone, *J. Immunol.,* 143, 1745, 1989.
108. **Mishra, G. C., Berton, M. T., Oliver, K. G., Krammer, P. H., Uhr, J. W., and Vitetta, E. S.,** A monoclonal anti-mouse LFA-1 alpha antibody mimics the biological effects of B cell stimulatory factor-1 (BSF-1), *J. Immunol.,* 137, 1590, 1986.
109. **Pircher, H., Groscurth, P., Baumhutter, S., Aguet, M., Zinkernagel, R. M., and Hengartner, H.,** A monoclonal antibody against altered LFA-1 induces proliferation and lymphokine release of cloned T cells, *Eur. J. Immunol.,* 16, 172, 1986.
110. **Van Noesel, C., Miedema, F., Brouwer, M., De Rie, M. A., Aarden, L. A., and Van Lier, R. A. W.,** Regulatory properties of LFA-1 alpha and beta chains in human T-lymphocyte activation, *Nature,* 333, 850, 1988.
111. **Carrera, A. C., Rincon, M., Sanchez-Madrid, F., Lopez-Botet, M., and De-Landazuri, M. O.,** Triggering of co-mitogenic signals in T cell proliferation by anti-LFA-1 (CD18, CD11a), LFA-3, and CD7 monoclonal antibodies, *J. Immunol.,* 141, 1919, 1988.
112. **Cambier, J. C. and Ransom, J. T.,** Molecular mechanisms of transmembrane signalling in B lymphocytes, *Annu. Rev. Immunol.,* 5, 175, 1987.
113. **Klaus, G. G. B., Bijsterbosch, M. K., O'Garra, A., Harnett, M. M., and Rigley, K. P.,** Receptor signalling and crosstalk in B lymphocytes, *Immunol. Rev.,* 99, 19, 1987.
114. **Brunswick, M., June, C. H., Finkelman, F. D., Dintzis, H. M., Inman, J. K., and Mond, J. J.,** Surface immunoglobulin-mediated B cell activation in the absence of detectable elevations in intracellular ionized calcium: a model for T cell independent B cell activation, *Proc. Natl. Acad. Sci. U.S.A.,* 86, 6724, 1989.

115. **Rigley, K. P.,** Receptor mediated signal transduction in B lymphocytes, in *Cytokines and Lymphocytes,* Callard, R., Ed., Academic Press, New York, 1990.
116. **Justement, L., Chezn, Z., Harris, L., Ransom, J., Sandoval, V., Smith, C., Rennick, D., Roehm, N., and Cambier, J.,** BSF1 induces membrane protein phosphorylation but not phosphoinositide metabolism, calcium mobilization, protein kinase C translocation, or membrane depolarisation in resting murine B lymphocytes, *J. Immunol.,* 137, 3664, 1986.
117. **Finney, M., Guy, G. R., Michell, R. H., Dugas, B., Rigley, K. P., Gordon, J., and Callard, R. E.,** Interleukin 4 activates human B lymphocytes via transient inositol lipid hydrolysis and delayed cAMP generation, *Eur. J. Immunol.,* 20, 151, 1990.
118. **Shields, J. G., Armitage, R. J., Jamieson, B. N., Beverley, P. C. L., and Callard, R. E.,** Increased expression of surface IgM but not IgD or IgG on human B cells in response to interleukin 4, *Immunology,* 66, 224, 1989.
119. **Foxwell, B. M. J., Woerly, G., and Ryffel, B.,** Identification of interleukin 4 receptor-associated proteins and expression of both high and low affinity binding on human lymphoid cells, *Eur. J. Immunol.,* 19, 1637, 1989.
120. **Cambier, J. C. and Lehmann, K. R.,** Ia-mediated signal transduction leads to proliferation of primed B lymphocytes, *J. Exp. Med.,* 170, 877, 1989.
121. **Nakagawa, T., Hirano, T., Nakagawa, N., Yoshizaki, K., and Kishimoto, T.,** Effect of recombinant IL 2 and gamma-IFN on proliferation and differentiation of human B cells, *J. Immunol.,* 134, 959, 1985.
122. **Boyd, A. W., Fisher, D. C., Fox, D. A., Schlossman, S. F., and Nadler, L. M.,** Structural and functional characterization of IL2 receptors on activated human B cells, *J. Immunol.,* 134, 2387, 1985.
123. **Defrance, T., Vanbervliet, B., Aubry, J. P., Takebe, Y., Arai, N., Miyajima, A., Yokota, T., Lee, F., Arai, K., De Vries, J., and Banchereau, J.,** B cell growth promoting activity of human IL-4, *J. Immunol.,* 139, 1135, 1987.
124. **Richard, Y., Leprince, C., Dugas, B., Treton, D., and Galanaud, P.,** Reactivity of Leu-1$^+$ tonsillar B cells to a high molecular weight B cell growth factor, *J. Immunol.,* 139, 1563, 1987.
125. **Kehrl, J. H., Miller, A., and Fauci, A. S.,** Effect of tumor necrosis factor alpha on mitogen-activated human B cells, *J. Exp. Med.,* 166, 786, 1987.
126. **Jelinek, D. F. and Lipsky, P. E.,** Enhancement of human B cell proliferation and differentiation by tumor necrosis factor-alpha and interleukin 1, *J. Immunol.,* 139, 2970, 1987.
127. **Kehrl, J. H., Alvarez-Mon, M., Delsing, G. A., and Fauci, A. S.,** Lymphotoxin is an important T cell-derived growth factor for human B cells, *Science,* 238, 1144, 1987.
128. **Romagnani, S., Giudizi, M. G., Biagiotti, R., Almerigogna, F., Mingari, C., Maggi, E., Liang, C.-M., and Moretta, L.,** B cell growth factor activity of interferon-gamma. Recombinant human interferon-gamma promotes proliferation of anti-μ-activated human B lymphocytes, *J. Immunol.,* 136, 3513, 1986.
129. **Defrance, T., Aubry, J.-P., Vanbervliet, B., and Banchereau, J.,** Human interferon-gamma acts as a B cell growth factor in the anti-IgM antibody to co-stimulatory assay but has no direct B cell differentiation activity, *J. Immunol.,* 137, 3861, 1986.
130. **Gordon, J., Guy, G., and Walker, L.,** Autocrine models of B-lymphocyte growth. II. Interleukin-1 supports the proliferation of transformed lymphoblasts but not the stimulation of resting B cells triggered through their receptors for antigen, *Immunology,* 57, 419, 1986.
131. **Kehrl, J. H., Roberts, A. B., Wakefield, L. M., Jakowlew, S., Sporn, M. B., and Fauci, A. S.,** Transforming growth factor beta is an important immunomodulatory protein for human B lymphocytes, *J. Immunol.,* 137, 3855, 1986.
132. **Petit-Koskas, E., Genot, E., Lawrence, D., and Kolb, J.-P.,** Inhibition of the proliferative response of human B lymphocytes to B cell growth factor by transforming growth factor-beta, *Eur. J. Immunol.,* 18, 111, 1988.
133. **Yamasaki, K., Taga, T., Hirata, Y., Yawata, H., Kawanishi, Y., Seed, B., Taniguchi, T., Hirano, T., and Kishimoto, T.,** Cloning and expression of the human interleukin-6 (BSF-2/IFN beta-2) receptor, *Science,* 241, 825, 1988.
134. **Hirano, T., Yasukawa, K., Harada, H., Taga, T., Watanabe, Y., Matsuda, T., Kashiwamura, S.-I., Nakajima, A., Koyama, K., Iwamatsu, A., Tsunasawa, S., Sakiyama, F., Matsui, H., Takahara, Y., Taniguchi, T., and Kishimoto, T.,** Complementary DNA for a novel human interleukin (BSF-2) that induces B lymphocytes to produce immunoglobulin, *Nature,* 324, 73, 1986.
135. **Billiau, A.,** BSF-2 is not just a differentiation factor, *Nature,* 324, 415, 1986.
136. **Clutterbuck, E., Shields, J. G., Gordon, J., Smith, S. H., Boyd, A., Callard, R. E., Campbell, H. D., Young, I. G., and Sanderson, C. J.,** Recombinant human interleukin 5 is an eosinophil differentiation factor but has no activity in standard human B cell growth factor assays, *Eur. J. Immunol.,* 17, 1743, 1987.
137. **Pene, J., Rousset, F., Briere, F., Chretien, I., Wideman, J., Bonnefoy, J. Y., and De Vries, J. E.,** Interleukin 5 enhances interleukin 4-induced IgE production by normal human B cells. The role of soluble CD23 antigen, *Eur. J. Immunol.,* 18, 929, 1988.

138. **Yokota, T., Coffman, R. L., Hagiwara, H., Rennick, D. M., Takebe, Y., Yokota, K., Gemmell, L., Shrader, B., Yang, G., Meyerson, P., Luh, J., Hoy, P., Pene, J., Briere, F., Spits, H., Banchereau, J., De Vries, J., Lee, F. D., Arai, N., and Arai, K.-I.,** Isolation and characterization of lymphokine cDNA clones encoding mouse and human IgA-enhancing factor and eosinophil colony-stimulating factor activities: relationship to interleukin 5, *Proc. Natl. Acad. Sci. U.S.A.*, 84, 7388, 1987.
139. **Karray, S., Defrance, T., Merle-Beral, H., Banchereau, J., Debre, P., and Galanaud, P.,** Interleukin 4 counteracts the interleukin 2-induced proliferation of monoclonal B cells, *J. Exp. Med.*, 168, 85, 1988.
140. **Defrance, T., Vanbervliet, B., Aubry, J.-P., and Banchereau, J.,** Interleukin 4 inhibits the proliferation but not the differentiation of activated human B cells in response to interleukin 2, *J. Exp. Med.*, 168, 1321, 1988.
141. **Ralph, P., Jeong, G., Welte, K., Mertelsmann, R., Rabin, H., Henderson, L. E., Souza, L. M., Boone, T. C., and Robb, R. J.,** Stimulation of immunoglobulin secretion in human B lymphocytes as a direct effect of high concentrations of IL 2, *J. Immunol.*, 133, 2442, 1984.
142. **Hirano, T., Taga, T., Nakano, N., Yasukawa, K., Kashiwamura, S., Shimizu, K., Nakajima, K., Pyun, K. H., and Kishimoto, T.,** Purification to homogeneity and characterization of human B-cell differentiation factor (BCDF or BSFp-2), *Proc. Natl. Acad. Sci. U.S.A.*, 82, 5490, 1985.
143. **Kishimoto, T. and Hirano, T.,** Molecular regulation of B lymphocyte response, *Annu. Rev. Immunol.*, 6, 485, 1988.
144. **Muraguchi, A., Hirano, T., Tang, B., Matsuda, T., Horii, Y., Nakajima, K., and Kishimoto, T.,** The essential role of B cell stimulatory factor 2 (BSF-2/IL6) for the terminal differentiation of B cells, *J. Exp. Med.*, 167, 332, 1988.
145. **Bich-Thuy, L. T., Queen, C., and Fauci, A. S.,** Interferon-gamma induces light chain synthesis in interleukin 2 stimulated human B cells, *Eur. J. Immunol.*, 16, 547, 1986.
146. **Jelinek, D. F., Splawski, J. B., and Lipsky, P. E.,** The roles of interleukin 2 and interferon-gamma in human B cell activation, growth and differentiation, *Eur. J. Immunol.*, 16, 925, 1986.
147. **Harfast, B., Huddlestone, J. R., Casali, P., Merigan, T. C., and Oldstone, M. B. A.,** Interferon acts directly on human B-lymphocytes to modulate immunoglobulin synthesis, *J. Immunol.*, 127, 2146, 1981.
148. **Splawski, J. B., Jelinek, D. F., and Lipsky, P. E.,** Immunomodulatory role of IL-4 on the secretion of Ig by human B cells, *J. Immunol.*, 142, 1569, 1989.
149. **Smeland, E. B., Blomhoff, H. K., Funderud, S., Shalaby, M. R., and Espevik, T.,** Interleukin 4 induces selective production of interleukin 6 from normal human B lymphocytes, *J. Exp. Med.*, 170, 1463, 1989.
150. **Snapper, C. M., Finkelman, F. D., and Paul, W. E.,** Regulation of IgG1 and IgE production by interleukin 4, *Immunol. Rev.*, 102, 51, 1988.
151. **Coffman, R. L., Seymour, B. W. P., Lebman, D. A., Hiraki, D. D., Christiansen, J. A., Shrader, B., Cherwinski, H. M., Savelkoul, H. F. J., Finkelman, F. D., Bond, M. W., and Mosmann, T. R.,** The role of helper T cell products in mouse B cell differentiation and isotype regulation, *Immunol. Rev.*, 102, 5, 1988.
152. **Takatsu, K., Tominaga, A., Harada, N., Mita, S., Matsumoto, M., Takahashi, T., Kikuchi, Y., and Yamaguchi, N.,** T cell-replacing factor (TRF)/interleukin 5 (IL-5): molecular and functional properties, *Immunol. Rev.*, 102, 107, 1988.
153. **Snapper, C. M., Peschel, C., and Paul, W. E.,** IFN-gamma stimulates IgG2a secretion by murine B cells stimulated with bacterial lipopolysaccharide, *J. Immunol.*, 140, 2121, 1988.
154. **Coffman, R. L., Lebman, D. A., and Shrader, B.,** Transforming growth factor beta specifically enhances IgA production by lipopolysaccharide-stimulated murine B lymphocytes, *J. Exp. Med.*, 170, 1039, 1989.
155. **Rothman, P., Lutzker, S., Cook, W., Coffman, R., and Alt, F. W.,** Mitogen plus interleukin 4 induction of $C_{epsilon}$ transcripts in B lymphoid cells, *J. Exp. Med.*, 168, 2385, 1988.
156. **Esser, C. and Radbruch, A.,** Rapid induction of transcription of unrearranged $s_{gamma}1$ switch regions in activated murine B cells by interleukin 4, *EMBO J.*, 8, 483, 1989.
157. **Harriman, G. R., Kunimoto, D. Y., Elliott, J. F., Paetkau, V., and Strober, W.,** The role of IL-5 in IgA B cell differentiation, *J. Immunol.*, 140, 3033, 1988.
158. **Beagley, K. W., Eldridge, J. H., Kiyono, H., Everson, M. P., Koopman, W. J., Honjo, T., and McGhee, J. R.,** Recombinant murine IL-5 induces high rate IgA synthesis in cycling IgA-positive Peyer's patch B cells, *J. Immunol.*, 141, 2035, 1988.
159. **Matsumoto, R., Matsumoto, M., Mita, S., Hitoshi, Y., Ando, M., Araki, S., Yamaguchi, N., Tominaga, A., and Takatsu, K.,** Interleukin-5 induces maturation but not class switching of surface IgA-positive B cells into IgA-secreting cells, *Immunology*, 66, 32, 1989.
160. **Schoenbeck, S., McKenzie, D. T., and Kagnoff, M. F.,** Interleukin 5 is a differentiation factor for IgA B cells, *Eur. J. Immunol.*, 19, 965, 1989.
161. **Del Prete, G., Maggi, E., Parronchi, P., Chretien, I., Tiri, A., Macchia, D., Ricci, M., Banchereau, J., De Vries, J. E., and Romagnani, S.,** IL-4 is an essential factor for the IgE synthesis induced in vitro by human T cell clones and their supernatants, *J. Immunol.*, 140, 4193, 1988.

162. **Pene, J., Rousset, F., Briere, F., Chretien, I., Bonnefoy, J.-Y., Spits, H., Yokota, T., Arai, N., Arai, K.-I., Banchereau, J., and De Vries, J. E.**, IgE production by normal human lymphocytes is induced by interleukin 4 and suppressed by interferons gamma and alpha and prostaglandin E2, *Proc. Natl. Acad. Sci. U.S.A.*, 85, 6880, 1988.

163. **Pene, J., Chretien, I., Rousset, F., Briere, F., Bonnefoy, J. Y., and De Vries, J. R.**, Modulation of IL-4 induced human IgE production in vitro by IFN-gamma and IL-5: the role of soluble CD23 (s-CD23), *J. Cell. Biochem.*, in press.

164. **Sarfati, M. and Delespesse, G.**, Possible role of human lymphocyte receptor for IgE (CD23) or its soluble fragments in the in vitro synthesis of human IgE, *J. Immunol.*, 141, 2195, 1988.

165. **Pene, J., Rousset, F., Briere, F., Chretien, I., Paliard, X., Banchereau, J., Spits, H., and De Vries, J. E.**, IgE production by normal human B cells induced by alloreactive T cell clones is mediated by IL-4 and suppressed by IFN-gamma, *J. Immunol.*, 141, 1218, 1988.

166. **Vercelli, D., Jabara, H. H., Arai, K., Yokota, T., and Geha, R. S.**, Endogenous interleukin 6 plays an obligatory role in interleukin 4 dependent human IgE synthesis, *Eur. J. Immunol.*, 19, 1419, 1989.

167. **Umetsu, D. T., Leung, D. Y. M., Siraganian, R., Jabara, H. H., and Geha, R. S.**, Differential requirements of B cells from normal and allergic subjects for the induction of IgE synthesis by an alloreactive T cell clone, *J. Exp. Med.*, 162, 202, 1985.

168. **Leung, D. Y. M., Young, M. C., Wood, N., and Geha, R. S.**, Induction of IgE synthesis in normal human B cells. Sequential requirements for activation by an alloreactive T cell clone and IgE-potentiating factors, *J. Exp. Med.*, 163, 713, 1986.

169. **Maggi, E., Del Prete, G., Macchia, D., Parronchi, P., Tiri, A., Chretien, I., Ricci, M., and Romagnani, S.**, Profiles of lymphokine activities and helper function for IgE in human T cell clones, *Eur. J. Immunol.*, 18, 1045, 1988.

170. **Paliard, X., Malefijt, R. D. W., Yssel, H., Blanchard, D., Chretien, I., Abrams, J., De Vries, J., and Spits, H.**, Simultaneous production of IL-2, IL-4, and IFN-gamma by activated human $CD4^+$ and $CD8^+$ T cell clones, *J. Immunol.*, 141, 849, 1988.

171. **Thyphronitis, G., Tsokos, G. C., June, C. H., Levine, A. D., and Finkelman, F. D.**, IgE secretion by Epstein-Barr virus-infected purified human B lymphocytes is stimulated by interleukin 4 and suppressed by interferon-gamma, *Proc. Natl. Acad. Sci. U.S.A.*, 86, 5580, 1989.

172. **Lundgren, M., Persson, U., Larsson, P., Magnusson, C., Edvard Smith, C. I., Hammarstrom, L., and Severinson, E.**, Interleukin 4 induces synthesis of IgE and IgG4 in human B cells, *Eur. J. Immunol.*, 19, 1311, 1989.

173. **Flores-Romo, L., Millsum, M. J., Gillis, S., Stubbs, P., Sykes, C., and Gordon, J.**, Engagement of surface IgM on cycling human B lymphocytes promotes a switch in cytokine-directed immunoglobulin isotype production, *Immunology*, in press.

174. **Wang, A.-C., Tung, E., and Fudenberg, H. H.**, The primary structure of a human IgG2 heavy chain: genetic, evolutionary, and functional implications, *J. Immunol.*, 125, 1048, 1980.

175. **Callard, R. E. and Turner, M. W.**, Cytokines and switching: evolutionary divergence in mouse and man, *Immunol. Today*, 11, 200, 1990.

176. **Delfraissy, J.-F., Wallon, C., and Galanaud, P.**, Interferon-alpha can synergize with interleukin 2 for human *in vitro* antibody response, *Eur. J. Immunol.*, 18, 1379, 1988.

177. **Callard, R. E., Smith, S. H., and Scott, K. E.**, Interleukin-4 is an inhibitor of specific antibody responses by human B cells, submitted.

178. **Llorente, L., Crevon, M.-C., Karray, S., Defrance, T., Banchereau, J., and Galanaud, P.**, Interleukin (IL)4 counteracts the helper effect of IL2 on antigen-activated human B cells, *Eur. J. Immunol.*, 19, 765, 1989.

179. **Kishimoto, T.**, Factors affecting B cell growth and differentiation, *Annu. Rev. Immunol.*, 3, 133, 1985.

180. **Rousset, F., Malefijt, R. D. W., Slierendregt, B., Aubry, J.-P., Bonnefoy, J.-Y., Defrance, T., Banchereau, J., and De Vries, J. E.**, Regulation of Fc receptor for IgE (CD23) and class II MHC antigen expression on Burkitt's lymphoma cell lines by human IL-4 and IFN-gamma, *J. Immunol.*, 140, 2625, 1988.

181. **Tony, H.-P. and Parker, D. C.**, Major histocompatibility complex-restricted, polyclonal B cell responses resulting from helper T cell recognition of anti-immunoglobulin presented by small B lymphocytes, *J. Exp. Med.*, 161, 223, 1985.

182. **Clarke, E. A. and Ledbetter, J. A.**, Structure, function, and genetics of human B cell associated surface molecules, *Adv. Cancer Res.*, in press.

183. **Stamenkovic, I. and Seed, B.**, CD19, the earliest differentiation antigen of the B cell lineage, bears three extracellular immunoglobulin-like domains and an Epstein-Barr virus-related cytoplasmic tail, *J. Exp. Med.*, 168, 1205, 1988.

184. **Tedder, T. F. and Isaacs, C. M.**, Isolation of cDNAs encoding the CD19 antigen of human and mouse B lymphocytes: a new member of the immunoglobulin superfamily, *J. Immunol.*, 143, 712, 1989.

185. **Brown, A. N., Thurstan, S. M., and Callard, R. E.,** Action of the B cell panel on B cell proliferation to anti-IgM plus Il4, and on specific antibody responses to influenza virus, in *Leucocyte Typing IV,* Knapp, W. et al., Eds., Oxford University Press, Oxford, 1989, 203.
186. **Shields, J. G., Rigley, K. P., and Callard, R. E.,** Regulation of human B cell proliferation by the CD19 cell surface glycoprotein, in *Leucocyte Typing IV,* Knapp W. et al., Eds., Oxford University Press, Oxford, 1989, 40.
187. **Einfeld, D. A., Brown, J. P., Valentine, M. A., Clark, E. A., and Ledbetter, J. A.,** Molecular cloning of the human B cell CD20 receptor predicts a hydrophobic protein with multiple transmembrane domains, *EMBO J.,* 7, 711, 1988.
188. **Tedder, T. F., Streuli, M., Schlossman, S. F., and Saito, H.,** Isolation and structure of a cDNA-encoding the B1(CD20) cell-surface antigen of human B lymphocytes, *Proc. Natl. Acad. Sci. U.S.A.,* 85, 208, 1988.
189. **Clark, E. A. and Ledbetter, J. A.,** Amplification of the immune response by agonistic antibodies, *Immunol. Today,* 7, 267, 1986.
190. **Clark, E. A. and Ledbetter, J. A.,** Activation of human B cells mediated through two distinct cell surface differentiation antigens, Bp35 and Bp50, *Proc. Natl. Acad. Sci. U.S.A.,* 83, 4494, 1986.
191. **Golay, J. T., Clark, E. A., and Beverley, P. C. L.,** The CD20 (Bp35) antigen is involved in activation of B cells from the Go to the G1 phase of the cell cycle, *J. Immunol.,* 135, 3795, 1985.
192. **Tedder, T. F., Forsgren, A., Boyd, A. W., Nadler, L. M., and Schlossman, S. F.,** Antibodies reactive with the B1 molecule inhibit cell cycle progression but not activation of human B lymphocytes, *Eur. J. Immunol.,* 16, 881, 1986.
193. **Smeland, E. B., Beiske, K., Ohlsson, R., Holte, H., Ruud, E., Blomhoff, H. K., Jefferis, R., and Godal, T.,** Activation of human B cells: alternate options for initial triggering and effects of nonmitogenic concentrations of anti-IgM antibodies on resting and activated cells, *J. Immunol.,* 138, 3179, 1987.
194. **Valentine, M. A., Cotner, T., Gaur, L., Torres, R., and Clark, E. A.,** Expression of the human B-cell surface protein CD20: alteration by phorbol 12-myristate 13-acetate, *Proc. Natl. Acad. Sci. U.S.A.,* 84, 8085, 1987.
195. **Moore, M. D., Cooper, N. R., Tack, B. F,. and Nemerow, G. R.,** Molecular cloning of the cDNA encoding the Epstein-Barr virus/C3d receptor (complement receptor type 2) of human B lymphocytes, *Proc. Natl. Acad. Sci. U.S.A.,* 84, 9194, 1987.
196. **Nemerow, G. R., McNaughton, M. E., and Cooper, N. R.,** Binding of monoclonal antibody to the Epstein-Barr virus (EBV)/CR2 receptor induces activation and differentiation of human B lymphocytes, *J. Immunol.,* 135, 3068, 1985.
197. **Frade, R., Crevon, M. C., Barel, M., Vazquez, A., Krikorian, L., Charriaut, C., and Galanaud, P.,** Enhancement of human B cell proliferation by an antibody to the C3d receptor, the gp 140 molecule, *Eur. J. Immunol.,* 15, 73, 1985.
198. **Carter, R. H., Spycher, M. O., Ng, Y. C., Hoffman, R., and Fearon, D. T.,** Synergistic interaction between complement receptor type 2 and membrane IgM on B lymphocytes, *J. Immunol.,* 141, 457, 1988.
199. **Gordon, J., Flores-Romo, L., Cairns, J. A., Millsum, M. J., Lane, P. J., Johnson, G. D., and MacLennan, I. C. M.,** CD23: a multi-functional receptor/lymphokine?, *Immunol. Today,* 10, 153, 1989.
200. **Delespesse, G., Sarfati, M., and Hofstetter, H.,** Human IgE-binding factors, *Immunol. Today,* 10, 159, 1989.
201. **Yokota, A., Kikutani, H., Tanaka, T., Sato, R., Barsumian, E. L., Suemura, M., and Kishimoto, T.,** Two species of human Fc epsilon receptor II (FceRII/CD23): tissue specific and IL-4 specific regulation of gene expression, *Cell,* 55, 611, 1988.
202. **Crow, M. K., Jover, J. A., and Friedman, S. M.,** Direct T helper-B cell interactions induce an early B cell activation antigen, *J. Exp. Med.,* 164, 1760, 1986.
203. **Stamenkovic, I., Clark, E. A., and Seed, B.,** A B-lymphocyte activation molecule related to the nerve growth factor receptor and induced by cytokines in carcinomas, *EMBO J.,* 8, 1403, 1989.
204. **Braesch-Andersen, S., Paulie, S., Koho, H., Nika, H., Aspenstrom, P., and Perlmann, P.,** Biochemical characteristics and partial amino acid sequence of the receptor-like human B cell and carcinoma antigen CDw40, *J. Immunol.,* 142, 562, 1989.
205. **Valle, A., Zuber, C. E., Defrance, T., Djossou, O., De Rie, M., and Banchereau, J.,** Activation of human B lymphocytes through CD40 and interleukin 4, *Eur. J. Immunol.,* 19, 1463, 1989.
206. **Liu, Y.-J., Joshua, D. E., Williams, G. T., Smith, C. A., Gordon, J. G., and MacLennan, I. C. M.,** The mechanism of antigen-drive selection in germinal centres, *Nature,* in press.
207. **Valentine, M. A., Clark, E. A., Shu, G. L., Norris, N. A., and Ledbetter, J. A.,** Antibody to a novel 95-kDa surface glycoprotein on human B cells induces calcium mobilization and B cell activation, *J. Immunol.,* 140, 4071, 1988.
208. **MacLennan, I. C. M., Oldfield, S., Liu, Y.-J., and Lane, P. J. L.,** Regulation of B-cell populations, in *Cell Kinetics of the Inflammatory Reaction,* Iverson, O. H., Ed., Springer-Verlag, New York, 1988, 37.

Chapter 8

CYTOKINES AND THEIR ROLE IN ANTIBODY CLASS EXPRESSION

Judy M. Teale and D. Mark Estes

TABLE OF CONTENTS

I. INTRODUCTION

For some time now it has been known that helper T cells are required for optimal isotype switching and the production of multiple antibody classes during an antigen specific response.[1-4] In recent years, the study of antibody class regulation has become an exciting research area since the molecular mechanisms associated with isotype regulation by T lymphocytes are rapidly being elucidated. Of primary importance is the accumulating evidence for the role of lymphokines produced by activated T cells in signaling B cells to synthesize specific antibody classes. Much of this chapter will be devoted to reviewing the information known about the role of soluble mediators in antibody class regulation.

A. INHERENT COMPLEXITIES

From the start it is important to point out some of the inherent complexities in the studies described. Most obvious is the fact that there are multiple cytokines and often multiple target cells for each cytokine. In addition, in many cases several cell types can produce a given cytokine. Since the bioassays used to analyze the influence of individual lymphokines on isotype expression often involve B cells, T cells, and macrophages, it is possible that the observed effects are not direct effects on the B cell itself. The addition of an exogenous lymphokine could potentially induce production of yet another cytokine which could then exert its effect on the B cell. This becomes especially significant since it is likely that we have not yet identified all of the relevant lymphokines. Moreover, several studies have demonstrated the importance of lymphokine concentration as well as the kinetics of addition of lymphokines. These problems are compounded by the fact that combinations of lymphokines can synergize with one another in inducing antibody synthesis. For example, the combination of IL-4 and IL-5 results in greater production of several isotypes.[5-8] In addition, IFN-γ is known to inhibit the ability of IL-4 to enhance IgG1 and IgE synthesis.[9-10]

Other complexities are associated with the finding that certain lymphokines can differentially affect a given B cell depending upon its maturational state.[11-13] Thus, IL-4 can affect unactivated B cells by inducing Ia expression, but can also affect activated B cells by inducing the specific expression of IgG1 or IgE.[9-15] Importantly, the effect on isotype expression of exogenously added lymphokines to cultures has often led to different results depending upon whether small, resting B cells or activated B cell blasts were used. For example, Tonkonogy et al. found that synergy of IL-4 and IL-5 in enhancing the production of all isotypes is only observed when large, presumably activated cells are used.[8] In addition, the activation requirements of resting and previously activated cells appear to be different with the resting cells requiring cognate interactions with the T cell.[8-19] Thus, in defining the exact mechanisms associated with antibody class regulation by cytokines, it is necessary that the maturational state of the B cell is known. In many of the earlier experiments, this was not the case.

Another potential problem in several of the studies to date that have examined the effect of cytokines in isotype regulation is the use of the mitogen lipopolysaccharide (LPS) as the B cell stimulus instead of an antigen-specific, T cell driven response. It is highly possible that LPS will result in different signals to the B cell which could promote the effect of certain cytokines and obscure the effects of others in a nonphysiological manner. This is exemplified by the finding that IgG1 enhancement by IL-4 is more apparent with LPS-stimulated B cells compared with antigen-stimulated B cells.

In addition, there is accumulating data for T cell heterogeneity among CD4+ helper T cells that results in activated T cells producing distinct lymphokine profiles. Given the role of certain lymphokines in the induction of specific isotypes, the particular helper T cells activated during an immune response could significantly influence the isotypes produced.

B. DISTINCT SUBSETS OF HELPER T CELLS

Evidence for distinct subsets of helper T cells was obtained when a large number of mouse helper T cell clones were analyzed for lymphokine production.[20-23] TH1 (T inflammatory) type T cell clones were found to produce IL-2, interferon gamma (IFN-γ), and lymphotoxin, whereas TH2 type T cell clones produce IL-4, IL-5, and IL-6.[23,24] Both types of T cells produce GM-CSF, TNF, IL-3, and preproenkephalin.[24] Subsequent studies have indicated distinct functional capabilities. TH1 cells seem to be more important for cell mediated immunity and are involved in the induction of both delayed type hypersensitivity and the cytotoxic T cell response.[23,24] Although TH1 cells have been shown to be effective in providing help to secondary antigen-specific responses with primed B cells, their ability to provide help to unprimed B cells has led to mixed results.[25,26] In contrast, TH2 cells are very effective at providing help for B cell antibody responses.[5,18,20,27,28]

It is still not certain how these different subsets of cells defined by *in vitro* propagated T cell clones relate to normal T cells *in vivo*. It is possible that the findings of distinct function described above relate to the maturational state of the helper T cell as opposed to truly different subsets.[29-31] Moreover, more recent experiments indicate that the classification of T cells into just two subsets based on lymphokine production is an oversimplification. Thus, some clones do not fall into the TH1, TH2 lymphokine pattern and produce IL-4 and IFN-γ.[32,33] In addition, the Pgp-1$^-$ subset of CD4+ T cells secretes substantial amounts of IL-2 but little or no IFN-γ.[34] Additional evidence for heterogeneity was observed when lymphokine profiles of mouse helper T cell clones were tested early (2 to 4 weeks) instead of late (months) after establishment in culture, and evidence for distinct lymphokine patterns produced by T cell precursors has emerged.[7,24,31] Such heterogeneity may help to explain the results obtained with human T cell clones in which the two distinct patterns of lymphokine profiles originally found with mouse TH1 and TH2 clones were not apparent.[35,36] Resolution of these issues may require definition of both lineage and activation-specific markers. Although markers such as the leukocyte common antigen in humans and mice, and the OX22 marker in the rat appear to distinguish between functional T cell subsets,[37,38] it is not clear whether these markers represent lineage or activation specific markers.

Despite the basis for different lymphokine profiles by individual T cells, the phenomenon appears to be biologically relevant. Thus, it seems clear from a number of infection models that distinct lymphokine patterns are induced.[40,41] The most convincing experiments have compared the lymphokines produced as a result of infection in both susceptible and resistant strains of mice. For example, when *Leishmania major* infects the susceptible strain, BALB/c; IL-4 was induced and an antibody response was apparent. In contrast, the nonsusceptible strain C57BL/6, which resolved the infection, produced IFN-γ with evidence for cell mediated immunity.[40] Apparently, for defense against this particular organism, cell-mediated immunity is more important. Thus, the specific lymphokines produced as a result of infection can affect the outcome of the infectious disease process.

C. DISTINCT SUBSETS OF T CELLS AND ISOTYPE REGULATION

T cell clones that were classified as either TH1 or TH2 were also shown to differ in their ability to induce the various isotypes.[5,10] The most compelling differences were observed with the isotypes IgE and IgG2a which appear to be specifically induced by the lymphokines IL-4 and IFN-γ, respectively.[10,39] Coffman et al. used a panel of TH1 and TH2 clones and found that the vast majority of IL-4-producing TH2 clones were capable of inducing IgE responses in contrast to none of the TH1 clones lacking detectable IL-4.[5] In contrast, TH1 clones were able to induce substantially more IgG2a relative to the TH2 clones. In another study, Stevens et al.[17] used isolated TNP-specific B cells and showed that a TH1 clone induced IgG2a whereas the TH2 clone induced IgG1. Both types of TH clones were shown to promote IgM and IgG3 production. In this system the regulation of isotype synthesis appeared to occur in at least two stages involving activation via cognate interaction between the T and B cell, and the interaction of T cell-derived lymphokines with the B cell.[18] The bottom line of these experiments is that the particular lymphokines produced by individual T cells substantially influence the production of specific antibody classes. Interestingly, though, the regulation of isotypes by cytokines is rarely absolute resulting in the presence or complete absence of individual isotypes.

As discussed above, however, the designation of TH1 and TH2 is likely to be an oversimplification and not as useful for an analysis of human systems. Therefore, the next section will emphasize the information known about particular cytokines and their potential role in antibody class regulation.

II. CYTOKINES AND ISOTYPE REGULATION

A. IL-1

IL-1 is one of the highly pleiotropic cytokines. First identified in 1972 by Gery et al.,[43] it is produced primarily by macrophages and monocytes. Subsequently, many different cell types have been shown to produce IL-1 or IL-1-like activities, including B cell lines and B lymphoblasts.[44-47] Expression of IL-1 or IL-1-like activities by B cells is especially relevant to their function as antigen presenting cells (APCs) since IL-1 may be a necessary stimulus for effective activation of T cells.

Among the effects of IL-1 on immune function is the ability to co-stimulate B cell differentiation and proliferation, primarily in the presence of IL-6.[48] IL-1 may also act indirectly on B cells through its ability to promote IL-2 and IL-2R transcriptional activity in certain populations of T cells, possibly by induction or enhancement of factors and receptors necessary for T cell activation.[49,51]

With regard to isotype expression, IL-1 may favor the proliferation of the helper/inducer T cell subpopulation (TH2) since T cell clones capable of providing antigen-specific B cell help can respond to the autocrine growth factor IL-4 only in the presence of IL-1.[52] Favoring the expansion of TH2 cells instead of TH1 cells would presumably promote IgG1, IgE, and IgA synthesis while minimizing IgG2a synthesis. IL-1 also acts synergistically both in enhancing IgM production in the presence of IL-6 and enhancing IgA production in the presence of IL-5.[53]

B. IL-2

Although IL-2 was originally described as a potent activator of T cells, subsequent studies have demonstrated that IL-2 plays a prominent role in B cell proliferation and differentiation. Although many of the effects of IL-2 are indirectly manifested through stimulation of other T-cell-derived cytokines, the potential exists for direct stimulation of B cells as the presence of IL-2 receptor (R)-like molecules has been demonstrated on both B cell lines and normal activated B cells.[54-58] Additionally, B cells shown to be positive for IL-2R expression, not only proliferate in the presence of exogenous IL-2, but synthesize and secrete antibody as well.[59-61] In accord with its role as a B cell differentiation factor, IL-2 induces J-chain transcription and synthesis in the BCL_1 B cell line and thus facilitates assembly of the pentameric form of IgM. J chain has been shown to be the limiting factor for high rate synthesis of IgM.[62,63]

The role of IL-2 in isotype expression is currently unresolved although several studies suggest that IL-2 is an essential cofactor for B cell differentiation and directed isotype switching. In studies of anti-Ig treated B cell blasts, Purkerson et al.[64] demonstrated that while IL-4 was solely capable of suppressing IgM expression, other lymphokines including IL-2 were necessary for secretion of IgG1. Additionally, in studies of IgE synthesis in humans, DeKruyff et al.[18] showed optimal expression of IgE in resting B cells required the presence of T cells, IL-4, low molecular weight B cell growth factor (LMW-BCGF), and Il-2. Importantly, IL-2 as the lone soluble factor was not sufficient to generate *in vitro* IgE responses under cognate conditions. Activated or low density B cells, on the other hand, were capable of producing IgE in the absence of T cells but in the presence of soluble IL-4 and IL-2. Utilizing activated cells, IL-2 was shown to be a potent B cell differentiation factor for IgG synthesis.[18,65,66] In addition, IL-2 appears to be a necessary cofactor for Ig synthesis by *Staphylococcal aureus* (SA)-stimulated B cells with little or no bias in specific isotype expression.[66]

In addition to indirect effects of IL-2 addition, a recent study indicates that IL-2 may regulate isotype production, possibly through non-T cells, by production of IFN-γ and IL-4.[67] Utilizing LPS-stimulated spleen cells derived from nude mice, IL-2-induced IgG2a secretion was blocked by both monoclonal and polyclonal antibodies to IFN-γ. In the case of the monoclonal antibody, blockade resulted in increased IgG1 production. Increased IgG1 production was countered by the addition of an anti-IL-4 monoclonal antibody. These results infer that IL-2 may regulate isotype production through an undefined T cell-independent mechanism involving both IFN-γ and IL-4.

C. IL-4

IL-4, first described in 1982, was orignally isolated from crude supernatants of the EL-4 thymoma and characterized as the specific factor which supported the growth of anti-Ig stimulated blasts.[68] Derived primarily from T cells, its activities have been demonstrated to affect a wide range of cell types including most cells of hematopoietic lineage.[69] Among its activities on B cells, IL-4 enhances the expression of class II major histocompatability antigens, stimulates IgG1 and IgE production by LPS blasts, is necessary for generation and maintenance of IgE responses *in vivo*, and stimulates Thy-1 expression on normal B cells.[9-12,14,70,71]

In terms of cytokines specifically inducing particular isotypes, the evidence is perhaps the strongest with IL-4 and IgE synthesis. Initial studies tested the effect of exogenous IL-4 on mitogen stimulated B cells. Culture of murine spleen cells with LPS results in activated B cells which then secrete largely IgM, IgG3 and IgG2b.[72,73] In contrast, spleen cells cultured with LPS and IL-4, selectively produce IgG1 and IgE while expression of IgG3 and IgG2b are substantially suppressed.[69] Importantly, the addition of anti-IL-4 profoundly inhibits IgE synthesis.[9,15,24] Comparable systems for activating human B cells and studying the direct effects of human IL-4 have not been fully delineated as LPS is not a potent activator of human B cells. A recent study utilizing IL-4, phorbol 12-myristate 13-acetate (PMA), and a mouse thymoma cell line in co-culture with human B cells resulted in increased amounts of IgE and IgG4 production by limiting dilution analysis.[65] Whether human IgG4 is analogous to murine IgG1 has not been clearly demonstrated although IgG4 and IgE appear to be co-regulated in humans.[74]

In both human and murine systems it is clear that IL-4 plays an obligatory role in IgE synthesis. Thus, in a variety of *in vitro* systems, it has been shown that the absence of IL-4 or its inhibition with anti-IL-4 results in essentially no IgE response.[9,36,75-79] This conclusion was confirmed in an *in vivo* model in which the IgE response normally found in animals infected with *Nippostrongylus brasiliensis* could be virtually eliminated by the *in vivo* administration of anti-IL-4.[70] IL-4 appears to affect IgE production by increasing the frequency of B cells that switch to IgE as well as increasing the clonal burst size.[80,81] The examination of single B cell clones stimulated with TH2 cells strongly suggested that IL-4 was inducing IgM bearing cells and their clonal progeny to subsequently switch to IgE synthesis.[80]

Although from both murine and human systems it seems clear that IL-4 is absolutely required for IgE production, there is evidence from studies with human B cells that IL-4 is not the sole lymphokine required. Thus, a recent study by Vercelli et al.[78] indicated that

IgE production could be inhibited by both anti-IL-4 and anti-IL-6 antibodies suggesting a requirement for IL-6.[78] In addition, IgE production was optimally enhanced by the presence of IL-5. Another study used cyclosporin A treated, allogeneic T cell clones to stimulate B cells and then examine the effects of lymphokines.[18] It has been shown that cyclosporin A treatment results in the inability of T cells to secrete their own lymphokines.[16] In this system an apparent requirement for low molecular weight B cell growth factor was observed.[18] It is uncertain at present if IgE is regulated in a fundamentally different manner in mice and humans, or if further details in IgE regulation remain to be elucidated.

The precise molecular mechanisms by which IL-4 specifically induces IgE synthesis remain unresolved. However, recent studies may suggest that IL-4 prepares the B cell for Ig rearrangement and the isotype switch (see below).

IL-4 is also known to enhance IgG1 production. Previous studies clearly showed that LPS activated B cells in the presence of IL-4 switch to the production of IgG1.[9,14,82-84] Moreover, a limiting dilution analysis indicated that IL-4 was inducing IgM bearing cells to switch to IgG1.[85] However, despite its strong IgG1 enhancing effects in mitogen driven systems, its role in the specific regulation of IgG1 synthesis remains uncertain. Thus, a number of TH1 clones that do not produce IL-4 induce substantial IgG1 production.[5] In addition, anti-IL-4 results in little or no inhibition of IgG1 responses either *in vitro* or *in vivo*.[5,7,28,86] Such studies indicate IL-4-independent mechanisms for IgG1 production. Additionally, studies of anti-Ig derived murine B cell blasts indicate that IL-4 alone cannot support proliferation and differentiation of B cells.[64] In order to obtain optimal stimulation and IgG1 production, the further addition of IL-2 and IL-5 was required.[64]

Interestingly, it has been reported that IL-4 can affect resting B cells and presumably influence their commitment to IgG1 production upon subsequent activation.[15] This would predict that increased IL-4 levels *in vivo* would influence the potential of the resident, primary B cell pool. Infection of mice with *Mesocestoides corti* results in an antibody response restricted to IgM and IgG1.[87-89] Although increased IL-4 appears to be associated with the infection, the resting B cells were unaltered in their potential to produce multiple isotypes in response to hapten-carrier conjugates.[90] Thus, no evidence was found for prior commitment to IgG1 synthesis. One of the major unresolved areas in isotype regulation by cytokines is when during B cell activation and clonal proliferation cytokines induce istotype-specific signals.

IL-4 induction of both IgE and IgG1 synthesis is inhibited by relatively low concentrations of IFN-γ.[10,91] The differences in the production of IL-4 and IFN-γ between TH1 and TH2 subsets of T cells can explain their differences in the ability to induce an IgE response. Additional evidence for this is the capacity to obtain an IgE response using TH1 cells if IL-4 and anti-IFN-γ are added to the cultures.[5]

Snapper et al.[92] have shown that IL-4 inhibits the enhanced expression of the IgG FcRII on LPS-activated B cells. Although the function of FcRII on B cells remains unclear, cross-linkage of surface Ig and FcRII with rabbit IgG specific for mouse Ig inhibits many B cell functions including activation and differentiation.[93-95] IL-4 has also been shown to regulate the expression of the low affinity receptor for IgE on lymphocytes adding yet another parameter through which IL-4 can potentially regulate IgE responses.[96-98]

D. IL-5

IL-5, first isolated in the early 1970s, is a potent activator of eosinophils. It is also known to induce Ig snythesis and proliferation of B cells.[99] This was first demonstrated by co-stimulation of mouse BCL_1 cells with dextran sulfate.[100,101] IL-5 appears to have the capability to act directly on B cells, as an IL-5 receptor has been identified on murine BCL1-B20 cells.[102]

IL-5 induces differentiation and Ig secretion in murine splenic B cells.[100-103] In studies utilizing LPS-activated splenic B cells in the presence of IL-5, several investigators demonstrated increased production of IgA. Thus early evidence indicated that IL-5 may act as a possible IgA switch factor.[107-110] However, several subsequent studies have demonstrated that IL-5 increased the IgA response by acting on surface IgA + B cells which are presumably already committed to IgA synthesis. Therefore, IL-5 appears to serve more as a terminal differentiation factor rather than as an isotype switch factor.[111-113,147]

IgG1 expression is affected by IL-5. In LPS-stimulated B cell cultures or cultures utilizing Percoll-fractionated B cell blasts, IL-4 and IL-5 synergize to enhance the IgG1 response.[8] In studies of LPS-stimulated anti-Ig blasts, a requirement was demonstrated for both IL-5 and IL-2 for optimal IgG1 expression.[64] As previously shown for IgA secretion, IL-5 may act as a differentiation factor for IgG1-committed cells.

IL-5 can also influence IgM expression when present together with IL-4. Stimulation of LPS or anti-Ig blasts in the presence of IL-4 alone results in reduced IgM secretion. This IL-4 dependent reduction in IgM expression was shown to be substantially overcome by the addition of IL-5.[8,64]

In contrast to the activities of IL-5 in the murine system, the effects of IL-5 on human B cell differentiation and proliferation are unresolved at present. Studies using human B cells have been unable to reproduce IL-5-induced effects on antibody production as observed in the murine system. Although experiments of Yokota et al.[10a] indicated an effect of recombinant human IL-5 on B cell antibody production, it is possible that the predominant effect of IL-5 in humans is its role in eosinophil development.[99,114]

While the precise mechanism of IL-5 action on B cells is poorly understood, IL-5 stimulation of a BCL_1 subline resulted in increased synthesis of μ_s mRNA and modest increases in J chain mRNA.[63] IL-5, together with IL-2 also appeared to augment J chain mRNA synthesis.[63] In addition, B cells cultured with LPS plus IL-4 showed increased steady state levels of mRNA for the secreted form of $\gamma 1$ and α.[141] Such activities help to explain increases in IgM, IgA, and IgG1 secretion. They are also consistent with the notion that IL-5 is important as a B cell terminal differentiation factor.

E. IL-6

As is apparent for other lymphokines, IL-6 has a wide variety of effects on immune responsiveness, although largely in concert with other soluble mediators. Many of the biological activities formerly ascribed to IL-1 have now been shown to be due to IL-6 or at least due to synergy with IL-6. Among its many activities, IL-6 has the ability to promote the growth and differentiation of B cells.[115,116,147] IL-6 has been shown to enhance the production of IgM by activated murine splenic B cells, particularly when IL-1 is present.[53] IL-6 also supports both Ig synthesis by human tonsillar B cells stimulated with PWM and the growth and Ig secretion of a human B-lymphoblastoid cell line.[115] In concert with IL-5, murine IL-6 has been shown to be a potent cofactor in IgA synthesis by T cell-depleted Peyer's patch (PP) B cells.[53,112]

IL-6 has been proposed to act as a relatively late acting lymphokine whose main function is directing terminal B cell differentiation. Purified IL-6 has been shown to increase Ig secretion without bias to particular isotypes.[119] However, introduction of a human IL-6 transgene into C57BL/6 mice resulted in high concentrations of human IL-6 accompanied by a 120- to 400-fold increase in polyclonal IgG1 expression in sera of all transgenic mice surveyed.[120] In IL-6 transgenic mice, a large proportion of IgM-expressing plasma cells present in lymph nodes appear to undergo a rapid class switch to IgG1, suggesting that novel mechanisms for substantial IgG1 production may exist.

Recent studies by Raynal et al.[121] indicate that IL-6 up-regulates the synthesis of immunoglobulin mRNAs, preferentially increasing mRNAs encoding the secretory form of immunoglobulin relative to the membrane form. This effect appears to be mediated either at the level of posttranscriptional processing and/or by affecting message stability.

IL-6 is also likely to function in a synergistic manner with IL-4 and affect isotype expression. In humans as in mice, it is clear that IL-4 plays an essential role in the regulation of IgE production. In humans, rIL-4 has been shown to induce IgE synthesis by normal peripheral blood mononuclear cells (PBMC).[78,79,122] Interestingly, a recent study of IL-4 driven IgE production by PBMC was strongly inhibited by an anti-IL-6 monoclonal antibody, indicating a potential role for this cytokine in human IgE production.[78] Isotype switching to IgE may be initiated by IL-4, as IL-4 is required for induction of responsiveness to IL-6 by increasing IL-6R expression on B cells.

F. INTERFERON-GAMMA (IFN-γ)

IFN-γ, derived primarily from T lymphocytes, serves as a potent regulator of B cell development and differentiation in addition to its other numerous biological activities. IFN-γ has been shown to up-regulate expression of secretory component necessary for IgM and IgA binding and transport, in addition to up-regulation of IgG(Fc) receptors on PMNs.[124,125]

IFN-γ appears to regulate both *in vitro* and *in vivo* antibody responses. Utilizing LPS-activated B cells *in vitro*, it was demonstrated that IFN-γ stimulates IgG2a secretion.[39] Marked increases in IgG2a production have also been demonstrated for resting B cells before LPS stimulation, suggesting that IFN-γ acts directly on B cells. Importantly, IFN-γ, as well as stimulating IgG2a synthesis, has also been shown to inhibit IgG1, IgG2b, IgG3, and IgE secretion.[39] Such findings indicate that IFN-γ may play an inhibitory role in antibody production.

The importance of IFN-γ in IgG2a synthesis has also been shown in several *in vivo* models. Previous studies indicated that the administration of anti-IgD results in relatively large increases in serum levels of both IgG1 and IgE.[86,126] In mice treated with a goat antibody to mouse IgD, administration of IFN-γ inhibited production of IgG1 and IgE and stimulated production of IgG2a.[91] In another model, monoclonal antibodies specific for IFN-γ were shown to block IgG2a responses elicited by the bacterium *Brucella abortus*, an organism which elicits primarily an IgG2a response in mice.[91]

Infectious disease models indicate the physiological importance of the IgG2a response and the presumed role for IFN-γ. For example, it has been shown by Coutelier et al.[127] that infection with a large number of both DNA and RNA viruses elicits predominantly an IgG2a response. Because interferons are typically produced as a result of viral infections, it was postulated that IFN-γ was responsible for the preferential expression of IgG2a antibodies. It has been suggested that IFN-γ is involved in humoral immune responses to a variety of pathogens that elicit predominantly IgG2a responses.[128,129] The advantages of an IgG2a response to microbial pathogens are evident when comparing the effector functions of the various isotypes. IgG2a can effectively fix complement. In addition, antibodies of the IgG2a subclass bind with high avidity to Fc receptors on both macrophages and NK cells and more effectively mediate antibody-dependent cellular cytotoxicity reactions relative to other isotypes.[130-133] These effector mechanisms should be effective in combating intracellular infections.

Studies of mouse TH1 clones are particularly relevant to understanding the functional consequences of IFN-γ production as detectable levels of IFN-γ are not produced by TH2 clones.[24] As indicated previously, TH1 clones induce substantial quantities of IgG2a relative to amounts produced by TH2 clones, and have also been shown to suppress TH2-induced *in vitro* responses as well.[24] In addition, differential selection of TH1 clones in the presence of IFN-γ *in vitro* implies that preferential expansion of this subset may also occur *in vivo*, selectively altering the isotype distribution indirectly through regulation of T cell responses.[24,31]

G. TRANSFORMING GROWTH FACTOR-BETA (TGF-β)

TGF-β was first described by its ability to induce abnormal growth patterns of non-malignant fibroblasts.[134] TGF-β is produced by a variety of cell types including activated B cells, T cells, and macrophages.[135-137] Additionally, TGF-β has largely a suppressive effect on lymphoid cells and has been shown to inhibit the production of many immunoglobulin isotypes in LPS-stimulated B cell cultures (IgM, IgG1, IgG2a, and IgG3) while not affecting others (IgG2b).[138,139] Consequently, TGF-β, like IFN-γ, may play an important role in negatively regulating antibody synthesis.

The predominance of IgA in mucosal secretions and various lymphoid tissues has led to a number of investigations of the cytokines present in these tissues as well as those which preferentially enhance IgA production *in vitro*. A number of factors have been shown to enhance IgA secretion including IL-5, IL-4, IL-2, and TGF-β. However TGF-β, unlike the other factors known to affect IgA production, appears to specifically induce surface IgA^- cells to switch to IgA production.[138,139] Therefore, TGF-β appears to be an IgA switch factor as opposed to a terminal differentiation factor like IL-5. Importantly, TGF-β increases IgA production approximately tenfold in LPS-stimulated cultures whereas other cytokines (IL-5, IL-2, and IL-4) elicit less than tenfold increases.[138] Additionally, increased production of IgA by TGF-β is further enhanced by IL-2 and to some extent by IL-5. Whether or not TGF-β is preferentially expressed in mucosal tissues needs to be explored. Thus, the physiological role of this cytokine has not been fully delineated.

III. POTENTIAL MOLECULAR MECHANISM FOR CYTOKINE-INDUCED ISOTYPE EXPRESSION

Precise mechanisms of the isotype switch process are still uncertain. It has been postulated that antibody class switch recombination can be controlled either by isotype specific recombinases or by relaxing chromatin structure of the immunoglobulin switch regions providing accessibility for switch recombinases.[140-143] Support for this latter model was provided through studies of B cell lines in which a preferential switch to certain isotypes occurs upon activation. Switching to these particular isotypes was shown to be preceded by transcription of the unrearranged constant region locus coding for the isotypes including the 5′ switch region.[141,143]

Although much remains to be learned about lymphokines and how they might specifically regulate isotype expression, recent molecular studies regarding germline transcripts provide some interesting clues.[141,142,144,145] Thus, it has been shown with both B cell lines and splenic B cells that the addition of IL-4 induces germline transcription of both the $C_{\gamma1}$ and C_ϵ constant region loci.[144,145] Moreover, this occurs within hours of the addition of IL-4.[145] It was postulated that IL-4 induces IgE and the IgG1 expression by inducing accessibility of the associated switch regions thus promoting the isotype switch to these constant region loci. Importantly, this phenomenon does not appear to be limited to IL-4 since it has been reported that TGF-β, which appears to induce IgA synthesis, results in the transcription of the unrearranged Cα locus prior to the appearance of detectable sIgA+ cells or secreted IgA.[138] Together, these data support the idea that cytokines induce isotype switching not by altering the specificity of the recombination but rather by changing the accessibility of specific C_H loci.

IV. SUMMARY AND CONCLUDING REMARKS

Substantial progress has been made in defining how antibody classes are produced and regulated. Upon B cell activation, either as a result of mitogen or antigen dependent T-B cell interactions, a number of cytokines have been shown to affect the expression of particular isotypes. As discussed above, it appears that in some cases the cytokine is directly inducing the isotype switch. The strongest case can be made for IL-4 and its induction of IgE synthesis.[24] Thus, IL-4 is an obligatory cytokine for IgE expression, and apparently promotes the isotype switch in surface IgM bearing B cells previously uncommitted to the production of IgE. Moreover, IL-4 induces germline transcripts of the epsilon constant region gene complex within hours after B cell activation providing potential molecular mechanisms for promotion of the isotype switch to IgE (see above).

Much remains to be learned about the process, however. It seems clear that there will not be a single, unique lymphokine that controls each of the antibody classes. Even in the more defined analyses of murine IgE regulation, IL-4 promotes the synthesis of both IgE and IgG1 and results in germline transcripts of $C_{\gamma 1}$ and C_{ϵ}. Moreover, for synthesis of IgE in the human there appear to be other obligatory signals apart from IL-4. Another unexplained observation is that IL-4 does not induce all activated B cells to switch to IgE.

Even further refinement will be required for a complete understanding of how the other isotypes are regulated. For example, although production of IFN-γ by TH1 cells has clearly been shown to promote IgG2a synthesis, the addition of anti-IFN-γ does not substantially alter the IgG2a response.[17] Moreover, the addition of IFN-γ to TH2 stimulated cultures fails to significantly enhance the IgG2a response. Thus, if there is an obligatory signal for IgG2a synthesis, it remains to be established. The regulation of IgG1 is also uncertain. Although early reports indicated the importance of IL-4 in specifically inducing IgG1 production in LPS blasts, it is likely that there are IL-4-independent mechanisms for IgG1 induction in that anti-IL-4 causes little or no inhibition of IgG1 *in vitro* or *in vivo* (see above). This is of particular significance since studies suggest that IL-4 promotes IgG1 by inducing the isotype switch in previously uncommitted surface IgM bearing cells rather than acting as a terminal differentiation factor.[85] The regulatory mechanisms of IgA synthesis also continue to unfold. In this case, IL-5 has been shown to act as a terminal differentiation factor and induce IgA secretion in cells already committed to IgA. In contrast, TGF-β appears to induce the isotype switch to IgA (see above). Thus, the vast majority of evidence points to a multifaceted regulation of individual isotypes that will require detailed studies for full understanding.

In more general terms, it is interesting that many of the cytokines that affect isotype regulation are produced by a variety of cells and influence a number of target cells apart from the B cell. Under these circumstances, it would be somewhat surprising that each of them would convey an isotype-specific signal. It seems more likely that each cytokine or combination of cytokines would induce a particular activation pathway that favors the production of one or more antibody classes.

The purification and characterization of cytokines that affect isotype expression have resulted in significant new advances in determining mechanisms of antibody class regulation. The road is now paved for defining cytokine-induced second messengers and the cellular processes associated with the isotype switch. Significant areas of research will include elucidating when during B cell activation lymphokines act and the precise mechanisms involved.

REFERENCES

1. **Taylor, R. B. and Wortis, H. H.,** Thymus dependence of antibody response: variation with dose of antigen and class of antibody, *Nature,* 220, 927, 1968.
2. **Torrigiani, G.,** Quantitative estimation of antibody in the immunoglobulin classes of the mouse. II. Thymus dependence of the different classes, *J. Immunol.,* 108, 161, 1972.

3. **Teale, J.**, The use of specific helper T cell clones to study the regulation of isotype expression by antigen stimulated B cell clones, *J. Immunol.*, 131, 2170, 1983.
4. **Teale, J. and Abraham, K.**, Regulation of antibody class expression, *Immunol. Today*, 8, 122, 1987.
5. **Coffman, R., Seymour, B., Lebman, D., Hiraki, D., Christiansen, J., Shrader, B., Cherwinski, H., Savelkoul, H., Finkelman, F., Bond, M., and Mosmann, T.**, The role of helper T cell products in mouse B cell differentiation and isotype regulation, *Immunol. Rev.*, 102, 5, 1988.
6. **Lebman, D. and Coffman, R.**, The effects of IL-4 and IL-5 on the IgA response by murine Peyer's patch B cell populations, *J. Immunol.*, 141, 2050, 1988.
7. **Swain, S. L., McKenzie, D. T., Dutton, R. W., Tonkonogy, S. L., and English, M.**, The role of IL4 and IL5: characterization of a distinct helper T cell subset that makes IL4 and IL5 (TH2) and requires priming before induction of lymphokine secretion, *Immunol. Rev.*, 102, 77, 1988.
8. **Tonkonogy, S., McKenzie, D., and Swain, S.**, Regulation of isotype production by IL-4 and IL-5. Effects of lymphokines depend on the state of activation of the responding B cells, *J. Immunol.*, 142, 4351, 1989.
9. **Coffman, R., Ohara, J., Bond, M., Carty, J., Zlotnik, A., and Paul, W.**, B cell stimulatory factor-1 enhances IgE response of lipopolysaccharide-activated B cells, *J. Immunol.*, 136, 4358, 1986a.
10. **Coffman, R. and Carty, J.**, A T cell activity that enhances polyclonal IgE production and its inhibition by interferon-γ, *J. Immunol.*, 136, 949, 1986b.
11. **Roehm, N., Liebson, J., Zlotnik, A., Kappler, J., Marrack, P., and Cambier, J.**, Interleukin-induced increase in Ia expression by normal mouse B cells, *J. Exp. Med.*, 160, 679, 1984.
12. **Noelle, R., Krammer, P., Ohara, J., Uhr, J., and Vitettta, E.**, Increased expression of Ia antigens on resting B cells: an additional role for B cell growth factor, *PNAS*, 81, 6149, 1984.
13. **Callard, R. and Smith, S.**, Interleukin 2 and low molecular weight B cell growth factor are T cell-replacing factors for different subpopulations of human B cells, *Eur. J. Immunol.*, 18, 1635, 1988.
14. **Isakson, P., Pure, E., Vitetta, E., and Krammer, P.**, T cell-derived B cell differentiation factor(s). Effect on the isotype switch of murine B cells, *J. Exp. Med.*, 155, 734, 1982.
15. **Snapper, C. M., Finkelman, F. D., and Paul, W. E.**, Differential regulation of IgG1 and IgE synthesis by interleukin 4, *J. Exp. Med.*, 167, 183, 1988b.
16. **Krusemeier, M. A. and Snow, E. C.**, Induction of lymphokine responsiveness of hapten-specific B lymphocytes promoted through an antigen-mediated T helper lymphocyte interaction, *J. Immunol.*, 140, 367, 1988.
17. **Stevens, T., Bossie, A., Sanders, V., Fernandez-Botran, R., Coffman, R., Mosmann, T., and Vitette, E.**, Subsets of antigen-specific helper T cells regulate isotype secretion by antigen-specific B cells, *Nature*, 334, 255, 1988.
18. **DeKruyff, R. H., Turner, T., Abrams, J. S., Palladino, M. A., and Umetsu, D. T.**, Induction of human IgE synthesis by CD4+ T cell clones. Requirement for interleukin 4 and low molecular weight B cell growth factor, *J. Exp. Med.*, 170, 1477, 1989.
19. **Noelle, R., McCann, J., Marshall, L., and Bartlett, W.**, Cognate interactions between helper T cells and B cells. III. Contact-dependent, lymphokine-independent induction of B cell cycle entry by activated helper T cells, *J. Immunol.*, 143, 1807, 1989.
20. **Kim, J., Woods, A., Backer-Donn, E., and Bottomly, K.**, Distinct functional phenotypes of cloned Ia-restricted helper T cells, *J. Exp. Med.*, 162, 188, 1985.
21. **Mosmann, T. R., Cherwinski, H., Bond, M. W., Giedlin, M. A., and Coffman, R. L.**, Two types of murine helper T cell clone. I. Definition according to profiles of lymphokine activities and secreted proteins, *J. Immunol.*, 136, 2348, 1986.
22. **Cherwinksi, H. C., Schumacher, J. H., Brown, K. D., and Mosmann, T. R.**, Two types of mouse helper T cell clone. III. Further differences in lymphokine synthesis between TH1 and TH2 clones revealed in RNA hybridization, functionally monospecific bioassays and monoclonal antibodies, *J. Exp. Med.*, 166, 1229, 1987.
23. **Bottomly, K.**, A functional dichotomy in CD4 T lymphocytes, *Immunol. Today*, 9, 268, 1988.
24. **Mosmann, T. and Coffman, R.**, TH1 and TH2 cells: different patterns of lymphokine secretion lead to different functional properties, *Annu. Rev. Immunol.*, 7, 145, 1989.
25. **Giedlin, M. A., Longenecker, B. M., and Mosmann, T. R.**, Murine T cell clones specific for chicken erythrocyte alloantigens, *Cell Immunol.*, 97, 357, 1986.
26. **Killar, L., MacDonald, G., West, J., Woods, A., and Bottomly, K.**, Cloned, Ia-restricted T cells that do not produce interleukin-4 (IL-4)/B cell stimulatory factor (BSF-1) failed to help antigen-specific B cells, *J. Immunol.*, 138, 1674, 1987.
27. **Tony, H. and Parker, D.**, Major histocompatability complex-restricted B cell responses resulting from helper T cell recognition and anti-immunoglobulin presented by small B lymphocytes, *J. Exp. Med.*, 161, 223, 1985.
28. **Boom, W. H., Liano, D., and Abbas, A. I.**, Heterogeneity of helper/inducer T lymphocytes. II. Effects of interleukin 4- and interleukin 2-producing T cell clones on resting B lymphocytes, *J. Exp. Med.*, 167, 1352, 1988.

29. **Swain, S. L., McKenzie, D. T., Weinberg, A. D., and Hancock, W.,** Characterization of T helper 1 and 2 cell subsets in normal mice: helper T cells responsible for IL-4 and IL-5 production are present as precursors which require priming before they develop into lymphokine-secreting cells, *J. Immunol.*, 141, 3445, 1988b.
30. **Powers, G. D., Abbas, A. K., and Miller, R. A.,** Frequencies of IL-2 and IL-4 secreting T cells in naive and antigen-stimulated lymphocyte populations, *J. Immunol.*, 140, 3352, 1988.
31. **Gajewski, T., Joyce, J., and Fitch, F.,** Antiproliferative effect of IFN-γ in immune regulation. III. Differential selection of TH1 and TH2 murine helper T lymphocyte clones using recombinant IL-2 and recombinant IFN-gamma, *J. Immunol.*, 143, 15, 1989.
32. **Gajewski, T. and Fitch, F.,** Anti-proliferative effect of IFN-gamma in immune regulation. I. IFN-gamma inhibits the proliferation of TH2 but not TH1 murine helper T lymphocyteclones, *J. Immunol.*, 140, 4245, 1988.
33. **Bass, H., Mossmann, T., and Stober, S.,** Evidence for mouse TH1 and TH2-like helper T cells *in vivo*. Selective reduction of TH1-like cells after total lymphoid irradiation, *J. Exp. Med.*, 170, 1495 1989.
34. **Butt, R. Cerottini, S., and MacDonald, H.,** Selectively increased production of interferon γ by subsets of Ly+2+ and L3T4+ T cells identified by expression of Pgp-1, *J. Immunol.*, 138, 3583, 1987.
35. **Paliard, X., de Waal Malefijt, R., Yssel, H., Blanchard, D., Chretien, I., Abrams, J., de Vries, J., and Spits, H.,** Simultaneous production of IL2, IL4 and IFNγ by activated human CD4+ and CD8+ T cell clones, *J. Immunol.*, 141, 849, 1988.
36. **Maggi, E., Del Prete, G., Macchia, D., Parronchi, P., Tiri, A., Chretien, I., Ricci, M., and Romagnani, S.,** Profiles of lymphokine activities and helper function for IgE in human T cell clones, *Eur. J. Immunol.*, 18, 1045, 1988.
37. **Arthur, R. and Mason, D.,** T cells that help B cell responses to soluble antigen are distinguishable from those producing interleukin 2 on mitogenic or allogenic stimulation, *J. Exp. Med.*, 163, 74, 1986.
38. **Rudd, C., Morimoto, C., Wong, L., and Schlossman, S.,** The subdivision of the T4 (CD4) subset on the basis of the differential expression of L-C/T200 antigens, *J. Exp. Med.*, 161, 1758, 1987.
39. **Snapper, C., Peschel, C., and Paul, W.,** IFN-gamma stimulates secretion by murine B cells stimulated with bacterial lipopolysaccharide, *J. Immunol.*, 140, 2121, 1988a.
40. **Scott, P., Natovitz, P., Coffman, R. L., Pearce, E., and Sher, A.,** Immunoregulation of cutaneous Leishmaniasis T cell lines which transfer protective immunity or exacerbation belong to different T_H subsets and respond to distinct parasite antigens, *J. Exp. Med.*, 168, 1675, 1988.
41. **Heinzel, F., Sadick, M., Holaday, B., Coffman, R., and Lockslay, R.,** Reciprocal expression of interferon γ or interleukin 4 during the resolution or progression of murine Leishmaniasis. Evidence for distinct helper T cell subsets, *J. Exp. Med.*, 169, 59, 1989.
42. **Gery, I. and Waksman, B.,** Protentiation of the T lymphocyte response to mitogens. II. The cellular source of potentiating mediator(s), *J. Exp. Med.*, 136, 143, 1972.
43. **Acres, R. B., Larsen, A., Gillis, S., and Conlon, P.,** Production of IL-1α and IL-1β by clones of EBV transformed human B cells, *Mol. Immunol.*, 24, 479, 1987.
44. **Pistoia, V., Cozzolino, F., Rubartelli, A., Torcia, M., Roncella, S., and Ferrarini, M.,** *In vitro* production of interleukin 1 by normal and malignant human B lymphocytes, *J. Immunol.*, 136, 1688, 1986.
45. **Rimsky, L., Wakasugi, H., Ferrara, P., Robin, P., Capdeveille, J., Tursz, T., Fradelizi, D., and Bertoglio, J.,** Purification to homogeneity and NH_2-terminal amino acid sequence of a novel interleukin 1 species derived from a human B cell line, *J. Exp. Med.*, 159, 1637, 1986.
46. **Valentine, M., Lotz, M., Dinarello, A., Carson, D., and Vaughan, J.,** Lymphoblastoid cell lines produce an interleukin-1-like activity that can be serologically distinct from macrophase interleukin 1, *Lymphokine Res.*, 5, 173, 1986.
47. **Hawrylowicz, C., Duncan, L., Fuhlbrigge, R., and Unanue, E.,** Regulation of antigen presentation. II. Anti-Ig and IL-2 induce IL-1 production by murine splenic B cells, *J. Immunol.*, 142, 3361, 1989.
48. **Vink, A., Coulie, P., Wauters, P., Nordan, R., and Van-Snick, J.,** B cell growth and differentiation activity in interleukin-HGF and related murine plasmacytoma growth factors. Synergy with interleukin-1, *Eur. J. Immunol.*, 18, 607, 1988.
49. **Garman, R., Jacobs, K., Clark, S., and Raulet, D.,** B cell stimulatory factor 2 (Beta-2 interferon) functions as a second signal for interleukin 2 production by mature murine T cells, *PNAS*, 84, 7629, 1987.
50. **Hagiwara, H., Huang, H., Arai, N., Herzenberg, L., Arai, K., and Zlotnik, A.,** Interleukin 1 modulates messenger RNA levels of lymphokines and of other molecules associated with T cell activation in the T cell lymphoma LBRM-33-1A5, *J. Immunol.*, 138, 2514, 1987.
51. **Lowenthal, J., Cerottini, J., and MacDonald, H.,** Interleukin 1-dependent induction of both interleukin 2 secretion and interleukin 2 receptor expression by thymoma cells, *J. Immunol.*, 137, 1226, 1988.
52. **Greenbaum, L. A., Horowitz, J. B., Woods, A., Pasqualine, T., Reich, E., and Bottomly, K.,** Autocrine growth of CD4+ T cells. Differential effects of IL-1 on helper and inflammatory cells, *J. Immunol.*, 140, 1555, 1988.

53. **Kunimoto, D. Y., Nordan, R., and Strober, W.,** IL-6 is a potent cofactor of IL-1 in IgM synthesis and of IL-5 in IgA synthesis, *J. Immunol.*, 143, 2230, 1989.
54. **Malek, T. R., Robb, R. J., and Shevach, J. M.,** Identification and initial characterization of a rat monoclonal antibody reactive with the murine interleukin 2 receptor-ligand complex, *PNAS,* 80, 5694, 1983.
55. **Zubler, R. H., Lowenthal, J. W., Erard, F., Hashimoto, N., Devos, R., and MacDonald, H. R.,** Activated B cells express receptor for, and proliferate in response to pure interleukin 2, *J. Exp. Med.*, 160, 1170, 1984.
56. **Mingari, M. C., Gerosa, F., Carra, G., Acolla, R. S., Moretta, Z., Zubler, T. H., Waldman, T. A., and Moretta, L.,** Human interleukin-2 promotes proliferation of activated B cells via surface receptors similar to those of activated T cells, *Nature,* 312, 641, 1984.
57. **Osawa, H. and Diamanstein, T.,** A rat monoclonal antibody that binds specifically to mouse T lymphoblasts and inhibits IL-2 receptor functions: a putative anti-IL2 receptor antibody, *J. Immunol.*, 132, 2445, 1984.
58. **Tsudo, M., Uchiyama, T., and Uchino, H.,** Expression of Tac antigen on activated normal human B cells, *J. Exp. Med.*, 160, 612, 1984.
59. **Teranishi, T., Hirano, T., Lin., B., and Onoue, K.,** Demonstration of the involvement of interleukin 2 in the differentiation of *Staphylococcus aureus* cowan I stimulated B cells, *J. Immunol.*, 133, 3062, 1983.
60. **Nakanishi, K., Malek, T., Smith, K., Hamaoka, T., Shevach, E., and Paul, W. E.,** Both interleukin 2 and a second T cell-derived factor in EL-4 supernatant have activity as differentiation factors in IgM synthesis, *J. Exp. Med.*, 160, 1605, 1984.
61. **Jelinek, D. F., Splawski, J. B., and Lipsky, P. E.,** The role of interleukin 2 and interferon-gamma in human B cell activation, growth, and differentiation, *Eur. J. Immunol.*, 16, 925, 1986.
62. **Blackman, M. A., Tigges, M. A., Minie, M. E., and Koshland, M. E.,** A model system for peptide hormone action in differentiation: interleukin 2 induces a B cell lymphoma to transcribe the J chain gene, *Cell,* 47, 609, 1986.
63. **Matsui, K., Nakanishi, K., Cohen, D., Hada, T., Furuyama, J., Hamaoka, T., and Higashino, K.,** B cell response pathways regulated by IL-5 and IL-2. Secretory IgM H chain mRNA and J chain mRNA expression are separately controlled events, *J. Immunol.*, 142, 2918, 1989.
64. **Purkerson, J., Newberg, M., Wise, G., Lynch, K., and Isakson, P.,** Interleukin 5 and interleukin 2 cooperate with interleukin 4 to induce IgG1 secretion from anti-Ig-treated B cells, *J. Exp. Med.*, 168, 1175, 1988.
65. **Lundgren, M., Persson, U., Larsson, P., Magnusson, C., Edvard Smith, C. I., Hammarstrom, L., and Severinson, E.,** Interleukin 4 induces synthesis of IgE and IgG4 in human B cells, *Eur. J. Immunol.*, 19, 1311, 1989.
66. **Splawski, J. B., Jelinek, D. F., and Lipsky, P. E.,** Immunomodulatory role of IL-4 on the secretion of Ig by human B cells, *J. Immunol.*, 142, 1569, 1989.
67. **Amigorena, S., Teillaud, J., and Fridman, W.,** IL2 regulates the isotypes production through the induction of IFN-gamma and IL4, in *Abstracts of the 7th International Congress of Immunology (Berlin),* Gustav Fischer, New York, 1989, 23.
68. **Howard, M., Farrar, J., Hilfiker, M., Johnson, B., Takatsu, K., Hamaoka, T., and Paul, W.,** Identification of a T cell-derived growth factor distinct from IL-2, *J. Exp. Med.*, 155, 214, 1982.
69. **Ohara, J. and Paul, W.,** Receptors for B cell stimulatory factor-1 expressed on cells of hematopoietic lineage, *Nature,* 325, 537, 1987.
70. **Finkelman, F., Katona, I., Urban, J., Holmes, J., Ohara, J., Tung, A., Sample, J., and Paul, W.,** IL-4 is required to generate and sustain *in vivo* IgE responses, *J. Immunol.*, 141, 2335, 1988a.
71. **Snapper, C., Hornbeck, P., Atasoy, U., Pereira, M., and Paul, W.,** Interleukin 4 induces membrane Thy-1 expression on normal B cells, *PNAS,* 85, 6107, 1988c.
72. **Andersson, J., Coutinho, A., and Melchers, F.,** The switch from IgM to IgG secretion in single mitogen-stimulated B-cell clones, *J. Exp. Med.*, 147, 1744, 1979.
73. **Yuan, D. and Vitetta, E.,** Structural studies of cell surface and secreted IgG in LPS-stimulated murine B cells, *Mol. Immunol.*, 20, 367, 1983.
74. **Hammarstrom, L. and Smith, C. I. E.,** Immunoglobulin subclass distribution of specific antibodies in allergic patients, *Allergy,* 42, 529, 1987.
75. **Vitetta, E., Ohara, J., Myers, S., Layton, C., Krammer, P., and Paul, W.,** Serological, biochemical, and functional identification of B cell stimulatory factor-1 and B cell differentiation factor for IgG1, *J. Exp. Biol.*, 162, 1726, 1985.
76. **Snapper, C. and Paul, W.,** B cell stimulatory factor-1 (Interleukin 4) prepares resting murine B cells to secrete IgG1 upon subsequent stimulation with bacterial lipopolysaccharide, *J. Immunol.*, 139, 10, 1987.
77. **Pene, J., Rousset, F., Briere, F., Chretien, I., Bonnefoy, J., and Vries, J.,** IgE production by normal human lymphocytes is induced by interleukin 4 and suppressed by interferons γ and α and prostaglandin E_2, *PNAS,* 85, 6880, 1988.

78. **Vercelli, D., Jabara, H. H., Arai, K., Yokata, T., and Geha, R. S.**, Endogenous interleukin 6 plays an obligatory role in interleukin 4-dependent human IgE synthesis, *Eur. J. Immunol.*, 19, 1419, 1989b.
79. **Del Prete, G., Maggi, E., Parronchi, P., Chrètien, I., Tiri, A., Macchia, D., Ricci, M., Banchereau, J., DeVries, J., and Romagnini, J.**, IL-4 is an essential factor for IgE synthesis induced *in vitro* by human T cell clones and their supernatants, *J. Immunol.*, 140, 4193, 1988.
80. **Lebman, D. A. and Coffman, R. L.**, Interleukin 4 causes isotype switching to IgE in T cell-stimulated clonal B cell cultures, *J. Exp. Med.*, 168, 853, 1988.
81. **Savelkoul, H. F. J., Lebman, D., Benner, R., and Coffman, R. L.**, Increase of precursor frequency and clonal burst size of murine immunoglobulin E-secreting cells by interleukin-4, *J. Immunol.*, 141, 749, 1988.
82. **Lee, F., Yokota, T., Otsuka, T., Meyerson, T., Villaret, D., Coffman, R., Mosmann, T., Renrick, D., Roehm, N., Smith, C., Zlotnick, A., and Arai, K.**, Isolation and characterization of a mouse interleukin cDNA clone that expresses B cell-stimulatory factor-1 activities and T cell- and mast cell-stimulating activities, *Proc. Natl. Acad. Sci. U.S.A.*, 83, 2061, 1986.
83. **Bergstedt-Lindquist, S., Sideras, P., MacDonald, H. R., and Severinson, E.**, Regulation of Ig class secretion by soluble products of certain T-cell lines, *Immunol. Rev.*, 78, 25, 1984.
84. **Noma, Y., Sideras, P., Naito, T., Bergstedt-Lindquist, S., Azuma, C., Severinon, E., Tanabe, T. Kinashi, T., Matsuda, F., Yaoita, Y., and Honjo, T.**, Cloning of cDNA encoding the murine IgG1 induction factor by a novel factor using SP6 promoter, *Nature,* 319, 640, 1986.
85. **Layton, J., Vitetta, E. S., Uhr, J. W., and Krammer, P. H.**, Clonal analysis of B cells induced to secrete IgG by T cell-derived lymphokines, *J. Exp. Med.*, 160, 1850, 1984.
86. **Finkleman, F., Snapper, C., Mountz, J., and Katona, I.**, Polyclonal activation of the murine system by a goat antibody to mouse IgD. IX. Induction of a polyclonal IgE response, *J. Immunol.*, 138, 2826, 1987.
87. **Chapman, C., Knopf, P., Hicks, J., and Mitchell, G.**, IgG1 hypergammaglobulinaemia in chronic parasitic infections in mice: magnitude of the response in mice infected with various parasites, *Aust. J.Exp. Med. Sci.*, 57, 369, 1979.
88. **Abraham, K. and Teale, J.**, Isotype restriction during infection of mice with the cestode *Mesocestoides corti*: role for immune suppression, *J. Immunol.*, 138, 1699, 1987a.
89. **Abraham, K. and Teale, J.**, The contribution of parasite-specific T cells to isotype restriction in *Mesocestoides corti* infected mice, *J. Immunol.*, 139, 2530 1987b.
90. **Desai, B., Abraham, K., and Teale, J.**, The isotype potential of B cells present in mice chronically infected with *Mesocestoides corti, Cell Immunol.*, in press.
91. **Finkelman, F., Katona, I., Mosmann, T., and Coffman, R.**, IFN-γ regulates the isotypes of Ig secreted during *in vivo* humoral immune responses, *J. Immunol.*, 140, 1022 1988b.
92. **Snapper, C., Hooley, J., Atasoy, U., Finkelman, F., and Paul, W. E.**, Differential regulation of murine B cell FcRII expression by CD4+ T helper subsets, *J. Immunol.*, 143, 2133, 1989.
93. **Sidman, C. and Unanue, E.**, Control of B lymphocyte function. I. Inactivation of mitogenesis by interactions with surface immunoglobulin and Fc-receptor molecules, *J. Exp. Med.*, 144, 882, 1976.
94. **Philips, N. and Parker, D.**, Cross-linking of B lymphocyte Fc-gamma receptors and membrane immunoglobulin inhibits anti-immunoglobulin-induced blastogenesis, *J. Immunol.*, 132, 627, 1984.
95. **Bijsterbosch, M. and Klavs, G.**, Crosslinking of surface immunoglobulin and F_c receptors on B lymphocytes inhibits stimulation of inositol phospholipid breakdown via the antigen receptors, *J. Exp. Med.*, 162, 1825, 1985.
96. **Conrad, D., Waldschmidt, T., Lee, W., Rao, M., Keegan, A., Noelle, R., Lynch, R., and Kehry, M.**, Effect of B cell stimulatory factor-1 (interleukin 4) on $F_{c\epsilon}$ and $F_{c\gamma}$ receptor expression on murine B lymphocytes and B cell lines, *J. Immunol.*, 139, 2290, 1987.
97. **DeFrance, T., Vanbervliet, B., Aubry, J., Takebe, Y., Arai, N., Miyajima, A., Yokota, T., Lee, F., Arai, K., de Vries, J., and Bancherean, S.**, B cell growth-promoting activity of recombinant human interleukin 4, *J. Immunol.*, 139, 1135, 1987.
98. **Hudak, S., Gollnick, S., Conrad, D., and Kehry, M.**, Murine B cell stimulatory factor 1 (interleukin 4) increases expression of the F_c receptor for IgE on mouse B cells, *Proc. Natl. Acad. Sci. U.S.A.*, 84, 4604, 1987.
99. **Sanderson, C., Campbell, H., and Young, I.**, Molecular and cellular biology of eosinophil differentiation factor (interleukin 5) and its effect on human and mouse B cells, *Immunol. Rev.*, 102, 29, 1988.
100. **Shimple, et al.**, A third signal in B cell activation given by TRF, *Transplant. Rev.*, 23, 176, 1975.
101. **Swain, S. and Dutton, R.**, Production of a B cell growth-promoting activity, (DC) BCGF, from a cloned T cell line and its assay on the BCL_1 B cell tumor, *J. Exp. Med.*, 156, 1821, 1982.
102. **Takatsu, K., Tominaga, A., Harada, N., Mita, S., Matsumoto, M., Takahashi, T., Kikuchi, Y., and Yamaguchi, N.**, T cell-replacing factor (TRF)/Interleukin 5 (IL-5): molecular and functional properties, *Immunol. Rev.*, 102, 107, 1988.

103. **Takatsu, K., Tanaka, A., Tominaga, A., Kumahara, Y., and Hamaoka, T.,** Antigen-induced T cell-replacing factor (TRF). III. Establishment of T cell hybrid clone continuously producing TRF and functional analysis of released TRF, *J. Immunol.*, 125, 2646, 1980.
104. **Harada, N., Kikuchi, Y., Tominaga, A., Takaki, S., and Takatsu, K.,** BCGFII activity on activated B cells of a purified murine T cell-replacing factor (TRF) from a T cell hybridoma (B151K12), *J. Immunol.*, 134, 3944, 1985.
105. **Kinashi, T., Harada, N., Severinson, E., Tanabe, T., Sideras, P., Konishi, M., Azuma, C., Tominaga, A., Bergstedt-Lindqvist, S., Takahashi, M., Matsuda, F., Yaoita, Y., Takatsu, K., and Honjo, T.,** Cloning of complementary DNA for T cell-replacing factor and identity with B cell-growth factor II, *Nature*, 324, 70, 1986.
106. **Moller, G.,** IL-4 and IL-5: biology and genetics, *Immunol. Rev.*, 102, 1988.
107. **Coffman, R., Shrader, B., Carty, J., Mosmann, T., and Bond, M.,** A mouse T cell product that preferentially enhances IgA production. I. Biologic characterization, *J. Immunol.*, 139, 3685, 1987.
108. **Bond, M., Shrader, B., Mosmann, T., and Coffman, R.,** A mouse T cell product that preferentially enhances IgA production. II. Physiochemical characterization, *J. Immunol.*, 139, 3685, 1987.
109. **Yokota, T., Coffman, R., Hagiwara, R., Rennick, D., Takebe, T., Yokota, K., Gemmel, L., Shrader, B., Yang, G., Meyerson, P., Luh, J., Hoy, P., Pene, J., Briere, F., Spits, H., Banchereau, J., de Vries, J., Lee, F., Arai, N., and Arai, K.,** Isolation and characterization of lymphokine cDNA clones encoding mouse and human IgA-enhancing factor and eosinophil colony-stimulating activities: relationship to interleukin 5, *PNAS*, 84, 7388, 1987.
110. **Murray, P., McKenzie, D., Swain, S., and Kagnoff, M.,** Interleukin 5 and interleukin 4 produced by Peyer's patch T cells selectively enhance immunoglobulin A expression, *J. Immunol.*, 139, 2669, 1987.
111. **Beagley, K. W., Eldridge, J. H., Kiyono, H., Everson, M. P., Koopman, W. J., Honjo, T., and McGhee, J. R.,** Recombinant murine IL-5 induces high rate IgA synthesis in cycling IgA-positive Peyer's patch B cells, *J. Immunol.*, 141, 2035, 1988.
112. **Beagley, K. W., Eldridge, J. H., Lee, F., Kiyono, H., Everson, M. P., Koopman, W., Hirano, T., Kishimoto, T., and McGhee, J. R.,** Interleukins and IgA synthesis. Human and murine interleukin 6 induce high rate IgA secretion in IgA-committed B cells, *J. Exp. Med.*, 169, 2133, 1989.
113. **Schoenbeck, S., McKenzie, D., and Kagnoff, M.,** Interleukin 5 is a differentiation factor for IgA B cells, *Eur. J. Immunol.*, 19, 965, 1989.
114. **Clutterbuck, E., Shields, J., Gordon, J., Smith, S., Boyd, A., Callard, R., Campbell, H., Young, I., and Sanderson, C.,** Recombinant human interleukin is an eosinophil differentiation factor but has no activity in standard human B cell growth factor assays, 1987.
115. **Hirano, T., Yasukawa, K., Harada, H., Taga, T., Watanabe, Y., Matsuda, T., Kashiwamura, S., Nakajima, K., Koyama, K., Iwamatsu, A., Tsunasawa, S., Sakiyama, F., Matsui, H., Takahara, T., Taniguichi, T., and Kishimoto, T.,** Complementary DNA for a novel human interleukin (BSF-2) that induced B lymphoytes to produce immunoglobulin, *Nature*, 324, 73, 1986.
109. **Yokota, T., Coffman, R., Hagiwara, R., Rennick, D., Takebe, T., Yokota, K., Gemmel, L., Shrader, B., Yang, G., Meyerson, P., Luh, J., Hoy, P., Pene, J., Briere, F., Spits, H., Banchereau, J., de Vries, J., Lee, F., Arai, N., and Arai, K.,** Isolation and characterization of lymphokine cDNA clones encoding mouse and human IgA-enhancing factor and eosinophil colony-stimulating activities: relationship to interleukin 5, *PNAS*, 84, 7388, 1987.
110. **Murray, P., McKenzie, D., Swain, S., and Kagnoff, M.,** Interleukin 5 and interleukin 4 produced by Peyer's patch T cells selectively enhance immonoglobulin A expression, *J. Immunol.*, 139, 2669, 1987.
Beagley, K. W., Eldridge, J. H., Kiyono, H., Everson, M. P., Koopman, W. J., Honjo, T., and McGhee, J. R., Recombinant murine IL-5 induces high rate IgA synthesis in cycling IgA-positive Peyer's patch B cells, *J. Immunol.*, 141, 2035, 1988.
112. **Beagley, K. W., Eldridge, J. H., Lee, F., Kiyono, H., Everson, M. P., Koopman, W., Hirano, T., Kishimoto, T., and McGhee, J. R.,** Interleukins and IgA synthesis. Human and murine interleukin 6 induce high rate IgA secretion in IgA-committed B cells, *J. Exp. Med.*, 169, 2133, 1989.
113. **Schoenbeck, S., McKenzie, D., and Kagnoff, M.,** Interleukin 5 is a differentiation factor for IgA B cells, *Eur. J. Immunol.*, 19, 965, 1989.
114. **Clutterbuck, E., Shields, J., Gordon, J., Smith, S., Boyd, A., Callard, R., Campbell, H., Young, I., and Sanderson, C.,** Recombinant human interleukin is an eosinophil differentiation factor but has no activity in standard human B cell growth factor assays, 1987.
115. **Hirano, T., Yasukawa, K., Harada, H., Taga, T., Watanabe, Y., Matsuda, T., Kashiwamura, S., Nakajima, K., Koyama, K., Iwamatsu, A., Tsunasawa, S., Sakiyama, F., Matsui, H., Takahara, T., Taniguichi, T., and Kishimoto, T.,** Complementary DNA for a novel human interleukin (BSF-2) that induced B lymphocytes to produce immunoglobulin, *Nature*, 324, 73, 1986.
116. **Van Snick, J. A., Cayphas, S., Vink, A., Uyttenhove, C., Coulie, P. G., Rubira, M. R., and Simpson, R. J.,** Purification and NH_2 terminal amino acid sequence of a T-cell-derived lymphokine with growth factor activity for B-cell hybridomas, *Proc. Natl. Acad. Sci. U.S.A.*, 83, 9679, 1986.

117. **Beagley, K. W., Eldridge, J. H., Lee, F., Kiyono, H., Everson, M. P., Koopman, W., Hirano, T., Kishimoto, T., and McGhee, J. R.,** Interleukins and IgA synthesis. Human and murine interleukin 6 induce high rate IgA secretion in IgA-committed B cells, *J. Exp. Med.*, 169, 2133, 1989b.
118. **Kishimoto, T. and Hirano, T.,** Molecular regulation of B lymphocyte responses, *Annu. Rev. Immunol.*, 6, 485, 1988.
119. **Muraguchi, A., Hirano, T., Tang, B., Matsuda, T., Horii, Y., Nakajima, K., and Kishimoto, T.,** The essential role of B cell stimulatory factor 2 (BSF2/IL-6) for the terminal differentiation of B cells, *J. Exp. Med.*, 167, 332, 1988.
120. **Suematsu, S., Matsuda, T., Aozasa, K., Akira, S., Nakano, N., Ohno, S., Miyazaki, J., Yamamura, K., Hirano, T., and Kishimoto, T.,** IgG_1 plasmacytosis in interleukin 6 transgenic mice, *PNAS*, 86, 7547, 1989.
121. **Raynal, M., Liu, Z., Hirano, T., Mayer, L., Kishimoto, T., and Chen-Kiang, S.,** Interleukin 6 induces secretion of IgG1 by coordinated transcriptional activation and differential mRNA accumulation, *PNAS*, 86, 8024, 1989.
122. **Jabara, H., Ackerman, S., Vercelli, D., Yokota, T., Arai, K., Abrams, J., Dvorak, A., Lavigne, M., Banchereau, J., De Vries, J., Leung, D., and Geha, R.,** Induction of interleukin 4-dependent IgE synthesis and interleukin 5-dependent eosinophil differentiation by supernatants of a human T cell clone, *J. Clin. Immunol.*, 8, 437, 1988.
123. **Sarfati, M. and Delepese, G.,** Possible role of human lymphocyte receptor for IgE (CD23) or its soluble fragments in the *in vitro* synthesis of human IgE, *J. Immunol.*, 141, 2195, 1988.
124. **Pace, J., Russell, S., Torres, B., Johnson, H., and Gray, P.,** Recombinant mouse interferon induces the priming step in macrophage activation for tumor cell killing, *J. Immunol.*, 130, 2011, 1983.
125. **Sollid, L., Kvale, D., Brandtzaeg, P., Markussen, G., and Thorsby, E.,** Interferon-gamma enhances the expression of secretory component, the ephithelial receptor for polymeric immunoglobulins, *J. Immunol.*, 138, 4303, 1987.
126. **Mountz, J., Smith, J., Snapper, C., Mushiaski, J., and Finkelman, T.,** Polyclonal activation of the murine system by antibody to IgD. VII. Stimulation of IgD secretion, *J. Immunol.*, 139, 2172, 1987.
127. **Coutelier, J., van der Logt, J., Heesen, F., Warnier, G., and Van Snick, J.,** IgG2a restriction of murine antibodies elicited by viral infections, *J. Exp. Med.*, 165, 164, 1987.
128. **Duran, L. and Metcalf, E.,** Clonal analysis of primary B cells responsive to the pathogenic bacterium *Salmonella typhimurium*, *J. Exp. Med.*, 165, 340, 1987.
129. **Winter, J., Duncan, J., Santisteban, C., Douglas, J., and Adams, L.,** Capacity of passively administered antibody to prevent establishment of *Brucella abortus* infection in mice, *Infect. Immun.*, 57, 3438, 1989.
130. **Matthews, T. J., Collins, J. J., Roloson, G. J., Thiel, H. J., and Bolognesi, D. P.,** Immunologic control of the ascites form of murine adenocarcinoma 755. IV. Characterization of the protective antibody in hyperimmune serum, *J. Immunol.*, 126, 2332, 1981.
131. **Herlyn, D. and Koprowski, H.,** IgG2a monoclonal antibodies inhibit tumor growth through interaction with effector cells, *PNAS*, 79, 4761, 1982.
132. **Unkeless, J., Scigliano, E., and Freedman, V.,** Structure and function of human and murine receptors for IgG, *Annu. Rev. Immunol.*, 6, 251, 1988.
133. **Johnson, W., Stephlewski, Z., Koprowski, H., and Adams, D.,** Destructive interactions between murine macrophage tumor cells and antibodies of the IgG_{2a} isotype, in *Mechanisms of Cell-Mediated Cytotoxicity*, Vol. II, Henkart, P. and Martz, E., Eds., Plenum Press, New York, 1985, 75.
134. **Sporn, M., Roberts, A., Wakefield, L., and Associan, R.,** Transforming growth factor-β: biochemical function and chemical structure, *Science*, 233, 532, 1986.
135. **Kehrl, J., Roberts, A., Wakefield, L., Jakolew, S., Sporn, M., and Fauci, A.,** Transforming growth factor-β is an important immunomodulatory protein for human B lymphocytes, *J. Immunol.*, 137, 3855, 1986a.
136. **Kehrl, J., Wakefield, L., Roberts, A., Jakolew, S., Alvarerz-Mon, M., Derynck, R., Sporn, M., and Fauci, A.,** Production of transforming growth factor-β by human T lymphocytes and its potential role in the regulation of T cell growth, *J. Exp. Med.*, 163, 1037, 1986b.
137. **Assoian, R., Fleurdlys, B., Stevenson, H., Miller, P., Madtes, D., Raines, E., Ross, R., and Sporn, M.,** Expression and secretion of type beta-transforming growth factor by activated human macrophages, *PNAS*, 84, 6020, 1987.
138. **Coffman, R., Lebman, D., and Shrader, B.,** Transforming growth factor-β specifically enhances IgA production by lipopolysaccharide-stimulated murine B lymphocytes, *J. Exp. Med.*, 170, 1039, 1989.
139. **Sonoda, E., Matsumoto, R., Hitoshi, Y., Ishii, T., Sugimoto, M., Araki, S., Tominaga, A., Yamaguchi, N., and Takatsu, K.,** Transforming growth factor-β induces IgA production and acts additively with interleukin 5 for IgA production, *J. Exp. Med.*, 170, 1415, 1989.
140. **Davis, M., Kim, S., and Hood, L.,** DNA sequences mediating class switching in α-immunoglobulins, *Science*, 209, 1360, 1980.

141. **Stavnezer-Nordgren, J. and Sirlin, S.,** Specificity of immunoglobulin heavy chain class switch correlates with activity of germline heavy chain genes prior to switching, *EMBO J.*, 5, 95, 1986.

142. **Lutzker, S., Rothman, P., Pollock, R., Coffman, R., and Alt, F.,** Mitogen- and IL-4-regulated expression of germ-line IgG2b transcripts: evidence for directed heavy chain class switching, *Cell,* 53, 177, 1988.

143. **Lutzker, S. and Alt, F. W.,** Structure and expression of germline γ_{2b} constant region, *Mol. Cell Biol.*, 8, 1849, 1988.

144. **Rothman, P., Lutzger, S., Cook, W., Coffman, R., and Alt, F.,** Mitogen plus interleukin 4 induction of C_ϵ transcripts in B lymphoid cells., *J. Exp. Med.*, 168, 2385, 1988.

145. **Esser, C. and Rabruch, A.,** Rapid induction of transcription of unrearranged sγ1 switch regions in activated murine B cells by interleukin 4, *EMBO J.*, 8, 483, 1989.

146. **Matsumoto, R., Matsumoto, M., Mita, S., Hitoshi, Y., Ando, M., Araki, S., Yamaguchi, N., Tominaga, A., and Takatsu, K.,** Interleukin 5 induces maturation but not class-switching of surface IgA-positive B cells into IgA-secreting cells, *Immunology,* 66, 32, 1989.

147. **Shields, J., Smith, S., Levinsky, R., DeFrance, T., De Vries, J., Banchereau, J., and Callard, R.,** The response of selected human B cell lines to B cell growth and differentiation factors, *Eur. J. Immunol.*, 17, 535, 1987.

Chapter 9

B-CELL RESPONSES TO NATURAL ANTIGENS: REPERTOIRE ANALYSES AND V GENE FAMILY UTILIZATION

Eleanor S. Metcalf

TABLE OF CONTENTS

I. INTRODUCTION

The B cells of an adult animal represent a complex, heterogeneous population which is responsive to a wide array of antigenic determinants. Studies have suggested that this B-cell repertoire consists of greater than 10^7 unique antibody specificities (clonotypes).[1-6] In a given individual, the extent and specificity of this expressed repertoire appears to be primarily a product of the germline genes which encode the immunoglobulin variable regions and of the environmentally induced regulatory processes which occur subsequent to antigen-specific receptor expression.[6,7] To date, the ligands which have been used to evaluate B-cell repertoire expression have been primarily simple, synthetic antigens such as dinitrophenyl (DNP),[2,8,9] fluorescein,[9] *p*-azophenylarsonate,[10,11] 4-hydroxy-3-nitro-phenylacetyl (NP),[12,13] and [(T,G)-A--L].[14,15] However, it is not yet clear to what extent the characteristics of the B cells specific for synthetic antigens and their immune responses are correlative with immune responses to physiologically relevant agents such as infectious disease organisms.

To date, there have been relatively few studies which have probed the mechanisms of immunity to infectious disease agents at the cellular level. Analyses of the B-cell repertoire specific for these "natural" antigens have been primarily restricted to a few environmental antigens such as inulin[16] and to purified cell surface components isolated from bacteria, such as phosphocholine (PC)[17-20] and dextran.[21-23] However, presentation of these molecules in soluble form parallels the presentation of the simple haptenic molecules described above, and thus it is not clear whether such responses give a true representation of the immune response which would occur if these same components were presented to the host in the context of the intact infectious disease organism.

Presently, analyses of the adult, murine B-cell responses after stimulation with intact bacteria and viruses have been limited to the bacteria *Salmonella typhimurium*,[24,25] *Streptococcus pneumoniae*,[26,27] *Proteus morganii*,[27,28] *Pasteurella pneumotropica*,[29,30] *Mycoplasma pulmonis*,[31,32] and to the viruses influenza[3,33-35] and SV-40.[36] In this review, the B-cell responses to these bacterial and viral organisms will be examined, to aid in the clarification of the role of genetics and the contribution of environmental selective processes to the expression of the B-cell repertoire specific for physiologically relevant antigens. An understanding of the variables which regulate B-cell repertoire expression for these antigens may permit manipulation of the immune system, either to enhance the immune response by the host or to inactivate host responses.

II. WHAT DO WE KNOW ABOUT THE PRIMARY REPERTOIRE SPECIFIC FOR NATURAL ANTIGENS?

One of the difficulties in evaluating the primary B-cell repertoire is that most of the studies have been done in conventionally, presumably naive, animals after *deliberate* exposure to antigen. The expectation is that the data derived from such experiments represent the potential of the nonimmune (pre-immune) repertoire of these naive animals. While this conclusion might be warranted for studies of synthetic antigens or simple haptenic determinants, such studies do not necessarily represent the potential of the B-cell repertoire for antigens which are subject to environmentally induced influences, such as natural antigens. Moreover, these studies do not reflect concern for the potential role of antigen selection in primary immune responses. Nevertheless, even given these constraints, significant data have accrued on the primary repertoire specific for a few natural antigens.

A. ESTIMATES OF REPERTOIRE SIZE

To date, the principle source of information on the phenotypic characteristics of *primary* B-cell repertoires specific for natural antigens has come from the B-cell limiting dilution assay, the splenic fragment culture system.[2,8] While this experimental system may not provide a complete picture of the repertoire for a given antigen, it nevertheless enables one to compare characteristics of antigen-specific B-cell repertoires. Studies of Cancro and colleagues[3,33] have shown that the frequency of B cells (an estimate of the size of the B-cell repertoire for a given antigen) for the hemagglutinin (HA) molecule(s) on the influenza virus in adult, nonimmune mice was, on the average, 13 per 10^6 B cells. This frequency is extremely small when compared with frequency of DNP-responsive B cells, which is 200 per 10^6 B cells.[2] Moreover, this repertoire size estimate is 15- to 20-fold lower than the frequency observed for several other chemically defined haptenic determinants (reviewed in References 2 and 7). Nevertheless, the frequencies of B cells responsive to other natural antigens are also several orders of magnitude lower than those for synthetic antigens. Studies from this laboratory[25] established that the frequency of primary B cells specific to the whole bacteria *S. typhimurium* (strain TML) was 9.6 per 10^6 B cells. This response may be construed to be lower if the frequency of one of the major antigenic determinants, lipopolysaccharide (LPS), is evaluated; the frequency of LPS-specific B cells was only 2.4 per 10^6 B cells. The frequency of PC-specific B cells in adult, nonimmune mice was also similar to that for other natural antigens, 19 per 10^6 B cells,[17] even though in the original studies PC was coupled to a purified protein, rather than the whole bacterial organism. Subsequently, Hurwitz et al.[26] showed similar results were obtained when strain R36A of *S. pneumoniae* was utilized in the splenic fragment culture system.

Cancro et al.[3] hypothesized that the relatively low frequencies observed for natural antigens, as compared to synthetic antigens, was a consequence of the similarity between complex antigens, such as bacteria and viruses, and self-antigens. Thus, they supposed that only a small number of determinants on the bacteria or viruses would be recognized as foreign. Furthermore, they suggested that it seemed likely that small moieties with conserved features, e.g., charge or hydrophobicity, may more easily generate K_as commensurate with triggering across available combining sites than would a more complex antigen. A second possible explanation for these findings is that clonotypes responsive to determinants on these natural antigens cross-react with antigens abundantly present in nature, and thus, cross-reactive clonotypes would be tolerized during ontogeny.[37-39] Our studies[37,38,40] and subsequently those of Teale et al.[39,41] have provided evidence for the existence of such mechanisms by showing that developing B cells mature through a stage in which they are extremely susceptible to tolerance induction (inactivation by antigen). These studies also suggested that the interaction of immature B cells and multivalent antigens (e.g., bacteria and viruses) would be tolerogenic unless concomitantly recognized by primed T cells.[37,38] Since primed T cells would not be present in the neonate as yet, most bacteria- or virus-specific clonotypes would be tolerized rather than stimulated by interaction with cross-reactive normal gut flora. Thus, only a limited subset of clonotypes would be available for activation by deliberate exposure to these organisms, and the frequency of these primary B cells specific for natural antigens would, not surprisingly, be low.

An elegant series of studies, primarily by Thompson and Cancro,[42,43] is consistent with these ideas. These investigators chronically or acutely exposed mice to influenza virus during the first week of life and then allowed the mice to mature. Analysis of the influenza-specific B-cell repertoires of these mice showed frequencies and fine specificity patterns which were characteristic of the age when immunization first occurred. Their results suggest that antigen exposure appears to fix some specificities while preventing the expression of other clonotypes. Thus, these studies are consistent with the idea that during development many clonotypes that are emergent within the bacterial-specific B-cell repertoire would be preserved upon contact with normal gut flora, while others were deleted.

Studies of the B-cell repertoire in germfree (GF) mice should ostensibly aid in the understanding of genetic and environmental selective processes which influence repertoire expression to natural antigens. However, to date, the results of B-cell repertoire analysis in GF mice in different laboratories appear to be inconsistent. Studies by Cancro et al.,[3] Press and Klinman,[9] and Sigal et al.,[17] using small numbers of animals, have indicated that the B-cell frequencies of HA, DNP, fluorescein, and PC in GF mice were comparable to those of adult, conventional (CV) mice, whereas Gearhart and Cebra[18] and Shahin and Cebra[16] showed that the frequencies to PC and inulin were much lower and of thc same order of magnitude as B cells from neonatal mice. No studies to date have analyzed the GF B-cell repertoire to an infectious disease organism per se. However, it might be predicted that if exposure to environmental antigens is the only factor which regulates B-cell repertoire expression, and assuming the mice are truly axenic, then the frequency of bacterial-specific B cells in CV and GF mice could provide the lower and upper limits, respectively, for the size of a given bacterial B-cell repertoire. In contrast, it is also possible that GF mice may have smaller numbers of antigen-specific B cells than CV mice,[16,18] which would suggest that exposure to the host's bacterial flora during development may expand the B-cell repertoire rather than limit it. Such a notion is supported by the studies of Vakil and Kearney[44] which demonstrated that injection of neonatal mice with anti-idiotype antibodies prior to immunization with *S. pneumoniae* R36A expanded the PC-specific repertoire in these mice, and the study by Stein and Sonderstrom[45] which showed that the administration of either idiotype or anti-idiotype within 24 h after birth can prime weanling mice for protection against *E. coli* K13 infection. These studies suggest idiotypic networks can play a role in B-cell repertoire expression.

B. V_H/V_L GENE FAMILY UTILIZATION IN PRIMARY RESPONSES

Immunoglobulin (Ig) molecules are composed of light chains (L) and heavy chains (H), each of which has variable (V) and constant (C) regions. During ontogeny, a B cell selects one of many variable (V_H), diversity (D_H), and joining (J_H) gene segments to synthesize a functional heavy chain variable region gene.[46,47] Murine V_H segments have been divided into nine families based on amino acid or nucleotide sequence homology.[48-54] In contrast, the L chain is composed of only 2 gene segments, V_L and J_L which encode approximately 20 families (18 V_κ, and 2 V_λ) (reviewed in References 55 and 56). Whether the expression of these V region gene segments is restricted for all antigens in the adult remains controversial, but there is evidence that V gene family utilization for natural antigens may be restricted.

In the primary antibody response to PC in BALB/c mice after immunization with *S. pneumoniae*, the majority of the antibodies are highly restricted with respect to idiotype expression and variable region gene usage.[57-59] A single V_H and V_L combination (V_H-1, V_κ22) defines the T15 idiotype which comprises greater than 90% of the PC response to *S. pneumoniae*[60-64] and is a member of the V_HS107 gene family. Limited heterogeneity exists as a consequence of non-T15 clones which utilize the T15 V_H gene product (V_H-1) in association with V_κ8 or V_κ24 light chains.[59] It appears that a V_H-1 homolog is present in all common laboratory mouse strains, since studies by Clarke et al.[65] have shown that the protein sequence of the V_H encoded by the BALB/c V_H-1 gene (T15^a) appears to differ by only four amino acids from the V_H encoded by the C57BL or CBA equivalent (T15^b, C3).

To date, the majority of studies show that antibodies secreted by primary B cells are unmutated (i.e., do not have any somatic mutations) (reviewed in References 5 and 7) and the T15-related antibodies isolated from *S. pneumoniae* immunized mice fall into this category.[64,65] Such a finding is particularly interesting in light of the study by Berek who showed that the PC response in *S. pneumoniae* immunized mice is a result of a specific VDJ combination[64] and produces a T-15-like antibody with a D segment identical to the myeloma protein T15. In contrast, antibodies to PC generated from other PC-carriers (e.g., PC-KLH) showed heterogeneity in their D regions. Berek's finding demonstrates that the response to *S. pneumoniae* is extremely restricted in V gene utilization and is consistent with the view that there may be a bias in the response to some antigens towards sequences which express germline genes with minimal junctional diversity.[5] Subsequent analyses of responses to natural antigens may show such restrictions in V_H and/or D_H gene utilization to be characteristic of immune responses to bacteria and viruses.

The conservation of the antibody response to PC may be related to its presence as an immunodominant determinant on a variety of bacterial surfaces and its similarity to normal membrane phospholipids. However, the ability of the host to generate a protective host response to PC in the context of *S. pneumoniae* may be multifaceted. Cerny and co-workers have demonstrated that antigen dose has a dramatic effect on the idiotypy of primary anti-*S. pneumoniae* response, when the R36A heat killed vaccine strain was employed as immunogen.[66] This study showed that the dominant idiotope was unaffected by dose of *S. pneumoniae*, but the expression of infrequent idiotopes was significantly decreased. This study suggests that the dose of the *S. pneumoniae* that the host receives may affect the type of host response that is stimulated and may ultimately affect the protective capacity of the host's response. Moreover, a second study suggests that continual exposure of the host to *S. pneumoniae*, and the subsequent production of T15 idiotype, may contribute to alterations in neurotransmission processes and possible disease, since T15 antibodies have been shown to cross-react with acetylcholinesterase and the acetylcholine receptor.[67]

To date, little is known about the V gene family usage for either *S. typhimurium* or influenza virus in *primary* responses. The recent study by Kienker et al.[68] suggests that the V_H gene families utilized by 1-week-old neonates in response to influenza virus is distinct from those V_H gene families used by 2-week-old mice. However, Teale and colleagues[69] have suggested that adult patterns of V_H are established by 1 to 2 weeks. Since other investigators have observed a few primary, adult anti-influenza antibodies composed of V_H genes from all families (see references for secondary influenza-specific repertoire below), then it might be predicted that the most frequently observed V_H genes utilized would be members of the largest V_H gene family (J558) (M. Cancro, personal communication).

III. WHAT DO WE KNOW ABOUT THE SECONDARY REPERTOIRE SPECIFIC FOR NATURAL ANTIGENS?

A. ESTIMATES OF REPERTOIRE SIZE

Presently, the only two natural antigens for which a secondary B-cell repertoire has been analyzed, at the individual B-cell level, are *S. typhimurium*[24] and influenza virus.[3,70] The frequency of secondary *S. typhimurium*-specific B cells is similar to that for the influenza HA-specific B cells.[70] Studies by Cancro et al.[3] compared the primary and secondary influenza-specific B-cell pools and showed that the HA-specific B-cell repertoire expands 10- to 20-fold post-immunization. Studies from this laboratory[25] suggest that the *S. typhimurium*-specific B-cell repertoire expands in a manner similar to influenza. In contrast, other studies have shown that the responses to synthetic antigens such as DNP, fluorescein, and (T,G)-A--L expands only 3- to 6-fold.[8,9,14]

In an earlier section of this review, it was proposed that the B-cell repertoire to natural antigens may be low because only a limited set of clonotypes would be available to respond upon deliberate immunization. However, the scenario for the secondary repertoire would be different. Secondary exposure to natural antigens would provide sufficient T-cell help for stimulation of newly generated B cells emerging from the bone marrow, and thus these B cells would not be susceptible to tolerance induction.[37,70] As a consequence, the secondary bacterial (and potentially viral) specific B-cell repertoires would theoretically include those B-cell subsets which would have been tolerized in the primary repertoire as well as those B-cell subsets which were not tolerized during development. Thus, the secondary B-cell repertoire specific for natural antigens might be expected to increase significantly after immunization, as has been shown to be the case for B-cell responses to *S. typhimurium* and influenza virus (see discussion above: References 3,24,25, and 34).

In contrast, the clonotypes for synthetic antigens, which are primarily non-self structures, permit the stimulation of multiple B-cell precursors after primary immunization and are not tolerized by this initial exposure to antigen. Thus, after secondary stimulation, the increase in B-cell frequency observed for synthetic antigens would seem less dramatic. Indeed, post-immunization, the frequency of HA-specific B cells increases from 1 in 10^5 to 1 in 4×10^3 splenic B cells[3,70,72] and the frequency of *S. typhimurium*-specific B cells increases from 1 in 10^5 to 1 in 6.1×10^3 splenic B cells[24,25] whereas the frequency of DNP-specific B cells increases from 1 in 5×10^3 to 1 in 1.3×10^3.[2] Although only two such antigen responses have been analyzed, it is intriguing that the size of the primary repertoires for these two natural antigens is so similar and that the expansion of these repertoires post-immunization is of the same order of magnitude, both of which are in contrast to the data accrued for synthetic antigens. Taken together, these data suggest that the rules and restrictions which define the B-cell repertoires specific for synthetic antigens may not be directly correlative or applicable to those specific for natural antigens. Thus, restraints should be applied when attempting to assign characteristics of host responses on the basis of data compiled solely from analyses of synthetic antigens.

B. ARE RESPONSES TO NATURAL ANTIGENS THYMUS-DEPENDENT OR THYMUS-INDEPENDENT?

The results to date suggest that the primary response to natural antigens can be both TI and TD. Both the primary response to influenza and *S. typhimurium* appear to be TD when analyzed in the splenic fragment culture system.[3,25] However, the response to PC, presented in the context of *S. pneumoniae*, appears to have a TI component or may be entirely TI.[26] It is not yet clear what the differences are between the stimulatory requirements of *S. typhimurium* and *S. pneumoniae*. While both sets of studies used killed preparations, the salmonella strain was acetone-killed and the streptococcus preparation was heat-killed. Moreover, in the salmonella studies, the organisms were injected, while in the other study, streptococci were added to the fragment cultures directly. Until comparable antigen preparations and stimulatory conditions are utilized, it will not be possible to directly assess the differences between T-cell requirements for these two responses.

A variety of studies have suggested that the secondary B-cell repertoire is predominantly T dependent, and that the generation of these B cells is enhanced by optimization of collaborative interactions between these B cells and helper T cells (reviewed in References 5 and 7). While such T dependence may be required for secondary B-cell responses to synthetic antigens, the secondary B-cell responses to natural antigens studied thus far suggest that the responses are T independent. Analysis of the secondary B-cell repertoire specific for influenza was carried out in unprimed mice,[34,71] i.e., in the absence of primed T cells. Although it is possible that the donor inoculum contained primed T cells in this study, the studies from this laboratory (Reference 24 and unpublished data) have shown that purified B-cell preparations from primed mice gave frequencies equivalent to those which used whole spleen cell preparations. Since the thymus-dependence/thymus-independence of the secondary responses for other natural antigens has not yet been analyzed critically, it is important to consider the possibility that the secondary responses to natural antigens may, at the very least, be less T-cell dependent than those responses to synthetic antigens. If this were so, then T-independence might serve as an advantage for the host because the development of protective humoral responses would not then require that both lymphoid cell arms of the immune system be intact. Finally, it should be mentioned that the categorization of B cells into T-dependent/T-independent subsets may be more apparent than real and may be a consequence of observed results to fractionated vs. whole antigen. Such conclusions await further experimental clarification.

C. V_H/V_L GENE FAMILY UTILIZATION IN SECONDARY RESPONSES

B cells in the primary repertoire generally express unmutated antibodies, as noted earlier.[5,7] In contrast, the hallmarks of the secondary repertoire are somatic hypermutation and recruitment of new clonotypes (see Volume 96 of *Immunologic Reviews*, 1987; and References 5,7,72, and 73). The targets of somatic hypermutation are the V,D,J immunoglobulin gene segments which have been re-arranged and their immediate 3′ and 5′ flanking regions. In general, point mutations are the main class of mutations observed and these mutations are introduced stepwise during clonal expansion.

The PC memory response to *S. pneumoniae* is similar, if not identical to the primary response.[74] Whereas the secondary immune response to PC-KLH is characterized by two groups of antibodies (Groups I and II),[75] immunization with *S. pneumoniae* (R36A) does not give rise to Group II antibodies.[72,74] The Group I antibodies utilize only genes from the S107 V_H family, as mentioned in a previous section, while the Group II antibodies can use a minimum of three V_H gene families (Q52, S107, and J558).[74,77] These data show that immunization with a whole organism in this case gives a more restricted response than does immunization with a synthetic antigen. While it has been shown that T15 antibodies can be protective in one experimental system,[78] it is not yet clear whether such a restriction in utilization of V region gene families is characteristic of the response when analyzed after a live infection with *S. pneumoniae*.

Claflin and Berry have recently studied the role of somatic hypermutation in the PC-specific antibody response to *S. pneumoniae* by analysis of sets of hybridoma antibodies derived from individual mice.[79]

The results of this study indicated that most of the monoclonal antibodies (mAbs) were encoded by the V1 gene of the T15 V_H family and $V_\kappa 22$, while a minority utilized V1 and the $V_k 8$ or $V_k 24$, all of which is reminiscent of the primary germline gene repertoire.[57-64] Six of the 17 cell lines which were sequenced had mutations. A comparison of the mutated and unmutated T15-positive lines indicated that the mutated lines had a decreased binding capacity to *S. pneumoniae*. Thus, these studies suggest that the antibody specificity for PC presented in the context of *S. pneumoniae* is encoded in the germline and that antigen-driven events selected *for* this germline encoded specificity[79] and against hypermutation for higher affinity antibodies. If these findings were expanded to include responses to all natural antigens, it could be proposed that germline V gene utilization by physiologically relevant antigens is very restricted and that the clonotypes which dominate the primary response also dominate the secondary response because these germline encoded specificities are the most protective for the host.

In contrast, it is clear that not all anti-PC secondary responses, when presented in the context of bacterial organisms, express germline specificities. Claflin et al.[28] have recently shown that the secondary response to PC in the context of *Proteus morganii* belongs to the M603 idiotype family and were derived from a single V_H/V_L pair, V_H1-DFL16-J_H1/V_κ8-J_κ5. Genealogic evidence suggests that somatic mutations at key amino acid positions are required for binding to *P. morganii*, and apparently the response in an individual mouse is derived stepwise by somatic mutation as a consequence of intraclonal diversity and selection from a small number of B-cell precursors.[80,81] Interestingly, a carrier element of the epitope seems to be an important contributor to the selection and expansion of these B cells.[81] These findings support the hypothesis that the co text in which the antigenic determinant is presented is important when considering the relevance of a particular response to protective immunity.

Similary, Staudt and Gerhard,[71] McKean et al.,[82] and Clark et al.,[83] in analyses of hybridoma antibodies specific for the HA of influenza, have shown that the memory response to HA is very restricted in V_H/V_L gene family utilization, and that somatic hypermutations are essential for the diverse response observed There is *interclonal* diversity in which HA antibodies use three different V_H families (S107, 7183, and 3660) and at least three different D_H and J_H regions. Different V_κ gene products are also used including V_κ21C and other V_κ21 subgroups, as well as V_κ8.[83] There is also *intraclonal* diversity in which all antibodies are derived from a single V_H/V_L pair,[83] reminiscent of the PC memory response to *P. morganii*.[29,80,81] Nevertheless, these investigators suggested that their results could be explained by a genealogic tree in which all of the antibodies arose from only two precursors by stepwise mutation and antigenic selection. Subsequent studies by Caton et al.[84] and Reale et al.[85] confirmed and extended these studies. Taken together with the studies of Claflin and colleagues with *P. morganii*, these data present very strong evidence for the role of somatic hypermutation in the response to bacterial and viral antigens. If microorganisms attempt to escape the host immune responses by mutation or alterations in cell surface determinants, then the capacity of the host to diversify rapidly through somatic mutation could potentially increase the host's advantage after infection.

Unlike the responses described above, the V_H/V_L gene family utilization of *S. typhimurium*-specific B cells has not yet been carried out. However, it is interesting to speculate that the response might be distinct from those outlined above because the characteristics of the primary and secondary B-cell repertoires to *S. typhimurium* appear to be more complex. In the analysis of the primary B-cell repertoire specific for *S. typhimurium*, it was shown that less than 30% of these primary B cells were directed against LPS[25] while greater than 80% of the secondary *S. typhimurium* specific B cells were LPS specific.[24] Moreover, of those primary B cells that were specific for LPS, the majority were specific for the outer region of the LPS molecule, the O antigen region, whereas the LPS-specific secondary B-cell clones were predominantly specific for the membrane proximal region of the LPS molecule, the KDO/lipid A region. These data suggest that there is an apparent preferential increase in those B cells specific for the KDO/lipid A region of the LPS molecule after immunization. Such findings may be a consequence of tolerization to environmentally cross-reactive determinants in the primary response and protection from these selective processes in the secondary response (see previous section and References 25 and 37). Alternatively, continued selection and expansion of these particular specificities could occur as a consequence of somatic hypermutation as described above for *P. morganii* and influenza. It may be of interest to analyze the V_H/V_L gene family utilization of antibodies produced after vaccination with various vaccines and correlate V_H/V_L gene usage with protection after challenge with a virulent bacteria. Such studies might provide a valuable approach to analysis of the components that are important in the development of effective vaccines.

D. SUSCEPTIBILITY TO INFECTION AND V_H/V_L GENE FAMILY UTILIZATION

Previous studies have shown that mutant CBA/N mice are susceptible to *S. typhimurium* infection as a consequence of a defective anti-salmonella antibody response.[86,87] Moreover, recent studies from this laboratory[88] suggested that susceptibility might be a result of an altered B-cell repertoire because *S. typhimurium*-specific B-cell clones from these mice expressed a skewed fine specificity pattern when compared with normal CBA/Ca mice. One possible explanation for these findings may be that the B-cell repertoire in CBA/N mice is abnormal or is composed of a skewed B-cell compartment which can not participate in all B-cell responses[89] (reviewed in Reference 90). A second possibility is that susceptibility of CBA/N mice to *S. typhimurium* may be related to an altered utilization pattern of V_H/V_L gene families. This contention is supported by studies by Slack et al.[91] who postulated asymmetric V_H expression among B-cell subsets after they demonstrated that CBA/N mice could not generate an IgG3 response to TNP-Ficoll. Clarke et al.[92] also showed that a CBA/N hybridoma induced by PC-KLH expressed a V_H segment identical to C3 (the C57BL equivalent to T15), while Primi and Cazenave showed that the subset of CBA/N B cells stimulated by LPS did not express the S107 and J606 V_H families;[93] however, Primi and Cazenave only analyzed three different V_H families and $V_\kappa 1$ and $V_\kappa 21D$-E light chains. Previous studies have suggested that immature B cells preferentially express the most D_H proximal V_H family, 7183 (reviewed in References 6 and 69). It is possible that since approximately 50% of the CBA/N splenic B cells are immature by the criteria of tolerance induction,[94] there may be significant expression of the $V_H 7183$ family in these mice. These studies should provide additional insights into the relationship between disease states and V_H/V_L gene family utilization.

IV. CONCLUSION

The review of the studies on natural antigens presented herein will hopefully contribute to the understanding of the genetic and environmental selective processes that are involved in antibody-mediated protection against infectious disease organisms. To date, there appear to be significant differences between the host humoral responses to these diverse natural antigens which need to be explored by further assessment of the humoral immune responses that they induce, as well as analyses of B-cell responses to other natural antigens. Despite these differences, the response to natural antigens as a whole is distinct from B-cell responses to synthetic antigens in significant ways. The studies carried out to date provide a foundation and framework on which future studies of biologically relevant organisms can be built and expanded. (Supported by USPHS grant AI22436 and USUHS grant CO7305.)

REFERENCES

1. **Kreth, H. W.,** The extent of diversity of anti-hapten antibodies in inbred mice: anti-NIP (4-hydroxy-5-iodo-3 nitrophenacetyl) antibodies in CBA/H mice, *Eur. J. Immunol.*, 3, 141, 1973.
2. **Klinman, N. R. and Press, J. L.,** The B-cell specificity repertoire: its relationship to definable subpopulations, *Transplant. Rev.*, 24, 41, 1975.
3. **Cancro, M. P., Gerhard, W., and Klinman, N. R.,** The diversity of the influenza specific primary B-cell repertoire in BALB/c mice, *J. Exp. Med.*, 147, 776, 1978.
4. **Sigal, N. H. and Klinman, N. R.,** The B cell clonotype repertoire, *Adv. Immunol.*, 26, 55, 1978.
5. **Berek, C. and Milstein,** The dynamic nature of the antibody repertoire, *Immunol. Rev.*, 105, 5, 1988.
6. **Malynn, B. A., Berman, J. E., Yancopoulos, G. D., Bona, C. A., and Alt, F. W.,** Expression of immunoglobulin heavy-chain variable region repertoire, *Curr. Topics Microbiol. Immunol.*, 135, 75, 1987.
7. **Klinman, N. R. and Linton, P. J.,** The clonotype repertoire of B cell subpopulations, *Adv. Immunol.*, 42, 1, 1988.
8. **Klinman, N. R.,** The mechanism of antigenic stimulation of primary and secondary precursor cells, *J. Exp. Med.*, 136, 241, 1972.
9. **Press, J. L. and Klinman, N. R.,** Frequency of hapten-specific B cells in neonatal and adult murine spleens, *Eur. J. Immunol.*, 4, 155, 1974.
10. **Sigal, N. H.,** The frequency of p-azophenylarsonate and 5-dimethylamino-naphthalene-sulfonyl specific B cells in neonatal and adult BALB/c mice, *J. Immunol.*, 119, 1129, 1977.
11. **Sigal, N. H.,** Regulation of azophenylarsonate-specific repertoire expression. I. Frequency of cross-reactive idiotype-positive B cells in A/J and BALB/c mice, *J. Exp. Med.*, 156, 1352, 1982.
12. **Stashenko, P. and Klinman, N. R.,** Analysis of the primary anti-(4-hydroxy-3-nitrophenyl) acetyl responsive B cells in BALB/c and B10.D2 mice, *J. Immunol.*, 125, 531, 1980.
13. **Riley, R. L. and Klinman, N. R.,** Differences in antibody repertoires for (4-hydroxy-3-nitrophenyl) acetyl (NP) in splenic vs. immature bone marrow precursor cells, *J. Immunol.*, 135, 3050, 1985.
14. **Press, J. L. and Giorgetti, C. A.,** Clonal analysis of the primary and secondary B cell responses of neonatal, adult, and *xid* mice to (T,G)-A--L, *J. Immunol.*, 137, 784, 1986.
15. **Busto, P., Gerstein, R., Dupre, L., Giorgetti, C., Selsing, E., and Press, J. L.,** Molecular analysis of heavy and light chains used by primary and secondary anti-(T,G)-A--L antibodies produced by normal and *xid* mice, *J. Immunol.*, 139, 608, 1987.

16. **Shahin, R. K. and Cebra, J. J.,** Rise in insulin-sensitive B cells during ontogeny can be prematurely stimulated by thymus-dependent and thymus-independent antigens, *Infect. Immun.*, 32, 211, 1981.
17. **Sigal, N. H., Gearhart, P. J., and Klinman, N. R.,** The frequency of phosphorylcholine-specific B cells in conventional and germ free BALB/c mice, *J. Immunol.*, 114, 1354, 1975.
18. **Gearhart, P. and Cebra, J.,** Differentiated B lymphocytes, *J. Exp. Med.*, 149, 216, 1979.
19. **Fung, J. and Kohler, H.,** Immune response to phosphorylcholine. VII. Evidence for three separate B cell subpopulations responding to T-independent and T-dependent PC-antigens, *J. Immunol.*, 125, 640, 1980.
20. **Klinman, N. R. and Stone, M. R.,** The role of variable region gene expression and environmental selection in determining the anti-phosphorylcholine B-cell repertoire, *J. Exp. Med.*, 158, 1948, 1984.
21. **Fernandez, C. and Moller, G.,** Immunological unresponsiveness to native dextran B512 in young animals of dextran high responder strains is due to lack of Ig receptor expression: evidence for non-random expression of V-genes, *J. Exp. Med.*, 147, 645, 1978.
22. **Ward, R. and Kohler, H.,** Regulation of clones responding to dextran B13555, *J. Immunol.*, 126, 146, 1981.
23. **Stohrer, R. and Kearney, J.,** Ontogeny of B cell precursors responding to 1, 3, dextran in BALB/c mice, *J. Immunol.*, 133, 2323, 1984.
24. **Metcalf, E. S., Gaffney, H., and Duran, L. W.,** The diversity of the secondary *Salmonella typhimurium*-specific B cell repertoire, *J. Immunol.*, 138, 3461, 1987.
25. **Duran, L. W. and Metcalf, E. S.,** Clonal analysis of primary B cells responsive to the pathogenic bacterium *Salmonella typhimurium, J. Exp. Med.*, 165, 340, 1987.
26. **Hurwitz, J. L., Torgart, V. B., Schweitzer, P. A., and Cebra, J. J.,** Patterns of isotype expression by B cell clones responding to thymus-dependent and thymus-independent antigens *in vitro, Eur. J. Immunol.*, 12, 342, 1982.
27. **Williams, K. R. and Claflin, J. L.,** Clonotypes of anti-phosphocholine antibodies induced with *Proteus morganii* (Potter). I. Structural and idiotypic similarities in a diverse repertoire, *J. Immunol.*, 125, 2429, 1980.
28. **Claflin, J. L., Berry, J., Flaherty, D., and Dunnick, W.,** Somatic evolution of diversity among anti-phosphocholine antibodies induced with *Proteus morganii, J. Immunol.*, 138, 3060, 1987.
29. **Marion, T. N., Dzierzak, E. A., Lee, H., Adams, R. L., and Janeway, C. A.,** Non-dinitrophenyl-binding immunoglobulin that bears a dominant idiotype (Id460) associated with antidinitrophenyl antibody is specific for an antigen on *Pasteurella pneumotropica, J. Exp. Med.*, 159, 221, 1984.
30. **Dzierzak, E. A., Janeway, C. A., Richard, N., and Bothwell, A.,** Molecular characterization of antibodies bearing Id-460. I. The structure of two highly homologous V_H genes used to produce idiotype positive immunoglobuins, *J. Immunol.*, 136, 1864, 1986.
31. **Rose, F. V. and Cebra, J. J.,** Isotype commitment of B cells and dissemination of the primed state after mucosal stimulation with *Mycoplasma pulmonis, Infect. Immun.*, 49, 428, 1985.
32. **Cebra, J. J., Gearhart, P. J., Halsey, J. F., Hurwitz, J. L., and Shahin, R. D.,** Role of environmental antigens in the ontogeny of the secretory immune response, *J. Reticuloendothel. Soc.*, 28, 61s, 1980.
33. **Cancro, M. P., Thompson, M., and Hilbert, D. M.,** Developmental aspects of B-cell repertoire phenotype, *Survey Immunol. Res.*, 2, 62, 1983.
34. **Gerhard, W., Braciale, T. J., and Klinman, N. R.,** The analysis of the monoclonal immune response to influenzae virus. I. Production of monoclonal anti-viral antibodies *in vitro, Eur. J. Immunol.*, 5, 720, 1975.
35. **Wylie, D. and Klinman, N. R.,** The murine B cell repertoire responsive to an influenza-infected syngeneic cell line, *J. Immunol.*, 127, 194, 1981.
36. **Froscher, B. G. and Klinman, N. R.,** Immunization with SV40-transformed cells yields mainly MHC-restricted monoclonal antibodies, *J. Exp. Med.*, 164, 196, 1986.
37. **Metcalf, E. S. and Klinman, N. R.,** *In vitro* tolerance induction of neonatal B cells, *J. Exp. Med.*, 143, 1327, 1976.
38. **Metcalf, E. S., Schrater, A. F., and Klinman, N. R.,** Murine models of tolerance induction in developing and mature B cells, *Immunol. Rev.*, 43, 143, 1979.
39. **Teale, J. and Klinman, N. R.,** Tolerance as an active process, *Nature*, 288, 385, 1980.
40. **Metcalf, E. S., Sigal, N. S., and Klinman, N. R.,** *In vitro* tolerance induction of neonatal and adult B cells as a probe for the study of B-cell diversification, *J. Exp. Med.*, 145, 1382, 1977.
41. **Teale, J. M., Layton, J., and Nossal, G. J. V.,** *In vitro* model for natural tolerance to self-antigens. Inhibition of the development of sIg negative lymphocytes into T-dependent responsive B cells by antigen, *J. Exp. Med.*, 150, 205, 1979.
42. **Thompson, M. A. and Cancro, M. P.,** Dynamics of B-cell repertoire formation: normal patterns of clonal turnover are altered by ligand interaction, *J. Immunol.*, 129, 2372, 1982.
43. **Thompson, M. A., Raychaudhuri, S., and Cancro, M. P.,** Restricted adult clonal profiles induced by neonatal immunization, *J. Exp. Med.*, 158, 112, 1983.

44. **Vakil, M. and Kearney, J. F.,** Functional characterization of monoclonal auto-anti-idiotype antibodies isolated from the early B cell repertoire of BALB/c mice. *Eur. J. Immunol.,* 16, 1151, 1986.
45. **Stein, K. and Sonderstrom, T.,** Neonatal administration of idiotype or anti-idiotype primes for protection against *E. coli* K13 infection in mice, *J. Exp. Med.,* 160, 1001, 1984.
46. **Tonegawa, S.,** Somatic generation of antibody diversity, *Nature,* 302, 575, 1983.
47. **Early, P., Huang, H., Davis, M., Calame, K., and Hood, L.,** An immunoglobulin heavy chain variable region gene is generated from three segments of DNA: V_H, D_H and J_H, *Cell,* 19, 981, 1980.
48. **Brodeur, P. and Riblett, R.,** The immunoglobulin heavy chain variable region (IgH-V) in the mouse. I. 100 IgH-V genes comprise 7 families of homologous genes, *Eur. J. Immunol.,* 14, 922, 1984.
49. **Winter, E. A., Radbruch, A., and Krawinkel, U.,** Members of novel V_H gene families are found in VDJ regions of polyclonally activated B-lymphocytes, *EMBO J.,* 4, 2861, 1985.
50. **Dildrop, R.,** A new classification of mouse V_H sequences, *Immunol. Today,* 5, 85, 1984.
51. **Kofler, R.,** A new murine Ig V_H gene family, *J. Immunol.,* 140, 4031, 1988.
52. **Rathburn, G., Capra, J. D., and Tucker, P.,** Organization of the murine immunoglobulin V_H complexity in inbred strains, *EMBO J.,* 6, 2391, 1987.
53. **Blankenstein, T. and Krawinkel, U.,** Immunoglobulin V_H region genes of the mouse are organized in overlapping clusters, *Eur. J. Immunol.,* 17, 1351, 1987.
54. **Brodeur, P., Osman, G. E., Mackle, J. J., Lalor, T. M.,** The organization of the mouse IgH-V locus: dispersion, interspersion, and the evalution of V_H gene family clusters, *J. Exp. Med.,* 168, 2261, 1988.
55. **D'Hoostelaere, L. A., Huppi, K., Mock, B., Mallett, C., and Potter, M.,** The Ig_κ L chain allelic groups among the Ig_κ haplotypes and Ig_κ crossover populations suggest a gene order, *J. Immunol.,* 141, 652, 1988.
56. **Kofler, R., Duchosal, M., and Dixon, F.,** Complexity polymorphism and connectivity of mouse V gene families, *Immunogenetics,* 29, 65, 1989.
57. **Claflin, J. L. and Davie, J. M.,** Clonal nature of the immune response to phosphoylcholine. IV. Idiotypic uniformity of binding site associated antigenic determinants among mouse anti-phosphyorylcholine antibodies, *J. Exp. Med.,* 140, 673, 1974.
58. **Lee, W., Cosenza, H., and Kohler, H.,** Clonal restriction of the immune response to phosphorylcholine, *Nature,* 247, 55, 1974.
59. **Perlmutter, R. M., Crews, S. T., Douglas, R., Sorenson, G., Johnson, N., Nivera, N., Gearheart, P. J., and Hood, L.,** The generation of diversity in phosphorylcholine-binding antibodies, *Adv. Immunol.,* 35, 1, 1984.
60. **Claflin, J. L., Hudak, S., Madalena, A., and Bender, T.,** Antigen-specific anti-PC antibodies: binding site studies, *J. Immunol.,* 134, 2536, 1985.
61. **Claflin, J. L. and Rudikoff, S.,** Uniformity in a clonal repertoire: a case for a germline basis of antibody diversity, *Cold Spring Harbor Symp. Quant. Biol.,* 41, 725, 1977.
62. **Cosenza, H. and Kohler, H.,** Specific inhibition of plaque formation to phosphorylcholine by antibody against antibody, *Science,* 176, 1027, 1972.
63. **Crews, S., Griffin, J., Huang, H., Calame, K., and Hood, L.,** A single VH gene segment encodes the immune response to phosphorylcholine: somatic mutation is correlated with the class of the antibody, *Cell,* 25, 59, 1981.
64. **Berek, C.,** The D segment defines the T15 idiotype: the immunoresponse of A/J mice to *Pneumococcus pneumoniae, Eur. J. Immunol.,* 14, 1043, 1984.
65. **Clark, S. H., Claflin, J. L., Potter, M., and Rudikoff, S.,** Polymorphisms in anti-phosphocholine antibodies reflecting evolution of immunoglobulin families, *J. Exp. Med.,* 157, 98, 1983.
66. **Stout, J. T., Strickland, F. M., and Cerny, J.,** Regulation of idiotope expression. I. The effect of antigen dose on expression of certain T15 idiotopes during primary IgM response to *S. pneumoniae* R36a, *J. Immunol.,* 134, 1926, 1985.
67. **Strickland, F. M., Hamilton, S. L., Blalock, E., and Cerny, J.,** Shared idiotype between phosphorylcholine-specific antibody and acetylcholinesterase detectable by a monoclonal antibody, *J. Immunol.,* 134, 1053, 1985.
68. **Kienker, L. J., Korostoff, J. M., and Cancro, M. P.,** Patterns of Ig V region expression among neonatal hemagglutinin-responsive B cells, *J. Immunol.,* 141, 3634, 1988.
69. **Jeong, H. D. and Teale, J. M.,** V_H Gene family repertoire of resting B cells. Preferential use of D-proximal families early in development may be due to distinct B cell subsets, *J. Immunol.,* 143, 2752, 1989.
70. **Metcalf, E. S. and Klinman, N. R.,** *In vitro* tolerance of bone marrow cells: a marker for B cell maturation, *J. Immunol.,* 118, 2111, 1977.
71. **Staudt, L. M. and Gerhard, W.,** Generation of antibody diversity in the immune response of BALB/c mice to influenza virus hemagglutinin, *J. Exp. Med.,* 157, 687, 1983.
72. **Rajewsky, K., Forster, I., and Cumano, A.,** Evolutionary and somatic selection of the antibody repertoire in the mouse, *Science,* 238, 1088, 1987.
73. **Kocks, C. and Rajewsky, K.,** Stable expression and somatic hypermutation of antibody V regions in B-cell developmental pathways, *Annu. Rev. Immunol.,* 7, 537, 1989.

74. **Stenzel-Poore, M. P., Bruderer, U., and Rittenberg, M.,** The adaptive potential of the memory response: clonal recruitment and epitope recognition, *Immunol. Rev.*, 105, 113, 1988.
75. **Chang, S. P., Brown, M., and Rittenberg, M. B.,** Immunologic memory to phosphorylcholine. II. PC-KLH induces two antibody populations that dominate different isotypes, *J. Immunol.*, 128, 702, 1982.
76. **Stenzel-Poore, M. P., Hall, T. J., Heusser, C. H., Faust, C. H., and Rittenberg, M. B.,** Immunologic memory to PC-KLH: participation of the Q52 V_H gene family, *J. Immunol.*, 139, 1698, 1987.
77. **Wicker, L. S., Guelde, G., Scher, I., and Kenny, J. J.,** The asymmetry in idiotype-isotype expression in the response to phosphocholine is due to divergence in the expressed repertoires of $Lyb5^+$ and $Lyb5^-$ B cells, *J. Immunol.*, 131, 2468, 1983.
78. **Briles, D. E., Forman, C., Hudak, S., and Claflin, J. L.,** Anti-phosphorylcholine antibodies of the T15 idiotype are optimally protective against *Streptococcus pneumoniae, J. Exp. Med.*, 156, 1177, 1982.
79. **Claflin, J. L. and Berry, J.,** Genetics of the phosphocholine-specific antibody response to *Streptococcus pneumoniae*. Germline but not mutated T15 antibodies are dominantly selected, *J. Immunol.*, 141, 4012, 1988.
80. **Dell, C. L., Lu, Y., and Claflin, J. L.,** Molecular analysis of clonal stability and longevity in B cell memory, *J. Immunol.*, 143, 3364, 1989.
81. **Claflin, J. L., George, J., Dell, C., and Berry, J.,** Patterns of mutations and selection in antibodies to the phosphocholine-specific determinant in *Proteus morganii, J. Immunol.*, 143, 3054, 1989.
82. **McKean, D., Huppi, K., Bell, M., Staudt, L., Gerhardt, W., and Weigert, M.,** Generation of antibody diversity in the immune response of BALB/c mice to influenza virus hemagglutinin, *Proc. Natl. Acad. Sci. U.S.A.*, 81, 3180, 1984.
83. **Clarke, S. H., Huppi, K., Ruezinsky, D., Staudt, L., Gerhard, W., and Weigert, M.,** Inter- and intraclonal diversity in the antibody response to influenza hemagglutinin, *J. Exp. Med.*, 161, 687, 1985.
84. **Caton, A. J., Brownlee, G. G., Staudt, L. M., and Gerhard, W.,** Structural and functional implications of a restricted antibody response to a defined antigenic region on the influenza virus hemagglutinin, *EMBO J.*, 5, 1577, 1986.
85. **Reale, M. A., Manheimer, A. J., Moran, T. M., Norton, G., Bona, C. A., and Schulman, J. L.,** Characterization of monoclonal antibodies specific for sequential influenza A/PR/8/34 virus variants, *J. Immunol.*, 137, 1352, 1986.
86. **O'Brien, A. D., Scher, I., Campbell, G. H., MacDermott, R. P., and Formal, S. B.,** Susceptibility of CBA/N mice to infection with *Salmonella typhimurium*: influence of the X-linked gene controlling B lymphocyte function, *J. Immunol.*, 123, 720, 1979.
87. **O'Brien, A. D., Scher, I., and Metcalf, E. S.,** Genetically conferred defect in anti-salmonella antibody formation renders CBA/N mice innately susceptible to *Salmonella typhimurium* infection, *J. Immunol.*, 126, 1368, 1981.
88. **Duran, L. W. and Metcalf, E. S.,** Antibody-defective, genetically susceptible CBA/N mice have an altered *Salmonella typhimurium*-specific B cell repertoire, *J. Exp. Med.*, 165, 29, 1987.
89. **Woodland, R. T. and Huber, B. T.,** Selective activation by thymus-dependent antigens of distinct B cell subpopulations expressing a major cross-reactive idiotype, *J. Immunol.*, 133, 1801, 1984.
90. **Scher, I.,** CBA/N immune defective mice: evidence for the failure of a B-cell subpopulation to be expressed, *Immunol. Rev.*, 64, 117, 1982.
91. **Slack, J., DerBalian, G., Nahm, M., and Davie, J. M.,** Subclass restriction of murine antibodies. II. The IgG plaque-forming cell response to thymus-independent type 1 and type 2 antigens in normal mice and mice expressing an X-linked immunodeficiency, *J. Exp. Med.*, 151, 853, 1980.
92. **Clark, S. H., Kenny, J. J., Sieckmann, D. G., and Rudikoff, S.,** Amino acid sequence of a phosphocholine-binding antibody from an immune defective CBA/N mouse employing the T15 V_H region associated with unusual D_H, J_H and V_κ segments, *J. Immunol.*, 132, 1544, 1984.
93. **Primi, D., Barbier, E., Drapier, A., and Cazenave, P.,** Paucity of phosphorylcholine-specific clones in B cells expressing the V_HT15 gene product, *J. Exp. Med.*, 164, 2107, 1986.
94. **Metcalf, E. S., Scher, I., and Klinman, N. R.,** Susceptibility to *in vitro* tolerance induction of adult B cells from mice with an X-linked B cell defect, *J. Exp. Med.*, 151, 486, 1980.

Chapter 10

HELPER T CELLS AND THE MATURATION OF THE B CELL RESPONSE TO ANTIGEN

Suzanne Fish and Tim Manser

TABLE OF CONTENTS

I. INTRODUCTION

In order for most humoral immune responses to occur, B cells and T cells must interact. Therefore, the potential exists for T cells to influence the expressed antibody repertoire. Different types of these helper T cells may be stimulated with different antigens. Susceptibility of different subsets of B cells to the influence of T_H cells may also vary. Consequently, selective stimulation of subsets of T cells or specific interactions among T_H and B cells may affect the expressed antibody variable region repertoire.

This review will begin with a discussion of interactions between T cell subsets and B cells. Then, a comparison of responses to T cell dependent antigens and T cell independent antigens will be made. The existence of different B cell subsets and the influence of T_H cells on their responses to antigen will be discussed. Finally, the maturation of the B cell response will be examined in terms of T_H cell influence on somatic mutation of antibody variable regions, affinity maturation and B cell memory development.

II. T CELL-B CELL INTERACTION DURING THE IMMUNE RESPONSE

T cell-B cell interaction may occur through several mechanisms. The best characterized mechanism is "cognate" interaction. T_H cells can recognize and respond to antigen which is presented on the major histocompatibility (MHC) proteins, I-A and I-E. For this recognition to occur, the nominal antigen must be "processed", or partially degraded within an antigen presenting cell (APC). When the protein complex of the resulting peptides and MHC proteins are expressed on the APC surface, the T cell receptor can then bind this complex. B cells can perform as APC. When this processed antigen is presented to T_H cells, T_H cells respond by proliferation and secretion of a number of cytokines which can, in turn, influence B cell proliferation, isotype switching and differentiation to secretory phenotype. By only stimulating certain B cells, T_H cells may directly influence or alter the B cell repertoire.

A. CARRIER SPECIFIC T_H CELLS

Carrier specific T_H cells are essential for the response of B cells to most antigens. This has been demonstrated by both *in vivo* and *in vitro* experiments. B cells cannot respond to most antigens purified away from carrier specific T_H cells. However, either carrier primed T cells or cloned, carrier specific T cell lines are able to restore the B cell response to T cell dependent antigens.[1-3]

B. IDIOTYPE SPECIFIC T_H CELLS

Idiotype specific T cells may play a part in the antigenic stimulation of B cells expressing antibodies which have particular idiotypic determinants, and thus directly affect the composition of the expressed antibody repertoire. Although the existence of both idiotype specific T helper and T suppressor cells has been documented, this review will concentrate on T helper cells.

Experiments in two antigenic systems, *p*-azophenylarsonate (Ars) and phosphorylcholine (PC), indicate that the presence of T_H cells specific for their respective major idiotypes, CRI and T15, may be necessary for the predominance of B cells expressing antibodies of these idiotypes.[4-7] Adoptive transfer experiments were used to show that anti-Ars antibody was always made in response to Ars-keyhole limpet hemocyanin (Ars-KLH) as long as the donor T cells had been primed to the carrier used in boosting.[4] However, antibody of the CRI idiotype was not made if idiotype specific T cells were depleted from this carrier primed population by absorption to plates coated with idiotype positive, anti-Ars antibodies. Conventional carrier specific T cells recognize processed forms of the carrier protein. Thus, it is not clear how idiotype specific T cells could have been derived in these experiments unless there is a cross reactive epitope present in both the carrier protein and idiotype positive antibody.

T cells primed *in vivo* and transferred into antigen primed athymic mice induce B cells producing idiotype positive antibodies in response to the PC epitope on *S. pneumoniae* R63a (Pn).[7] T cells remaining after depletion of idiotype specific T cells through panning on plates coated with idiotype positive antibodies did not stimulate B cells to respond to Pn. The transfer of T cells enriched for idiotype specificity resulted in an enhanced response of B cells producing idiotype positive antibody. Unfortunately, these experiments are complicated by the use of a TI antigen to study the effect of T cells on a B cell response.

Addition of T cells enriched for idiotype specific T cells through binding to plates coated with idiotype positive antibody has been shown to greatly augment the amount of $T15^+$ antibodies produced by naive B cells *in vitro* in response to PC-ovalbumin.[8] An idiotype specific T_H cell line also augments the response of B cells producing idiotype positive antibodies.[9]

Idiotype specific T cells can be enriched or depleted (see above) through direct binding to intact antibody.[4,7,8,10] This binding obviously does not require that the antibody be "processed". Presentation of antigen to carrier specific T cells is very sensitive to changes in MHC. Some results indicate that recognition of antibody by idiotype specific T_H cells is not restricted by MHC.[5,9] In these cases, it is not known how the antibody is being recognized by the T cells. F_c receptors are unlikely because these T cells do not bind to idiotype negative antibody. Therefore, while T_H cells that can directly recognize antibody idiotypes appear to exist, their mechanism of action and their role in humoral responses remain unclear.

In some cases, idiotype specific T cells can recognize their respective idiotypes after processing by APC.[11] Unfortunately, these idiotype specific T_H cells have been elicited by injecting idiotypic antibody, not by injecting the antigen which normally elicits idiotype positive antibody. Experiments using B lymphoma cells indicate that B cells may process and present their own immunoglobulin in the context of MHC antigens.[12] These points suggest that at least some idiotype specific T cells may recognize antibody in the same manner as carrier specific T cells recognize their cognate antigen.

The experiments described above indicate that idiotype specific T cells can influence the stimulation of B cells expressing specific variable regions. In contrast, experiments using the hapten (4-hydroxy-3-nitrophenyl)acetyl (NP) suggest that idiotypic T_H cells are not required for expression of two major idiotopes of this system (Ac38 and Ac146).[13] Two different carrier specific T_H cell lines were able to stimulate naive B cells to secrete idiotype positive antibody in an adoptive host. The ability of the T_H cell lines to bind directly to idiotype positive antibody was not tested. The studies described above (using Ars and Pn) indicate idiotype specific T_H cells can be elicited by carrier protein alone.[4,9] Although it is unlikely that the two T cell lines used in the experiments with NP were also idiotype specific, this remains a formal possibility. The ability of idiotype specific T_H cells to influence the variable region repertoire expressed by B cells thus appears to vary with the system used for study.

Interaction between B cells and T cells is clearly required for the stimulation of B cells by TD antigens. Carrier specific T_H cells play a pivotal role in this interaction through antigen presentation in the context of MHC proteins and subsequent production of B cell growth and differentiation factors. Although idiotype specific T_H cells have the potential to influence the B cell repertoire, the mechanism and outcome of interaction between these T cells and B cells is not clear.

III. TI VS. TD RESPONSES

One way to examine the effect of T cells on a humoral immune response is to use antigens that can elicit a response in the absence of T cells. (Such an antigen is termed T cell independent (TI).) As discussed above, idiotype specific T_H cells may influence the response of certain B cells to T cell dependent (TD) antigens. However, TI forms of an epitope are often capable of stimulating B cells producing idiotype positive antibody characteristic of TD responses to the same epitope. B cells expressing antibody with similar idiotypic determinants are stimulated with either a TI or TD form of dextran (dextran alone or dextran-Limulus hemocyanin).[14] Similar idiotype expression is found in euthymic mice in response to TD and TI forms of NP.[13] Even more convincingly, in athymic mice where few mature T cells exist, the predominant TD idiotype is expressed in response to a TI form of NP.[13] Immune responses to the TD antigen, Ars-KLH and the TI antigen, Ars-*Brucella abortus* (Ars-Bruc) elicit antibody encoded by the same combination of variable region gene segments.[15] Analysis of hybridomas isolated after immunization with these two forms of Ars indicates that T cells are not directly selective in stimulating B cells expressing particular variable region genes. Nevertheless, TI forms of a hapten do not always elicit antibody of the idiotype seen in a response to a TD form. NP-Ficoll does not elicit the major idiotype characteristic of the TD response. Hybridomas have been isolated which express only one (the V_H gene segment) of the variable region gene segments necessary for the presence of idiotypic determinants.[16] The V_L which encodes idiotype positive antibodies was not expressed by these hybridomas. Thus, the results obtained from different experimental systems using TD and TI antigens differ regarding the importance of T cells for the expression of certain variable regions by B cells.

Ideally, the use of TI antigens could yield insights regarding the role that T_H cells play in alteration of the expressed antibody repertoire. Results of experiments using TI and TD antigens have, nevertheless, failed to clarify this role. Mature T cells present in euthymic mice probably do affect the response even to "TI" antigens. Use of athymic mice in combination with TI antigens may allow an assessment of the changes that can occur in the variable region repertoire expressed by B cells in the absence of the influence of mature T cells.

IV. DIFFERENTIAL T_H INTERACTION WITH B CELL SUBSETS

As summarized above, T_H cells may influence the ability of B cells to respond to certain antigens. In addition, different subsets of B cells may express different variable regions (i.e., different repertoires) as well as have different requirements for stimulation. Ly1 B cells are one subgroup of B cells which appear to be very different from "conventional" B cells. Ly1 B cells express a limited number of variable region genes and are predominantly localized to the peritoneal cavity.[17] The specificities of Ly1 B cells resident in the peritoneum appear to be the same in the presence or absence of T cells. This indicates that these cells may predominantly respond to TI antigens.[18,19] Thus, T cells probably have little influence on the limited repertoire of variable regions expressed by Ly1 B cells.

Differential stimulation of two other subsets of B cells, $Lyb5^-$ and $Lyb5^+$, (also referred to as B1 and B2, respectively), may also play a role in the control of the expressed B cell repertoire. These two subsets of cells are defined by the CBA/N strain of mice which carries the *xid* gene. Normal mice have both subsets of B cells; *xid* mice have only $Lyb5^-$ B cells. TI antigens are categorized into two groups, TI-1 and TI-2, on the basis of the ability of the antigen to elicit an immune response from $Lyb5^-$ B cells in CBA/N mice.[20] Only $Lyb5^+$ B cells are able to respond to TI-2 antigens. TD antigens can be divided into two groups in a similar manner.[21] In this case it is not clear whether the designation of TD antigens into these two groups is due to a defect in T cells of the CBA/N mice or their lack of $Lyb5^+$ B cells. There also may be a difference in the requirement for MHC restriction in the interaction of the different B cell subgroups with T cells.[22,23] Some experiments have suggested that $Lyb5^+$ B cells do not require MHC restricted interaction with T cells for activation, while $Lyb5^-$ B cells do require MHC restricted interaction with T cells for activation.

As defined by idiotype, the repertoire of B cells from normal and *xid* mice able to respond to LPS 0113, 2,4-dinitrophenyl (DNP), (T,G)-A—L, and Ars are similar.[24-27] An attempt to determine which subset of B cells in normal mice produced these idiotypes was only made in the experiments using Ars. In the other systems, it is possible that normal $Lyb5^-$ cells and *xid* $Lyb5^-$ B cells were being compared instead of $Lyb5^+$ and $Lyb5^-$ B cells. Further complications with these experiments result from the fact that *xid* mice lack Ly1 B cells. In response to another antigen, PC, $Lyb5^-$ B cells from *xid* mice were not able to produce antibodies of the major idiotype, T15.[28] This was confirmed through sequencing of the variable region genes. A hybridoma from an *xid* mouse (thus derived from an $Lyb5^-$ B cell) expressed the correct heavy chain variable region gene segment for idiotype expression but the incorrect D, J_H and V_K gene segments.[29] Therefore, it is unclear whether $Lyb5^+$ and Lyb^- subsets of B cells are capable of expressing the same variable regions or whether each subset is restricted in the expression of certain variable regions.

Different subsets of B cells may express different variable regions and may have different requirements for interaction with T cells. Ly1 B cells may not require an interaction with T cells to respond to antigen. $Lyb5^+$ and $Lyb5^-$ B cells may both require T_H cells for stimulation. $Lyb5^-$ B cells appear to require MHC restricted interaction with T_H cells while $Lyb5^+$ B cells do not appear to require MHC restricted interaction.

V. MATURATION OF THE B CELL RESPONSE

Many changes occur in the expressed antibody variable region repertoire during a response to antigen. These include somatic mutation, affinity maturation and development of the memory repertoire. The role that T_H cells might play in these changes will now be discussed.

A. T_H AND SOMATIC MUTATION OF ANTIBODY VARIABLE REGIONS

Somatic mutation of variable region genes serves to increase the diversity of the variable region repertoire. Several haptens conjugated to TD carriers elicit B cell responses in which the mutation of expressed variable region genes occurs at a high rate.[30-34] An antigenic response is necessary for this mutation. Hybridomas derived from naive B cells express unmutated variable region genes, as do those derived from naive B cells which have been driven to proliferate *in vitro* by the mitogens LPS and dextran sulfate.[35,36] Early in the primary response to TD antigens, hybridomas can be isolated which also express variable region genes in the unmutated form. It is later in the primary response that mutated forms of variable region genes are observed.

Results of experiments using TI antigens are ambivalent with regard to whether T cells are necessary for the somatic mutation of variable region genes. Somatic mutation is observed among the variable regions expressed by hybridomas isolated after a secondary immunization of Ars-Bruc.[15] The amount of mutation observed is about half of that seen in response to Ars-KLH. Because these experiments were performed in euthymic mice, it is likely that T cells influenced this immune response. T_H cells may augment the amount of somatic mutation that took place in this response.

Conflicting results have been observed in another antigenic system. A tertiary response to the TI antigen, NP-Ficoll, resulted in isolation of hybridomas expressing unmutated variable regions.[16] Priming with NP-Ficoll followed by boosting with NP-chicken gamma globulin (CGG), also gives rise to hybridomas expressing variable region genes in the germline form.[37] This would indicate that the participation of T_H cells is necessary for somatic mutation to occur in B cells.

It is therefore not clear whether T cells are necessary for the somatic mutation of variable region genes. Although B cells may not require T cells for induction of mutation, T cells may augment this process.

B. T_H AND AFFINITY MATURATION

As an immune response to a TD antigen progresses, the average affinity of serum antibody for the antigen increases. An increase in affinity can also be observed in monoclonal antibodies produced by hybridomas derived from B cells at various times of the immune response. These changes in affinity reflect changes in the expressed antibody repertoire. Are T_H cells necessary for the change in the B cell repertoire which is represented by this observed change in antibody affinity?

Similar affinities were determined for antibodies derived from a TD response to dinitrophenyl-fowl gamma globulin (DNP-FGG) and a TI response to DNP-flagellin.[38] A difference in affinity was not found on comparison of antibodies derived from DNP-KLH and DNP-Ficoll responses either.[39] Monoclonal antibodies derived from a secondary response to Ars-Bruc exhibit an increase in affinity for Ars similar to that found for Ars-KLH.[15] These experimental systems all show that TI antigens are capable of eliciting an immune response which involves an increase in antibody affinity as is seen with TD antigens. Therefore, affinity maturation per se appears not to require T_H cells and thus the requisite changes in antibody repertoire do not require T cells.

These results are in conflict with reports that responses with little T cell help produce antibody of low infinity.[40-42] These experiments examined the stimulation of B cells with TD antigen in the absence of T cells. Unfortunately, it is difficult to examine affinity maturation if little antibody is produced. The experiments previously discussed utilized TI antigens to which B cells can respond directly. The idea that T_H cells may influence the affinity of antibodies elicited by a TI antigen is also challenged by the findings that *in vitro* stimulation of B cells by the TI antigen, TNP-Ficoll, with or without T cells present results in IgG2a antibodies of similar affinity.[43] Therefore, T_H cells probably do not affect affinity maturation, except for augmenting greater increases in affinity through somatic mutation.

C. T_H AND B CELL MEMORY DEVELOPMENT

Often, a shift in the B cell repertoire occurs between the primary response and the secondary response. This shift appears to be due to the development of memory cells. NP or oxazolone (OX) conjugated to TD carriers elicit a major cross reactive idiotype in the primary response. However, after a second injection, the expressed antibody repertoire is very heterogeneous. The antibodies of the predominant, primary idiotype are only a small fraction of the antibodies of the secondary response. In the Ars system, the secondary response is dominated by antibodies encoded by a single combination of variable region gene segments and the primary response initially consists of a small amount of antibody of this type. These shifts in variable region usage may be due to the preferential expansion of B cell clones with certain variable regions, or the selective entry of B cells into the memory population. Control of somatic mutation and the potential for affinity maturation may also play roles in a repertoire shift. Idiotypic suppression may be involved in some systems. What role do T_H cells play in this repertoire shift?

TI forms of haptens have been used to explore the role of T cells in this repertoire shift. Unfortunately, the secondary response to TI antigens is usually poor. This has discouraged detailed molecular studies of this type. Responses to TI antigens have mostly been studied with respect to the isotypes produced. When experiments have been done, conflicting results with respect to the role of T cells in the repertoire shift of B cell memory have been obtained. The response to Ars conjugated *Brucella abortus*, a TI antigen, appears to follow the repertoire pattern of the response to Ars conjugated to KLH, a TD antigen.[15] On the other hand, NP-Ficoll, a TI antigen, does not stimulate B cells expressing the major idiotype of the primary response elicited by a TD form of NP.[16]

The control of somatic mutation may play a central role in the development of memory B cells. Two different experimental systems suggest that somatic mutation stops or occurs at a very slow rate in memory B cells. Adoptive transfer experiments indicate that clonal expansion of memory B cells can occur without somatic mutation.[44,45] One such experiment involved the transfer of primed cells from individual mice into several irradiated adoptive recipients.[45] Clonally related B cells, which shared all detected somatic mutations, were isolated from different adoptive recipients. Stimulation of memory B cells with an antigen structurally related to the priming antigen also resulted in isolation of clonally related hybridomas which shared all detected somatic mutations.[46] These results indicate that mutation had occurred in a precursor B cell and then stopped, as proliferation and expansion of the clone continued. T cells might play a role in regulating mutation in B cells and thereby affect the expressed antibody repertoire. If the signal to mutate is not removed at some point, then the variable region of the B cell clone may eventually incur so much mutation that it loses all affinity for the antigen. The consequence of this may be the ''death'' of the B cell clone. It is possible that this occurs in B cells producing antibodies of the dominant idiotype during the primary responses to NP and OX, thus causing a repertoire shift.

The development of memory B cells results in a change in the B cell repertoire that is revealed upon further immunization. Comparison of immune responses elicited with TI and TD forms of an epitope yields conflicting results on the role of T_H cells in the repertoire shift between primary and secondary B cells. Control of the development of memory cells may be through the control of somatic mutation. Depletion of a B cell clone may occur through mutational ''death'' before differentiation to memory can occur. The role T_H cells play in the termination of somatic mutation is deserving of future experiments.

VI. CONCLUSION

B and T cells must interact for most humoral immune responses to take place. The variable region structural outcome of the response may be affected by which subsets of T cells and B cells are involved. There is also potential for specific aspects of the maturation of the B cell response to be affected by the involvement of T_H cells. Many approaches have been used to study these interactions: adoptive transfer of purified sets of B and T cells into euthymic or athymic mice; comparison of responses to TI antigens and TD forms of the same epitope; and studies of the response of different B cell subsets defined by the *xid* mutation. Unfortunately, a unified view of the role of T cells in the regulation of the maturation of the antibody variable region repertoire during an immune response has yet to emerge.

In conclusion, their very requirement for the production of most humoral immune responses suggests that T cells may influence the composition of the expressed antibody repertoire. Further, several modes of T cell-B cell interaction have been characterized that would allow such influence to be exerted. However, little experimental evidence has been forwarded in support of such an influence, and the few experiments regarding this issue that have been conducted have often yielded conflicting results. Further experimentation regarding this issue is sorely needed.

REFERENCES

1. **Tada, T., Takemori, T., Okumura, K., Nonaka, M., and Tokuhisa, T.,** Two distinct types of helper T cells involved in the secondary antibody response: independent and synergistic effects of Ia^- and Ia^+ helper T cells, *J. Exp. Med.,* 147, 446, 1976.
2. **Keller, D. M., Swierkosz, J. E., Marrack, P., and Kappler, J. W.,** Two types of functionally distinct, synergizing helper T cells, *J. Immunol.,* 124, 1350, 1980.
3. **Hiramatsu, K., Miyatani, S., Kim, M., Yamada, S., Okumura, K., and Tada, T.,** Unique T cell Ia antigen expressed on a hybrid cell line producing antigen-specific augmenting T cell factor, *J. Immunol.,* 127, 1118, 1981.
4. **Woodland, R. and Cantor, H.,** Idiotype-specific T helper cells are required to induce idiotype-positive B memory cells to secrete antibody, *Eur. J. Immunol.,* 8, 600, 1978.
5. **Bottomly, K. and Janeway, C. A.,** Selected populations of alloreactive T cells contain helper T cells but lack Thld, an antigen-specific helper T cell required for dominant production of the T15 idiotype, *Eur. J. Immunol.,* 11, 270, 1981.
6. **Cerny, J., Cronkhite, R., and Stout, J. T.,** Rapid changes in the regulatory potential of autologous anti-idiotopic T cells during an antigen-driven primary response, *J. Immunol.,* 136, 3597, 1986.
7. **Cerny, J. and Caulfield, M. J.,** Stimulation of specific antibody-forming cells in antigen-primed nude mice by the adoptive transfer of syngeneic anti-idiotypic T cells, *J. Immunol.,* 126, 2262, 1981.
8. **Dunn, E. B. and Bottomly, K.,** T15-specific helper T cells: analysis of idiotype specificity by competitive inhibition analysis, *Eur. J. Immunol.,* 15, 728, 1985.
9. **Dunn, E. B., Kim, J., and Bottomly, K.,** A cloned T cell line that selectively augments antibody responses of phosphorylcholine-specific B cells bearing the T15 idiotype, *J. Mol. Cell. Immunol.,* 2, 209, 1986.
10. **Jayaraman, S., Swierkosz, J. E., and Bellone, C. J.,** T cell replacing factor substitutes for an $I\text{-}J^+$ idiotype-specific T helper cell, *J. Exp. Med.,* 155, 641, 1982.
11. **Hannestad, K., Kristoffersen, G., and Briand, J. P.,** The T lymphocyte response to syngeneic lambda2 light chain idiotypes. Significance of individual amino acids revealed by variant lambda2 chains and idiotype-mimicking chemically synthesized peptides, *Eur. J. Immunol.,* 16, 889, 1986.
12. **Weiss, S. and Bogen, B.,** B-lymphoma cells process and present their endogenous immunoglobulin to major histocompatibility complex-restricted T cells, *Proc. Natl. Acad. Sci. U.S.A.,* 86, 282, 1989.
13. **Smith, F. I., Tesch, H., and Rajewsky, K.,** Heterogeneous and monoclonal helper T cells induce similar anti-(4-hydroxy-3-nitrophenyl)acetyl (NP) antibody populations in the primary adoptive response, *Eur. J. Immunol.,* 14, 195, 1984.
14. **Stohrer, R. and Kearney, J. F.,** Fine idiotype analysis of B cell precursors in the T-dependent and T-independent responses to alpha 1 −>3 dextran in Balb/c mice, *J. Exp. Med.,* 158, 2081, 1983.
15. **Fish, S. and Manser, T.,** Influence of the macromolecular form of a B cell epitope on the expression of antibody variable and constant region structure, *J. Exp. Med.,* 166, 711, 1987.
16. **Maizels, N. and Bothwell, A.,** The T-cell-independent immune response to the hapten NP uses a large repertoire of heavy chain genes, *Cell,* 43, 715, 1985.
17. **Andrade, L., Freitas, A. A., Heutz, F., Poncet, P., and Coutinho, A.,** Immunoglobulin V_H expression in $Ly\text{-}1^+$ and conventional B lymphocytes, *Eur. J. Immunol.,* 19, 1117, 1989.
18. **Heutz, F., Poncet, P., Coutinho, A., and Portnoi, D.,** Ontogenic development of autoantibody repertoires in spleen and peritoneal cavity of normal mice: examples of T cell-dependent and -independent reactivities, *Eur. J. Immunol.,* 19, 1195, 1989.
19. **Forster, I. and Rajewsky, K.,** Expansion and functional activity of $Ly\text{-}1^+$ B cells upon transfer of peritoneal cells into allotype-congenic, newborn mice, *Eur. J. Immunol.,* 17, 521, 1987.
20. **Mosier, D. E. and Subbarao, B.,** Thymus-independent antigens: complexity of B-lymphocyte activation revealed, *Immunol. Today,* 3, 217, 1982.
21. **Press. J. L.,** The CBA/N defect defines two classes of T cell-dependent antigens, *J. Immunol.,* 126, 1234, 1981.
22. **Singer, A., Morrissey, P. J., Hathcock, K. S., Ahmed, A., Scher, I., and Hodes, R. J.,** Role of the major histocompatibility complex in T cell activation of B cell subpopulations, *J. Exp. Med.,* 154, 501, 1981.
23. **Singer, A. and Hodes, R. J.,** Mechanisms of T cell-B cell interaction, *Annu. Rev. Immunol.,* 1, 211, 1983.
24. **Hiernaux, J. R., Jones, J. M., Rudbach, J. A., Rollwagen, F., and Baker, P. J.,** Antibody response of immunodeficient (xid) CBA/N mice to *Escherichia coli* 0113 Lipopolysaccharide, A thymus-independent antigen, *J. Exp. Med.,* 157, 1197, 1983.
25. **Dzierzak, E. A., Janeway, C. A., Rosenstein, R. W., and Gottlieb, P. D.,** Expression of an idiotype (Id-460) during in vivo anti-dinitrophenyl antibody responses, *J. Exp. Med.,* 152, 720, 1980.

26. **Busto, P., Gerstein, R., Dupre, L., Giorgetti, C. A., Selsing, E., and Press, J. L.,** Molecular analysis of heavy and light chains used by primary and secondary anti-(T,G)-A—L antibodies produced by normal and *xid* mice, *J. Immunol.*, 139, 608, 1987.
27. **Woodland, R. T. and Huber, B. T.,** Selective activation by thymus-dependent antigens of distinct B cell subpopulations expressing a major cross-reactive idiotype, *J. Immunol.*, 133, 1801, 1984.
28. **Wicker, L. S., Guelde, G., Scher, I., and Kenny, J. J.,** The asymmetry in idiotype-isotype expression in the response to phosphocholine is due to divergence in the expressed repertoires of Lyb-5^+ and Lyb-5^- B cells, *J. Immunol.*, 131, 2468, 1983.
29. **Clarke, S. H., Kenny, J. J., Sieckmann, D. G., and Rudikoff, S.,** Amino acid sequence of a phosphocholine-binding antibody from an immune defective CBA/N mouse employing the T15 V_H region associated with unusual D_H, J_H, and V_K segments, *J. Immunol.*, 132, 1544, 1984.
30. **Allen, D., Cumano, A., Dildrop, R., Kocks, C., Rajewsky, K., Rajewsky, N., Roes, J., Sablitzky, F., and Siekevitz, M.,** Timing, genetic requirements and functional consequences of somatic hypermutation during B-cell development, *Immunol. Rev.*, 96, 141, 1987.
31. **Berek, C. and Milstein, C.,** Mutation drift and repertoire shift in the maturation of the immune response, *Immunol. Rev.*, 96, 23, 1987.
32. **Malipiero, U. V., Levy, N. S., and Gearhart, P. J.,** Somatic mutation in anti-phosphorylcholine antibodies, *Immunol. Rev.*, 96, 59, 1987.
33. **Manser, T., Wysocki, L. J., Margolies, M. N., and Gefter, M. L.,** Evolution of antibody variable region structure during the immune response, *Immunol. Rev.*, 96, 141, 1987.
34. **Hartman, A. B. and Rudikoff, S.,** V_H genes encoding the immune response to beta-(1,6)-galactan: somatic mutation in IgM molecules, *EMBO J.*, 3, 3023, 1984.
35. **Manser, T.,** Mitogen-driven B cell proliferation and differentiation are not accompanied by hypermutation of immunoglobulin variable region genes, *J. Immunol.*, 139, 234, 1987.
36. **Wysocki, L., Manser, T., and Gefter, M. L.,** Somatic evolution of variable region structures during an immune response, *Proc. Natl. Acad. Sci. U.S.A.*, 83, 1847, 1986.
37. **Maizels, N., Lau, J. C., Blier, P. R., and Bothwell, A.,** The T-cell independent antigen NP-ficoll, primes for a high affinity IgM anti-NP response, *Mol. Immunol.*, 25, 1277, 1988.
38. **Huchet, R. and Feldmann, M.,** Studies on antibody affinity in mice, *Eur. J. Immunol.*, 3, 49, 1973.
39. **Burkly, L. C., Nikolic, J., and Wortis, H. H.,** T cell regulation of light chain expression: preferential enhancement of Igκ in a primary thymus-dependent response does not require affinity-based selection, *J. Immunol.*, 135, 1573, 1985.
40. **Cunningham, A. J.,** Evolution in microcosm: the rapid somatic diversification of lymphocytes, *Cold Spring Harbor Symp. Quant. Biol.*, 41, 761, 1976.
41. **Walters, C. S., Moorhead, J. W., and Claman, H. N.,** Immunologic reactions to haptens on autologous carriers, *J. Immunol.*, 113, 654, 1974.
42. **Gershon, R. K. and Paul, W. E.,** Effect of thymus-derived lymphocytes on amount and affinity of anti-hapten antibody, *J. Immunol.*, 106, 872, 1971.
43. **Mongini, P. K. A., Paul, W. E., and Metcalf, E. S.** T cell regulation of immunoglobulin class expression in the antibody response to trinitrophenyl-ficoll, *J. Exp. Med.*, 155, 884, 1982.
44. **Siekevitz, M., Kocks, C., Rajewsky, K., and Dildrop, R.,** Analysis of somatic mutation and class switching in naive and memory B cells generating adoptive primary and secondary responses, *Cell*, 48, 757, 1987.
45. **Dell, C. L., Lu, Y., and Claflin, J. L.,** Molecular analysis of clonal stability and longevity in B cell memory, *J. Immunol.*, 143, 3364, 1989.
46. **Fish, S., Zenowich, E., Fleming, M., and Manser, T.,** Molecular analysis of original antigenic sin, *J. Exp. Med.*, 170, 1191, 1989.

Chapter 11

T CELL CONTROL OF MEMORY DEVELOPMENT

Thomas L. Feldbush, David Leitenberg, and Helen Braley-Mullen

TABLE OF CONTENTS

I. INTRODUCTION

A. HISTORICAL PERSPECTIVE

A major defining characteristic of the immune system is its ability to ''remember'' prior exposure to an antigen, and to protect against subsequent exposure. This was evident as long as 2000 years ago when it was noted that survivors of various infectious epidemics were not re-afflicted with the same disease. In the late 19th century Edward Jenner applied this observation by using cowpox pustules to immunize against the similar but much more severe disease small-pox. Starting at the turn of this century more systematic observations were made which distinguished the secondary or memory response from the primary response to antigen. Traditionally, the secondary response has been described as occurring in a more rapid and more vigorous manner than the initial response. In terms of antibody titer, the memory response is often several logs greater in magnitude than the primary response, usually appears more rapidly after immunization, and the amount of antigen needed to generate a secondary response is usually much lower.[1-9] The nature of the antibody produced after secondary challenge often changes as well, from IgM to IgG,[6-9] and the affinity of this antibody for antigen is usually increased.[10-12] Taken together these observations illustrate the profound differences in quantity and character between the primary and secondary responses to antigen.

Adoptive transfer techniques have been instrumental in identifying the nature and origin of the cells responsible for the secondary humoral response. In these experiments, cells from an immunized animal are transferred into an irradiated recipient which is then challenged with the immunizing antigen and the ability to mount an anamnestic response is analyzed. These experiments established that lymphocytes could transfer to a naive animal the ability to respond in a secondary manner.[9,13,14] A rather extensive body of information on the site and kinetics of memory B cell development has accumulated. The memory B cells were originally characterized as predominantly long-lived small lymphocytes[9,15-21] which arise within the follicular region[22-24] of lymphoid tissue draining the site of antigen injection.[25-27] These memory cells are the progeny of a precommitted lymphocyte pool not involved in primary antibody formation[28-31] and are continually recruited by available antigen[32-36] from both the recirculating population and the tissue-fixed cells.[27] Once generated, the memory cells may either become recirculating cells or perhaps remain sessile within the draining node.[15,17,25,26,33,36,38-40]

Following immunization with particulate antigen, memory T cells can be seen within 2 to 3 days, while memory B cells are detectable by day 4.[35,41] Soluble antigen priming is slightly slower with memory T cells appearing on days 3 to 4 and memory B cells by day 8.[27] Memory B cells begin leaving the draining lymph node by 11 days after immunization[21] with some of these cells subsequently homing into distal nodes resulting in systemic immunity.[36,39,42,43]

After generation, the memory cells acquire the ability to recirculate and appear to enter a quiescent phase in which they can remain essentially unchanged for long periods of time.[21,44-48] Upon restimulation by antigen these cells proliferate and differentiate into plasma cells or they proliferate to expand the memory cell pool[46,49-54] with the available antigen concentration apparently regulating which pathway predominates.[49]

II. MEMORY B CELL SUBSETS

Two of the major subsets of memory B cells probably represent stages in the antigen-driven maturation of these cells; the immature memory B cell and the long-lived mature memory cell subset. Strober[55] was the first to describe a large, apparently immature, memory cell subpopulation found in the thoracic duct lymph (TDL) 4 to 8 weeks following priming. This cell population was four to five times more active on a per cell basis in adoptive transfer experiments than the small memory cell population. Sixteen weeks after priming, the response within the large cell fraction waned while the small cell fraction maintained its activity. Using tritiated thymidine suicide, Strober demonstrated that these large cells were short-lived, since a 48-h exposure to [^{3}H] thymidine before transfer to adoptive recipients markedly reduced the subsequent response to antigen. The small cell population was resistant to this treatment.

Hobbs and Feldbush[56] confirmed the appearance of large memory cells in the TDL early after priming. In addition, they analyzed the size distribution of the memory population within the draining lymph node from day 5 to day 77 after immunization. At 5 days the memory B cell population was restricted to the large cell fraction and with time (77 days) shifted, leaving only 10% of the activity in the large cell fraction and 46% residing in the small cell fraction. In addition, they demonstrated that the shift in the size distribution was the result of a memory cell maturation process rather than the delayed appearance of an unrelated small memory cell. This was demonstrated by transferring only the large cell fraction to an intermediate host. After 7 or 11 days TDL were collected, the cells separated by $1 \times g$ velocity sedimentation, and the large, medium, and small cell fractions transferred to final irradiated hosts. The antibody responses of these final hosts indicated that after 7 days 53% of the memory activity was found in the medium fraction, and after 11 days the majority (60%) of the memory response was in the small fraction.

Unlike memory T cells which have been shown to display unique membrane antigens (for review see Reference 57), memory B cells appear, at least so far, to lack such markers. However, several investigators have used lectins or monoclonal antibodies to partially identify mature and immature memory B cell subsets. Butcher, et al.[58,59] and Reichert et al.[60] described a large B cell which develops in the germinal centers of Peyer's patches and binds large amounts of peanut agglutinin (PNA^{hi}). This PNA^{hi} population is distinct from other B cells, plasma cells, and most T cells which bind low levels of PNA (PNA^{low}). Using both immunofluorescent staining and panning, these PNA^{hi} cells were found to be negative for membrane IgD (mIgD), weakly positive for mIgM or mIgG (10 to 30%), but strongly positive for surface IA-encoded determinants. Coico et al.[61] isolated PNA^{hi} B cells from murine spleens at various times after immunization and tested their ability to adoptively transfer secondary responses. Ten days after immunization the PNA^{hi} B cells were enriched for memory cells, while 14 weeks later only PNA^{low} memory cells were found. Fyfe et al[62,63] used three monoclonal antibodies to subdivide the PNA^{hi} memory B cell into three additional subsets: one monoclonal called 5B5 recognizes a globotriaosyl ceramide (a glycosphingolipid) expressed on some B cells and lymphoma cells; the second mAb designated HK23 recognizes a 90 kDa protein found on B and T cells and the third recognized CD23, commonly called Blast-2. One subset was found to be PNA^{hi}, 5B5$^+$, HK23$^-$, CD23$^-$, sIgM^{lo} and sIgD^{lo}; the second subset was PNA^{hi}, HK23$^+$, 5B5$^-$, CD23$^-$, and sIgM^{hi} and the third subset was PNA^{hi}, CD23$^+$, 5B5$^-$, and HK23^{lo}. While no monoclonal antibodies have yet been described which recognize molecules found on all memory B cells, at least three have been shown to react only with virgin B cells. Bruce et al.[64] described a monoclonal antibody, J11d, which bound to >90% thymus cells and most B cells. Using this mAb to pretreat B cells before adoptive transfer substantially reduced primary responses and LPS reactivity but failed to inhibit secondary responses. Kroese et al.[65] described a monoclonal antibody, HIS22, which bound to a 210 kDa molecule found on about 33% of pre-B cells and 80% of Ig$^+$ spleen cells but was either absent or very low on germinal center B cells. Likewise, Marshall-Clarke et al.[66] reported on a mAb, NIM-R2, which stained B cells responsive *in vitro* to primary stimulation with DNP-LPS, TNP-BA and TNP-Ficoll but was lacking on cells which transferred secondary responsiveness to CGG.

The large, predominantly PNA^{hi} memory cell population has also been divided into two subsets based upon its expression of mIgD. Lafrenz et al.[67] and Herzenberg et al.[68] have found that most or all immature memory cells are mIgD+ when sorted by FACS and tested in adoptive recipients. As described above, Butcher et al.[58,59] and Reichert et al.[60] have found that early after immunization a large immature memory cell population appears which lacks mIgD and binds PNA. It is possible that these mIgD-large cells are rapidly dividing precursors whose lack of surface delta reflects their proliferative state. Kanowith-Klein et al.[69] and Ashman[70] have described such a phenomenon where 5 days after *in vitro* stimulation, B cells lose mIgD and display a cyclical modulation of surface IgM and IgG. By FACS analysis and immunoprecipitation, other investigations have confirmed this loss of mIgD following *in vitro* activation of B cells.[71-74] The weak surface expression of IgM and IgG by PNA^{hi} cells and the increased expression of Ia antigen[75] would be consistent with this hypothesis. It is even possible that the mIgD− cells are precursors of the mIgD+ cells, but this hypothesis must await formal proof. Regardless of their surface phenotype, all large memory cells lack complement receptors.[76]

Mason and Gowans[19] were the first to suggest that large memory cells arise in antigen-draining lymph nodes and circulate through the thoracic duct to other lymphoid tissues, but are unable to recirculate. This hypothesis was tested directly[77] and it was shown that large memory cells could be recovered from lymph nodes draining the site of antigen injection, but could not be recovered from the thoracic duct lymph following adoptive transfer.[56] Furthermore, as described above, Hobbs and Feldbush[56] were able to show that the large memory cells differentiate into smaller, more mature cells in the process of becoming a recirculating population. Butcher et al.[58,59] and Reichert et al.[60] have confirmed these observations using a monoclonal antibody to a putative receptor for the post-capillary venule. This monoclonal antibody is called Mel-14 and fails to bind to the PNA^{hi} memory B cell subset.[60]

The phenotypic characteristics of the small, mature memory cells are less well characterized. It is generally accepted that the mature memory B cells are small,[9,14-17,19-21] quiescent,[9,14-17,19-21] PNA^{low},[58-61] $Ia^{intermediate}$,[58-60] recirculating,[15] and may either display or lack complement receptors.[76] Since all these cells are recirculating,[15] we would assume that they are Mel-14 positive. The controversy surrounds the surface Ig (IgM, IgG, and IgD) displayed on these populations.

When mature B (IgM+, IgD+) cells become activated *in vitro* or *in vivo* (including germinal center B cells) they become IgD−.[61,69,78-80] It is not clear however if these cells re-express IgD upon returning to a resting state or if they remain IgD−. Several investigators have adoptively transferred IgD+ or IgD− cells into irradiated recipients and examined the ability of these subpopulations to transfer a secondary response.[67,68,81] Herzenberg et al. and Black et al. observed that early after priming (<10 weeks), both IgD+ and IgD− subsets could transfer a memory response. However, the affinity of the IgD− transferred response was always much higher; and after 10 week post-immunization the memory response was almost exclusively a property of the IgD− subset.[68,81] In similar experiments Lafrenz et al. found that early after immunization IgD+ cells restored the majority of the secondary response, and at later time points the responses restored by the two subsets were equivalent.[67] Lafrenz et al. also found that the affinity of the secondary response was independent of the reconstituting cell population.[67] One possible explanation for these disparate results is that Herzenberg et al. and Black et al. measured affinity only of the IgG2a isotype,[68,81] while Lafrenz et al. measured affinity without respect to antibody isotype.[67]

The presence of surface IgM and IgG on memory B cells has also been a matter of some controversy. Since the humoral memory response is associated with the secretion of IgG, investigators have hypothesized that only B cells with surface IgG can adoptively transfer a secondary response.[19] It has been clear for some time that all B cells upon initial maturation bear surface IgM[82] and that these cells can mature into IgG bearing cells. However, it is not clear if the ability to transfer a secondary response occurs only after the B cell is surface IgG+, or if a surface IgM+ cell can transfer the ability to generate an IgG secreting secondary response. Mason has shown that when IgG and IgM positive cells are separated by cell sorting techniques only the IgG bearing cells could transfer a secondary response.[83] In support of these results Coffman and Cohn showed that only when IgG bearing cells were protected from radiolabelled antigen-induced suicide could an IgG secondary response be transferred, while protection of surface IgM+ cells did not prevent the ability of primed cells to adoptively transfer a response.[84]

Other investigators have reported results which indicate that both IgM+ and IgG+ cells can transfer a memory response. For example, Yuan et al. specifically depleted IgM+ cells or IgG+ cells with antibody and complement, and found that depletion of IgG+ cells diminished the secondary response by 70% while depletion of both IgM and IgG positive cells completely inhibited the ability to restore the response.[85] Zan-Bar et al. separated IgG+ and IgM+ cells from a primed donor by flow cytometry and found that both populations could restore a secondary IgG response,[86] and finally, Strober found that the large B cells described previously were surface IgM positive and could adoptively transfer a 2-mercaptoethanol resistant antibody (presumptively IgG) response.[55] These studies indicated that surface IgM positive cells upon secondary stimulation could differentiate into IgG secreting cells. This was more directly examined using limiting dilution analysis in splenic focus assays.[87] In these experiments cells from primed animals were separated according to surface immunoglobulin phenotype and transferred in limiting number to carrier primed recipients and splenic fragments challenged with specific antigen. These studies showed that 90% of the surface IgM positive clones could produce antigen-specific IgG secreting progeny upon secondary challenge.[87]

III. INDUCTION OF MEMORY

Since, as mentioned above, the secondary immune response differs in magnitude, kinetics, and affinity from the primary response, any theory of memory B cell development must account for these differences. Burnet's landmark clonal selection theory can account for these differences and is the predominantly held theory today.[88] Burnet proposed that lymphocytes bearing surface receptors specific for the immunizing antigen are stimulated to proliferate upon contact with antigen thus dramatically increasing the clone size of B cells which can respond upon secondary stimulation. In addition, Burnet hypothesized that somatic mutation could occur so that cells with a higher affinity for the antigen would be preferentially selected, especially late in the response when antigen concentration may become limiting. Thus when these cells become restimulated during secondary stimulation, the response is larger due to the increased clone size, and since the cells are more efficient at binding antigen they respond more quickly and with a higher affinity.

Two general hypotheses have evolved to explain the steps involved in the generation of the memory B cell population. One theory proposes that mature virgin B lymphocytes which mature into the primary antibody forming cell and those which mature into memory cells are one and the same; the determining factor in which event predominates is the microenvironment in which the B cell proliferation and differentiation occurs and/or the sum of the regulatory controls applied by accessory cells (including both T cells and antigen presenting cells). One example of this hypothesis, which has never been formally tested, is the X-Y-Z scheme of Sercarz and colleagues.[89,90] In this model the resting lymphocyte (X) is stimulated by antigen to become a proliferating memory cell (Y) which becomes an antibody secreting cell (Z) upon re-encounter with the same antigen. Antibody production during the primary response would only occur if the cell had two distinct contacts with antigen, while long-term memory would occur if the cell had a single contact with antigen, and then upon secondary stimulation would respond very vigorously.

A second theory of memory B cell development holds that the cells which produce antibody during primary and secondary responses arise from distinct precursors.[91] This theory implies that during initial contact with antigen both cell subsets are stimulated, but in different ways: one differentiates into antibody secreting cells, while one clonally expands to become the memory cell population.

No one has yet been able to stimulate memory B cell development *in vitro*. This single fact probably explains why we have so little data directly supporting either theory. Klaus and Humphrey[92] recently reviewed evidence supporting a role for complement in memory B cell development. In this model, initial antigen immunization leads to the formation of antigen-antibody-complement complexes which are bound by the follicular dendritic cells (FDC) of primary follicles. These complexes then stimulate the formation of germinal centers leading to memory B cell development. In support of this model, Klaus and Humphrey[92] point out: (1) depletion of circulating C3 in mice results in no antigen-antibody deposition on FDC nor memory B cell formation; (2) ligation of the complement receptors CR2 and CR1 with antibody leads to enhanced secretion of Ig by PWM-stimulated B cells and synergizes with BCGF to augment B cell proliferation, and (3) C3 fragments have been shown to modulate B cell responses. The key factor in this model is that the germinal center microenvironment selectively supports memory cell induction rather than AFC formation. Gordon et al[93] reviewed evidence supporting a role for CD23 (Blast-2) in memory cell development. In this model, B cells are activated by antigen and T cells resulting in the expression of CD23. The CD23 is then released and acts in an autocrine manner, stimulating the activated B cell to proliferate. These proliferating B cells then develop into a germinal center where the cells continue to proliferate and switch to IgG or IgA expression. These switched cells become CD23 negative. In the absence of a positive signal many of these proliferating $CD23^-$ cells spontaneously die. However, some may re-encounter antigen (retained for a long time by FDC) and in the presence of CD23 are restimulated to proliferate and survive as memory B cells. Again microenvironment is perceived to be important since FDC can retain antigen as antigen-antibody-complement complexes for long periods of time and CD23 can display mostly BCGF activity in germinal centers.

In the last few years investigators have used molecular genetics to trace the development of specific B cell clones during primary and secondary responses to antigen. In these experiments B cell hybridomas are made from animals at various intervals after primary, secondary, and even tertiary immunization. The sequences of the rearranged immunoglobulin genes are then analyzed and compared to known germline sequences to determine if any somatic mutation occurred, and if the clonotypes isolated from primary and secondary responses arose from a common progenitor. Essentially, these experiments have shown that high affinity memory B cells can arise from both mechanisms outlined above; they can result from both germline and somatically mutated cells which participated in the primary response and are selected by FDC retained antigen and/or from novel clones which do not participate significantly in the primary response.[94-96]

Somatic mutation is rarely seen in hybridomas isolated 7 days after primary immunization,[97] but if hybridomas are isolated 14 days after immunization or after secondary challenge, significant somatic mutation occurs.[98-105] Because of these kinetics, somatic mutation is thought to occur independently of the early burst of proliferation that is seen in the draining lymph node which occurs within 7 days after immunization.[6] Somatic mutation also appears to be independent of immunoglobulin isotype switching since IgM bearing cells can be mutated[103,106] and IgG bearing cells can still have a germline sequence.[104] Instead, there appears to be a period from day 7 to 13 in which high rate somatic mutation (frequency of 10^{-3} to 10^{-4}) occurs[107,108] but after generating memory cells little further change occurs.[104] These mutational events are not random, but appear to be directed toward the V regions of both L and H chains and away from the constant and framework regions.[94,107-109] This may simply be a product of selection since mutations in the structural regions tend to be deleterious, but also may indicate a regulated, specific mechanism in which only the areas important in binding antigen are affected. These types of experiments, in which clones from primary and secondary responses are compared, have also directly demonstrated specific mutations which increase the affinity of the antibody produced during a secondary response.[94-96] These results indicate that somatic mutation is an important factor in the maturation of the immune response and can be shown to be responsible for increasing affinity for antigen in the secondary response.

Just as compelling are the experiments which have shown that clones distinct from those isolated during the primary response make up a significant part, and sometimes the majority of the memory response.[94,95,100,101,104,105,110] A distinct clonal origin of these cells (and not simply extensive somatic mutation) derived during a secondary response is inferred because these clones often have different heavy and light chain combinations and/or because the sequences (N sequences) involved in joining the V-D-J genes during rearrangement are different. One example that has been extensively studied is the immune response to phosphocholine (PC). The antibodies produced during the primary and secondary response to PC can be divided into two groups. Group I antibodies are of the TEPC15 idiotype, will bind free PC, and make up greater than 90% of the primary response.[110,111] Group II antibodies, on the other hand, do not have TEPC15 iditotye, will not bind free PC (need aromatic ring in conjunction with PC, i.e., nitrophenyl PC), generally make up 50 to 80% of the secondary response, and have higher affinity for the immunizing antigen than Group I antibodies.[110,111] By sequence analysis these antibodies can be shown to belong to different gene families,[112,113] showing that these differences are due to the presence of different B cell subsets being active in the primary and secondary responses. Further evidence that different B cell subsets are involved in the secondary response is found with the immunodeficient *xid* mouse strain. These mice lack the Lyb5 B cell subset and are unable to make Group I antibodies, but respond normally to secondary stimulation producing large amounts of Group II antibodies.[114]

As mentioned above, the finding that different clones of B cells are involved in the primary and secondary response is not limited to the PC system, but is found in many other systems, such as the nitrophenyl, phenyloxazalone, and arsonate systems.[94,95,105] One possible explanation for this phenomenon is that the clones which appear in the secondary response are rare representatives of the naive repertoire, but have high affinity for the antigen prior to immunization, and/or are particularly amenable to further somatic mutation thus enhancing their selective advantage. However, it is difficult to understand how this could explain the wholesale change in cellular repertoire often observed without invoking other forms of regulation. This probably involves complex interactions with other cells, including antigen presenting cells, T cells, regulatory B cells, and many soluble factors which may be produced by these cells.

In summary it seems clear that the development of memory is a complex event involving antigen selection of existing clones, somatic mutation and the expression of new B cell clones. Furthermore, the microenvironment of the germinal centers plays an important role in supporting these processes and allowing the development of high affinity memory B cells. However, the molecular and cellular mechanisms involved in these processes are poorly understood.

IV. ROLE OF T CELLS IN THE GENERATION OF MEMORY B CELLS

A. STUDIES WITH T CELL DEPENDENT ANTIGENS

Although it is well established that antigen-specific T_H are required for the induction of antibody responses to most antigens, the question as to whether T_H are also required for the generation of memory B cells has not been clearly resolved. Thus, B cells which initially encounter antigen under conditions where T_H are either absent or unable to be activated by the antigen may or may not exhibit an altered response when the same antigen is subsequently injected in the presence of T cells. The changes induced in B cells in the absence of T cells include both specific unresponsiveness (tolerance)[115-119] and differentiation to memory cells.[120-126] Alternatively, other reports indicate that B cells exposed to antigen in the absence of T cells undergo no apparent change, i.e., they are neither tolerant nor do they exhibit memory.[118,123,127-132] Important variables that may influence the results obtained in such experiments include the extent of T cell depletion, the nature of the T cell depleted environment, the nature of the antigen and the immunization protocol including such variables as antigen dose, adjuvant used and route of injection. In addition, the B cell subpopulations being analyzed as well as the isotype and/or avidity of the antibody analyzed could have also influenced the results of these studies.

Several studies have been done to directly assess the possible influence of some of these variables on the generation of memory B cells in T cell deprived animals. For example, Lafrenz and Feldbush[129] addressed the possibility that small numbers of residual T cells might contribute to the development of B cell memory in T-deprived animals by reconstituting T-deprived rats with graded numbers of T cells at the time of priming. Their results indicated that little or no memory to DNP-BGG developed in the absence of added T cells and that memory cell development increased with increasing numbers of exogenous T cells. Most importantly, the increase in memory cell development with added T cells was shown to be due to the selective generation of B cells producing high affinity antibody.[129] These results indicate that T cells are required for the generation of memory B cells and that this requirement is most pronounced for B cells capable of secreting high affinity IgG antibody. Similarly, Okumura et al.[123] showed that the development of high affinity memory B cells was T cell dependent whereas development of low affinity memory B cells was relatively T cell independent. Taken together, these results suggest that some reports of T cell independent induction of memory B cells could be due either to the presence of some residual T cells in the T cell-depleted environment and/or to the assessment of low affinity rather than high affinity B cell responses.

There is also evidence which indicates that the requirement for T cells for the generation of B memory cells may vary depending on the B cell subpopulation being analyzed. Johnson et al. showed that the polypeptide antigen TTGG-A-L required the activity of specific T_H to generate memory in B cells from *xid* mice whereas T cells were not required for the induction of memory to TTGG-A-L in normal B cells.[124] While these studies demonstrate that B cell subsets can have different T cell requirements for induction of memory, T_H have been shown to be required for generation of memory in the normal B cell subset in several other studies.[118,130,133,134]

In another study,[131] several variables which might influence experiments designed to determine if T cells were required for the generation of memory B cells were assessed using TNP-KLH as the antigen. These experiments demonstrated that neither the method used for T cell depletion (athymic nude mouse, adult thymectomized irradiated and bone marrow reconstituted (ATxBM) mouse or antilymphocyte serum (ALS)-treated mouse) nor the type of adjuvant given with the priming antigen apparently influenced the results since the generation of B memory cells was found to be highly T cell dependent in all cases. However, in the same study, the *dose* of antigen was found to quite markedly influence the results in that significant B memory cell development occurred when T cell-depleted animals were primed with relatively high doses (100 to 500 μg) of TNP-KLH but not when mice were primed with a lower dose (10 μg) of antigen.[131] In a separate study in which T cell depleted mice were primed with various doses of type III pneumococcal polysaccharide (S3) coupled to erythrocytes (S3-RBC), high doses of antigen were shown to induce tolerance in S3-

specific B cells whereas lower doses of antigen induced neither B cell tolerance nor specific B cell priming; both doses of antigen induced specific B cell priming in the presence of T cells.[118] That the antigen used could also influence results obtained in these types of experiments was shown by the fact that high doses of S3-RBC which induced tolerance in *S3-specific* B cells when injected in the absence of T cells[118] induced memory in *RBC-specific* B cells.[135] Lower doses of antigen required the presence of T cells for induction of memory in both S3- and RBC-specific B cells.[118] Overall, these results indicate that the induction of memory in B cells is highly T cell dependent when relatively low doses of antigen are used for priming. However, an apparent T cell independent induction of memory B cells can be observed when high doses of antigen are used for priming. High doses of some antigens induce B cell tolerance in the absence of T cells[115-119] which can in some cases be prevented by use of an appropriate adjuvant.[119]

In summary, when the variables which can influence the results of these types of experiments are taken into account, it may be possible to come to some general consensus regarding the requirement for T cells for the generation of memory B cells by TD antigens. For example, in those studies in which the generation of memory B cells was found to be T cell *independent*, relatively high doses of antigen were used for priming[118,120-123,126,135] and/or the antigens were given with adjuvants which may be able to replace T cell function or permit the activation of very small numbers of residual T cells.[119,120,126] In contrast, when memory B cell generation was found to be T cell *dependent*, the priming dose of antigen has generally been lower.[118,128,131] When very high doses of antigen[115,116,119] or optimal doses of highly tolerogenic antigens such as S3[118] or DNP-GL[117] have been used, B cell tolerance was induced in the absence of T cells.[115-119] Lower doses of antigen will primarily activate B cells with high avidity for the antigen while higher doses of antigen can activate both high and low avidity B cells.[136] Moreover, it has been shown that T cells are required for the maturation and activation of high affinity B cells.[129,137,138] Thus, the apparent T cell independence of memory B cell generation with high doses of priming antigen may be explained if the memory cells being assayed have low avidity for antigen whereas memory cells assayed after priming with low doses of antigen would be likely to have high avidity for antigen and to require T cells for their generation.[123,129,137] Overall, the majority of the studies done to assess the T cell requirement for memory B cell generation would be consistent with the idea that T cells are essential for the generation of memory B cells capable of differentiating into B cells producing high affinity IgG antibody. On the other hand, generation of memory in B cells able to produce IgM antibody[118,128,132] or low affinity IgG antibody[123,129] may not require T cells or, alternatively, may require fewer T cells which could be present in T-deprived animals.

B. "T INDEPENDENT" ANTIGENS

An alternative approach to studying the role of T cells in the development of memory B cells has been to utilize so-called thymus independent (TI) antigens which were initially thought to be capable of activating B cells without a requirement for helper T cells.[139] It is now generally accepted that optimal B cell activation by most, if not all, TI antigens requires lymphokines and possibly other signals produced by CD4+ T cells.[140-143] However, it is probable that *antigen-specific* T-B collaboration, as occurs with thymus dependent (TD) antigens,[144,145] does not occur in most primary responses to TI antigens.[141,143] A number of studies have shown that most TI antigens are incapable of generating memory in B cells.[130,139,146] However, limited memory responses, generally restricted to IgM responses, have been reported by some investigators.[130,147-150] Memory responses such as those obtained with TD antigens with affinity maturation and increased production of antibody of the IgG isotype have only rarely been observed using TI antigens (see below). Early studies suggested that ineffective or inadequate T cell stimulation might explain the general inability of TI polysaccharide antigens to induce memory responses since TI antigens coupled to carrier TD antigens (which could activate helper T cells) could prime B cells for the production of polysaccharide-specific IgG antibody after secondary challenge with the polysaccharide-TD carrier.[118,135] Adoptive transfer experiments clearly established that the TD form of antigen was required for the *generation* of memory B cells,[118,130,135,151] i.e., the above-mentioned results did not simply reflect a requirement for T_H activation for *expression* of secondary IgG responses. Moreover, nonspecific T cell activation, e.g., by Con A, failed to provide the necessary signal(s) for generation of memory B cells[151] suggesting that memory B cell generation might require an antigen-specific T cell signal. Based on the above types of studies using polysaccharides coupled to TD carriers, it was suggested that T cell activation was required for the generation of memory in polysaccharide-specific B cells. If this is true, it follows that if conditions could be found such that "TI" antigen could generate memory in B cells, the antigen should stimulate helper T cells.[130] Evidence to support this suggestion was first obtained in experiments using the type 2 TI antigen polyvinylpyrrolidone (PVP); PVP could prime mice for a PVP-specific secondary IgG response when mice were primed with a low dose of (0.0025 μg) of antigen which was able to induce only a very marginal primary IgM antibody response.[152] In contrast, priming with doses of PVP (0.25 μg) which induced optimal primary IgM responses did not lead to the generation of B memory cells and 0.25 μg PVP prevented the generation of memory by the TD form of antigen, PVP-HRBC.[152] Further studies indicated that low doses of PVP activated PVP-specific CD4+ T cells which could provide help to PVP-specific memory B cells for induction of IgG antibody responses.[153,154] Although PVP-specific T_H were also activated by higher (0.25 μg) doses of PVP, the higher amount of PVP preferentially activated T suppressor cells (T_S) which prevented the expression of T_H activity in unfractionated T cell populations.[153] Elimination of T_S at the time of priming with 0.25 μg PVP resulted in the generation of PVP-specific B memory cells[155] indicating that expression of T_H activity was required for B memory generation. Based on these results, it was suggested that the inability of type 2 "TI" antigens to generate IgG memory in B cells is due to the fact that commonly employed doses of these antigens, i.e., doses that elicit optimal primary IgM responses, preferentially activate T_S. These T_S can interfere with the expression of the antigen-specific T_H activity required for generation of IgG memory in B cells.[153-155] Indeed, further studies have shown that other type 2 TI antigens (pneumococcal capsular polysaccharides) also activate both T_H and T_S in a dose-dependent fashion. Very low doses of these antigens activate T_H which provide help to polysaccharide-primed B cells for IgG responses.[156] In contrast, doses of antigen that are optimally immunogenic for primary IgM responses interfere with the induction and expression of B cell memory by TD forms of the antigens[151,157-159] and these antigen doses preferentially activate T_S which mask the functional expression of T_H activity.[153,155,156] As was the case with PVP, elimination of T_S at the time of priming with an optimal dose of S3 leads to the generation of S3-specific IgG memory.[156] There are also several other reports which indicate that the generation of IgG memory by type 2 TI antigens is T cell dependent[133,134] although in these studies the antigen specificity of the T_H involved in B memory generation is unknown.

The studies just described suggest that the generation of long-term IgG memory in B cells requires the activity of T_H even using "TI" antigens which probably can to some extent directly activate primary B cells to produce IgM antibody. There are also several reports which indicate that both type 1 and type 2 TI antigens can generate a limited degree of IgM memory in the apparent absence of T_H activation[130,147-150] and some of these antigens can also apparently prime B memory cells which will produce hapten-specific IgG antibody after secondary immunization.[160-163] When type 2 TI antigens are used for priming it has generally been necessary to use the hapten coupled to either a TD carrier or to a type 1 TI carrier[161,163] to elicit the IgG response. The reason for this has not been determined but it is apparently not due to activation of T_S by the type 2 antigen.[163]

In the most extensively studied model of IgG B memory generation by TI antigens, mice primed with the type 2 antigens DNP-Ficoll, DNP-dextran or DNP-S3 produce a very substantial IgG response when challenged with the type 1 antigen, TNP-LPS (References 162,163, Braley-Mullen, unpublished results). As noted above, the memory cells activated by the type 2 antigen cannot be stimulated to produce IgG when a type 2 antigen is used for challenge. The memory cells can be generated in athymic nude mice and in irradiated recipients of T cell depleted spleen cells suggesting that the generation of memory in this model is T cell independent (References 162,163, Braley-Mullen, unpublished results). It is difficult to reconcile these observations with those discussed above which generally support the idea that generation of IgG memory requires T cells. However, the IgG memory responses generated by TI-2 priming followed by challenge with TNP-LPS, while very substantial in magnitude,[162,163] do differ from most classical memory responses in that the memory persists for only a few weeks after priming (Reference 164, Braley-Mullen, unpublished results). It is therefore possible that the memory cells generated by TI-2 priming in this situation may be distinct from the memory cells generated under conditions where T cell help is available although further studies would be needed to determine if this is true.

V. ROLE OF LYMPHOKINES IN MEMORY B CELL DIFFERENTIATION

The role of various lymphokines in stimulating virgin B cell proliferation and differentiation has been studied by a great number of investigators. Essentially, three different approaches have been used. (1) Different lymphokines have been used to co-stimulate polyclonal B cell growth and/or differentiation in conjunction with mitogens such as LPS, SAC, anti-Ig, PWM or combinations of agents such as calcium ionophore plus phorbol esters. (2) B cell lymphomas which generally require no additional mitogenic signals have been treated with various lymphokines and differentiation was measured by Ig secretion. (3) More recently, B cells have been mixed with antigen specific T helper clones which are known to secrete certain sets of lymphokines and the ability to stimulate B cell differentiation is determined. It is not the purpose of this chapter to review this extenstive literature on regulation of virgin B cell differentiation but one can probably conclude that various B cell subsets (i.e., those which secrete IgE) respond differently to each lymphokine or combination of lymphokines. Since memory B cells represent a separate subset of the B cell population, it would not be surprising if they too required a particular lymphokine or unique cocktail of lymphokines.

The simple experiment, polyclonal activation of memory B cells, has not been performed since most investigators have found that these cells are refractory to activation with such mitogens as LPS[165-168] and SAC.[169] Results supporting the unresponsiveness of primed B cells to polyclonal activators have also been obtained in our laboratory.[170] Dextran sulfate plus LPS stimulation of highly purified, DNP-primed, lymph node-derived, B cells, did not typically lead to the expression of IgG anti-DNP PFC, while an IgM response similar to that obtained from an unprimed cell population was seen. However, under the same culture conditions but utilizing antigen plus lymphokine stimulation (see below), large amounts of antigen-specific IgG antibody were produced demonstrating the presence of functional memory B cells.[170,171]

In a few instances, polyclonal activation of memory B cells has been reported to occur; however, in each case a rather unconventional method was employed making interpretation of the data very difficult.[167,172-174] For example, Zubler and Glasebrook[167] demonstrated that IgG antigen-specific AFC arose from primed B cells in the presence of LPS and a source of nonspecific T cell factors, provided that hapten was included in the culture. In the absence of hapten, little anti-TNP-specific IgG was seen, though LPS yielded a considerable polyclonal IgM response by the reverse plaque assay.

While neither polyclonal activation of memory B cells nor lymphoma models has provided us with useful information about lymphokine requirements for memory B cell differentiation, T helper clone dependent, antigen-specific B cell activation has been informative. In these experiments antigen-specific T helper cell clones are used in conjunction with antigen to stimulate B cell growth and differentiation. Mossman and colleagues have shown that T helper cell clones can be divided into two subgroups (Th1 and Th2) based on their pattern of lymphokine secretion.[175,176] Both groups are known to produce IL-3 and granulocyte-monocyte colony-stimulating factor (GM-CSF), but only Th1 clones secrete IL-2, IFNγ and lymphotoxin, while Th2 clones uniquely secrete IL-4 and IL-5. These T helper cell subsets have been assessed for their ability to help B cell responses.

The ability of Th1 clones to provide help to B cells is somewhat controversial. In some studies, these clones have been able to provide help to antigen specific B cells,[177,178] in others they only stimulated bystander B cell activation,[179] while in still others, no help has been observed.[180] A partial explanation for this heterogeneity may be in the relative amounts of lymphokines secreted. Some clones may secrete more lymphotoxin than others, and since lymphotoxin in high concentration is toxic, the B cells in close proximity (antigen-specific antigen presenting B cells) may be preferentially killed.[181] Similarly, IFNγ at high concentrations also has inhibitory effects in murine systems, while at lower concentrations it can have stimulatory effects.[182,183] In fact, antibody to IFNγ when added to previously inactive Th1 stimulated cultures can reverse this inhibition.[183] Thus it appears from these studies that there is considerable heterogeneity among Th1 clones, perhaps due to varying amounts of inhibitory lymphokines being produced or due to the absence of other unknown lymphokines.

Most groups agree that Th2 clones can provide cognate help to antigen-specific B cells, stimulating antigen-specific primary antibody production.[177,179,180] IL-4 and IL-5 have been shown to be important in this response, since antibodies to these molecules separately or in combination inhibit immunoglobulin secretion.[180,184] It should be noted, however, that this does not rule out the possibility that other lymphokines may also be involved. In fact, DeKruyff et al., in one of the only studies using primed animals, have recently analyzed a large number of Th2 clones and found that many of them were not able to provide antigen-specific help, although they all secreted IL-4 and IL-5.[178] These results suggest that the secretion of IL-4 and IL-5 in and of itself is not sufficient to stimulate all antigen-specific antibody responses, and that other lymphokines may also be involved.

In support of this hypothesis are the earlier results of Kim et al.[185] Prior to Mossman's classification of T helper cells as Th1 and Th2, Kim et al. described three types of cloned helper T cells, types 1,2, and 3. Type 1 and 2 clones, when stimulated with specific carrier and PC as the hapten, stimulate PC-specific antibody production. Type 3 clones on the other hand stimulate an antigen-dependent polyclonal B cell activation (bystander effect), but do not stimulate antigen-specific antibody production. Although Type 1 and 2 clones both stimulate PC-specific B cells, type 1 clones stimulate an idiotypically heterogeneous response which is typical of memory B cell stimulation (60% TEPC15 negative), while type 2 clones stimulate an idiotypically restricted response (90% TEPC15 positive) typical of the *in vivo* primary response (see memory B cell development section). These results indicate that different T helper cells can preferentially stimulate different B cell subsets. Subsequently, it has been shown that Type 1 and 2 clones both secrete IL-4 and IL-5, while Type 3 clones secrete IL-2 and IFNγ, corresponding to the aforementioned Th2 and Th1 clones, respectively. Since both type 1 and type 2 clones secrete IL-4 and IL-5, the basis for their ability to stimulate different B cell subsets remains obscure, perhaps indicating the presence of other lymphokines which are involved in regulating different B cell subsets, particularly memory and naive B cells.

There are very few reports in the literature which directly address which lymphokines stimulate memory B cell differentiation. The experiments mentioned earlier by DeKruyff et al.[178] using a panel of Th1 and Th2 clones are among the first to measure antigen-specific IgG responses from primed B cells. These investigators found that neither all Th1 (IL-2 producing) nor Th2 (IL-4 producing) clones could provide help for secondary B cell responses, although some clones from both subsets did provide help. These experiments indicate that the lymphokines which induce memory B cell differentiation can be produced by both T cell subsets and are not an exclusive property of a single subset of T helper cells defined by IL-4 and IL-2 secretion.

In order to directly assess the requirement for particular lymphokines in memory B cell differentiation, we recently developed a system for antigen-induced, lymphokine-dependent B cell activation which does not require cognate T cell recognition.[171] In this model system highly-purified, low-density rat B cells are isolated from 4-week primed antigen-draining lymph nodes and cultured *in vitro*. When stimulated with antigen (DNP-KLH) plus a partially purified mixture of lymphokines (50 to 80% ammonium sulfate fraction of a Con-A induced rat spleen cell supernatant), antigen-specific IgG antibody is secreted; this is used as a measure of memory cell differentiation since only primed B cells secrete such an antibody. When the same cell populations are stimulated with mitogen (LPS plus Dxs) plus lymphokines, polyclonal IgG and IgM secretion is seen and this was used as a measure of virgin B cell differentiation. Thus, with the same cell preparation, both memory and virgin B cell responses can be measured. The results of a large number of experiments performed using purified, partially purified or recombinant lymphokines are summarized in Table 1 (Leitenberg and Feldbush, submitted). The lymphokines in the Con-A supernatant (CSN) were fractionated by gel filtration and compared to unfractionated CSN for their relative ability to support both virgin and memory B cell differentiation (BCDF). Memory BCDF activity was found only in the relatively high molecular weight fractions corresponding to approximately 55 to 65 kDa, while virgin BCDF activity was found mostly in the lower molecular weight fractions corresponding to approximately 15 to 25 kDa. When the column fractions were tested for IL-2, IL-4 and IL-5 biologic activity (T cell blast proliferation, Ia induction and BCL-1 proliferation, respectively) most was found in the 15 to 25 kDa fractions but some IL-5 activity was also associated with the 55 to 65 kDa fraction. Furthermore, memory B cell differentiation was not induced by various sources of purified and recombinant IL-1, IL-2, IL-4, IL-5, IL-6, IFNγ and TGFβ. Preliminary experiments did indicate that the combination of IL-6 and high molecular weight BCGF (HMW-BCGF) may stimulate modest levels of memory B cell differentiation, but more work is necessary to confirm this observation. Thus, within the parameters of our system (large memory B cells stimulated with antigen plus various mouse or human purified or recombinant interleukins or fractionated rat cytokines) unique factors are required for rat memory B cell differentiation.

It is unclear why lymphokines such as IL-1, IL-4, IL-6, and TNF-α which have previously been shown to promote virgin B cell differentiation in some systems, had essentially no effect on the large virgin B cells used in our experiments. The most probable explanation for this is the different model systems which were used. Differences range from the source of B cells used (spleen, lymph node, peripheral blood, low-density, high-density, highly or partially purified), to the various mitogens which are used such as LPS, anti-Ig, and SAC. Another important factor is that species differences between rat, human, and murine B cells may exist. In fact, recombinant IL-5 which is generally thought to be a potent B cell growth factor in the murine system had no detectable activity on human B cells using a variety of different activation protocols.[186] This is also true in the experiments described here, where recombinant IL-5 had no effect on polyclonal B cell differentiation unless IL-2 was also present.

Differences also exist between our results on lymphokine requirements for memory B cell differentiation and the results of several other investigators. For example, there are several reports which state that IL-5 can induce the differentiation of hapten-primed B cells

TABLE 1
Cytokine Requirements for Memory and Virgin B Cell Differentiation[a]

Factors	Relative amount of response[b] Polyclonal IgM and IgG	DNP-Specific IgG
CSN	4+	4+
IL-1	0	0
IL-2	3+	0
IL-4	0	0
IL-5	1+	0
IL-6	0	0
IFNγ	0	0
IL-2 + IL-5	4+	0
IL-2 + HMW-BCGF	0	1+
IL-4 + IL-5	3+	0
IL-5 + IL-6	0—1+	0
IL-5 + IFNγ	0	0
TNF	0	0
TNF + IL-5	0—1+	0
IL-6 + HMW-BCGF	0	1+
TGFβ	0	0
TGFβ + CSN	0—1+	0
CSN		
55—65 kDa	0—1+	2+
15—25 kDa	3+	0

[a] Large B cells were isolated from the antigen-draining lymph nodes of 4-week primed donor rats. 0.125×10^6 cells/ml were stimulated with 10 μg/ml LPS plus D×S or 1 μg/ml antigen (DNP-KLH) plus various concentrations of the partially purified, purified or recombinant cytokines listed. On day 7 of culture the supernatants were harvested and the level of total IgM and IgG or antigen-specific IgG determined by ELISA {776}.

[b] The amount of antibody measured in cultures receiving CSN were set equal to 4+ and all other responses were compared to those controls. In a typical experiment, 0.125×10^6 cells/ml will secrete 100 μg of total IgM, 50 μg of total IgG and 4 μg of DNP-specific IgG.

into antigen-specific, IgG secreting plaque-forming cells. Initially this activity was described as a product of the T cell hybridoma, B151-K12,[187-189] and later upon cloning of the active product, was termed IL-5.[190] When assayed in our system neither supernatants from B151-K12 cells or recombinant IL-5 was able to support memory B cell differentiation, although upon the addition of IL-2, significant virgin B cell differentiation was observed. One possible reason for this apparent discrepancy is that the mice used in the aforementioned study were initially primed 6 to 8 weeks previously, but were then boosted 2 to 5 days prior to sacrifice.[187] It is possible that the late antigen challenge *in vivo* triggers the cells to progress to a level of differentiation in which only a late-acting differentiation factor is required *in vitro*. Furthermore, since no control experiments were done with animals which were not immunized 6 to 8 weeks previously, the differentiation observed in these experiments may be due to virgin B cells initially stimulated 3 days earlier *in vivo* which are responding to regulatory lymphokines in the normal course of the primary response.

Another lymphokine which has been implicated in memory B cell differentiation is IL-

6.[191] In these experiments splenocytes were harvested from animals primed 3 days previously with sheep red blood cells (SRBC) and cultured for an additional 4 days with IL-6 and antigen. In this system IL-6 was found to augment the PFC response approximately fivefold, while, interestingly, in results similar to ours, IL-2 alone or in conjunction with IL-6 had no effect.[191] However, as with the experiments with IL-5, it is unlikely that the responses measured in this system represent a secondary response, but instead are simply an extension of the primary response initiated by the immunization 3 days previously. This is supported by the studies which show that a memory response is not generated until after germinal center development, approximately 10 days after priming.[61]

There is also a report in which IL-2 is used to provide help in stimulating human B cells to secrete influenza-specific antibody.[192] In this system peripheral blood-derived B cells were cultured with IL-2 and various influenza virus stains and 7 days later specific antibody was measured by ELISA. IL-2 was able to significantly enhance specific antibody formation in these cultures, although the magnitude of the response was relatively small.[192] Since the immunoglobulin isotype of the response was not determined, it is difficult to compare these results with those obtained in our laboratory in which IL-2 consistently failed to support antigen-specific IgG secretion. In addition, because human peripheral blood is used as a B cell source, and since non-exposure to influenza virus would be a rarity, unprimed donors cannot be used to determine whether these results represent a primary or a secondary response.

A recent report by Ambrus et al.,[193] and personal communication, has implicated high molecular weight-BCGF (HMW-BCGF) in memory B cell activation. In this system human B cells were initially stimulated with HMW-BCGF in conjunction with SAC, the HMW-BCGF was removed and the cells were recultured in supernatant from a mixed lymphocyte reaction (MLR) and immunoglobulin secretion was measured. These experiments suggested that HMW-BCGF was able to selectively expand the memory B cell subset since tetanus toxid specific IgG was preferentially produced in these cultures, while parallel cultures expanded with IL-2 produced nonspecific immunoglobulin. This was of particular interest since the molecular weight of HMW-BCGF is similar to that of the memory BCDF activity isolated by gel-filtration in our laboratory. When HMW-BCGF was used in our system little or no memory BCDF activity was observed; however, the addition of IL-6 to HMW-BCGF stimulated cultures did induce modest levels of memory B cell differentiation. It must be emphasized however that these results are extremely preliminary in nature and further work is necessary to confirm this observation. In addition, even if these results are reproducible, it is still not clear if the memory BCDF activity partially purified from the CSN represents rat HMW-BCGF and IL-6. Further purification and characterization of the memory BCDF is required.

VI. T CELL MEMORY

Although considerable evidence indicates that immunologic memory is characteristic of both T cells and B cells, the molecular and cellular basis of T cell memory is still poorly understood. It is clear that both CD4+ and CD8+ T cells, following initial activation by antigen, exhibit immunologic memory as defined by accelerated kinetics and higher levels of activity following restimulation with an appropriate antigen.[194-196] Antigen activation is known to lead to clonal expansion of CD4+ and CD8+ T cells resulting in an increased frequency of antigen-specific T cells. While it has been known for some time that clonal expansion of T cells occurs during the generation of memory,[194-196] qualitative differences between memory and naive T cells have only recently begun to be defined. The various phenotypic markers that may distinguish naive and memory T cells have been discussed in depth in recent reviews[196,197] and will thus be summarized only briefly here.

A. Pgp-1

Several recent studies suggest that the 95 kDa cell surface glycoprotein Pgp-1 or Ly24 may be a marker for memory T cells in the mouse. Only a minor subpopulation of peripheral T cells in some strains of mice express a high density of Pgp-1.[196] Pgp-1 − CD8 + T cells have been shown to become Pgp-1 + after antigen or mitogen stimulation *in vitro* and this marker is expressed constitutively thereafter.[198] Pgp-1 has also been shown to identify recently activated CD4 + and CD8 + T cells *in vivo*[196,198,199] and Pgp-1 + CD8 + T cells were highly enriched for cells having high affinity T cell receptors.[196] Pgp-1 + and Pgp-1 − T cells also differ in lymphokine production: both subpopulations produce IL-2 while the Pgp-1 + T cells produce more IFNγ, IL-4, and IL-5 than do Pgp-1 − T cells.[196,200,201] Since IL-2 producing T cells may be precursors of T cells producing other cytokines,[200] the differential cytokine production by Pgp-1 + vs. Pgp-1 − T cells is consistent with the idea that the former subset may represent T cells which have previously encountered antigen. Pgp-1 also appears to be increased on a subset of human T cells which may represent memory T cells and which produce increased amounts of IFNγ.[202] While the above-mentioned studies indicate that CD4 + and CD8 + T cells having the properties of memory cells can be identified by their expression of Pgp-1, this may be true for only some strains of mice such as C_{57}BL/6. In other strains of mice such as Balb/c, the vast majority of peripheral T cells express Pgp-1[196,203] so that phenotypic identification of memory T cells in some strains of mice will need to rely on other markers (see below).

B. CD45R

Subsets of CD4 + T cells in rats, mice, and humans can be defined based on their reactivity with monoclonal antibodies (mAb) directed against determinants on various isoforms of the leukocyte common antigen, T200 or CD45.[197,200,204-207] Different isoforms of the CD45 protein are expressed by different lymphocytes; mAb that recognize determinants on one of the variably expressed exons are referred to as CD45 restricted (CD45R) determinants.[200] The expression of CD45R determinants on T cell subsets has been studied most extensively in humans. One subset of T cells, originally described functionally as "suppressor-inducers", expresses the 220 kDa form of CD45R and is defined by the mAbs 2H4 and HB-11.[197,204,208] The other subset, originally described functionally as "helper-inducers", expresses lower molecular weight forms of CD45 defined by the mAbs 4B4 (CDw29) or UCHL1.[197,208] Following activation by antigen or mitogen, expression of 2H4 and HB-11 is lost[197,200,208] and these cells convert to cells having increased expression of CDw29 and UCHL1 as well as Pgp-1, CD2, LFA-1, and LFA-3;[196,197,208] conversion in the opposite direction has not been observed.[196,197] Moreover, the phenotypic conversion of the 2H4 +/HB-11 + subset is accompanied by a change in function as the activated UCHL1 + cells can provide help to B cells.[208] The CDw29 subset has properties consistent with memory cells including increased proliferation to recall antigens, increased production of IFNγ and apparently higher affinity antigen receptors.[197] Similarly, expression of the CD45R markers OX-22 in the rat and 23G2 in the mouse are lost upon activation[200,205,206] and the CD45R low or negative cells in these species are the major subsets providing B cell help and producing lymphokines such as IL-4 and IL-5.[200,205,207] Thus, it is likely that expression of distinct CD45 epitopes may distinguish memory from naive T cells in man as well as in the mouse and rat. There is also some evidence to suggest that CD45 epitopes can be differentially expressed on distinct subsets of CD4 + T cells defined by their function and the types of lymphokines they produce,[205-207] i.e., the murine Th1 and Th2 subsets described by Mosmann.[200]

C. OTHER MARKERS

As mentioned above, several cell surface antigens in addition to Pgp-1 and CD45R have been shown to be increased following T cell activation. These include CD-2, LFA-3[197] and Ta_1 or CDw26.[209] Recent studies in the murine system using the anti-T cell mAbs 3G11 and 6C10 also demonstrate both functional and phenotypic alteration in CD4+ T cell subsets following antigen activation.[210] It is not yet known whether any of the markers discussed herein will be found to be unique to all memory T cells in all species or whether they may in fact be found to define several memory cell subsets[210] or activation states of T cells. Clearly, however, the availability of several specific markers for distinguishing memory and naive T cells will be of considerable importance for furthering our understanding of the cellular and molecular events underlying selection and activation of clonal T cell populations.

REFERENCES

1. **Glenny, A. T. and Sudmersen, H. J.,** Notes on the production of immunity to diphtheria toxin, *J. Hyg.*, 20, 176, 1989.
2. **Burnet, F. M. and Fenner, F.,** *The Production of Antibodies,* 2nd ed., Macmillan, Melbourne, 1953.
3. **Freund, J.,** The response of immunized animals to specific and non-specific stimuli, in *The Nature and Significance of the Antibody Response,* Pappenheimer, A. M., Ed., Columbia University Press, New York, 1953.
4. **Dixon, F. J., Maurer, P. H., Weigle, W. O., and Deichmiller, M. P.,** Rates of antibody synthesis during first, second, and hyperimmune response of rabbits to bovine gamma globulin, *J. Exp. Med.*, 103, 425, 1956.
5. **Uhr, J. W., Finkelstein, M. S., and Baumann, J. B.,** Antibody formation. III. The primary and secondary antibody response to bacteriophage ϕX 174 in guinea pigs, *J. Exp. Med.*, 115, 655, 1962.
6. **Thorbecke, G. J., Asofsky, R. M., Hochwald, G. M., and Siskind, G. W.,** Gamma globulin and antibody formation in vitro. III. Induction of secondary response at different intervals after the primary; the role of secondary nodules in the preparation for the secondary response, *J. Exp. Med.*, 116, 295, 1962.
7. **Bauer, D. C. and Stavitsky, A. B.,** On the different forms of antibody synthesized by rabbits during the early response to a single injection of protein and cellular antigen, *Proc. Natl. Acad. Sci. U.S.A.*, 47, 1667, 1961.
8. **Uhr, J. W. and Finkelstein, M. S.,** Antibody formation. IV. Formation of rapidly and slowly sedimenting antibodies and immunological memory to bacteriophage ϕX 174, *J. Exp. Med.*, 117, 457, 1963.
9. **Gowans, J. L. and Uhr, J.,** The carriage of immunological memory by small lymphocytes in the rat, *J. Exp. Med.*, 124, 1017, 1966.
10. **Eisen, H. N. and Siskind, G. W.,** Variations in affinities of antibodies during the immune response, *Biochemistry,* 3, 996, 1964.
11. **Steiner, L. A. and Eisen, H. N.,** Sequential changes in the relative affinity of antibodies synthesized during the immune response, *J. Exp. Med.*, 126, 1161, 1967.
12. **Davie, J. M. and Paul, W. E.,** Receptors on immunocompetent cells. V. Cellular correlates of the maturation of the immune response, *J. Exp. Med.*, 135, 660, 1972.
13. **Roberts, J. C. and Dixon, F. J.,** The transfer of lymph node cells in the study of the immune response to foreign proteins, *J. Exp. Med.*, 102, 379, 1955.
14. **Ellis, S. T. and Gowans, J. L.,** The role of lymphocytes in antibody formation. V. Transfer of immunological memory to tetanus toxoid; the origin of plasma cells from small lymphocytes, stimulation of memory cells in vitro and the persistence of memory after cell-transfer, *Proc. R. Soc. London Ser. B*, 183, 125, 1973.
15. **Gowans, J. L.,** Lymphocytes, *Harvey Lect.*, 64, 87, 1970.
16. **Bosman, C. and Feldman, J.,** Cytology of immunologic memory: a morphological study of lymphoid cells during the ananmestic response, *J. Exp. Med.*, 128, 293, 1968.
17. **Hunt, S. V., Ellis, S. T., and Gowans, J. L.,** The role of lymphocytes in antibody formation. IV. Carriage of immunological memory by lymphocyte fractions separated by velocity sedimentation and on glass bead columns, *Proc. R. Soc. London Ser. B,* 182, 211, 1972.
18. **Gearhart, P. J. and Cebra, J. J.,** Most B cells that have switched surface immunoglobulin isotypes generate clones of cells that do not secrete IgM, *J. Immunol.*, 127, 1030, 1981.
19. **Mason, D. W. and Gowans, J. L.,** Subpopulations of B lymphocytes and the carriage of immunologic memory, *Ann. Inst. Pasteur,* 127C, 657, 1976.

20. **Shortman, K., Fidler, J. M., Schlegel, R. A., et al.,** Subpopulations of B lymphocytes: physical separation of functionally distinct stages of B cell differentiation, in *Contemporary Topics Immunobiology,* Weigle, W. O., Ed., Plenum Press, New York, 1976, 1.
21. **McGregor, D. D. and Mackaness, G. B.,** The properties of lymphocytes which carry immunologic memory of Phi-X174, *J. Immunol.,* 114, 336, 1975.
22. **Durkin, H. G. and Thorbecke, G. J.,** Relationship of germinal centers in lymphoid tissue to immunologic memory. V. The effect of prednisolone administered after the peak of the primary response, *J. Immunol.,* 106, 1079, 1971.
23. **Durkin, H. G. and Thorbecke, G. J.,** Homing of immunologically committed lymph node cells to germinal centers in rabbits, *Nature (London) New Biol.,* 238, 53, 1972.
24. **Grobler, P., Buerki, H., Cottier, H., Hess, M. W., and Stoner, R. D.,** Cellular bases for relative radioresistance of the antibody-forming system at advanced stages of the secondary response to tetanus toxoid in mice, *J. Immunol.,* 112, 2154, 1974.
25. **Smith, J. B., Cunningham, A. J., Lafferty, K. J., and Morris, B.,** The role of the lymphatic system and lymphoid cells in the establishment of immunological memory, *Aust. J. Exp. Biol. Med. Sci.,* 48, 57, 1970.
26. **Stavitsky, A. B., and Folds, J. D.,** The differential localization of antibody synthesis and of immunologic memory in lymph nodes draining and not draining the site of primary immunization with hemocyanin, *J. Immunol.,* 108, 152, 1972.
27. **Mohr, R. and Krawinkel, U.,** Helper T cell kinetics and investigation of antigen receptor expression on early and memory T helper cells, *Immunology,* 31, 249, 1976.
28. **Cunningham, A. J.,** Studies on the cellular basis of IgM immunological memory, *Immunology,* 16, 621, 1969.
29. **Kappler, J. W., Hoffmann, M., and Dutton, R. W.,** Regulation of the immune response. I. Differential effect of passively administered antibody on the thymus-derived and bone marrow-derived lymphocytes, *J. Exp. Med.,* 134, 577, 1971.
30. **Axelrad, M. A.,** Suppression of memory by passive immunization late in the primary response, *J. Exp. Med.,* 133, 857, 1971.
31. **Safford, J. W. and Tokeida, S.,** Antibody mediated suppression of the immune response: effect on the development of immunologic memory, *J. Immunol.,* 107, 1213, 1971.
32. **Steiner, L. A. and Eisen, H. N.,** Symposium on in vitro studies of the immune response, *Bacteriol. Rev.,* 30, 383, 1966.
33. **Feldbush, T. L. and Gowans, J. L.,** Antigen modulation of the immune response. The effect of delayed challenge on the affinity of anti-DNP-BGG antibody produced in adoptive recipients, *J. Exp. Med.,* 134, 1453, 1971.
34. **Hamaoka, T., Kitagawa, M., Matsuoka, Y., and Yamamura, Y.,** Antibody production in mice. I. The analysis of immunologic memory, *Immunology,* 17, 55, 1969.
35. **Cunningham, A. J. and Sercarz, E. E.,** The asynchronous development of immunological memory in helper (T) and precursor (B) cell lines, *Eur. J. Immunol.,* 1, 413, 1971.
36. **Greene, E. J., Tew, J. G., and Stavitsky, A. B.,** The differential localization of the in vitro spontaneous antibody and proliferative responses in lymphoid organs proximal and distal to the site of primary immunization, *Cell. Immunol.,* 18, 476, 1975.
37. **Strober, S.,** Maturation of B lymphocytes in the rat. II. Subpopulations of virgin B lymphocytes in the spleen and thoracic duct lymph, *J. Immunol.,* 114, 877, 1975.
38. **Phillips, M. E., Quagliata, F., and Thorbecke, G. J.,** Effect of thoracic duct drainage on the primary and secondary immune responses in rats. Induction of immunological tolerance, *Immunology,* 22, 565, 1972.
39. **Thorbecke, G. J. and Bell, M. K.,** The proliferative and anamnestic antibody response of rabbit lymphoid cells in vitro. II. Effect of passive antibody on immunologic memory in lymph nodes contralateral to the site of antigen injection, *J. Immunol.,* 111, 1043, 1973.
40. **Lerner, A., Yamada, T., and Miller, R. A.,** Pgp-1^{hi} T lymphocytes accumulate with age in mice and respond poorly to concanavalin A., *Eur. J. Immunol.,* 19, 977, 1989.
41. **Black, S. J. and Inchley, J.,** Characteristics of immunological memory in mice. I. Separate early generation of cells mediating IgM and IgG memory to sheep erythrocytes, *J. Exp. Med.,* 140, 333, 1974.
42. **Weissman, I. L., Peacock, M., and Eltringham, J. R.,** Regional lymph node irradiation: effect on local and distant generation of antibody forming cells, *J. Immunol.,* 110, 1300, 1973.
43. **Jacobsen, E. B. and Thorbecke, G. J.,** The proliferative and anamnestic antibody response of rabbit lymphoid cells in vitro. I. Immunological memory in the lymph nodes draining and contralateral to the site of a primary antigen injection, *J. Exp. Med.,* 130, 1969, 1969.
44. **Fecsik, A. I., Butler, W. T., and Coons, A. H.,** Studies of antibody production. XI. Variation in the secondary response as a function of the length of the interval between two antigenic stimuli, *J. Exp. Med.,* 120, 1041, 1964.

45. **Nettesheim, P., Makinodan, T., and Williams, M. L.,** Regenerative potential of immunocompetent cells. I. Lack of recovery of secondary antibody-forming potential after x-irradiation, *J. Immunol.,* 99, 150, 1967.
46. **Feldbush, T. L.,** Antigen modulation of the immune response. The decline of immunologic memory in the absence of continuing antigenic stimulation, *Cell Immunol.,* 8, 435, 1973.
47. **Strober, S.,** Immune function. Cell surface characteristics and maturation of B cell subpopulations, *Transplant. Rev.,* 24, 84, 1975.
48. **Cramer, M. and Braun, D. G.,** Immunological memory: stable IgG patterns determine in vivo responsiveness at the clonal level, *Scand. J. Immunol.,* 4, 63, 1975.
49. **Feldbush, T. L. and van der Hoven, A.,** Antigen modulation of the immune response. IV. Selective trigger of memory cells, *Cell. Immunol.,* 25, 152, 1976.
50. **Askonas, B. A., Cunningham, A. J., Dreth, H. W., Roelants, G. E., and Williamson, A. R.,** Amplification of B cell clones forming antibody to the 2,4-dinitrophenyl group, *Eur. J. Immunol.,* 2, 494, 1972.
51. **Williamson, A. R. and Askonas, B. A.,** Senescence of an antibody-forming cell clones, *Nature,* 238, 337, 1972.
52. **McMichael, A. J. and Williamson, A. R.,** Clonal memory. I. Time-course of proliferation on B memory cells, *J. Exp. Med.,* 139, 1361, 1974.
53. **Klaus, G. G. B. and Willcox, H. N. A.,** B cell tolerance induced by polymeric antigens. III. Dissociation of antibody formation and memory generation in tolerant mice, *Eur. J. Immunol.,* 5, 699, 1975.
54. **Nakashima, I. and Kato, N.,** Maintenance and amplification of cell-associated immunological memory in in vivo culture system, *Cell. Immunol.,* 20, 156, 1975.
55. **Strober, S.,** Maturation of B lymphocytes in rats. III. Two subpopulations of memory B cells in the thoracic duct lymph differ by size, turnover rate, and surface immunoglobulin, *J. Immunol.,* 117, 1288, 1976.
56. **Hobbs, M. V. and Feldbush, T. L.,** Antigen modulation of the immune response. VI. Rate of large memory cell appearance in lymph nodes and thoracic duct lymph, *Cell Immunol.,* 50, 30, 1980.
57. **Hogg, N.,** The leukocyte integrins, *Immunol. Today,* 10, 111, 1989.
58. **Butcher, E. C., Reichert, R. A., Coffman, R. L., Nottenburg, C., and Weissman, I. L.,** Surface phenotype and migratory capability of Peyer's patch germinal center cells, *Adv. Exp. Med. Biol.,* 149, 765, 1982.
59. **Butcher, E. C., Rouse, R. V., Coffman, R. L. Nottenburg, C. N., Hardy, R. R., and Weissman, I. L.,** Surface phenotype of Peyer's patch germinal center cells: implications for the role of germinal centers in B cell differentiation, *J. Immunol.,* 129, 2698, 1982.
60. **Reichert, R. A., Gallatin, W. M., Weissman, I. L., and Butcher, E. C.,** Germinal center B cells lack homing receptors necessary for normal lymphocyte recirculation, *J. Exp. Med.,* 157, 813, 1983.
61. **Coico, R. F. Bhogal, B. S., and Thorbecke, G. J.,** Relationship of germinal centers in lymphoid tissue to immunologic memory. VI. Transfer of B cell memory with lymph node cells fractionated according to their receptors for peanut agglutinin, *J. Immunol.,* 131, 2254, 1983.
62. **Fyfe, G., Cebra-Thomas, J. A., Mustain, E., Davie, J. M., Alley, C. D., and Hahm, M. H.,** Subpopulations of B lymphocytes in germinal centers, *J. Immunol.,* 139, 2187, 1987.
63. **Nahm, M. H., Thomas, J. C., Mustain, E., Davie, J. M., and Fyfe, G.,** Subpopulations among germinal center B cells (abstr.), *Fed. Proc.,* 46, 6036, 1987.
64. **Bruce, J., Symington, F. W., McKearn, T. J., and Sprent, J.,** A monoclonal antibody discriminating between subsets of T and B cells, *J. Immunol.,* 127, 2496, 1981.
65. **Kroese, F. G. M., Wubbena, A. S., Opstelten, D., Deenen, G. J., Schwander, E. H., Leij, L. D., Vos, H., Poppema, S., Volberda, J., and Nieuwenhuis, P.,** B lymphocyte differentiation in the rat: production and characterization of monoclonal antibodies to B lineage-associated antigens, *Eur. J. Immunol.,* 17, 921, 1987.
66. **Marshall-Clarke, S., Chayen, A., and Parkhouse, R. M. E.,** Monoclonal antibody NIM-R2 shows differential reactivity with virgin and memory B cells, *Eur. J. Immunol.* 12, 733, 1982.
67. **Lafrenz, D., Strober, S., and Vitetta, E.,** The relationship between surface immunoglobulin isotype and the immune function of murine B lymphocytes. V. High affinity secondary antibody reponses are transferred by both IgD-positive and IgD-negative memory B cells, *J. Immunol.,* 127, 867, 1981.
68. **Herzenberg, L. A., Black, S. J., and Tokuhisa, T.,** Memory B cells at successive stages of differentiation. Affinity maturation and the role of IgD receptors, *J. Exp. Med.,* 151, 1071, 1980.
69. **Kanowith-Klein, S., Vitetta, E. S., and Ashman, R. F.,** The isotype cycle: successive changes in surface immunoglobulin classes expressed by the antigen-binding B-cell population during the primary in vivo immune response, *Cell. Immunol.,* 62, 377, 1981.
70. **Ashman, R. F.,** The immunological role of antigen-binding cells, *Immunol. Today,* 3, 349, 1982.
71. **Preudhomme, J. L.,** Loss of surface IgD by human B lymphocytes during polyclonal activation, *Eur. J. Immunol.,* 7, 191, 1977.

72. **Pernis, B., Brouet, J. C., and Seligmann, M.,** IgD and IgM on the membrane of lymphoid cells in macroglobulinemia. Evidence for identity of membrane IgD and IgM antibody activity in a case with anti-IgG receptors, *Eur. J. Immunol.*, 4, 776, 1974.
73. **Sitia, R., Abbott, J., and Hammerling, U.,** The ontogeny of B lymphocytes. V. Lipopolysaccharide-induced changes in IgD expression on murine B lymphocytes, *Eur. J. Immunol.*, 9, 859, 1979.
74. **Bourgois, A., Kitajiama, K., Hunter, J. R., and Askonas, B. A.,** Surface immunoglobulins of lipopolysaccharide-stimulated spleen cells. The behavior of IgM, IgD and IgG, *Eur. J. Immunol.*, 7, 151, 1977.
75. **Monroe, J. G. and Cambier, J. C.,** Level of mIa expression in mitogen-stimulated murine B lymphocytes is dependent on position in cell cycle, *J. Immunol.*, 130, 636, 1983.
76. **Feldbush, T. L.,** Separation of memory cell subpopulations by complement receptors: in vivo analysis, *Eur. J. Immunol.*, 10, 443, 1980.
77. **Feldbush, T. L. and Stewart, N.,** Antigen modulation of the immune response. V. Generation of large memory cells in antigen draining lymph nodes, *Cell. Immunol.*, 37, 336, 1978.
78. **Kraal, G., Weissman, I. L., and Butcher, E. C,** Germinal centre B cells: antigen specificity and changes in heavy chain class expression, *Nature,* 298, 377, 1982.
79. **Kraal, G., Hardy, R. R., Gallatin, W. M., Weissman, I. L., and Butcher, E. C.,** Antigen-induced changes in B cell subsets in lymph nodes: analysis by dual fluorescence flow cytometry, *Eur. J. Immunol.*, 16, 829, 1986.
80. **Hayakawa, K., Hardy, R. R., Honda, M., Herzenberg, L. A., Steinberg, A. D., and Herzenberg, L. A.,** Ly-1 B cells: functionally distinct lymphocytes that secrete IgM autoantibodies, *Proc. Natl. Acad. Sci. U.S.A.*, 81, 2494, 1984.
81. **Black, S. J., van der Loo, W., Loken, M. R., and Herzenberg, L. A.,** Expression of IgD by murine lymphocytes, *J. Exp. Med.*, 147, 984, 1974.
82. **Lawton, A. R., Asofsky, R., Hylton, M. D., and Cooper, M. D.,** Suppression of immunoglobulin class synthesis in mice. I. Effects of treatment with antibody to mu chain, *J. Exp. Med.*, 135, 277, 1972.
83. **Mason, D. W.,** The class of surface immunoglobulin on cells carrying IgG memory in rat thoracic duct lymph: the size of the subpopulation mediating IgG memory, *J. Exp. Med.*, 143, 1122, 1976.
84. **Coffman, R. L. and Cohn, M.,** The class of surface immunoglobulin on virgin and memory B lymphocytes, *J. Immunol.*, 118, 1806, 1977.
85. **Yuan, D., Vitetta, E. S., and Kettman, J. R.,** Cell surface immunoglobulin. XX. Antibody responsiveness of subpopulations of B lymphocytes bearing different isotypes, *J. Exp. Med.*, 145, 1421, 1977.
86. **Zan-Bar, I., Vitetta, E. S., Assisi, F., and Strober, S.,** The relationship between surface immunoglobulin isotype and immune function of murine B lymphocytes. III. Expression of a single predominant isotype on primed and unprimed B cells, *J. Exp. Med.*, 147, 1374, 1978.
87. **Teale, J. M., Lafrenz, D., Klinman, N. R., and Strober, S.,** Immunoglobulin class commitment exhibited by B lymphocytes separated according to surface isotype, *J. Immunol.*, 126, 1952, 1981.
88. **Burnet, F. M.,** *The Clonal Selection Theory of Acquired Immunity,* Vanderbilt University Press, Nashville, TN, 1959.
89. **Sercarz, E. E. and Coons, A.,** *Mechanisms of Immunologic Tolerance,* Czechoslovak Academy of Science Press, Prague, 1962, 73.
90. **Sercarz, E. E. and Byers, V. S.,** The X-Y-Z scheme of immunocyte maturation. III. Early IgM memory and the nature of the memory cell, *J. Immunol.*, 98, 836, 1967.
91. **Nossal, G. J. V.,** The mechanisms of action of antigen, *Aust. Ann. Med.*, 14, 321, 1965.
92. **Klaus, G. G. B. and Humphrey, J. H.,** A re-evaluation of the role of C3 in B-cell activation, *Immunol. Today,* 7, 163, 1986.
93. **Gordon, J., Flores-Romo, L., Cairns, J. A., Millsum, M. J., Lane, P. J., Johnson, G. D., and MacLennan, I. C. M.,** CD23: a multi-functional receptor/lymphokine?, *Immunol. Today,* 10, 153, 1989.
94. **Allen, D., Cumano, A., Dildrop, R., Kocks, C., Rajewsky, K., Rajewsky, N., Roes, J., Sablitzky, F., and Siekevitz, M.,** Timing, genetic requirements and functional consequences of somatic hypermutation during B-cell development., *Immunol. Rev.*, 96, 5, 1987.
95. **Berek, C. and Milstein, C.,** Mutation drift and repertoire shift in the maturation of the immune response, *Immunol. Rev.*, 96, 23, 1987.
96. **Fish, S., Zenowich, E., Fleming, M., and Manser, T.,** Molecular analysis of original antigenic SIN. I. Clonal selection, somatic mutation, and isotype switching during a memory b cell response, *J. Exp. Med.*, 170, 1191, 1989.
97. **Kaartinen, M., Griffiths, G. M., Markham, A. F., and Milstein, C.,** mRNA sequences define an unusually restricted IgG response to 2-phenyloxazolone and its early diversification, *Nature,* 304, 320, 1983.
98. **Bothwell, A. L. M., Paskind, M., Reth, M., Imanishi-Kari, T., Rajewsky, K., and Baltimore, D.,** Heavy chain variable region contribution to the NP^6 family of antibodies: somatic mutation evident in a IgG2a variable region, *Cell,* 24, 625, 1981.

99. **Cumano, A. and Rajewsky, K.,** Structure of primary anti-(4-hydroxy-3-nitrophenyl)acetyl (NP) antibodies in normal and idiotypically suppressed C57BL/6 mice, *Eur. J. Immunol.* 15, 512, 1985.
100. **Griffiths, G. M., Berek, C., Kaartinen, M., and Milstein, C.,** Somatic mutation and the maturation of immune response to 2-phenyl oxazolone, *Nature,* 312, 271, 1984.
101. **Berek, C., Griffiths, G. M., and Milstein, C.,** Molecular events during maturation of the immune response to oxazolone, *Nature,* 316, 412, 1985.
102. **Gearhart, P., Johnson, N. D., Douglas, R., and Hood, L.,** IgG antibodies to phosphoryl choline exhibit more diversity than their IgM counterparts, *Nature,* 291, 29, 1981.
103. **Mckean, D., Huppi, K., Bell, M., Staudt, L., Gerhard, W., and Weigert, M.,** Generation of antibody diversity in the immune response of Balb/c mice to influenza virus hemagglutinin, *Proc. Natl. Acad. Sci. U.S.A.,* 81, 3180, 1984.
104. **Siekevitz, M., Kocks, C., Rajewsky, K., and Dildrop, R.,** Analysis of somatic mutation and class switching in naive and memory B cells generating adoptive primary and secondary responses, *Cell,* 48, 757, 1987.
105. **Wysocki, L. J., Manser, T., and Gefter, M. L.,** Somatic evolution of variable region structures during an immune response, *Proc. Natl. Acad. Sci. U.S.A.,* 83, 1847, 1986.
106. **Rudikoff, S., Pawlita, M., Pumphrey, J., and Heller, M.,** Somatic diversification of immunoglobulins. *Proc. Natl. Acad. Sci. U.S.A.,* 81, 2162, 1984.
107. **Levy, N. S., Malipiero, U. V., Lebecque, S. G., and Gearhart, P. J.,** Early onset of somatic mutation in immunoglobulin V_H genes during the primary immune response, *J. Exp. Med.,* 169, 2007, 1989.
108. **French, D. L., Laskov, R., and Scharff, M. D.,** The role of somatic hypermutation in the generation of antibody diversity, *Science,* 244, 1152, 1989.
109. **Gearhart, P. J. and Bogenhagen, D. F.,** Clusters of point mutations are found exclusively around rearranged V genes, *Proc. Natl. Acad. Sci. U.S.A.,* 80, 3439, 1983.
110. **Stenzel-Poore, M. P., Bruderer, U., and Rittenberg, M. B.,** The adaptive potential of the memory response: clonal recruitment and epitope recognition, *Immunol. Rev.,* 105, 113, 1988.
111. **Wicker, L. S., Guelde, G., Scher, I., and Kenny, J. J.,** Antibodies from the Lyb5− B cell subset predominate in the secondary IgG response to phosphocholine, *J. Immunol.,* 129, 950, 1982.
112. **Chang, S. P., Perlmutter, R. M., Brown, M., Heusser, C. H., Hood, L., and Rittenberg, M. B.,** Immunologic memory to phosphorylcholine. IV. Hybridomas representative of Group I (T15-like) and Group II (non-T15-like) antibodies utilize distinct V_H genes, *J. Immunol.,* 132, 1550, 1984.
113. **Stenzel-Poore, M. P., Hall, T. J., Heusser, C. H., Faust, C. H., and Rittenberg, M. B.,** Immunologic memory to PC-KLH: participation of the Q 52 V_H gene family, *J. Immunol.* 139, 1698, 1987.
114. **Wicker, L. S., Guelde, G., Scher, I., and Kenny, J. J.,** The asymmetry in idiotype-isotype expression in the response to phosphocholine is due to divergence in the expressed repertoires of Lyb-5+ and Lyb-5− B cells, *J. Immunol.,* 131, 2468, 1983.
115. **Mitchell, G. R., LaFleur, L., and Anderson, K.,** Evidence for readily induced tolerance to heterologous erythrocytes in nude mice, *Scand. J. Immunol.,* 3, 39, 1974.
116. **Etlinger, H. M. and Chiller, J. M.,** Induction of tolerance in athymic mice with an antigen which is highly immunogenic in euthymic mice, *Cell. Immunol.,* 33, 297, 1977.
117. **Hamaoka, T., Inada, T., Yamashita, Y., and Kitagawa, M.,** Preventive effect of hapten-reactive thymus-derived helper lymphocytes on the tolerance induction in hapten-specific precursors of antibody-forming cells, *J. Immunol.,* 114, 1771, 1975.
118. **Braley-Mullen, H.,** Secondary IgG responses to type III pneumococcal polysaccharide. III. T cell requirement for development of B memory cells, *Eur. J. Immunol.,* 7, 775, 1977.
119. **Schrader, J. W.,** Induction of immunological tolerance to a thymus-dependent antigen in the absence of thymus-derived cells, *J. Exp. Med.,* 139, 1303, 1974.
120. **Schrader, J. W.,** The role of T cells in IgG production: thymus dependent antigens induce B-cell memory in the absence of T cells, *J. Immunol.,* 114, 1665, 1975.
121. **Roelants, G. E. and Askonas, B. A.,** Immunological B memory in thymus deprived mice, *Nature (London) New Biol.,* 239, 63, 1972.
122. **Diamantstein, T. and Blitstein-Willinger, E.,** T cell independent development of B memory cells, *Eur. J. Immunol.,* 4, 830, 1974.
123. **Okamura, K., Metzler, C. M., Tsu, T. T., and Herzenberg, L. A.,** Two stages of B-cell memory cell development with different T-cell requirements, *J. Exp. Med.,* 144, 345, 1976.
124. **Johnson, D. R., Faherty, D. A., and Zauderer, M.,** Different T cell requirements for specific memory induction in normal xid B cells, *J. Immunol.,* 137, 2796, 1986.
125. **Sokol, S. Y. and Borisova, T. K.,** Ir gene influence on DNP-specific B-memory-cell formation, *Cell. Immunol.,* 110, 449, 1987.
126. **Hosokawa, T., Amagi, T., and Muramatsu, S.,** Studies on B cell memory. I. Generation and exhaustion of B cell memory by thymus-dependent antigen in T cell depleted mice, *Immunology,* 38, 283, 1979.

127. **Schlegal, R. A.,** Antigen-initiated B lymphocyte differentiation. Kinetics and T lymphocyte dependence of the primary and secondary in vitro immune responses to the hapten 4-hydroxy-3-iodo-5-nitrophenyl acetic acid presented on the carrier polymerized bacterial flagellin, *Aust. J. Exp. Biol.*, 52, 471, 1974.
128. **Dresser, D. W. and Popham, A. M.,** The influence of T cells on the initiation and expression of immunological memory, *Immunology,* 38, 265, 1979.
129. **Lafrenz, D. E. and Feldbush, T. L.,** Role of T cells in the development of memory B cells. Quantitative and qualitative analysis, *Immunology,* 44, 177, 1981.
130. **Braley-Mullen, H.,** Antigen requirements for induction of B-memory cells: studies with dinitrophenyl coupled to T-dependent and T-independent carriers, *J. Exp. Med.,* 147, 1824, 1978.
131. **Braley-Mullen, H.,** The role of T cells in the induction of B memory cells: influence of variations in experimental design, *Cell. Immunol.*, 68, 234, 1982.
132. **Davie, J. M. and Paul, W. E.,** Role of T lymphocytes in the humoral immune response. I. Proliferation of B lymphocytes in thymus deprived mice, *J. Immunol.*, 113, 1438, 1974.
133. **Schuler, W., Schuler, A., and Kolsch, E.,** Immune response against the T-independent antigen (1->3) dextran. II. Occurrence of memory cells in the course of immunization with the native polysaccharide is T cell dependent, *Eur. J. Immunol.*, 14, 578, 1984.
134. **Hosokawa, T., Tanaka, Y., Aoike, A., Kawai, K., and Muramatsu, S.,** Studies on B-cell memory. III. T-dependent aspect of B-memory generation in mice immunized with T-independent type-2 (TI-2) antigen, *Immunology,* 53, 97, 1984.
135. **Braley-Mullen, H.,** Secondary IgG responses to type III pneumococcal polysaccharide. II. Different cellular requirements for induction and elicitation, *J. Immunol.*, 116, 904, 1976.
136. **Kim, Y. T. and Siskind, G. W.,** Studies on the control of antibody synthesis. VI. Effect of antigen dose and time after immunization on antibody affinity and heterogeneity in the mouse, *Clin. Exp. Immunol.*, 17, 329, 1974.
137. **Gershon, R. K. and Paul, W. E.,** Effect of thymus derived lymphocytes on amount and affinity of anti-hapten antibody, *J. Immunol.*, 106, 872, 1971.
138. **DeKruyff, R. and Siskind, G. W.,** Studies on the control of antibody synthesis. XIV. Role of T cells in regulating antibody affinity, *Cell. Immunol.*, 47, 134, 1979.
139. **Basten, A. and Howard, J. G.,** Thymus independence, *Contemp. Topics Immunobiol.*, 2, 265, 1973.
140. **Letvin, N. C., Benacerraf, B., and Germain, R. W.,** B lymphocyte responses to TNP-Ficoll: requirement for T lymphocytes and Ia bearing cells, *Proc. Natl. Acad. Sci. U.S.A.*, 78, 5113, 1981.
141. **Mond, J. J. and Brunswick, M.,** A role for IFN and NK cells in immune responses to T cell regulated antigens types 1 and 2, *Immunol. Rev.*, 99, 105, 1987.
142. **Pike, B. L. and Nossal, G. J. V.,** A reappraisal of "T independent" antigens. I. Effect of lymphokines on the response of single adult hapten-specific B lymphocytes, *J. Immunol.*, 132, 1687, 1984.
143. **Murray, P. D. and Kagnoff, M. F.,** Regulation of the anti-α(1,3) dextran response: two populations of dextran-reactive B cells that differ in their T cell requirements for induction to antibody synthesis, *J. Immunol.*, 138, 2439, 1987.
144. **Singer, A. and Hodes, R. J.,** Mechanisms of T cell-B cell interaction, *Annu. Rev. Immunol.*, 1, 211, 1983.
145. **Vitetta, E. S., Bossie, A., Fernandez-Botran, R., Myers, C. D., Oliver, K. S., Sanders, V. L., and Stevens, T. L.,** Interaction and activation of antigen-specific T and B cells, *Immunol. Rev.*, 99, 193.
146. **Mosier, D. E. and Subbarao, B.,** T-independent antigens: complexity of B lymphocyte activation revealed, *Immunol. Today,* 3, 217, 1982.
147. **del Guercio, P., Thobie, N., and Poirier, M. F.,** IgM anamnestic immune response to the haptenic determinant DNP on a thymus independent carrier, *J. Immunol.*, 112, 427, 1974.
148. **Lake, J. P. and Reed, N. D.,** Characterization of antigen specific immunologic paralysis induced by a single low dose of polyvinylpyrrolidone, *J. Reticuloendothel. Soc.*, 20, 307, 1976.
149. **Hosokawa, T.,** Studies on B cell memory. II. T-cell independent antigen can induce B-cell memory, *Immunology,* 38, 291, 1979.
150. **Schott, C. F., Lizzio, E. F., Inman, J. K., and Merchant, B.,** Immune memory to a nonmitogenic, thymic independent antigen in mice: variation among inbred strains and possible relationship to oncogenesis, *J. Immunol.*, 127, 139, 1981.
151. **Braley-Mullen, H.,** Antigen requirements for priming of IgG producing B memory cells specific for type III pneumococcal polysaccharide, *Immunology,* 40, 521, 1980.
152. **Lite, H. S. and Braley-Mullen, H.,** Induction of IgG memory responses with polyvinylpyrrolidone (PVP) is antigen dose dependent. *J. Immunol.*, 126, 928, 1981.
153. **Braley-Mullen, H.,** Regulation of IgG memory responses by helper and suppressor T cells activated by the type 2 antigen, polyvinylpyrrolidone *J. Exp. Med.*, 161, 1357, 1985.
154. **Van Buskirk, A. and Braley-Mullen, H.,** Characterization of helper T cells induced by the type 2 antigen polyvinylpyrrolidone. *J. Immunol.*, 138, 1031, 1987.

155. **Van Buskirk, A. and Braley-Mullen, H.,** Suppression of IgG memory responses by T cells activated with the type 2 antigen polyvinylpyrrolidone (PVP), *Cell. Immunol.,* 107, 121, 1987.
156. **Milligan, G. N. and Braley-Mullen, H.,** Regulation of IgG responses by helper and suppressor T cells activated by pneumococcal capsular polysaccharides, *Cell. Immunol.,* 119, 222, 1989.
157. **Leeman, S. P., Romano, T. J., Mond, J. J., Heidelberger, M., and Thorbecke, G. J.,** Induction of primary and inhibition of secondary antibody response to hapten by hapten conjugates of type III pneumococcal polysaccharide, *Cell. Immunol.,* 15, 321, 1975.
158. **Mitchell, G. F., Humphrey, J. H., and Williamson, A. R.,** Inhibition of secondary anti-hapten responses with the hapten coupled to type 3 pneumococcal polysaccharide, *Eur. J. Immunol.,* 2, 460, 1972.
159. **Milligan, G. N., Fairchild, R. L., Sterner, K. E., and Braley-Mullen, H.,** Types 6 and 19 pneumococcal polysaccharides coupled to erythrocytes elicit pneumococcal cell-wall specific primary IgM responses and capsular polysaccharide-specific secondary IgG responses, submitted.
160. **Shinohara, N. and Tada, T.,** Hapten-specific IgM and IgG antibody responses in mice against a thymus-independent antigen (DNP-Salmonella), *Int. Arch. Allergy Appl. Immunol.,* 47, 762, 1974.
161. **Umetsu, D. R., Chapman-Alexander, D. M., and Thorbecke, G. J.,** Cross-priming of murine B cells with TNP conjugates of hemocyanin and ficoll: characteristics of primed B cells responding to both antigens, *J. Immunol.,* 123, 396, 1979.
162. **Motta, I., Portnoi, D., and Truffa-Bachi, P.,** Induction and differentiation of B memory cells by a thymus-independent antigen, trinitrophenylated lipopolysaccharide, *Cell. Immunol.,* 57, 327, 1981.
163. **Colle, J-H., Motta, I., and Truffa-Bachi, P.,** Generation of immune memory by haptenated derivatives of thymus-independent antigens in C57BL/6 mice. I. The differentiation of memory B lymphocytes into antibody-secreting cells depends on the nature of the thymus-independent carrier used for memory induction and/or revelation, *Cell. Immunol.,* 75, 52, 1983.
164. **Le Moal, M. A. and Truffa-Bachi, P.,** Immune memory expression to TNP-Ficoll in CB20 mice: evidence for a multigenic control, *Cell. Immunol.,* 95, 425, 1985.
165. **Yokochi, T., Nakashima, I., Fujii, Y., and Kato, N.,** Suppression of B-memory cell function by polyclonal lymphocyte activators, *Immunobiology,* 158, 454, 1981.
166. **Shiozawa, C., Sawada, S., Inazawa, M., and Diener, E.,** Triggering of affinity enriched B cells. II. Composition and B cell triggering properties of affinity purified antigen specific, T cell derived helper factor, *J. Immunol.,* 132, 1892, 1984.
167. **Zubler, R. H. and Glasebrook, A. L.,** Requirement for three signals in "T-independent" (lipopolysaccharide-induced) as well as in T-dependent B cell responses, *J. Exp. Med.,* 155, 666, 1982.
168. **Tittle, T. V. and Rittenberg, M. B.,** IgG B memory cell subpopulations: differences in susceptibility to stimulation by TI-1 and TI-2 antigens, *J. Immunol.,* 124, 202, 1980.
169. **Falkoff, R. J. M., Zhu, L. P., and Fauci, A. S.,** Separate signals for human B cell proliferation and differentiation in response to *S. aureus:* evidence for a two-signal model of B cell activation, *J. Immunol.,* 129, 97, 1982.
170. **Burg, D. L. and Feldbush, T. L.,** Polyclonal activation of primed rat B cells, *Cell. Immunol.,* 98, 351, 1986.
171. **Leitenberg, D., Burg, D., and Feldbush, T. L.,** Lymphokine induction of memory B cell differentiation: differential stimulation of large virgin and memory B cell differentiation, *Cell. Immunol.,* 114, 83, 1988.
172. **Shenk, R. R., Weissberger, H. Z., and Dickler, H. B.,** Anti-idiotype stimulation of antigen-specific antigen-independent antibody responses in vitro. II. Triggering of B lymphocytes by idiotype plus anti-idiotype in the absence of T lymphocytes, *J. Immunol.,* 132, 2709, 1984.
173. **Nariuchi, H., Kakiuchi, T., and Ueno, J.,** Effect of priming with antigen on the expression of C3 receptors on murine B cells and the responses to TI-2 antigens, *J. Immunol.,* 134, 748, 1985.
174. **Fauntleroy, M. B., Rudbach, J. A., Prescott, B., and Baker, P. J.,** Increased activation of antigen-primed or memory B cells by bacterial lipopolysaccharide, *Cell. Immunol.,* 107, 263, 1987.
175. **Mosmann, T. R., Cherwinski, H., Bond, M. W., Giedlin, M. A., and Coffman, R. L.,** Two types of murine helper T cell clones. I. Definition according to profiles of lymphokine activities and secreted proteins, *J. Immunol.,* 236, 2348, 1986.
176. **Cherwinski, H. M., Schumacher, J. H., Brown, K. D., and Mosmann, T. R.,** Two types of mouse helper T cell clone III. Further differences in lymphokine synthesis between Th1 and Th2 clones revealed by RNA hybridization, functionally monospecific bioassays, and monoclonal antibodies, *J. Exp. Med.,* 166, 1229, 1987.
177. **Stevens, T. L., Bossie, A., Sanders, V. M., Fernandez-Botran, R., Coffman, R. L., Mosmann, T. R., and Vitetta, E. S.,** Regulation of antibody isotype secretion by subsets of antigen-specific helper T cells, *Nature,* 334, 255, 1988.
178. **Dekruyff, R. H., Ju, S., Hunt, A. J., Mosmann, T. R., and Umetsu, D. T.,** Induction of antigen-specific antibody responses in primed and unprimed B cells, *J. Immunol.,* 142, 2575, 1989.
179. **Killar, L., MacDonald, G., West, J., Woods, A., and Bottomly, K.,** Cloned, Ia-restricted T cells that do not produce interleukin 4(IL4)/B cell stimulatory factor 1(BSF-1) fail to help antigen-specific B cells, *J. Immunol.,* 138, 1674, 1987.

180. **Boom, W. H., Liano, D., and, Abbas, A. K.,** Heterogeneity of helper/inducer T lymphocytes. II. Effects of interleukin 4 and interleukin 2 producing T cell clones on resting B lymphocytes, *J. Exp. Med.*, 167, 1350, 1988.
181. **Tite, J. P., Powell, M., and Ruddle, N.,** Protein-antigen specific Ia restricted cytolytic T cells: analysis of frequency, target cell susceptibility, and mechanism of cytolysis, *J. Immunol.*, 135, 25, 1985.
182. **Snapper, C. M. and Paul, W. E.,** Interferon- and B cell stimulatory factor-1 reciprocally regulate Ig isotype production, *Science*, 236, 944, 1987.
183. **Reynolds, D. S., Boom, W. H., and Abbas, A. K.,** Inhibition of B lymphocyte activation by interferon-γ, *J. Immunol.*, 139, 767, 1987.
184. **Rasmussen, R., Takatsu, K., Harada, N., Takahashi, T., and Bottomly, K.,** T cell-dependent hapten-specific and polyclonal B cell responses require release of interleukin 5, *J. Immunol.*, 140, 705, 1988.
185. **Kim, J., Woods, A., Becker-Dunn, E., and Bottomly, K.,** Distinct functional phenotypes of cloned Ia-restricted helper T cells, *J. Exp. Med.*, 162, 188, 1985.
186. **Clutterbuck, E., Shields, J. G., Gordon, J., Smith, S. H., Boyd, A., Callard, R. E., Campbell, H. D., Young, I. G., and Sanderson, C. J.,** Recombinant human interleukin 5 is an eosinophil differentiation factor but has no activity in standard human B cell growth factor assays, *Eur. J. Immunol.*, 17, 1743, 1987.
187. **Takatsu, K., Harada, N., Hara, Y., Takahama, Y., Yamada, G., Dobashi, K., and Hamaoka, T.,** Purification and physicochemical characterization of murine T cell replacing factor (TRF), *J. Immunol.*, 134, 382, 1985.
188. **Takatsu, K., Tanaka, K., Tominaga, A., Kumahara, Y., and Hamaoka, T.,** Antigen-induced T cell replacing factor (TRF). III. Establishment of T cell hybrid clone continuously producing TRF and functional analysis of released TRF, *J. Immunol.*, 125, 2646, 1980.
189. **Ono, S., Hayashi, S.-I., Takahama, Y., Dobashi, K., Katoh, Y., Nakanishi, K., Paul, W. E., and Hamaoka, T.,** Identification of two distinct factors, B151-TRF1 and B151-TRF2, inducing differentiation of activated B cells and small resting B cells into antibody-producing cells, *J. Immunol.*, 137, 187, 1986.
190. **Kinashi, T., Harada, N., Severinson, E., Tanabe, T., Sideras, P., Konishi, M., Azuma, C., Tominaga, A., Bergstedt-Lindquist, S., Takahashi, M., Matsuda, F., Yaoita, Y., Takatsu, K., and Honjo, T.,** Cloning of complementary DNA encoding T-cell replacing factor and identity with B-cell growth factor II, *Nature*, 324, 70, 1986.
191. **Takatsuki, F., Okano, A., Suzuki, C., Chieda, R., Takahara, Y., Hirano, T., Kishimoto, T., Hamuro, J., and Akiyama, Y.,** Human recombinant IL-6/B cell stimulatory factor 2 augments murine antigen-specific antibody responses in vitro and in vivo, *J. Immunol.*, 141, 3072, 1988.
192. **Callard, R. E., Smith, S. H., Shields, J. G., and Levinsky, R. J.,** T cell help in human antigen-specific antibody responses can be replaced by interleukin 2, *Eur. J. Immunol.*, 16, 1037, 1986.
193. **Ambrus, J. L., Chesky, L., Chused, T., McFarlane, P., Mostowski, H., and Fauci, A. S.,** Subpopulations of activated human B lymphocytes distinguished by differences in intracellular signaling in response to human B cell growth factors, *FASEB J.*, 3, A499, 1989.
194. **Katz, D. H. and Benacerraf, B.,** The regulatory influence of activated T cells on B cell responses to antigen, *Adv. Immunol.*, 15, 1, 1972.
195. **Mitchell, G. F., Chan, E. L., Nobel, M. S., Weissman, I. L., Mishell, R. I., and Herzenberg, L. A.,** Immunologic memory in mice. III. Memory to heterologous erythrocytes in both T and B cell populations and requirement for T cells in expression of B cell memory, *J. Exp. Med.*, 135, 165, 1972.
196. **Cerottini, J. C. and MacDonald, H. R.,** The cellular basis of T-cell memory, *Annu. Rev. Immnol.*, 7, 77, 1989.
197. **Sanders, M. E., Makgoba, M. W., and Shaw, S.,** Human naive and memory T cells: reinterpretation of helper-inducer and suppressor-inducer subsets, *Immunol. Today*, 9, 195, 1988.
198. **Budd, R. C., Cerottini, J.-C., Horvath, C., Bron, C., Perdrazzini, T., Howe, R. C., and MacDonald, H. R.,** Distinction of virgin and memory T lymphocytes. Stable acquisition of the Pgp-1 glycoprotein concomitant with antigenic stimulation, *J. Immunol.*, 138, 3120, 1987.
199. **Butterfield, K., Eathman, C. G., and Budd, R. C.,** A subset of memory CD4+ helper T lymphocytes identified by expression of Pgp-1, *J. Exp. Med.*, 169, 1461, 1989.
200. **Mosmann, T. R. and Coffman, R. L.,** Heterogeneity of cytokine secretion patterns and functions of helper T cells, *Adv. Immunol.*, 46, 111, 1989.
201. **Budd, R. C., Cerottini, J. C., and MacDonald, H. R.,** Selectively increased production of interferon-c by subsets of Lyt2+ and L3T4+ T cells identified by expression of Pgp-1, *J. Immunol.*, 138, 3583, 1987.
202. **Sanders, M. E., Makgoba, M. W., Sharrow, S. O., Stephany, D., Springer, T. A., Young, H. A., and Shaw, S.,** Human memory T lymphocytes express increased levels of three cell adhesion molecules (LFA-3, CD2 and LFA-1) and three other molecules (UCHL1, CDw29 and Pgp-1) and have enhanced IFNγ production, *J. Immunol.*, 140, 1401, 1988.

203. **Lynch, F. and Ceredig, R.,** Mouse strain variation in Ly-24 (Pgp-1) expression by peripheral T cells and thymocytes: implications for T cell differentiation, *Eur. J. Immunol.,* 19, 223, 1989.
204. **Morimoto, C., Letvin, N. L., Distaso, J. A., Adrich, W. R., and Schlossman, S. F.,** The isolation and characterization of the human suppressor inducer T cell subset, *J. Immunol.,* 134, 1508, 1985.
205. **Powrie, F. and Mason, D.,** Phenotypic and functional heterogeneity of CD4+ T cells, *Immunol. Today,* 9, 274, 1988.
206. **Birkeland, M. L., Metlay, J., Sanders, V. M., Fernandez-Botran, R., Vitetta, E. S., Steinman, R. M., and Pure, E.,** Epitopes on CD45R (T200) molecules define differentiation antigens on murine B and T lymphocytes, *J. Mol. Cell. Immunol.,* 4, 71, 1988.
207. **Bottomly, K.,** A functional dichotomy in CD4+ T lymphocytes, *Immunol. Today,* 9, 268, 1988.
208. **Clement, L. T., Yamashita, N., and Martin, A. M.,** The functionally distinct subpopulations of human CD4+ helper inducer T cells defined by anti-CD45R antibodies derive sequentially from a differentiation pathway that is regulated by activation-dependent post-thymic differentiation, *J. Immunol.,* 141, 1464, 1988.
209. **Hafler, D. A., Chofflon, M., Benjamin, D., Dang, N. H., and Breitmeyer, J.,** Mechanisms of immune memory. T cell activation and CD3 phosphorylation correlates with Ta_1 (CDw26) expression, *J. Immunol.,* 142, 2590, 1989.
210. **Hayakawa, K. and Hardy, R. R.,** Phenotypic and functional alteration of $CD4^+$ T cells after antigen stimulation. Resolution of two populations of memory T cells that both secrete interleukin 4, *J. Exp. Med.,* 169, 2245, 1989.

Chapter 12

B CELL TOLERANCE

Thomas J. Waldschmidt

TABLE OF CONTENTS

I. INTRODUCTION

Immunologic tolerance, the phenomenon whereby an organism discriminates self from non-self, is the foundation for a properly functioning immune system. Each individual must invoke a variety of processes which render the immune system tolerant to autoantigens. Failure to induce self-tolerance can lead to the state of autoimmunity with its deleterious and often fatal results. This is evidenced by a host of clinical conditions, including rheumatoid arthritis, systemic lupus erythematosus, myasthenia gravis, autoimmune anemia, and a number of vasculitides.

A significant amount of energy has been spent investigating immune tolerance, emphasizing both whole animal and *in vitro* experimental systems. The results of this work have suggested that tolerance is enforced by a number of mechanisms which work in concert to prevent the overt expression of autoimmune responses. Although tolerance must be invoked in both the T and B cell compartments, the intent of this chapter is to review the various models proposed to account for B cell tolerance.

A. SELF-ANTIGENS

When considering the enormous number and variety of antigens within an organism, the complexity of the task for the immune system appears overwhelming. Self-antigens are composed of all classes of macromolecules, and exist as soluble, cell-bound, or intracellular entities. Some are distributed throughout the entire organism while others are restricted to specific tissues. Self-antigens can even be sequestered in specialized organs, and effectively hidden from the immune system. There is also a tremendous variance in the available concentration of these antigens, as some circulate at milligram levels or are displayed on all cells of the body, and others are only found in trace amounts. In addition, while some self-antigens present a pattern of repeating epitopes, many others are univalent in that their surface is contoured by a series of unique determinants. Given this spectrum of self-antigens, it is clear that exquisitely sensitive mechanisms are required to silence the array of potentially self-reactive lymphocytes.

B. RELEVANT ASPECTS OF B CELL PHYSIOLOGY

Before critically examining the various experimental models of B cell tolerance, it is necessary to consider our current understanding of B cell maturation, differentiation, and repertoire expression. The maturational scheme of B cells is well understood, and follows a progression of steps in which the developing B cell must productively assemble its immunoglobulin heavy and light chain variable region genes, express these genes in the form of surface antigen receptors, and migrate to the peripheral lymphoid organs.[1] This maturation is accompanied by the expression of both IgM and IgD forms of the antigen receptor, and the appropriate cell-interaction and lymphokine receptors. Although significant maturational activity occurs during the neonatal period, B cells must continually be generated throughout the lifetime of an organism, since virgin B cells are only short-lived. In both the human and the mouse, fetal and neonatal B cell maturation occurs in the liver, spleen, and bone marrow, whereas in the adult, the bone marrow appears to be the primary source of new B cells.

Once a B cell has matured and settled into the appropriate compartments of the peripheral lymphoid organs, it is now competent to differentiate into an antibody secreting cell. This process has become better understood in recent years, and requires that B cells initially recognize the antigen for which it is specific, and present this antigen to the corresponding helper T cell. Once focused onto the B cell, the helper T cell can deliver the necessary cognate and lymphokine signals required to drive the B cell through its differentiative sequence (Chapters 8 to 11). Although this is a complex series of events, one can condense the process by stating that ultimate commitment to antibody secretion requires two key elements, that of antigen and that of the helper T cell. It is generally assumed that during a normal immune response, both of these elements must be present in order to effect terminal B cell differentiation, even though each has been shown to independently initiate early activation events. This simplification of T cell-dependent B cell differentiation is merely a restatement of the original two-signal hypothesis proposed by Bretscher and Cohn.[2]

Due to an impressive series of molecular studies, immunologists also have an accurate description of how the B cell antigen receptor repertoire is generated.[3] At early stages of B cell maturation, selected V_H, D_H, and J_H segments, as well as V_L and J_L segments, are recombined into complete heavy and light chain variable region genes. These assembled genes are eventually expressed as the polypeptide regions which shape the antigen binding pocket of the immunoglobulin molecule. This recombination mechanism is capable of generating an enormous degree of diversity, as the specific V, D, and J elements selected by individual B cells are drawn from a sizable pool of related but different segments. Although this diversity generating mechanism is desirable, it cannot distinguish amongst the potential specificities it may create. Thus it is immunologically blind, and may generate an anti-self specificity as easily as an anti-viral specificity. In addition to the recombinatorial mechanism described above, a second mechanism is also known to contribute to the B cell repertoire. While the first mechanism is operative early in B cell development and recombines germ line gene segments, the second acts late in immune responses, and produces new specificities by mutating the variable region genes of differentiating cells.[4] It is now clear that this somatic mutation mechanism is intended to contribute antibodies with higher affinity to an ongoing immune response. As with the recombinatorial process however, somatic mutation has the potential to create antibodies specific for self-antigens. It appears therefore, that with the pressure to create and add diversity to the B cell antigen receptor repertoire, the immune system has devised molecular mechanisms with no obvious capacity to distinguish self from non-self.

II. GENERAL RULES FOR B CELL TOLERANCE

Given the complexity of self-antigens and the various aspects of B cell maturation and repertoire expression, the mechanisms which account for B cell tolerance must satisfy a number of conditions.

A. Tolerance mechanisms must be highly operative during periods of B cell maturation when antigen receptors are initially being expressed. Since the recombinatorial process which generates the range of immunoglobulin specificities becomes visible when IgM molecules are first displayed, it is expected that the silencing of self-reactive clones would be most efficient during this period. These mechanisms must function not only in the neonatal period when intense maturational activity is evident, but throughout adult life as B cells continue to mature in the bone marrow.

B. Tolerance mechanisms must also function to neutralize auto-reactive B cell clones generated by somatic mutation during secondary responses.

C. Tolerance mechanisms must cope with all self-antigens which are accessible to circulating antibodies. These include antigens which are not present in compartments where B cell repertoires are generated and expressed.

D. Tolerance mechanisms must be affinity selective. It is well established that immunoglobulin molecules can exhibit a broad range of cross-reactivity. A given antibody which binds one antigen with high affinity might bind a panel of other antigens with moderate to low affinity. Accordingly, it has been recognized that mechanisms which silence anti-self B cells must not eliminate the entire affinity range of self-reactivity (high, moderate, and low), for fear of depleting a significant portion of the protective B cell repertoire.[5] Thus, B cell tolerance must be neither too zealous nor too permissive, but must strike a balance between purging potentially destructive clones, and allowing for an acceptable level of self recognition.

As will be discussed in the forthcoming section, silencing of auto-reactive B cells is accounted for by a number of cooperating mechanisms. Although individually these mechanisms may not fulfill all of the four rules, on a collective basis, tolerance mechanisms must take all of these conditions into account.

III. EXPERIMENTAL MODELS OF B CELL TOLERANCE

Numerous experimental models have been designed to study the mechanisms of B cell tolerance. Most studies have been performed with mice, and have utilized both *in vitro* and *in vivo* systems. Owing to the inherent difficulties in examining responses to self-antigens, many investigators have used foreign or exogenous antigens as tolerogens. Among these antigens are various proteins, a number of highly polymeric thymus-independent antigens, synthetic polymers, and haptenated derivatives of all of these. When studied *in vitro*, the various antigens are typically exposed to the target B cells for a defined period of time, after which the cells are washed and challenged with an immunogenic form of the test antigen. *In vivo* analysis of tolerance has been performed by introducing the antigen systemically, allowing for exposure to the B cell compartments, and again, subsequent challenge with an immunogenic form of the antigen. In both types of experiments, a significant reduction or elimination of the antibody response is viewed as tolerance induction. In addition to studies using exogenous antigens as tolerogens, there are a number of extremely elegant studies in which researchers have taken advantage of allotypic differences in self-antigens,[6] the availability of mice with loss mutations,[7,8] and the construction of transgenic mice.[9-12] Although a number of strong similarities have been derived from the various experimental systems, it is not surprising that important differences have also been reported.

Based on the accumulated body of experimental evidence, two general mechanisms have been proposed to account for B cell tolerance; that of central tolerance and that of peripheral tolerance. Central tolerance operates through the direct inactivation or deletion of the auto-reactive B cells. This process is initiated by the interaction of antigen with the antigen-specific receptors, and is therefore clonal in nature. As a consequence of central tolerance, the auto-reactive B cell is lost to the repertoire, along with its potential to produce auto-antibodies. Peripheral tolerance functions through effector cells or regulatory molecules which suppress or neutralize self-reactive B cells once they have arisen. In this model, anti-self B cells are continually present in an individual and must be muted by regulatory mechanisms. Should the appropriate regulation be lost however, these B cells have the real potential to differentiate and produce auto-antibodies.

IV. CENTRAL MECHANISMS OF B CELL TOLERANCE

Since the antigen receptors, or immunoglobulin, expressed on B cells recognize native determinants on antigens, the idea of a central inactivating event mediated by surface immunoglobulin is simple and attractive. A number of models for central B cell tolerance have been proposed, with abundant data to support them. These models can be grouped into the two categories of clonal deletion and clonal anergy. Both mechanisms propose that interaction with antigen during an early stage of maturation leads to clonal inactivation. They differ in that clonal deletion predicts the physical elimination of the B cell, while clonal anergy suggests that the clone is only functionally impaired. Most of the discussion in this section will deal with experimental systems that have utilized exogenous antigens as tolerogens. Whole animal models examining tolerance to authentic or transgenic self-antigens will be analyzed in subsequent sections.

A. CLONAL DELETION

Clonal deletion predicts that B cells pass through a stage early in ontogeny in which interaction with antigen leads to cell death. Once the B cell passes out of this stage, it can no longer be tolerized but is intended for antibody production. This model, originally described by Burnet,[13] and refined by Lederberg,[14] has found an experimental basis in the extreme sensitivity of immature B cells to tolerance induction.[15-32,110] Upon exposure to a variety of exogenous tolerogens, immature B cells have been reported to be up to 1000 times more tolerizable than mature B cells.[16,24,25] Even among immature B cells (B cells which only display the IgM form of the antigen receptor), there is a hierarchy of susceptibility. Those cells which are emerging from the pre-B cell stage, and are in the process of acquiring surface IgM, are more easily tolerized than a cell already expressing high levels of IgM.[17,18] It should be noted however, that even though mature B cells are described as being resistant to tolerance, this is only a quantitative description since supraoptimal doses of tolerogen have been reported to inactivate mature cells. This aspect will be addressed later.

1. The Participation of Surface Immunoglobulin

Implicit in the model of clonal deletion is the participation of the surface antigen receptors, which bind the specific antigen and initiate the events leading to elimination. This is supported by a variety of experiments which demonstrate the specificity of unresponsiveness, in which the induction of tolerance to a given antigen does not effect immune responses to non-related antigens.[15,17,18,21,22,24,25,30,31,110] In addition, there is need for crosslinking of the B cell surface receptors,[16,21,29-31] and the requirement for minimum affinity thresholds.[30,31]

These attributes of tolerance induction are identical to those observed for surface immunoglobulin mediated activation, in which intracellular metabolic events are initiated as a consequence of antigen receptor crosslinking. Indeed, experiments reported by Klinman's group have demonstrated that active events are involved in antigen specific tolerance, as inhibitors of metabolism and inhibitors of protein synthesis prevented unresponsiveness.[30-32] This idea is also supported by tumor models of B cell clonal deletion. Several groups have shown that a number of B cell tumors with an immature phenotype (IgM+, IgD−) will invoke mechanisms leading to cell death when their surface immunoglobulin is crosslinked.[33-36] What is interesting is that the early activation events initiated by the antigen receptors on these tumors are identical to those observed in normal mature B cells.[36] Incorporating these findings into the scheme of clonal deletion, one could postulate that during the window of tolerance susceptibility early in development, B cells use the same transmembrane signal transduction process utilized by mature B cells, but couple it to a suicide rather than activation pathway.

2. The Roles of Surface IgM and IgD

As the individual B cell matures, clonal deletion predicts that it passes from a vulnerable to tolerance resistant stage. Investigators have sought to define this boundary, and have focused on the appearance of surface IgD. During B cell maturation, IgM is initially displayed on the surface followed by IgD. When fully mature, the conventional B cell expresses high levels of IgD and moderate levels of IgM. Accordingly, Vitetta and Uhr proposed that antigen interacting with surface IgM delivers a tolerizing signal to the B cell, whereas IgD transmits an activation signal.[37] Thus, an immature IgM+, IgD− B cell would exhibit tolerance susceptibility, while an IgM+, IgD+ cell would be resistant.

Various laboratories have tested the "IgD" hypothesis, and reported conflicting results.[22,38-42] In a series of experiments performed by Vitetta et al., adult B cells were stripped of their surface IgD by anti-delta antisera or by papain, and were found to have an increased sensitivity to tolerance induction *in vitro*.[38,39] In studies carried out by Scott et al., B cells treated *in vivo* or *in vitro* with anti-delta antisera were also found to have an increased sensitivity to tolerance induction.[40] In contrast however, Layton et al. reported different results when examining the tolerance susceptibility of IgD+ and IgD− B cells purified from neonatal mice.[41,42] In these experiments, no significant difference was found between maturing cells which expressed or lacked surface IgD. When monitoring the loss of tolerance susceptibility and the acquisition of IgD, Metcalf and Klinman also failed to observe a clear correlation between these two events.[22] Thus, although an attractive hypothesis, the experimental results could not unequivocally define IgD as the pivotal isotype in B cell tolerance.

In more recent experiments, Tisch et al.[43] and Ales-Martinez et al.[44] have provided evidence that IgD and IgM may indeed transmit different signals to maturing B cells. In these studies, complete IgD genes were transfected into the immature B cell tumors which, as described above, will die subsequent to their surface IgM being crosslinked. The resulting tumors, which expressed both IgM and IgD, still exhibited growth arrest when treated with anti-IgM reagents. When their surface IgD was crosslinked however, the tumors responded by increasing their proliferative rate. These observations suggest that in this particular system, surface IgM and IgD mediate fundamentally different events, and lend support to the original Vitetta and Uhr hypothesis. It is important to note that treatment of *mature* B cells with low concentrations of anti-IgM reagents will lead to activation and proliferation, rather than cell death. Thus, if IgM is intended to confer tolerance susceptibility to B cells, it must do so only at an early stage of maturation.

3. One Signal vs. Two Signals

In proposing a slightly different model of self-nonself discrimination, Bretscher and Cohn took issue with several tenets of the clonal deletion hypothesis.[2] Their basic premise, in contemporary terms, was that B cells require two signals to become activated; signal 1 coming from the antigen-antigen receptor interaction, and signal 2 being delivered by a second immunocyte recognizing a distal portion of the antigen (helper T cell). They further argued that regardless of the maturational stage of the B cell, encounter with antigen in the *absence* of a second activating signal would lead to a tolerant state. If both signals were present however, activation and antibody production would ensue, again, regardless of the stage of maturation. Thus, they differed with the conventional clonal deletion model by stating that an immature B cell could undergo differentiation leading to antibody production, and that a mature B cell could become tolerized.

As discussed in the introduction, their prediction that B cells require two basic signals for activation is essentially true, as both antigen and T helper cells are needed to effect differentiation. Evidence also exists which supports the tolerance aspects of their theory. In studying the tolerogenic potential of immature B cells, Klinman's group routinely used the splenic fragment culture system.[22,23] The utility of this system is that primed helper T cells are present in the same environment where B cells undergo tolerance induction. These studies demonstrated that hapten-specific tolerance could be easily induced to the DNP moiety if it were attached to a carrier other than the protein to which the T cells were primed. If the DNP were conjugated to the priming carrier however, an excellent anti-DNP response occurred (i.e., if the priming carrier were KLH, then DNP-MGG would induce profound tolerance, whereas DNP-KLH resulted in an anti-DNP response). Other groups have reported similar results using either helper T cells,[45-47] or mitogens as the second signal.[48] These results are a direct confirmation of the Bretscher and Cohn hypothesis in that signal 1 (antigen) in the presence of signal 2 (cognate interaction with helper T cell) will result in the activation of immature B cells, whereas signal 1 by itself will lead to tolerance. Thus, immature B cells are not automatically destined for elimination should they interact with their antigen.

The second prediction of the Bretscher and Cohn hypothesis, the tolerizability of mature B cells, is somewhat controversial. As previously stated, virtually all experimental systems of tolerance have demonstrated the exquisite sensitivity of immature B cells to tolerance induction. The various studies differ however, as to whether mature B cells can also be rendered tolerant. Some reports claim that mature B cells are resistant,[22,23] while others claim that these cells can be tolerized if the tolerogen is present in high concentrations.[16,19-21,24-26,29,47,49-51,109,110] The various conclusions may reflect the use of different tolerogens, and the different stimulation regimens used to assess the degree of tolerance induction (T cell dependent vs. independent antigens). Thus, although a consensus was not achieved, the possibility was presented that mature B cells can be tolerized.

B. CLONAL ANERGY

When testing the clonal deletion hypothesis, most investigators assumed that the observed unresponsiveness was due to physical elimination of the tolerized B cell. Given the fact that these systems used whole populations of B cells, it would have been technically challenging to follow the fate of the specific antigen-binding B cells. Although this was attempted in some experiments,[52-55] most investigators had no proof that the particular B cells in question were actually deleted subsequent to tolerogen exposure.

In order to directly approach this question, Scott's group[56-59] and Nossal's group[60-63] took advantage of techniques which allowed for the purification of antigen-specific B cells. The design of their experiments was straightforward in that mature or immature mice were systemically tolerized to a haptenated protein (fluorescein-IgG). The mice were subsequently tested for their ability to respond to the hapten, and for the presence of hapten-binding B cells. Although the animals were tolerant, the investigators were able to isolate normal numbers of fluorescein-binding B cells, as compared to control mice. Moreover, these purified hapten-specific B cells were unable to produce anti-fluorescein antibodies when appropriately stimulated *in vitro*. These results indicate that rather than dying, the immature B cells stored the negative signal delivered via the antigen-antigen receptor interaction, and continued their existence in an ''anergic'' state. Thus, B cell tolerance was proposed to occur through clonal anergy, in which the immature B cell experiences a functional rather than physical deletion.[60]

C. CONCLUDING REMARKS ON CENTRAL B CELL TOLERANCE

1. A Composite Model of Central B Cell Tolerance

The original Burnet/Lederberg model provided an invaluable conceptual framework from which a large number of experiments were designed. These studies concluded that immature B cells may indeed be programmed for elimination owing to the tremendous ease of antigen specific inactivation. Clonal anergy proposed an alternative outcome to the process of tolerance, in that the immature B cell, should it interact with its antigen, continue its existence functionally incapacitated rather than die. The Bretscher and Cohn hypothesis furthered the field by encouraging experiments which demonstrated the pivotal role of helper T cells in B cell tolerance. These studies showed that although easy to tolerize, immature B cells could be activated if antigen was delivered in the presence of T cell help. Moreover, some experiments indicated that adult B cells could also be tolerized, albeit with some difficulty. Collectively, one can combine these three models, and propose the following scenario: All B cells, regardless of their stage of maturation, have the potential to be tolerized at the clonal level. The induction of tolerance is the result of the B cell interacting with its antigen in the absence of specific helper T cells. Although all B cells can be tolerized, immature B cells are designed to be extremely sensitive to antigen-induced inactivation, whereas mature B cells are relatively resistant. The cellular basis for this difference is not readily apparent, but may lie in the density and isotype of the surface antigen receptors, or in differential signal transduction pathways coupled to the surface immunoglobulin. Finally, several experimental systems suggest that the intracellular events invoked during tolerance induction lead to an anergic state. It should be noted that the mechanism of clonal deletion has not been entirely disqualified by these latter experiments, and may still be operative in certain situations.

The fourth hypothesis discussed in this section, the Vitetta and Uhr "IgD" theory, is not easy to integrate into the composite model. The IgD hypothesis maintains that IgD serves as a triggering receptor while IgM functions as a tolerizing receptor. While some studies were consistent with this notion,[38-40,43,44] others demonstrated that IgD− B cells could be easily activated,[22,41,42,64] and IgD+ B cells could be tolerized.[42] Thus, a *strict* application of this proposal cannot be incorporated into the overall model. Since immature and mature B cells differ markedly in their IgM and IgD expression, and exhibit large differences in their tolerance susceptibility, there is still the strong possibility that the balance of these two isotypes plays a determining role in tolerance induction.

2. Does the Central Inactivation Model Obey the Rules for B Cell Tolerance?

As discussed in Section II, there are several rules which immune tolerance must obey in order to effectively eliminate the potential for anti-self reactions. Clearly, the requirement for tolerance mechanisms to be highly operative during early periods of maturation is met by the central tolerance model. It has been demonstrated repeatedly that B cells are most sensitive to inactivation when their surface IgM is first being expressed. Also, immature B cells recovered from both neonatal mice and adult bone marrow display extreme sensitivity to tolerance induction. Thus, the first rule is easily satisfied.

The second rule demanded that tolerance mechanisms neutralize any anti-self reactive clones generated during the somatic mutation process active in secondary immune responses. This implies that central tolerance must be able to inactivate a mature, differentiating B cell. The composite central tolerance model allows for mature B cells to undergo clonal inactivation, provided that higher concentrations of the tolerizing antigen are present and helper T cells are absent. Most of the studies which demonstrated the ability of mature B cells to undergo tolerance induction examined virgin IgM+, IgD+ B cells, and not IgG (or other isotype) producing cells. There is evidence however, that switched B cells can be tolerized. Both Weigle's group and Klinman's group demonstrated that IgG bearing B cells,[65,66] or B cells undergoing a secondary response in splenic fragment cultures can indeed be tolerized.[67-69] Klinman further postulated that switched B cells experience a second window of tolerance susceptibility in order to insure the elimination of anti-self clones generated through mutation.[68] These experiments thus support the contention that mature cells can be inactivated, and further suggest that nature had compensated for the somatic mutation process by introducing a second phase of heightened tolerance susceptibility. On the basis of these results therefore, it is possible that central tolerance can cope with switched, mutated B cells.

The third rule, that tolerance mechanisms consider all self-antigens accessible to circulating antibodies, is the most challenging for central tolerance to satisfy. Certainly, central tolerance will effectively eliminate those B cells which are specific for self-antigens present in compartments of B cell maturation. This should purge the repertoire of B cells with specificity to common circulating and cell surface antigens. After migration to the peripheral lymphoid organs, the mature B cells may encounter an additional set of antigens not present during maturation. Again, since the composite model allows for elimination of mature B cells, these antigens have the opportunity to induce tolerance. This set of antigens must be in sufficient concentration however, since mature B cells are relatively resistant to inactivation.

Regardless of the putative efficiency of central tolerance there is still a definite subset of self-antigens to which tolerance will not have been induced. These include antigens residing in locations other than the sites of B cell maturation and the peripheral lymphoid organs. They also include antigens present in these anatomic locations, but in concentrations too low to be effective tolerogens. There is also the special problem of soluble univalent antigens. A number of studies have demonstrated that antigens must crosslink the surface immunoglobulin in order to induce tolerance.[16,21,29-31] If a given self-antigen is composed entirely of unique determinants however, clonal inactivation initiated by surface receptor crosslinking is not possible. In order to circumvent this problem investigators have suggested that accessory cells may serve to present these antigens in a multivalent array.[70] There is no strong evidence for this *in vivo*, and univalent antigens continue to pose a conceptual problem for the proponents of central tolerance. Thus, self-reactive B cells which never interact with their specific antigen, or encounter it in too low a concentration or in monomeric form, may escape central inactivation. Accordingly, the central tolerance model can only partially fulfill the third rule.

The final rule instructed that tolerance must not be too stringent, and must eliminate anti-self B cells on the basis of affinity. This restriction was added in order to prevent the elimination of the entire affinity range of self-reactivity, and with it, a potentially dangerous reduction of B cells directed against pathogens. Since the events of central tolerance are mediated by a receptor-ligand interaction, (surface immunoglobulin and antigen), it should be quite possible for B cells to obey this requirement. Indeed, a number of investigators have reported that B cells are tolerized with an affinity bias. In a variety of systems, limiting amounts of tolerogen preferentially tolerized the higher affinity B cell population, while the lower affinity clones remained responsive.[71-73] In addition, experiments performed by Teale and Klinman not only demonstrated the need for sufficient affinity interactions to effect clonal inactivation, but further showed that a weak interaction with tolerogen will spare the B cell from elimination.[30,31] In their studies, tolerance was prevented in a population of B cells specific for the DNP hapten, if a TNP-conjugated protein was present in competing concentrations. Thus, the poor interaction of the DNP-specific cells with TNP actually prevented tolerance in these B cells. Collectively, these studies provide an experimental verification of the fourth rule, and further suggest that low affinity, crossreactive clones are protected from central inactivation.

Extrapolation of the preceding results predicts that mature, low affinity anti-self B cells should continually be present *in vivo*. When stimulating B cells with polyclonal activators *in vivo* or *in vitro*, many investigators have reported the presence of autoreactive antibodies.[74-81] Upon careful examination, many of these antibodies are low affinity and polyreactive in nature,[79-81] a finding consistent with a selective elimination of high affinity anti-self B cells. Thus, the mechanism of central tolerance appears to have incorporated the conditions of the fourth rule.

3. The Preferential Use of IgG Based Tolerogens

In designing experimental systems of central B cell tolerance, many investigators utilized hapten-protein conjugates as tolerogens.[82-85] The advantage of these conjugates is obvious, in that one can easily follow the immune response to a well-defined determinant. Thus, B cells could be tolerized to a given hapten-protein conjugate, challenged with the same hapten bound to a thymus independent or dependent antigen, and the secretion of anti-hapten antibodies determined with a plaque-forming cell or solid-state assay. When choosing a hapten-carrier conjugate, investigators sought to use carriers which are inert or non-immunogenic when presented to the immune system. Proteins thought to fill this criterion were isologous serum proteins, or deaggregated heterologous proteins. Among the various proteins tested, IgG molecules from both categories became the carriers of choice.

In keeping with the idea that the protein carrier is an inert portion of the tolerogen, it is expected that it will not contribute to the tolerogenic potential of the molecule. In many systems however, it was easily demonstrated that haptenated immunoglobulin molecules were more potent in their tolerogenic potential than similarly haptenated non-immunoglobulin conjugates. The non-immunoglobulin molecules required higher concentrations to induce inactivation,[49-51,89] or were not tolerogenic when directly compared to their IgG counterparts.[20,29,86-88,90-93] Moreover, in experiments performed by Waters and co-workers, the superior ability of IgG based tolerogens was clearly demonstrated with an *in vivo* model.[91-93] When haptenated IgG, albumin, or synthetic peptide tolerogens were introduced into immature animals in utero or as neonates, only those animals which received the IgG based conjugate displayed a significant degree of tolerance. This was also reported in Nossal's model of clonal anergy, in which haptenated albumin failed to induce the level of anergy observed with haptenated IgG.[5]

Apart from its antigen binding capacity, IgG molecules have a variety of other effector functions. Among these is the ability to bind to Fc receptors, including those on B cells. This latter fact becomes important, since IgG Fc receptors on B cells are known to participate in the down-regulation of B cell responses. When crosslinked with surface immunoglobulin, Fc receptors actively suppress both B cell proliferation and differentiation.[95-108] This mechanism has been proposed as a model for B cell suppression mediated by immune-complexes,[95-99] or anti-idiotypic antibodies.[100-102] Since haptenated IgG molecules could likewise crosslink surface immunoglobulin and Fc receptors, they might also induce inactivation through an Fc receptor dependent mechanism. This has been verified in a number of laboratories, in which Fab'2 fragments of IgG, or immunoglobulins which fail to bind to Fc receptors (murine IgG3) were extremely poor tolerogens when lightly hapenated.[29,94,109-111] In the same experiments, tolerance was easily induced using similarly haptenated intact IgG. Since the model of central tolerance only allows for the interaction between the antigen and antigen receptor, it is necessary to cautiously interpret studies employing Fc receptor binding tolerogens. This caveat does not negate the model of central B cell tolerance however, since a number of studies have carefully documented the phenomenon of central inactivation using Fc receptor independent carriers.[29,109,110]

V. PERIPHERAL MODELS OF B CELL TOLERANCE

As opposed to central tolerance, peripheral tolerance involves the neutralization of self-reactive B cells through mechanisms which leave the cell intact, but prohibit its functional expression. Investigation has delineated a number of mechanisms which fall into this category, and mediate tolerance in both active and passive manners.

A. ABSENCE OF HELPER T CELLS

The two signal hypothesis of Bretscher and Cohn was an important contribution to the field of central B cell tolerance. The suggestion that B cell activation will fail if signal 1 (antigen) is present without signal 2 (helper T cell), can also be applied to peripheral tolerance. A number of recent experiments have shown that clonal inactivation appears to be a dominant mechanism for eliminating self-reactive T cells.[112-114] This observation is important when considering B cells which escape central tolerance. As discussed in Section IV.C.2, a subset of autoreactive B cells is likely to survive inactivation owing to the location, concentration, and valency of their respective self-antigens. Should these B cells be appropriately stimulated, the secretion of autoantibodies will follow. Applying the Bretscher and Cohn model however, these self-reactive B cells will never be activated without the concommitant presence of the corresponding helper T cell. Thus, the central elimination of self-reactive helper T cells is a critical peripheral mechanism in maintaining tolerance in the B cell compartment.

B. SUPPRESSOR CELLS

Although never clearly defined, suppressor T cells have been postulated to play a supporting role in the maintenance of tolerance. Similar to B cells, suppressor cells are thought to recognize native determinants on antigens, and as such, are well suited to aid in B cell tolerance. Several studies have provided evidence for ''infectious'' tolerance mediated by suppressor cells.[8,115-123] In these experiments, cells from animals made tolerant to a soluble protein were able to suppress an antibody response to the same protein when adoptively transferred. Although normal antigen responsive B cells were present in the host animals, their activation was effectively inhibited. The difficulty with these experiments is in determining whether the suppression was acting at the level of the B cell or the helper T cell. It is ultimately not important however, since the goal of B cell tolerance is to prevent the secretion of anti-self antibodies. Thus, by preventing the activation of helper T cells or by the direct inhibition of B cells, suppressor T cells may strongly contribute to peripheral tolerance.

C. RECEPTOR CELL BLOCKADE

Receptor cell blockade, like central B cell tolerance, is mediated through the interaction of antigen with the surface antigen receptors. As opposed to the functional or physical elimination of the B cell, this form of tolerance causes only a temporary paralysis. In models of receptor blockade, highly polyvalent antigens are typically used as tolerogens, and when bound to the B cell, are proposed to occupy and immobilize a large number of receptors. This rigid antigen-receptor lattice is thought to prevent any further B cell activation. Once the antigen is removed however, the B cell is free to respond to antigenic challenge. As such, receptor blockade does not induce a central inactivating event, and must be considered peripheral in nature.

In examining receptor cell blockade, investigators have used polymerized flagellin,[124-126] type III pneumococcal polysaccharide,[127] dextran,[128] levan,[128] and haptenated derivatives of all of these.[129-135] When used in low concentrations, these antigens are usually immunogenic, but become tolerogenic when added in high amounts.[124,129-131] In addition, the high dose unresponsiveness can be reversed if the antigen is removed by incubation in antigen-free medium, or by proteolytic treatment.[126,135] Together, these studies indicate that in sufficient concentration, highly polyvalent antigens can ''overwhelm'' the surface immunoglobulin and inactivate the B cell.

Receptor blockade is not limited to resting B cells, as this form of tolerance can also affect antibody secreting cells. This phenomenon, termed effector cell blockade, has been demonstrated with normal secreting spleen cells,[136,137,140,142,143] myeloma cells,[144] and hybridoma cells.[138,139,141] In these studies, interaction of the antibody forming cells with tolerogen resulted in a specific reduction of antibody secretion. This form of blockade is very similar to receptor cell blockade in that multivalent antigens are required,[136,142] and the suppression is reversible with enzymatic treatment or removal of the antigen.[137,139,140,143,144] It should be noted however, that with both receptor and effector cell blockade, prolonged exposure to the polyvalent tolerogen may lead to an irreversible inactivating event.[135,143]

Given the requirement for rigid, highly polyvalent, and generally non-degradable tolerogens to induce blockade, it is likely that this form of peripheral tolerance is restricted to a small subset of self-antigens. One could postulate that a number of highly repetitive structures, such as sulfated carbohydrate molecules, may paralyze mature or secreting B cells in a similar fashion.

D. IgG Fc RECEPTOR MEDIATED SUPPRESSION

The ability of the B cell IgG Fc receptor to antagonize a number of B cell activities has been well described.[95-108] As discussed in Section IV.C.3, the IgG Fc receptor initiates its inhibitory function only when crosslinked with surface immunoglobulin. *In vivo*, this crosslinking is thought to be mediated by anti-idiotypic antibodies or by IgG immune complexes.[95-102] The expression of anti-idiotypic antibodies early in ontogeny has been postulated and experimentally proven to have a profound effect on the establishment of the B cell repertoire.[145] Moreover, in studies performed by several groups, anti-idiotypic antibodies administered to adult and neonatal mice required an Fc portion in order to be effective, thus implicating the need for Fc receptor binding.[100-102] The formation of autoreactive IgG and self-antigen immune complexes could also function to silence autoreactive B cells. The IgG autoantibodies could be present in mature organisms through the inadvertent activation of B cells which have escaped central tolerance, and could also be present early in ontogeny through maternal transmission. Since IgG based tolerogens were repeatedly shown to be highly effective in inducing tolerance (Section IV.C.3), both anti-idiotypic antibodies and immune complexes may provide important ancillary mechanisms of B cell inactivation.

E. CONCLUDING REMARKS ON PERIPHERAL B CELL TOLERANCE

Close scrutiny of central tolerance revealed that a proportion of self-reactive B cells is likely to escape this inactivation mechanism. The presence of peripheral tolerance mechanisms is necessary therefore to mute this population of anti-self B cells, and prevent the secretion of autoantibodies. Of the various peripheral mechanisms discussed, the absence of autoreactive helper T cells is clearly the major safeguard against the activation of these B cells. One can be quite forgiving of the incomplete nature of central B cell tolerance, given the extreme efficiency of central T cell inactivation. Thus, an organism can retain a significant element of self-reactive B cells, yet be phenotypically tolerant. This was demonstrated early on by experiments documenting the phenomenon of low zone tolerance,[72,73] and by studies performed by Weigle and co-workers,[146] in which mice phenotypically tolerant to HGG displayed a phase of tolerance in which HGG reactive B cells were clearly present.

Even with efficient tolerance in the T cell compartment, there are still conditions whereby autoreactive B cells could be activated and begin to secrete autoantibodies. The necessary T cell help could be provided through a rare failure of T cell tolerance, a fortuitous cross-reaction with an exogenous antigen, or by way of a bystander effect.[147-149] Under these circumstances, the anti-self B cells can be controlled by the recruitment of suppressor T cells, immune complexes and anti-idiotypic antibodies, and even by receptor blockade.

When considering the rules for B cell tolerance described in Section II, it is evident that the combination of central and peripheral tolerance mechanisms is necessary to satisfy all of the listed conditions. Although central tolerance met the conditions of rule 1 (heightened inactivation during the immature stage) and rule 4 (inactivation based on affinity), this form of tolerance failed to satisfy rule 3 (inactivation against all accessible self-antigens) and has not been thoroughly investigated in regards to rule 2 (inactivation of anti-self clones derived from somatic mutation). Peripheral tolerance can be envisioned to effectively neutralize the autoreactive B cells escaping central tolerance or being generated in secondary responses. Thus, *both* peripheral and central mechanisms are required to ensure the maintenance of self-tolerance.

VI. TOLERANCE MODELS USING TRUE SELF-ANTIGENS

The majority of our present knowledge on the subject of B cell tolerance has been derived from experiments using exogenous antigens as tolerogens. Although these experiments were essential in constructing the various models of central and peripheral tolerance, the tolerizing antigens were ultimately artificial. As such, these antigens may have introduced additional variables not present in the natural condition. An example of this is the Fc receptor binding capacity of the IgG based tolerogens. It is necessary therefore, to validate the various central and peripheral tolerance models using experimental systems based on true self-antigens. This has been accomplished in several studies, whereby clever use was made of altered self-antigens, alloantigens, congenic strains of mice, and the construction of transgenic animals.

A. TOLERANCE TO RARE SOLUBLE SELF-ANTIGENS

A straightforward approach to test for the presence and thoroughness of B cell tolerance has been to immunize animals with antigens which differ only slightly from a known self-antigen. When doing so, one is able to generate helper T cells against the portion of the antigen which is foreign, and in the process, provide the necessary help to reveal any B cells specific for epitopes shared between the immunizing and self antigen. In early studies by Rouslahti et al., animals were immunized with altered homologous or native heterologous alpha fetoprotein.[150,151] Using these immunization regimens, antibodies specific for self-alpha fetoprotein were easily detectable. Immunization with native homologous alpha fetoprotein did not result in auto-reactive antibodies. These results were interpreted to mean that tolerance in the helper T cell compartment was complete, hence the need for neo-determinants to provide the appropriate T cell help. B cell tolerance, on the other hand, was incomplete or lacking.

Using a different approach, Mitchison's group took advantage of the F1 and F2 allelic forms of the murine liver F protein.[6] As with alpha fetoprotein, mice challenged with autologous F protein failed to mount an antibody response to the self-antigen. If immunized with the allogeneic form however, autoantibodies against the syngeneic form of the F protein were elicited. Again, it was concluded that tolerance to self-F protein was maintained in the helper T cell compartment, and was not effectively induced in the B cell population. Thus, T cells specific for the foreign determinants on the immunizing allo-F protein were able to activate and reveal B cell clones directed against the native F antigen.

Studies performed by Weigle and colleagues using thyroglobulin provided similar results.[152] When administered to mice without adjuvant, self-thyroglobulin failed to provoke an immune response, as expected. When given with complete adjuvant, mice were found to produce significant titers of antibody against native thyroglobulin. In common with the alpha fetoprotein and F protein experiments, tolerance appeared to be deficient or lacking in the B cell compartment. These studies differed however, in that tolerance was also incomplete within the helper T cell subset.

Taken together, these experiments are noteworthy since they provide an interesting challenge to the notion of B cell tolerance. It is important to realize however, that all of these proteins are soluble antigens normally found in extremely low concentrations (10^{-9} to 10^{-10} M). As discussed in Section IV.C.2, it is anticipated that self-antigens of this type would have great difficulty in tolerizing B cells directed against them. Also, in all of the above studies, the affinity of the autoreactive antibodies was not carefully monitored. Hence, the highest affinity, and potentially most destructive B cell clones could have been centrally inactivated. Regardless of this possibility, these experiments underscore the importance of tolerance in the T cell compartment in preventing the expression of self-reactive B cells.

B. TOLERANCE TO COMMON SOLUBLE SELF-ANTIGENS

The above studies suggested that B cell tolerance to rare soluble proteins is incomplete at best. It would be of interest therefore, to ascertain whether soluble self-proteins circulating in higher concentrations are capable of inducing central B cell tolerance. Using a C5 sufficient, C5 deficient congenic pair of mice, Borel and colleagues were able to test the tolerogenic potential of a more abundant self-antigen.[7,8] The complement component C5 is normally present at concentrations of 50 to 85 μg/ml (10^{-6} to 10^{-7} *M*), a concentration at which many of the experimental tolerogens can induce significant B cell inactivation. The two strains of mice used by these investigators are genetically identical except for a mutation at the C5 locus (MUB1) resulting in the absence of circulating C5 protein in the deficient mice. Accordingly, the deficient animals' T cells and B cells will have never seen the C5 protein and will not be tolerant to it, whereas the sufficient strain is tolerant. Using an adoptive transfer technique, experiments were designed to test whether tolerance to C5 in the wild type mice is maintained in both the T cell and B cell compartments. As expected, irradiated *deficient* mice produced anti-C5 antibodies when given spleen cells from syngeneic C5 *deficient* mice, and immunized with C5 protein. Irradiated *deficient* mice given *sufficient* spleen cells produced no autoantibodies. When similarly transferring *sufficient* T cells and *deficient* B cells, and immunizing with C5, no antibody response was observed, indicating that the T cell subset is indeed tolerant. Surprisingly, transfer of *deficient* T cells and *sufficient* B cells allowed for a significant anti-C5 response, demonstrating the presence of self-reactive, non-tolerant B cells in C5 *sufficient* mice. Thus, even with a soluble antigen present in microgram concentrations, B cell tolerance is still lacking or incomplete. However, as with the thyroglobulin, alpha fetoprotein, and F protein experiments, the affinity of the autoantibodies was not determined. Hence, one could again postulate that the high affinity anti-C5 B cell clones were eliminated in the wild type mice, with the remaining moderate to low affinity clones kept in check by the absence of helper T cells.

C. TOLERANCE TO COMMON MULTIVALENT SELF-ANTIGENS

When analyzing central B cell tolerance to exogenous tolerogens, it was commonly observed that receptor crosslinking is necessary to induce inactivation. The failure to readily observe B cell tolerance to rare and common soluble self-antigens may reflect their univalent character, and corresponding inability to crosslink surface antigen receptors. It is not surprising therefore, that B cells directed against these antigens are found in the peripheral lymphoid tissue (as predicted in Section IV.C.2). When considering multivalent self-antigens however, it is anticipated that the majority of B cells directed against these molecules will be purged. In examining tolerance to single stranded DNA, Cowdery et al.[153] demonstrated that the majority of B cells specific for this repeating antigen were indeed eliminated in normal mice. Using a splenic fragment culture system, they observed that bone marrow cells depleted of immunoglobulin bearing B cells (leaving the pre-B cell population intact) contained a significant number of anti-DNA specific precursors. When similarly testing spleen cells from mature animals, the anti-DNA precursor frequency was markedly reduced. Thus, during the pre-B cell to B cell transition, with the accompanying expression of the surface antigen receptors, central tolerance was enforced in those B cells specific for a multivalent self-antigen.

D. THE USE OF TRANSGENIC MODELS TO STUDY B CELL TOLERANCE

One of the persistent difficulties in studying B cell tolerance has been to monitor the fate of the antigen-specific B cells, especially *in vivo*. The studies from Nossal's and Scott's laboratories attempted to circumvent this problem by purifying hapten-specific B cells from mice rendered tolerant to a hapten-carrier conjugate.[56-63] Although crucial in defining the model of clonal anergy, this technique did not allow for direct detection and study of B cells specific for self-antigens. This technical problem has been recently overcome with the utilization of transgenic mice. In studies by two different laboratories, the rearranged heavy and light chain genes encoding a self-reactive antibody were incorporated into the genome of mice. The resulting transgenic animals contain a high frequency of B cells bearing the specific antibody, thereby allowing for easy examination of autoreactive B cells.

Goodnow et al. analyzed B cell tolerance by developing two different strains of transgenic mice.[9,10] The first strain contained and expressed the gene for hen egg lysozyme. Although derived from another species, the transgenic lysozyme must be considered a self-antigen, as it is carried in the genome, found in the circulation, and induces self-tolerance. In the second strain of mice, the investigators introduced the genes for high affinity anti-lysozyme antibodies. The genetic construct given to these mice included the anti-lysozyme heavy chain variable region and both the IgM and IgD constant regions, with the result that upwards of 90% of the B cells displayed anti-lysozyme antibodies of both isotypes. The resulting transgenic animals were also observed to contain significant numbers of anti-lysozyme secreting B cells and high titers of anti-lysozyme antibodies.

The availability of these two strains provided the ideal opportunity to study the fate of antigen-specific B cells when encountering their antigen *in vivo*. This was performed by crossing the lysozyme and anti-lysozyme mice, and analyzing the F_1 offspring. In the resulting double-transgenic mice, the titers of anti-lysozyme antibody, and the numbers of anti-lysozyme plaque-forming cells were dramatically reduced. However, when examining the B cell population, normal numbers of lysozyme binding cells were found. Adoptive transfer experiments established that although present in the secondary lymphoid organs, these cells were not functional as they responded poorly when transferred to normal mice. Thus, tolerance was induced in the B cell population, with the result that the B cells were rendered anergic. These investigators also observed that the lysozyme-binding B cells in double-transgenic mice displayed normal levels of the IgD transgene, but abnormally low levels of the IgM transgene.

The double-transgenic experiments clearly demonstrated that tolerance to a soluble univalent self-antigen can be induced in the B cell compartment. In addition, these results appear to be an affirmation that clonal anergy is operative as a silencing mechanism among self-reactive B cells. The use of a high affinity transgenic antibody is also noteworthy, as it supports the notion that tolerance to univalent antigens might be limited to clones with highest affinity. It would be of interest to test a low affinity antibody in a similar system. Finally, the observation of depressed surface IgM levels is interesting and needs to be explored, since this has not been observed in other systems of clonal anergy.[160]

In a separate set of experiments, Nemazee and Burki constructed transgenic mice bearing the gene for an antibody directed against allogeneic H-2 class I antigens.[11,12] Specifically, mice of the H-2^d haploytpe were given the genes for an IgM anti-H-2^k class I antibody. The resulting mice displayed high serum levels of the transgenic antibody, and up to 50% of the splenic B cells displayed the anti-H-2^k antigen receptor. To investigate B cell tolerance, the transgenic mice were mated with normal mice bearing H-2^k class I antigens. In the H-2^d/H-2^k F_1 mice obtained from this cross, serum titers of the anti-H-2^k transgenic antibody were reduced to background levels, indicating that tolerance had been induced. Close inspection of the B cell compartment revealed the loss of all B cells expressing the anti-H-2^k transgene. The same result was obtained in chimeric mice, in which bone marrow cells from the anti-H-2^k transgenic animals were injected into irradiated H-2^k recipients. Thus, as opposed to the lysozyme transgenic system, the anti-class I self-reactive B cells underwent clonal deletion.

Both sets of transgenic experiments were valuable in demonstrating the ability of physiologic self-proteins to induce central tolerance in the B cell population. The two systems differed however, in the mechanism which caused the B cell inactivation. Whereas lysozyme induced clonal anergy, interaction with the H-2^k molecule resulted in clonal deletion. The difference between these two sets of observations is likely to reflect the biochemical characteristics of the tolerizing antigen. As discussed, lysozyme is a soluble univalent molecule, and should theoretically have difficulty crosslinking surface antigen receptors. Class I antigens on the other hand, are expressed at high density on most cells of the body, and on a collective basis, should effect excellent receptor crosslinking. Thus, one could postulate that clonal deletion will result from a high degree of receptor crosslinking, whereas inefficient or minimal crosslinking will lead to an anergic state.

VII. B CELL SUBPOPULATIONS AND TOLERANCE SUSCEPTIBILITY

Although often considered as a homogeneous group of cells, B cells can be grouped into different subsets based on a number of parameters. These subsets are defined by the state of maturation, state of activation and differentiation, anatomic location, whether the cells respond to TI or TD antigens, and the assignment to the conventional or Ly1/sister B cell lineages. Given these multiple subsets, it is quite likely that the requirements for tolerance induction may differ from cell type to cell type. It has already been established that immature B cells are much more sensitive to inactivation than mature B cells.

An additional distinction has been observed with B cells derived from the Ly1/sister lineage (reviewed in References 81, 154 to 156). In the mouse, B cells comprising this lineage are distinguished by their expression of the Ly1 and Mac1 markers, in addition to high levels of IgM and low levels of IgD. These cells predominate in the peritoneum of adult animals, and have the unique capacity for self-renewal. More important, studies by Herzenberg and co-workers have demonstrated a strong correlation between Ly1/sister B cells and autoimmunity.[154] Not only are these cells overexpressed in certain strains of autoimmune prone mice, but when purified and stimulated, Ly1 B cells appear to be a major source of IgM autoantibodies. A similar association of CD5 B cells with autoimmunity has been reported in the human.[81,157-159] Although the contribution of these IgM autoantibodies to the pathogenesis of autoimmune disease is debatable, these observations suggest that Ly1 B cells may exhibit a higher tolerance threshold when compared to conventional lineage B cells. Given this precedent, it is possible that self-reactive B cells from the various subsets may differ greatly in the manner in which they are silenced. Thus, some B cells may be intended to experience deletion, while others undergo anergy, and yet others may be resistant to central inactivation and require peripheral forms of tolerance.

VIII. CONCLUDING REMARKS

The induction and maintenance of B cell tolerance is clearly a required immune process. If this were not so, the devastating effects of autoimmune disease would never allow for the survival of an organism. The numerous studies utilizing experimental and authentic tolerogens have collectively led to a basic understanding of the processes which silence self-reactive B cells. The data clearly indicate that multiple mechanisms of both central and peripheral tolerance simultaneously operate to eliminate and mute autoreactive cells. The data also suggest that the mechanism employed to silence a given self-reactive B cell, depends greatly upon the self-antigen recognized by the clone. Thus, accessible antigens expressed in a multivalent array, such as cell surface molecules and those with repeating epitopes, may eliminate their respective B cells through clonal deletion. Conversely, soluble univalent antigens are likely to have difficulty deleting B cells directed against them. As such, these antigens may only be able to silence the higher affinity clones, and may do so by inducing a state of anergy. Finally, a proportion of soluble antigens, and self-antigens which are inaccessible to B cells, are unable to centrally inactivate their specific B cells, and must rely upon peripheral tolerance to prevent a damaging autoantibody response. Thus, helper T cell deletion, suppressor cells, Fc receptor feedback inhibition, and receptor blockade serve as additional layers of protection.

Although the combination of central and peripheral tolerance is extremely efficient in preventing pathologic autoimmune processes, several of the experimental systems suggested that self-reactive B cells are normally present. This should not be alarming however, as these cells typically express receptors with lower affinity towards self, and are likely to be present by design. Not only might the various B cell subsets have different thresholds for inactivation, but as discussed, central tolerance must preserve the low affinity autoreactive clones to retain a sufficient protective repertoire. It is the aberrant presence of activated, high affinity B cell clones which initiate the destructive inflammatory cascades observed in autoimmune diseases. Thus, it is necessary to utilize our gathered understanding of normal mechanisms of B cell tolerance, and attempt to discover the conditions which lead to their failure.

REFERENCES

1. **Kincade, P. W. and Gimble, J. M.,** B Lymphocytes, in *Fundamental Immunology,* Paul, W., Ed., Raven Press, New York, 1989, 41.
2. **Bretscher, P. and Cohn, M.,** A theory of self-nonself discrimination, *Science,* 169, 1042, 1970.
3. **Max, E. E.,** Immunoglobulins: molecular genetics, in *Fundamental Immunology,* Paul, W., Ed., Raven Press, New York, 1989, 235.
4. **Kocks, C. and Rajewsky, K.,** Stable expression and somatic hypermutations of antibody V regions in B-cell developmental pathways, *Annu. Rev. Immunol.,* 7, 537, 1989.
5. **Nossal, G. J. V.,** Cellular mechanisms of immunologic tolerance, *Annu. Rev. Immunol.,* 1, 33, 1983.
6. **Mitchison, N. A., Lukic, M., and Nardi, N.,** Proteins at the margin between being tolerated and being noticed, in *9th International Convocation in Immunology,* Milgrom, F., Abeyounis, C. J., and Albani, B., Eds., S. Karger, Basel, 1986, 184.
7. **Harris, D. E., Cairns, L., Rosen, F. S., and Borel, Y.,** A natural model of immunologic tolerance. Tolerance to murine C5 is mediated by T cells, and antigen is required to maintain unresponsiveness, *J. Exp. Med.,* 156, 567, 1982.
8. **Cairns, L., Rosen, F. S., and Borel, Y.,** Mice naturally tolerant to C5 have T cells that suppress the response to this antigen, *Eur. J. Immunol.,* 16, 1277, 1986.
9. **Goodnow, C. C., Crosbie, J., Adelstein, S., Lavoie, T. B., Smith-Gill, S. J., Brink, R. A., Pritchard-Briscoe, H., Wotherspoon, J. S., Loblay, R. H., Raphael, K., Trent, R. J., and Basten, A.,** Altered immunoglobulin expression and functional silencing of self-reactive B lymphocytes in transgenic mice, *Nature,* 334, 676, 1988.
10. **Goodnow, C. C., Crosbie, J., Jorgensen, H., Brink, R. A., and Basten, A.,** Induction of self-tolerance in mature peripheral B lymphocytes, *Nature,* 342, 385, 1989.
11. **Nemazee, D. A. and Buerki, K.,** Clonal deletion of B lymphocytes in a transgenic mouse bearing anti-MHC class I antibody genes, *Nature,* 337, 562, 1989.
12. **Nemazee, D. and Buerki, K.,** Clonal deletion of autoreactive B lymphocytes in bone marrow chimeras, *Proc. Natl. Acad. Sci. U.S.A.,* 86, 8039, 1989.
13. **Burnet, F. M.,** A modification of Jerne's theory of antibody production using the concept of clonal selection, *Aust. J. Sci.,* 20, 67, 1957.
14. **Lederberg, J.,** Genes and antibodies: do antigens bear instructions for antibody specificity or do they select cell lines that arise by mutation, *Science,* 129, 1649, 1959.
15. **Nossal, G. J. V. and Pike, B. L.,** Evidence for the clonal abortion theory of B-lymphocyte tolerance, *J. Exp. Med.,* 141, 904, 1975.
16. **Nossal, G. J. V. and Pike, B. L.,** Mechanisms of clonal abortion tolerogenesis I. Response of immature hapten specific B lymphocytes, *J. Exp. Med.,* 148, 1161, 1978.
17. **Pike, B. L. and Nossal, G. J. V.,** Mechanisms of clonal abortion tolerogenesis. III. Antigen abrogates functional maturation of surface immunoglobulin-negative adult bone marrow lymphocytes, *Eur. J. Immunol.,* 9, 708, 1979.
18. **Teale, J. M., Layton, J. E., and Nossal, G. J. V.,** In vitro model for natural tolerance to self-antigens, *J. Exp. Med.,* 150, 205, 1979.
19. **Kay, T. W., Pike, B. L., and Nossal, G. J. V.,** Mechanisms of clonal abortion tolerogenesis. IV. Verification of the hypothesis in living mice, *J. Immunol.,* 124, 1579, 1980.
20. **Stocker, J. W. and Nossal, G. J. V.,** Induction of B lymphocyte tolerance by an oligovalent antigen. I. Influence of the read-out system, *Cell. Immunol.,* 16, 162, 1975.

21. **Stocker, J. W.,** Tolerance induction in maturing B cells, *Immunology*, 32, 283, 1977.
22. **Metcalf, E. S. and Klinman, N. R.,** In vitro tolerance induction of neonatal murine B cells, *J. Exp. Med.*, 143, 1327, 1976.
23. **Metcalf, E. S. and Klinman, N. R.,** In vitro tolerance induction of bone marrow cells: a marker for B cell maturation, *J. Immunol.*, 118, 2111, 1977.
24. **Cambier, J. C., Kettman, J. R., Vitetta, E. S., and Uhr, J. W.,** Differential susceptibility of neonatal and adult murine spleen cells to in vitro induction of B-cell tolerance, *J. Exp. Med.*, 144, 293, 1976.
25. **Cambier, J. C., Uhr, J. W., Kettman, J. R., and Vitetta, E. S.,** B cell tolerance. I. Analysis of hapten-specific unresponsiveness induced in vitro in adult and neonatal murine spleen cell populations, *J. Immunol.*, 119, 2054, 1977.
26. **Venkataraman, M. and Scott, D. W.,** Cellular events in tolerance. VI. Neonatal vs. adult B cell tolerance: differences in antigen-binding cell patterns and lipopolysaccharide stimulation, *J. Immunol.*, 119, 1879, 1977.
27. **Elson, C. J.,** Tolerance in differentiating B lymphocytes, *Eur. J. Immunol.*, 7, 6, 1977.
28. **Bruyns, C., Urbain-Vansanten, G., Planard, C., De Vos-Cloetens, C., and Urbain, J.,** Ontogeny of mouse B lymphocytes and inactivation by antigen of early B lymphocytes, *Proc. Natl. Acad. Sci. U.S.A.*, 73, 2462, 1976.
29. **Pike, B. E., Battye, F. L., and Nossal, G. J. V.,** Effect of hapten valency and carrier composition on the tolerogenic potential of hapten-protein conjugates, *J. Immunol.*, 126, 89, 1981.
30. **Teale, J. M. and Klinman, N. R.,** Tolerance as an active process, *Nature*, 288, 385, 1980.
31. **Teale, J. M. and Klinman, N. R.,** The molecular requisites of B cell tolerance induction and stimulation, in *B Lymphocytes in the Immune Response: Functional, Developmental, and Interactive Properties*, Klinman, N., Mosier, D., Scher, I., and Vitetta, E., Eds., Elsevier North Holland, Amsterdam, 1981, 251.
32. **Teale, J. M. and Klinman, N. R.,** Membrane and metabolic requirements for tolerance induction of neonatal B cells, *J. Immunol.*, 133, 1811, 1984.
33. **Boyd, A. W. and Schrader, J. W.,** The regulation of growth and differentiation of a murine B cell lymphoma, *J. Immunol.*, 126, 2466, 1981.
34. **Pennell, C. and Scott, D.,** Lymphoma models for B cell activation and tolerance. IV. Growth inhibition by anti-Ig of CH31 and CH33 B lymphoma cells, *Eur. J. Immunol.*, 16, 1577, 1986.
35. **Scott, D. W., Livnat, D., Pennell, C. A., and Keng, P.,** Lymphoma models for B cell activation and tolerance. III. Cell cycle dependence for negative signalling of WEHI-231 B lymphoma cells by anti-μ, *J. Exp. Med.*, 164, 156, 1986.
36. **Fahey, K. A. and DeFranco, A. L.,** Cross-linking membrane IgM induces production of inositol tris-phosphate and inositol tetrakisphosphate in WEHI-231 B lymphoma cells, *J. Immunol.*, 138, 3935, 1987.
37. **Vitetta, E. S. and Uhr, J. W.,** Immunoglobulin-receptors revisited, *Science*, 189, 964, 1975.
38. **Vitetta, E. S., Cambier, J. C., Ligler, F. S., Kettman, J. R., and Uhr, J. W.,** B-cell tolerance. IV. Differential role of surface IgM and IgD in determining tolerance susceptibility of murine B cells, *J. Exp. Med.*, 146, 1804, 1977.
39. **Cambier, J. C., Vitetta, E. S., Kettman, J. R., Wetzel, G. M., and Uhr, J. W.,** B-cell tolerance. III. Effect of papain-mediated cleavage of cell surface IgD on tolerance susceptibility of murine B cells, *J. Exp. Med.*, 146, 107, 1977.
40. **Scott, D. W., Layton, J. E., and Nossal, G. J. V.,** Role of IgD in the immune response and tolerance. I. Anti-delta pretreatment facilitates tolerance induction in adult B cells in vitro, *J. Exp. Med.*, 146, 1473, 1977.
41. **Layton, J. E., Teale, J. M., and Nossal, G. J. V.,** Cloning of B cells positive or negative for surface IgD II. Triggering and tolerance in the T-dependent splenic focus assay, *J. Immunol.*, 123, 709, 1979.
42. **Layton, J. E., Pike, B. L., Battye, F. L., and Nossal, G. J. V.,** Cloning of B cells positive or negative for surface IgD I. Triggering and tolerance in T-independent systems, *J. Immunol.*, 123, 702, 1979.
43. **Tisch, R., Roifman, C. M., and Hozumi, N.,** Functional differences between immunoglobulins M and D expressed on the surface of an immature B-cell line, *Proc. Natl. Acad. Sci. U.S.A.*, 85, 6914, 1988.
44. **Ales-Martinez, J. E., Warner, G. L., and Scott, D. W.,** Immunoglobulins D and M mediate signals that are qualitatively different in B cells with an immature phenotype, *Proc. Natl. Acad. Sci. U.S.A.*, 85, 6919, 1988.
45. **Hamaoka, T., Inada, T., Yamashita, U., and Kitagawa, M.,** Preventive effect of hapten-reactive thymus-derived helper lymphocytes on the tolerance induction in hapten-specific precursors of antibody-forming cells, *J. Immunol.*, 114, 1771, 1975.
46. **Hamaoka, T. and Katz, D.,** Immunological tolerance in bone marrow-derived lymphocytes. II. Effect of allogeneic cell interactions and enzymatic digestion with trypsin of inactivated hapten-specific precursors of antibody-forming cells, *J. Exp. Med.*, 139, 1446, 1974.
47. **Chen, S. and Cerny, J.,** Studies on hapten-specific B cell tolerance in vitro. II. Regulation of central tolerance in T-independent B cells by antigen-activated syngeneic T cells, *J. Immunol.*, 125, 1970, 1980.

48. **Golub, E. S. and Weigle, W. O.,** Studies on the induction of immunologic unresponsiveness. I. Effect of endotoxin and phytohemagglutinin, *J. Immunol.,* 98, 6, 1967.
49. **Kettman, J.,** In vitro induction of specific unresponsiveness against the 2,4,6-trinitrophenyl determinant. I. The induction of unresponsiveness, *J. Immunol.,* 112, 1139, 1974.
50. **Schrader, J. W.,** The in vitro induction of immunological tolerance in the B lymphocyte by oligovalent thymus-dependent antigens, *J. Exp. Med.,* 141, 962, 1975.
51. **Chen, S. and Cerny, J.,** Studies on hapten-specific B cell tolerance *in vitro.* I. The inductive phase of central B cell tolerance, *J. Immunol.,* 125, 1962, 1980.
52. **Sjoberg, O. and Moller, E.,** Antigen binding cells in tolerant animals, *Nature,* 228, 780, 1970.
53. **Cooper, M. G., Ada, G. L., and Langman, R. E.,** The incidence of hemocyanin-binding cells in hemocyanin-tolerant rats, *Cell. Immunol.,* 4, 289, 1972.
54. **Aldo-Benson, M. and Borel, Y.,** The tolerant cell: direct evidence for receptor blockade by tolerogen, *J. Immunol.,* 112, 1793, 1974.
55. **Louis, J., Chiller, J. M., and Weigle, W. O.,** Fate of antigen-binding cells in unresponsive and immune mice, *J. Exp. Med.,* 137, 461, 1973.
56. **Venkataraman, M. and Scott, D. W.,** Cellular events in tolerance: decrease in tolerant spleens of PFC precursors stimulated in vitro by specific antigen or mitogen, *Cell. Immunol.,* 47, 323, 1979.
57. **Venkataraman, M. and Scott, D. W.,** Cellular events in tolerance. VIII. Analysis of the proliferative capacity of antigen-binding cells isolated from normal or tolerant spleens, *J. Immunol.,* 124, 607, 1980.
58. **Pillai, P. S. and Scott, D. W.,** Cellular events in tolerance. IX. Maintenance of immunological tolerance in the presence of normal B-cell precursors and in the absence of demonstrable suppression, *Cell. Immunol.,* 77, 69, 1983.
59. **Chace, J. H. and Scott, D. W.,** Activation events in hapten-specific B cells from tolerant mice, *J. Immunol.,* 141, 3258, 1988.
60. **Nossal, G. J. V. and Pike, B. L.,** Clonal anergy: persistence in tolerant mice of antigen-binding B lymphocytes incapable of responding to antigen or mitogen, *Proc. Natl. Acad. Sci. U.S.A.,* 77, 1602, 1980.
61. **Pike, B. L., Kay, T. W., and Nossal, G. J. V.,** Relative sensitivity of fetal and newborn mice to induction of hapten-specific B cell tolerance, *J. Exp. Med.,* 152, 1407, 1980.
62. **Pike, B. L., Boyd, A. W., and Nossal, G. J. V.,** Clonal anergy: the universally anergic B lymphocyte, *Proc. Natl. Acad. Sci. U.S.A.,* 79, 2013, 1982.
63. **Pike, B. L., Abrams, J., and Nossal, G. J. V.,** Clonal anergy: inhibiton of antigen-driven proliferation among single B lymphocytes from tolerant animals, and partial breakage of anergy by mitogens, *Eur. J. Immunol.,* 13, 214, 1983.
64. **Waldschmidt, T. J., Layton, J. E., McFadden, S. F., Uhr, J. W., and Vitetta, E. S.,** $sIgD^-$ and $sIgD^+$ B cells responding to thymus-independent antigens have a different lymphokine requirement, *J. Immunol.,* 135, 865, 1985.
65. **Walker, S. M. and Weigle, W. O.,** In vitro tolerance induction of primed, IgD-negative murine spleen cells, *J. Exp. Med.,* 153, 653, 1981.
66. **Walker, S. M. and Weigle, W. O.,** Primed lymphoid cell tolerance. II. *In vivo* tolerization of highly tolerogen-sensitive hapten-primed, potentially IgG-producing B cells, *Cell. Immunol.,* 90, 331, 1985.
67. **Neil, G. A. and Klinman, N. R.,** Repertoire expression in surface immunoglobulin-negative bone marrow B cell precursors, *Int. Rev. Immunol.,* 2, 307, 1987.
68. **Linton, P. J. and Klinman, N. R.,** The generation of secondary B cells *in vitro, Fed. Proc.,* 45, 378, 1986.
69. **Linton, P. J., Gilmore, G. L., and Klinman, N. R.,** The secondary B cell lineage, *J. Cell. Biochem.,* 12B, 73, 1988.
70. **Phipps, R. P. and Scott, D. W.,** A novel role for macrophages: the ability of macrophages to tolerize B cells, *J. Immunol.,* 131, 2122, 1983.
71. **Theis, G. A. and Siskind, G. W.,** Selection of cell populations in induction of tolerance: affinity of antibody formed in partially tolerant rabbits, *J. Immunol.,* 100, 138, 1968.
72. **Rajewsky, K. and Brenig, C.,** Paralysis to serum albumins in T and B lymphocytes in mice. Dose dependence, specificity and kinetics of escape, *Eur. J. Immunol.,* 4, 120, 1974.
73. **Andersson, B., Celada, F., and Asjo, B.,** Studies on high and low dose paralysis. Different effect on antibody avidity and on immunocompetent subpopulations of lymphocytes in the mouse, *Eur. J. Immunol.,* 4, 793, 1974.
74. **Steele, E. J. and Cunningham, A. J.,** High proportion of Ig-producing cells making autoantibody in normal mice, *Nature,* 274, 483, 1978.
75. **Dresser, D. W.,** Most IgM-producing cells in the mouse secrete autoantibodies (rheumatoid factor), *Nature,* 274, 480, 1978.
76. **Fournie, G. F., Lambert, P. H., and Miescher, P. A.,** Release of DNA in circulating blood and induction of anti-DNA antibodies by bacterial lipopolysaccharides, *J. Exp. Med.,* 140, 1109, 1974.

77. **Izui, S., Zaldivar, N. M., Scher, I., and Lambert, P. H.,** Mechanism for induction of anti-DNA antibodies by bacterial lipopolysaccharides in mice. I. Anti-DNA induction by LPS without significant release of DNA in circulating blood, *J. Immunol.*, 119, 2151, 1977.
78. **Fernandez, C., Hammarstrom, L., Moller, G., Primi, D., and Smith, E.,** Immunologic tolerance affects only a subpopulation of antigen-specific B lymphocytes: evidence against clonal deletion as the mechanism of tolerance induction, *Immunol. Rev.*, 43, 3, 1979.
79. **Prabhakar, B. S., Saegusa, J., Onodera, T., and Notkins, A. L.,** Lymphocytes capable of making monoclonal autoantibodies that react with multiple organs are a common feature of the normal B cell repertoire, *J. Immunol.*, 133, 2815, 1984.
80. **Ternynch, T. and Avrameas, S.,** Murine natural monoclonal autoantibodies: a study of their polyspecificities and their affinities, *Immunol. Rev.*, 94, 99, 1986.
81. **Casali, P. and Notkins, A. L.,** Probing the human B-cell repertoire with EBV: polyreactive antibodies and CD5+ B lymphocytes, *Annu. Rev. Immunol.*, 7, 513, 1989.
82. **Havas, H. F.,** The effect of the carrier protein on the immune response and on the induction of tolerance in mice to the 2,4-dinitrophenyl determinant, *Immunology,* 17, 819, 1969.
83. **Borel, Y.,** Induction of immunological tolerance by a hapten (DNP) bound to a non-immunogenic protein carrier, *Nature New Biology,* 230, 180, 1971.
84. **Katz, D. H., Davie, J. M., Paul, W. E., and Benacerraf, B.,** Carrier function in anti-hapten antibody responses. IV. Experimental conditions for the induction of hapten-specific tolerance or for the stimulation of anti-hapten anamnestic responses by ''nonimmunogenic'' hapten-polypeptide conjugates, *J. Exp. Med.*, 134, 201, 1971.
85. **Walters, C. S., Moorhead, J. W., and Claman, H. N.,** Immunity and tolerance to a hapten (NIP) coupled to an isologous carrier (mouse gamma globulin), *J. Exp. Med.*, 136, 546, 1972.
86. **Golan, D. T. and Borel, Y.,** Nonantigenicity and immunologic tolerance: the role of the carrier in the induction of tolerance to the hapten, *J. Exp. Med.*, 134, 1046, 1971.
87. **Galanaud, P., Aldo-Benson, M., and Borel, Y.,** Carrier-determined tolerance in vitro, *J. Immunol.*, 114, 141, 1975.
88. **Galanaud, P., Crevon, M.-C., Erard, D., and Dormont, J.,** Hapten-IgG-induced suppression of the in vitro antibody response. Role of hapten density and of antigenic challenge, *Cell. Immunol.*, 29, 72, 1977.
89. **Waldmann, H. and Pope, H.,** Can B-cell tolerance be induced by oligovalent thymus dependent antigens, *Immunology,* 32, 657, 1977.
90. **Lee, W. Y. and Sehon, A. H.,** Suppression of reaginic antibody formation. I. Induction of hapten-specific tolerance, *J. Immunol.*, 114, 829, 1975.
91. **Diener, E., Waters, C. A., and Singh, B.,** Restraints on current concepts of self-tolerance, in *T and B Lymphocytes: Recognition and Function, ICN-UCLA Symposia on Molecular and Cellular Biology,* Vol. 16, Bach, F., Bonavida, B., Vitetta, E., and Fox, C. F., Eds., Academic Press, New York, 1979, 209.
92. **Diener, E., Diner, U. E., von Borstel, R. C., and Waters, C. A.,** Can experimental B cell tolerance serve as a model for self tolerance, *Ann. N.Y. Acad. Sci.*, 392, 35, 1982.
93. **Waters, C. A. and Diener, E.,** Tolerance induction during ontogeny. II. Distinct unresponsive states in immature mice question the generality of clonal abortion, *Eur. J. Immunol.*, 13, 928, 1983.
94. **Grey, H. M., Anderson, C. L., Heussor, C. H., Bonthistle, B. K., Von Eschen, K. B., and Chiller, J. M.,** Structural and functional heterogeneity of Fc receptors, in *Cold Spring Harbor Symp. Quant. Biol.*, 41, 315, 1976.
95. **Chan, P. L. and St. C. Sinclair, N. R.,** Regulation of the immune response. V. An analysis of the function of the Fc portion of antibody in suppression of an immune response with respect to interaction with components of the lymphoid system, *Immunology,* 21, 967, 1971.
96. **Stockinger, B. and Lemmel, E. M.,** Fc receptor dependency of antibody-mediated feedback regulation: On the mechanism of inhibition, *Cell. Immunol.*, 40, 395, 1978.
97. **Kolsch, E., Oberbarnscheidt, J., Bruner, K., and Heuer, J.,** The Fc receptor. Its role in the transmission of differentiation signals, *Immunol. Rev.*, 49, 61, 1980.
98. **Taylor, R. B. and Tite, J. P.,** Immunoregulatory effects of a covalent antigen-antibody complex, *Nature,* 281, 488, 1979.
99. **Tite, J. P. and Taylor, R. B.,** Immunoregulation by covalent antigen-antibody complexes. II. Suppression of a T-cell independent anti-hapten response, *Immunology,* 38, 325, 1979.
100. **Pawlak, L. L., Hart, D. A., and Nisonoff, A.,** Requirements for prolonged suppression of an idiotypic specificity in adult mice, *J. Exp. Med.*, 137, 1442, 1973.
101. **Eichman, D.,** Idiotype suppression. I. Influence of the dose and of the effector functions of anti-idiotypic antibody on the production of an idiotype, *Eur. J. Immunol.*, 4, 296, 1974
102. **Kohler, H., Richardson, B. C., Rowley, D. A., and Smyk, S.,** Immune response to phosphorylcholine. III. Requirement of the Fc portion and equal effectiveness of IgG subclasses in anti-receptor antibody-induced suppression, *J. Immunol.*, 119, 1979, 1977.

103. **Sidman, C. L. and Unanue, E. R.,** Control of B-lymphocyte function. I. Inactivation of mitogenesis by interactions with surface immunoglobulin and Fc-receptor molecules, *J. Exp. Med.*, 144, 882, 1976.
104. **Scribner, D. J., Weiner, H. L., and Moorhead, J. W.,** Anti-immunoglobulin stimulation of murine lymphocytes, *J. Immunol.*, 121, 377, 1978.
105. **Tony, H.-P. and Schimpl, A.,** Stimulation of murine B cells with anti-Ig antibodies: dominance of a negative signal mediated by the Fc receptor, *Eur. J. Immunol.*, 10, 726, 1980.
106. **Phillips, N. E. and Parker, D. C.,** Fc-dependent inhibition of mouse B cell activation by whole anti-μ antibodies, *J. Immunol.*, 130, 602, 1983.
107. **Phillips, N. E. and Parker, D. C.,** Cross-linking of B lymphocyte Fcγ receptors and membrane immunoglobulin inhibits anti-immunoglobulin-induced blastogenesis, *J. Immunol.*, 132, 627, 1984.
108. **Phillips, N. E. and Parker, D. C.,** Subclass specificity of Fcγ receptor-mediated inhibition of mouse B cell activation, *J. Immunol.*, 134, 2835, 1985.
109. **Waldschmidt, T. J., Borel, Y., and Vitetta, E. S.,** The use of haptenated immunoglobulins to induce B cell tolerance *in vitro*. The roles of hapten density and the Fc portion of the immunoglobulin carrier, *J. Immunol.*, 131, 2204, 1983.
110. **Waldschmidt, T. J. and Vitetta, E. S.,** The use of haptenated-immunoglobulin molecules to induce tolerance in B cells from neonatal mice, *J. Immunol.*, 134, 1436, 1985.
111. **Borel, Y.,** Isologous IgG-induced immunologic tolerance to haptens: a model of self versus non-self recognition, *Transplant. Rev.*, 31, 3, 1976.
112. **Kappler, J. W., Staerz, U., White, J., and Marrack, P. C.,** Self-tolerance eliminates T cells specific for Mls-modified products of the major histocompatibility complex, *Nature,* 332, 35, 1988.
113. **Teh, H. S., Kisielow, P., Scott, B., Kishi, H., Uematsu, Y., Blüthmann, H., and von Boehmer, H.,** Thymic major histocompatibility complex antigens and the αβ T-cell receptor determine the CD4/CD8 phenotype of T cells, *Nature,* 335, 229, 1988.
114. **Lo, D., Burkly, L. C., Flavell, R. A., Palmiter, R. D., and Brinster, R. L.,** Tolerance to class II MHC in transgenic mice, *Semin. Immunol.*, 1, 147, 1989.
115. **Benjamin, D. C.,** Evidence for specific suppression in the maintenance of immunologic tolerance, *J. Exp. Med.*, 141, 635, 1975.
116. **Doyle, M. V., Parks, D. E., and Weigle, W. O.,** Specific suppression of the immune response by HGG-tolerant spleen cells. I. Parameters affecting the level of suppression, *J. Immunol.*, 116, 1640, 1976.
117. **Waters, C. A., Pilarski, L. M., Wegmann, T. G., and Diener, E.,** Tolerance induction during ontogeny. I. Presence of active suppression in mice rendered tolerant to human γ-globulin in utero correlates with the breakdown of the tolerant state, *J. Exp. Med.*, 149, 1134, 1979.
118. **Nakano, K. and Cinader, B.,** Suppressor cells and their precursors in A/J mice, tolerant to heterologous gamma globulin, *Scand. J. Immunol.*, 11, 291, 1980.
119. **Acres, R. B. and Cunningham, A. J.,** Neonatally induced tolerance to soluble protein antigens. I. Analysis of B-cell and T helper cell function in mice tolerant to bovine serum albumin or fowl γ-globulin, *Cell. Immunol.*, 73, 12, 1982.
120. **Acres, R. B. and Cunningham, A. J.,** Neonatally induced tolerance to soluble protein antigens. II. A role for the thymus in prolonging tolerance, *Cell. Immunol.*, 73, 31, 1982.
121. **Loblay, R. H., Fazekas de St. Groth, B., Pritchard-Briscoe, H., and Basten, A.,** Suppressor T cell memory. II. The role of memory suppressor T cells in tolerance to human gamma globulin, *J. Exp. Med.*, 157, 957, 1983.
122. **Fazekas de St. Groth, B., Basten, A., and Loblay, R.,** Induction of memory and effector suppressor T cells by perinatal exposure to antigen, *Eur. J. Immunol.*, 14, 228, 1984.
123. **Lukic, M. L. and Mitchison, N. A.,** Self- and allo-specific suppressor T cells evoked by intravenous injection of F protein, *Eur. J. Immunol.*, 14, 766, 1984.
124. **Diener, E. and Feldmann, M.,** Relationship between antigen and antibody-induced suppression of immunity, *Transplant. Rev.*, 8, 76, 1972.
125. **Diener, E. and Paetkau, V.,** Antigen recognition: early surface-receptor phenomena induced by binding of a tritium-labeled antigen, *Proc. Natl. Acad. Sci. U.S.A.*, 69, 2364, 1972.
126. **Diener, E. and Feldmann, M.,** Mechanisms at the cellular level during induction of high zone tolerance *in vitro, Cell. Immunol.*, 5, 130, 1972.
127. **Siskind, G. W., Paterson, D. Y., and Thomas, L.,** Induction of unresponsiveness and immunity in newborn and adult mice with pneumococcal polysaccharide, *J. Immunol.*, 90, 929, 1963.
128. **Howard, J. G. and Hale, C.,** Lack of neonatal susceptibility to induction of tolerance by polysaccharide antigens, *Eur. J. Immunol.*, 6, 486, 1976.
129. **Feldmann, M.,** Induction of immunity and tolerance in vitro by hapten protein conjugates. I. The relationship between the degree of hapten conjugation and the immunogenicity of dinitrophenylated polymerized flagellin, *J. Exp. Med.*, 135, 735, 1972.

130. **Desaymard, C. and Feldmann, M.,** Role of epitope density in the induction of immunity and tolerance with thymus-independent antigens. I. Studies with 2,4-dinitrophenyl conjugates *in vitro, Eur. J. Immunol.,* 5, 537, 1975.
131. **Desaymard, C. and Howard, J. G.,** Role of epitope density in the induction of immunity and tolerance with thymus-independent antigens. II. Studies with 2,4-dinitrophenyl conjugates *in vivo, Eur. J. Immunol.,* 5, 541, 1975.
132. **Desaymard, C., Pearce, B., and Feldmann, M.,** Role of epitope density in the induction of tolerance and immunity with thymus-independent antigens. III. Interaction of epitope density and receptor avidity, *Eur. J. Immunol.,* 6, 646, 1976.
133. **Humphrey, J. H.,** Tolerogenic or immunogenic activity of hapten-conjugated polysaccharides correlated with cellular localization, *Eur. J. Immunol.,* 11, 212, 1981.
134. **Abbas, A. K., Klaus, G. G. B., and McElroy, D. J.,** Mechanisms of B cell tolerance. I. Dynamic nature of the induction of hapten-specific unresponsiveness by hapten-conjugated pneumococcal polysaccharide, *Eur. J. Immunol.,* 7, 382, 1977.
135. **Klaus, G. G. B., Abbas, A. K., and McElroy, P. J.,** Mechanisms of B cell tolerance. II. Evidence for inhibition of antigen receptor expression as a mechanism of unresponsiveness, *Eur. J. Immunol.,* 7, 387, 1977.
136. **Schrader, J. W. and Nossal, G. J. V.,** Effector cell blockade. A new mechanism of immune hyporeactivity induced by multivalent antigens, *J. Exp. Med.,* 139, 1582, 1974.
137. **Schrader, J. W.,** Effector-cell blockade. II. A demonstration of the reversible masking of an immune response by blockade of antibody-forming cells, *Eur. J. Immunol.,* 5, 808, 1975.
138. **Boyd, A. W. and Schrader, J. W.,** Mechanism of effector-cell blockade. I. Antigen-induced suppression of Ig synthesis in a hybridoma cell line, and correlation with cell-associated antigen, *J. Exp. Med.,* 151, 1436, 1980.
139. **Boyd, A. W. and Schrader, J. W.,** Mechanism of effector-cell blockade. II. Colchicine fails to block effector-cell blockade but enhances its reversal, *Cell. Immunol.,* 73, 179, 1982.
140. **Boyd, A. W. and Schrader, J. W.,** Mechanism of effector cell blockade. III. The increased sensitivity of neonatal PFC to effector cell blockade, *Cell. Immunol.,* 73, 192, 1982.
141. **Boyd, A. W. and Schrader, J. W.,** Mechanism of effector cell blockade. IV. Induction by monoclonal anti-μ or anti-idiotypic antibody, role of secreted IgM and mechanism of decreased secretion, *Mol. Immunol.,* 21, 119, 1984.
142. **Klaus, G. G. B.,** B cell tolerance induced by polymeric antigens. IV. Antigen-mediated inhibition of antibody-forming cells, *Eur. J. Immunol.,* 6, 200, 1976.
143. **Klaus, G. G. B.,** B cell tolerance induced by polymeric antigens. VI. Kinetics and reversibility of the inhibition of antibody-forming cells by antigen, *Eur. J. Immunol.,* 6, 389, 1976.
144. **Abbas, A. K. and Klaus, G. G. B.,** Inhibition of antibody production in plasmacytoma cells by antigen, *Eur. J. Immunol.,* 7, 667, 1977.
145. **Kearney, J. F.,** Idiotypic networks, in *Fundamental Immunology,* Paul, W. E., Ed., Raven Press, New York, 1989, 663.
146. **Weigle, W. O., Chiller, J. M., and Habicht, G. S.,** Effect of immunological unresponsiveness on different cell populations, *Transplant. Rev.,* 8, 3, 1972.
147. **Whalen, B. J., Tony, H.-P., and Parker, D. C.,** Characterization of the effector mechanism of help for antigen-presenting and bystander resting B cell growth mediated by IA-restricted Th2 helper T cell lines, *J. Immunol.,* 141, 2230, 1988.
148. **Julius, M. H. and Rammensee, H.-G.,** T helper cell-dependent induction of resting B cell differentiation need not require cognate cell interactions, *Eur. J. Immunol.,* 18, 375, 1988.
149. **Riedel, C., Owens, T., and Nossal, G. J. V.,** A significant proportion of normal resting B cells are induced to secrete immunoglobulin through contact with anti-receptor antibody-activated helper T cells in clonal cultures, *Eur. J. Immunol.,* 18, 403, 1988.
150. **Ruoslahti, E., Pihko, H., Becker, M., and Makela, O.,** Rabbit alpha-fetoprotein: normal levels and breakage of tolerance with haptenated homologous alpha-fetoprotein, *Eur. J. Immunol.,* 5, 7, 1975.
151. **Ruoslahti, E. and Wigzell, H.,** Breakage of tolerance to alpha-fetoprotein in monkeys, *Nature,* 255, 716.
152. **Romball, C. G. and Weigle, W. O.,** T cell competence to heterologous and homologous thyroglobulins during the induction of experimental autoimmune thyroiditis, *Eur. J. Immunol.,* 14, 887, 1984.
153. **Cowdery, J. S., Jacobi, S. M., Pitts, A. K., and Tyler, T. L.,** Defective B cell clonal regulation and autoantibody production in New Zealand black mice, *J. Immunol.,* 138, 760, 1987.
154. **Herzenberg, L. A., Stall, A. M., Lalor, P. A., Sidman, C., Moore, W. A., Parks, D. R., and Herzenberg, L. A.,** The LY-1 B cell lineage, *Immunol. Rev.,* 93, 81, 1986.
155. **Hardy, R. R. and Hayakawa, K.,** Development and physiology of LY-1 B and its human homolog, LEU-1 B, *Immunol. Rev.,* 93, 53, 1986.
156. **Hayakawa, K. and Hardy, R. R.,** Normal autoimmune and malignant $CD5^+$ B cells: the LY-1 B lineage, *Annu. Rev. Immunol.,* 6, 197, 1988.

157. **Hardy, R. R., Hayakawa, K., Shimizu, M., Yamasaki, K., and Kishimoto, T.,** Rheumatoid factor secretion from human Leu-1^+ B cells, *Science,* 236, 81, 1987.
158. **Casali, P., Burastero, S. E., Nakamura, M., Inghirami, G., and Notkins, A. L.,** Human lymphocytes making rheumatoid factor and antibody to ssDNA belong to Leu-1^+ B-cell subset, *Science,* 236, 77, 1987.
159. **Burastero, S. E., Casali, P., Wilder, R. L., and Notkins, A. L.,** Monoreactive high affinity and polyreactive low affinity rheumatoid factors are produced by $CD5^+$ B cells from patients with rheumatoid arthritis, *J. Exp. Med.,* 168, 1979, 1988.
160. **Chace, J. H. and Scott, D. W.,** The tolerance defect. Properties and fate of tolerant B cells in adult mice, in *The Year in Immunology 1988. Cellular, Molecular, and Clinical Aspects,* Cruse, J. M. and Lewis, R. E., Jr., Eds., 1989, 181.

Chapter 13

REGULATION OF NATURAL AUTOANTIBODIES: BENIGN VS. DISEASE-CAUSING SELF-REACTIVITIES

Keith A. Krolick

TABLE OF CONTENTS

I. INTRODUCTION

Autoimmune diseases are a family of immunopathological disorders characterized on the basis of lost tolerance to self antigens. The breakdown of self-tolerance can result in the expression of autoreactivity in both the humoral and cellular compartments of the immune system. Furthermore, examples of such disorders have been described in which particular organ systems or tissues are affected in a relatively restricted manner, sometimes involving a single, identifiable target antigen (e.g., rheumatoid arthritis, myasthenia gravis). In contrast, certain autoimmune diseases (e.g., systemic lupus erythematosus), can include multiple demonstrable autoreactivities to a wide range of autoantigens and are considered to be "systemic" in nature; the exact identification of all antigenic specificities in such a system is often difficult. Whether organ specific or systemic, antibody-mediated or cell-mediated, an active area of investigation for many years has been focused on attempts to identify mechanisms by which normal maintenance of self-tolerance is bypassed, thus allowing the expression of immunopathological autoreactivities.

There appear to be a myriad of possible factors that can affect the development and eventual nature of autoimmune diseases; the complexity of this picture is so extensive that is has been referred to as a "mosaic of autoimmunity".[1] At the top of an undoubtedly incomplete list of influences would probably be genetic factors. For example, genes of the major histocompatibility complex (MHC) encode for cell surface proteins involved in a variety of immunoregulatory functions that can influence the nature of possible immune responses in an individual (i.e., helper vs. suppressor activities); there appear to be correlations (some strong and some rather weak) between the expression of particular MHC gene products and the predisposition of certain individuals to develop particular autoreactivities.[2] Although it is comforting to implicate genes that are known to influence the regulation of the immune system, it is important to superimpose over this body of information the fact that other complex genetic links to certain autoimmune diseases have been reported that are not associated with the MHC,[4-8] including the expression of genes encoding for immunoglobulin products that may also influence the demonstration of autoreactivities (reviewed in Reference 3 and discussed below). Moreover, it is not uncommon to find autoimmune disease activities that are associated with other abnormalities of the immune system. For example, certain immunodeficiencies, such as those involving IgA, are disproportionately detected in patients with SLE, RA, idiopathic thrombocytopenia purpura, dermatomyositis, and Sjögren's syndrome.[9-12] Conversely, relatively healthy individuals with benign IgA deficiencies often produce autoantibodies. Additionally, deficiencies in the complement system have also been reported to be associated with certain autoimmune diseases.[13]

A wide variety of non-immunological factors must also be incorporated into mechanisms involved in the induction of autoimmunity. For example, hormonal influences on the development of autoimmune disease suggest immuno-endocrine relationships; direct and indirect effects of sex hormones[14-16] and hormones of thymic origin[17] have been reported. Exposure to certain drugs has been long implicated as a determining factor in the development of an autoimmune state,[18-20] while dietary influences[21,22] may also play a role. This list of factors is admittedly incomplete.

Finally, exposure to infectious agents must be included as a possible influence on the induction and nature of potential autoreactivities.[23] Micro-organisms could induce the production of autoantibodies by mechanisms dependent on molecular mimicry, increased expression of products of the MHC, direct polyclonal B cell activation, or crossreactive interactions between infection-driven anti-idiotypic networks and self-antigens (e.g., anti-viral antibodies that induce anti-idiotypic antibodies that crossreact with self).[24-29]

The purpose of the discussion presented below is to describe factors that influence the development of B cell (antibody)-mediated autoimmunity within the context of prototypic models of autoimmune disease (both in humans and in experimental animal systems). The premise on which this discussion is based is that the induction of antibody-mediated autoimmune disease symptoms depends, at least in part, on the exact nature of the self-reactive antibody that is present. Thus, many instances of autoantibody production can be demonstrated in the absence of disease, and are even a common feature of "naturally" circulating serum immunoglobulins; antibody with disease-causing potential is likely a restricted subset of the total autoantibody repertoire. Mechanisms leading to the induction of an immunopathological state can only be fully appreciated when regulatory factors are understood that selectively expand the disease-causing antibody subset in certain individuals. Thus, investigations in recent years have focused on examining the nature and the regulation of the autoantibody subset of the "naturally" circulating serum antibodies as a potential source of disease-causing reactivities; other studies have attempted to evaluate the possible role of B cells with the $CD5^+$ phenotype as producers of both benign and disease-causing autoantibodies. This is where the present discussion begins.

II. HUMORAL AUTOIMMUNITY IN HEALTH

A. "NATURAL" AUTOANTIBODIES

In recent years there has been an apparent erosion of the original "rules" defining self-tolerance.[30-32] Numerous reports describe the presence in the sera of many normal, healthy individuals of naturally occurring (i.e., no intentional immunization) antibodies (nAb)[33,34] with reactivities toward a variety of self antigens.[35-39] The nAbs with identifiable binding specificities tend to be of the IgM class, polyreactive, generally of low titer and affinity, and of a relatively short half-life in serum. Importantly, autoreactive nAb most often do not express any disease-causing potential. In contrast, autoreactive antibodies that result in the induction of disease symptoms are often found to be IgG in isotype, monoreactive (or at least restricted with regard to its fine specificity), and generally of higher titer and affinity. Thus, regulatory activities that influence the serum titer, specificity, affinity, and/or isotype of self-reactive antibodies will likely influence the probability of disease onset.

There has been considerable discussion concerning whether the production of self-reactive nAb activity is the result of random polyclonal, mitogen-like stimulation (polyclonal B cell activation)[40-52] or immunologically specific triggering events.[53-59] Although it has been observed that under certain circumstances, intentional *in vivo* polyclonal B cell activation can lead to a disproportionately increased expansion of autoreactive B cells, it is also possible that the additional expansion of particular B cell clones with disease-causing specificities may be under more selective control (see below). Thus, polyclonal mitogen-like stimulation and antigen-specific stimulation may not be mutually exclusive explanations for the process leading to disease induction; polyclonal activators may drive the machinery, while antigen steers it.

The evaluation of the molecular biology of nAb production involving primarily IgM antibodies, has demonstrated restricted V_H gene usage with little if any somatic mutation.[60,61] This is reminiscent of mitogen driven antibody responses, and suggests that nAb are encoded by germline genes (demonstrable in both normal and autoimmune individuals) in antibody producing cells that have not been driven or selected by specific antigen.[62-70] Of possible further importance, immunoglobulin enhancer sequences do not appear to vary between normal and autoimmune mouse strains.[71] Thus, it has been concluded that it is unlikely that immunological defects involving nAb leading to autoimmune disease will be traced to immunoglobulin structural or enhancer elements found outside the normal repertoire of natural autoantibody genes. Further examination of RNA dot-blots of *in situ* hybridization analyses of V_H gene usage by autoimmune mouse strains has lead to further suggestions that in addition to the potential self-reactivities being present in the germline, they may be heavily represented in the 3′-proximal genes (e.g., V_H genes of the 7183, J558, and Q52 families);[3,61,72-76] this could contribute directly to a skewed expression of particular fine-specificities in natural autoantibodies and have some influence on their potential role in disease induction (see below). It has also been proposed that underlying the restricted usage of genes encoding for autoantibody variable regions are immunoregulatory networks based on restricted idiotype (Id) usage.[72,74,77]

Idiotype (Id) "connectivity" has been described as being high among natural antibodies and could have some bearing on the behavior of autoantibodies contained within this pool.[78-80] In a series of studies in which it was assumed that the B cells reactive with self-antigens that contribute to the nAb pool are large, "spontaneously" cycling (activated) cells,[81,82] it was observed that the expression (presence in serum following antigen stimulation) of an antibody with a particular anti-dextran idiotype (called 17-9), differed depending on whether it originated from the spontaneously cycling B cells or the resting B cells.[83] Thus, the resting 17-9 Id^+ B cells in C57BL/6 mice could be antigen activated to produce about half of the total anti-dextran antibody in the response. The already "naturally" activated 17-9 Id^+ anti-dextran B cell population in BALB/c mice, however, appeared to be unresponsive to antigen stimulation and therefore contributed no anti-dextran antibody to the response; this is particularly intriguing in light of the fact that BALB/c mice do express low serum levels of 17-9 Id^+ nAb. Further, the production of 17-9 Id in the C57BL/6 response is inhibited if the mice are pre-primed with anti-Id (17-9 specific) before antigen immunization. The pre-priming does not stimulate antidextran antibody production, but only activates the 17-9 Id^+ B cells into cycle so as to resemble conditions found in the unresponsive BALB/c strain. Thus, it appears that Ids expressed in the nAb pool, derived from spontaneously activated B cells, are under some strict regulation (perhaps idiotype-specific) that limits their participation in antigen-driven responses. Such idiotype "connectivity" which could form the basis of regulating self-reactive B cells, appears to be even stronger in newborns (compared to adults) in which a larger fraction of B cells appears to be in cycle;[84] a randomly chosen panel of anti-Id antibodies (with specificity for a variety of hapten-reactive Ids), were used to show that nAbs from neonatal mice demonstrate significantly more idiotope sharing (i.e., crossreactivity) then do nAb obtained from adult mice. Furthermore, some of the idiotopes associated with the nAb produced by the cycling B cells of the neonatal mice were characteristically not used in responses by adults of that particular mouse strain. Therefore, it appears that under normal circumstances a portion of the autoreactivities associated with the nAb pool are not capable of being selectively expanded by antigen-driven mechanisms; this notion and additional explanations will have to be incorporated into models of autoimmune disease induction that depend on selective production of disease-causing antibodies.

B. ROLE OF CD5⁺ B CELLS IN THE PRODUCTION OF NATURAL AUTOANTIBODIES

The $CD5^+$ B cell subset[85-89] has been the focus of attention with regard to its possible participation in the production of nAb. This B cell subset, sometimes called the Leu-1^+ subset in humans or the Ly-1^+ subset in mice, makes up about 10 to 25% of B cells,[90] and the majority of B cells in fetal spleen[91] (which may explain the high incidence of nAb production in newborns). Many of the $CD5^+$ B cells are activated and cycling,[39,82] and have been identified as a source of spontaneously produced natural antibodies;[92,93] the antibodies produced by $CD5^+$ B cells have many of the nAb characteristics mentioned above including the tendency to be polyreactive, low affinity IgM antibodies. Further, the nAb of $CD5^+$ B cell origin appear to demonstrate restricted V_H gene usage[94] and to contain an autoreactive subset that is affected by little, if any, somatic mutation.[95,96] Although it has been proposed that the nAb pool produced by $CD5^+$ B cells may play a role as the first line of defense against infection, it is also this pool of antibodies, containing a self-reactive subset, that may also play a role in autoimmune disease.

III. HUMORAL AUTOIMMUNITY IN DISEASE

Despite the evidence against frequent antigen-driven stimulation of natural autoantibody production mentioned above, evidence also exists that some of the autoantibody reactivities contained within the nAb population can, in fact, be produced and expanded following certain selective, perhaps antigen-specific events.[53-59,97-99] At the antibody level it has been noted that although the *potential* specificity repertoire of self-reactive nAb appears the same when one compares normal individuals to autoimmune individuals, individuals with autoimmune disease often demonstrate restricted, or at least skewed representation at the *expressed* level. This may be a major determining factor with regard to subsequent disease induction. In other words, it is possible that a very small subset of specific, self-reactive antibodies contained within a more complex spectrum of self-specificities is directly responsible for symptoms of disease, and the vast majority of remaining antibodies reactive with the same antigen have no disease-inducing potential. Consequently, if the appropriate circumstances were to occur and the disease-causing antibody subset were to be selectively produced, disease induction should follow; if selective production of any of the antibodies in the non-disease-causing subset were to occur, disease induction should not follow.

Evidence from experimental animal systems suggests that the precursor frequencies of autoreactive B cells in immunologically normal strains of mice are very similar to those determined for autoimmune strains of mice;[46,48] in both cases, $CD5^+$ B cells may make up a prominent proportion of the autoreactive cells.[100] Furthermore, naturally circulating murine autoantibodies and conventional antibodies produced in response to immunization may exhibit similar degrees of antigenic crossreactivity,[101] although they can often be of different fine-specificity.[102,103] Thus, the potential antibody repertoire appears to be similar when comparing individuals that produce natural autoantibody, spontaneously produced autoantibody, or artificially induced autoantibody. However, a subset of the potential autoantibody repertoire seems to be more directly involved in disease induction. V_H gene usage for the production of self-reactive antibodies in autoimmune strains of mice may be highly restricted;[3] for example, anti-DNA antibodies from lupus-prone mice have been observed to be encoded by a small number of V_H gene families,[73-76,100] (some of which can be traced back to $CD5^+$ B cells). Furthermore, the degree of restriction placed on V_H gene usage appears to increase as the disease progresses.[76] Although restricted V_H gene expression and crossreactive idiotypic associations can be observed in both normal and autoimmune mouse strains,[74,77,79,104-107] anti-DNA antibodies with disease-causing activities may be distinguishable from benign anti-DNA antibodies on the basis of detectable V_H gene mutations, as well as redirected, non-random patterns of antigen-binding fine-specificities. Therefore, with regard to disease onset it is not enough that the autoantibody response be restricted to a small number of B cell clones,[108] but must be restricted to a particular subset of autoreactive B cell clones. The possible non-random expansion of particular anti-DNA fine-specificities mentioned above may be due to non-random mechanisms of clonal selection that are normally down-regulated;[54] depending on which clones of B cells are allowed to expand, however, the resulting anti-DNA antibodies may or may not have disease-causing potential.

Another example of the fine-specificity of autoantibodies playing a determining role in the induction of autoimmune disease involves the rat model of myasthenia gravis.[109] In this system, it is clear that the ability of autoantibodies to recognize conformationally intact antigenic determinants on the acetylcholine receptor (AChR) is required for the induction of neuromuscular disease; antibodies produced by immunization with purified AChR of *denatured* conformation, even though crossreactive with AChR of *native* conformation, are incapable of inducing disease symptoms.[110-112] Interestingly, antibodies produced against the denatured form of the autoantigen were observed to interfere (protect?) with the disease-causing antibodies raised against the native form of the antigen.[112]

Examples are also found in human patient systems of distinctions between disease-causing and benign autoantibody reactivities. In one case, the *potential* fine specificity repertoire of anti-thyroglobulin antibody from Graves' disease patients compared to normal healthy individuals appeared to be similar.[55] However, the *expressed* Graves' antibody demonstrated antigen binding fine specificity that significantly favored one particular antigenic determinant.

From other studies, rheumatoid arthritis (RA) patients possess a rheumatoid factor (RF)-producing B cell population, possibly derived from the $CD5^+$ B cell subset.[87,92,113-116] Patient anti-IgG antibodies are often of defined, restricted specificity for particular regions of the target antigen,[117-120] and can be further defined by associations with certain crossreactive idiotopic determinants.[121-123] Furthermore, EB virus-transformed B cells from RA patients appear to produce two main types (specificities) of rheumatoid factor with distinct reactivities. One RF is polyreactive, of low binding affinity, and found in the serum of both normal subjects and RA patients, and a second RF that is monoreactive, of high binding affinity, and found in the serum of only RA patients.[114,124] In light of the overlapping binding specificities shared by RF and the Fc-binding proteins of certain bacteria, particularly Staphylococcal strains, it is interesting to speculate that the origin of the RF is as the anti-idiotypic internal image of anti-bacterial antibodies with specificity for Staph Protein A. Therefore, in this case there could be an indirect role for the antigen-driven expansion of a subset of the normal antibody repertoire that possesses disease-causing potential; superimposed over this apparently simple antigen/antibody system, however, the immune network may provide unexpected reactivities and exact binding specificities may at times appear ambiguous. For example, unusual crossreactivities have been described in which both anti-DNA activity and RF activity can be demonstrated by the same autoantibodies.[125,126]

One other example of biased antibody binding fine-specificity that may influence the severity of autoimmune disease symptoms involves anti-DNA antibodies produced in lupus patients; anti-DNA antibodies can be found in the nAb pool in healthy individuals, but when compared to anti-DNA antibodies in patients with disease symptoms a different reactivity profile is observed.[127,128] Naturally occurring or artificially induced (via immunization) anti-DNA antibody tends to be most reactive with single-stranded DNA, while the antibody obtained from lupus serum tends to be most reactive with native double-stranded DNA; furthermore, although lupus autoantibodies can demonstrate polyreactivity, there appear to be patterns of anti-DNA antibody crossreactivity among different organisms that can distinguish between nAb and lupus antibody.[128]

Although anti-DNA antibodies can be distinguished with regard to binding specificities that may correspond with disease state, there is a complex overlap of the antibody repertoires expressed by patients and their healthy counterparts. Moreover, lupus antibodies have been often observed to cross-react with numerous non-DNA antigens that possess phosphorylcholine moieties or various antigens that happened to have appropriately spaced phosphodiester bonds.[127] These overlapping antibody repertoires have been further evaluated by the identification of recurrent idiotypic determinants shared by benign and disease-causing anti-DNA antibodies.[4,129-136] Of possible additional importance are demonstrable crossreactive idiotypes between antibodies reactive with DNA and various micro-organisms such as certain species of *Klebsiella* associated with ankylosing spondylitis.[27] These overlapping specificity repertoires that include crossreactivities with antigens often associated with a variety of micro-organisms are suggestive of the possible involvement of bacterial infection in the generation of anti-DNA responses.

To conclude, antibodies with reactivities for self appear to be included in the normal specificity repertoire. Even in health, a small number of "spontaneously" cycling B cells, many that may be CD5 positive, are constantly producing self-reactive antibodies that make up a subset of the naturally circulating immunoglobulins found in the serum of normal individuals. It is not clear whether the normal production of "natural" autoantibodies is the result of an error in nature or if there is some selective advantage given to organisms that make such antibodies; there has been some discussion regarding the head-start that naturally circulating antibodies would have on infectious micro-organisms, some that may coincidentally crossreact with host constituents.[137] The rapid removal of infectious agents that mimic host antigens by low-affinity IgM antibodies might be a crucial mechanism for avoiding the production of high titered, high-affinity, IgG anti-self responses that could be destructive to the host. Thus, under most circumstances natural autoantibodies may not only be non-destructive, but in a sense protective. However, circumstances may change that give disease-causing potential to the usually benign autoantibody subset. This may include the expansion of a very particular subset of autoreactive B cell clones (possibly only one clone is necessary) whose antibody always had immunopathological potential, but due to immunoregulatory constraints contributed to only a very small portion of total natural antibody production; if the immunoregulatory state of an individual were to deteriorate, a disproportionate expansion of the disease-causing B cell clones might occur. In contrast, new immunopathological potential (e.g, altered fine-specificity) may be given to normally harmless, naturally occurring autoantibodies if produced by expanded clones of B cells following somatic mutational events involving V_H genes. Thus, general characteristics of *disease-causing* autoantibodies might include (1) they would make up a subset of the total natural autoantibody repertoire, (2) they would often demonstrate a shift away from the low-affinity polyreactivity of the average naturally circulating autoantibody to higher affinity monoreactivity with highly defined fine-specificity, and (3) they may be associated with particular antibody isotypes (e.g., IgG) required for the effector mechanisms that lead to disease symptoms. The discussion below describes possible instances in which T lymphocytes may influence the activities of autoreactive B cells. Attention is focused on immunoregulatory biases that may preferentially expand autoreactive B cell clones of particular fine-specificity and/or isotype.

IV. REGULATION OF AUTOANTIBODIES AND DETERMINATION OF DISEASE-CAUSING POTENTIAL

With regard to the involvement of T cells in determining the nature of autoantibody production in an individual, and the eventual influence that T cell regulation has on determining the severity of antibody-mediated disease symptoms that may result, the following discussion is divided into three areas: (1) general lymphokine influences, (2) influences of T cells demonstrating helper activity, and (3) influences of T cells demonstrating suppressive activity.

A. LYMPHOKINES

The role of T cells in the induction of antibody-mediated autoimmune disease may be felt at the level of the growth and differentiation factors that they produce or are affected by. Interleukin-2 (IL-2) has received the most attention in this regard.[138] A reduced level of IL-2 during ontogeny has been proposed as a requirement for effective establishment of tolerance against self; increased levels can interfere with tolerance induction and once established, tolerance can be broken by appropriate stimulation with the lymphokine.[140-142] Studies of T cell responses to mitogens such as concanavalin A (Con A) or phytohemagglutinin (PHA) hints at growth factor defects in various animal and human autoimmune disease states; impaired T cell proliferative function has been observed in lupus-prone mice,[143] diabetic BB rats,[144] and in patients with SLE,[145] rheumatoid arthritis,[146] Sjogren's syndrome,[147] and active multiple sclerosis.[148] Evaluation of data that has been collected over the past few years indicates that IL-2 involvement in autoimmune disease can be identified both in the production of the growth factor, and in its reception by IL-2 receptor-bearing target cells.

Anti-DNA antibody production in both SLE patients[149] and in lupus-prone MRL-lpr/lpr mice[150] appears to be enhanced in the presence of additional levels of IL-2. This may be of physiological importance taken together with reports of IL-2 hyperproduction.[151,152] However, there is some confusion concerning this issue, since in similar systems there have been descriptions of decreased responsiveness to IL-2[153-155] that correspond to decreased levels of IL-2 receptors on target cells.[154,156,157] Explanations for declining IL-2 responsiveness may also involve the presence of detectable concentrations of soluble IL-2 receptor chains in the serum of affected individuals that might interfere with proper IL-2 binding;[158,159] T cells with suppressive activity may also play a role.[160] Once removed from such *in vivo* antagonizing factors, T cells from SLE patients appear to be able to revert to normal IL-2 responsiveness.[149,161] To further complicate this picture of abnormal IL-2 functions are reports that other lymphokine systems such as IL-3 and IL-5 may also be impaired in SLE and/or its models.[162,163] These observations contradict what one might predict concerning diseases that are based on abnormal immunological hyperresponsiveness.

The role of lymphokines in other autoimmune disease systems also lacks in clarity. Although IL-2 secretion is down in rheumatoid arthritis, the number of IL-2 receptor-bearing (Tac$^+$) T cells is elevated and so is the concentration of soluble IL-2 receptor in the serum of patients.[164-167] Additional abnormal cytokine activities associated with rheumatoid arthritis have been reported for IL-1[168] and IL-6.[169] In multiple sclerosis, there is no agreement concerning whether IL-2 production is up or down,[170-172] although it is clearer that the number of IL-2 receptor-bearing cells and the number of receptors per cell are both up in multiple sclerosis patients.[173,174] It is also clear that production of tumor necrosis factor is important in disease progression,[175,176] but probably plays no immunoregulatory role in the autoimmune response. Finally, one system that demonstrates a virtual across-the-board increase in IL-2-dependent activities and appears to be more consistent with the (perhaps naive) notion that autoimmune disease should be accompanied by only hyperresponsiveness, is the spontaneous autoimmune thyroiditis model in obese strain chickens. This model of Hashimoto's thyroiditis demonstrates a significant elevation of various T cell functions including mitogen responsiveness and stimulated and spontaneous IL-2 secretion, increased numbers of IL-2 receptor-bearing cells, and increased numbers of IL-2 receptors per cell; furthermore, T cells extracted from diseased thyroid tissue appear to be in an appropriate state of activation.[177-180] Interestingly, genetic studies suggest that the basis for this abnormal immunological state is probably not linked to the major histocompatibility complex.[181]

B. HELPER T CELL ACTIVITY IN AUTOANTIBODY-MEDIATED DISEASE

Various studies implicate T cells as having a role in the generation of autoantibodies; further, depending on the nature of the help available to the autoreactive B cells, the resulting autoantibody production may or may not lead to disease induction. First, with regard to the generation of B cells capable of producing natural autoantibody activity it is clear that there is often a T cell influence. For example, splenic lymphocytes from athymic nude mice do not produce detectable LPS-stimulated autoantibody reactivity directed against bromelain-treated syngeneic red blood cells (BMRC).[182] Adoptive transfer of CD4$^+$, but not CD8$^+$, T cells early in ontogeny leads to the acquisition of this autoreactive activity in nude mice with B cell precursor frequencies comparable to those observed in normal control mice. However, this observation is complicated by the fact that T cell dependence on the generation of autoreactivity appears to also depend on the physiological compartment examined. In other words, in contrast to the results reported for the splenic B cell population, when nude mice are evaluated for the ability to produce anti-BMRC antibodies, B cells obtained from the peritoneal cavity are capable of producing the autoreactivity.[183] The explanation for differences in the T cell dependence of B cells from these two locations (spleen vs. peritoneal cavity) to express autoreactivities is not well understood, but may involve the increased frequencies of Ly-1 ($CD5^+$) B cells found in the peritoneal cavity of the mice. Consistent with this explanation, normal mice demonstrate the presence of anti-BMRC B cell precursors in the peritoneal cavity earlier (by 3 weeks of age) than are demonstrated in the spleen (reaching adult levels only by 6 to 10 weeks), corresponding to the early prominence of Ly-1 B cells early in life; the kinetics of generating significant frequencies of BMRC-reactive B cells in the spleen can be accelerated in neonates by transfer of adult T cells. It has also been noted that not all autoantibody activities are equally sensitive to regulation by T cells; precursor frequencies of DNA-reactive B cells in nude mice appear to be similar to those found in their euthymic counterparts.

In addition to the T cell dependence of generating the precursors of natural autoantibody as described above, additional studies have demonstrated the direct T cell dependence of the antigen-specific stimulation of autoreactive B cells to secrete autoantibody.[53,184] For example, *in vitro* exposure to human thyroglobulin (Tg) does not result in the production of anti-Tg antibodies by human peripheral blood mononuclear cells obtained from healthy donors as might have been predicted by the previous demonstration of mitogen-stimulated generation of the ‘‘natural’’ self-reactivity. However, if appropriate T cell help is provided, anti-Tg antibody (predominantly of the IgM isotype) is secreted into culture fluids. This can be accomplished by (1) stimulation of B cells with Tg-agarose beads plus a B cell growth factor derived from a human T cell clone, or (2) stimulation of B cells from tetanus toxoid-immune individuals with Tg-tetanus toxoid conjugates.

With regard to the generation of disease-inducing auto-reactivities, T cells have been discussed in terms of ‘‘selecting’’ particular subsets of the total natural autoantibody repertoire. T cell influence on the disease-causing potential of autoantibodies could be directed at the isotypic and/or idiotypic nature of antibodies produced. For example, in systems involving SLE patients and lupus-prone strains of mice, it has been shown that the fine-specificity of anti-DNA antibodies as well as the isotype of those antibodies can determine resulting disease severity, with IgG anti-dsDNA antibodies being much more prominent in disease then IgM anti-ssDNA antibodies, even though the latter appears to dominate natural autoantibody responses stimulated experimentally in healthy individuals.[98,99,185] Generation of IgG anti-DNA antibody responses is clearly T cell dependent,[186-190] as shown in studies in which $L3T4^+$ T cells were deleted *in vivo*.[186] The age-dependent shift from IgM to IgG production in lupus-prone mouse strains (but not in normal mouse strains) appears to be determined by $L3T4^+$ T cells.[191] An additional isotype-related abnormality that has been demonstrated in lupus-prone NZB mice that may bear on explanations for isotype shifts that are not observed in healthy mouse strains, involves the defective regulation of IgA anti-erythrocyte antibody responses.[193] In contrast to the IgA-binding factor produced by IgA Fc receptor-bearing T cells in normal mice that can suppress IgA secretion by normal or NZB B cells, the IgA-binding factor from the lupus-prone mouse is incapable of down-regulating IgA production; moreover, it was also observed that the defective binding factor could be shown to have stimulating activity under conditions in which suppression was expected.

With regard to biased idiotype expression in disease, it has been suggested that sustained release of B cell growth factors combined with selection and expansion by T cells of B cells expressing a particular crossreactive idiotype is responsible for the generation of at least a portion of the disease-causing anti-DNA antibodies,[107,193-195] and that in disease, idiotype specific T cells may preferentially recognize IgG-bearing B cells over IgM-bearing B cells; there have been previous reports of such idiotype/isotype relationships in other systems.[196]

Another antibody-mediated autoimmune disease in which helper T cells may play a role in determining the nature of autoantibodies produced and subsequent expression of disease symptoms is myasthenia gravis.[197-204] It is clear that some but not all antibodies reactive with the acetylcholine receptor (AChR) can cause neuromuscular dysfunction.[205-207] Thus, studies have focused on the importance of fine-specificities possessed by anti-AChR antibodies in disease induction. For example, it has been observed that disease activity can be demonstrated in Lewis rats following the induction of an oligoclonal subset of AChR-reactive antibodies that react with conformational determinants possessed by the AChR, while reactivities directed at denatured, linear determinants of the same molecule are incapable of disease-induction.[112] Although the exact disease-causing antibody fine-specificity has not been identified, the influence of T cell specificity for AChR has been examined with regard to its influence over the nature of anti-AChR antibodies that are produced. Thus, it has been observed that particular helper T cells with restricted reactivities directed at the alpha subunit of the AChR are the most prominent among the AChR-responsive T cells and are capable of driving the most heterogeneous anti-AChR antibody response (when compared to T cells with reactivities directed at the other subunits, beta, gamma, or delta);[200,202] moreover, T cells reactive with a small 17-amino acid synthetic peptide associated with the alpha subunit appear to represent a large portion of total AChR specific T cell reactivity,[208] as well as having the ability to help an anti-AChR antibody response that is virtually identical to the response helped by the entire unrestricted T cell anti-AChR repertoire with regard to characteristics of clonotypic anti-AChR antibody heterogeneity, affinity, and isotype distribution.[209] Finally, there are genetic associations (perhaps MHC-linked) in the rat model of myasthenia gravis with patterns of T cell fine specificity.[208,210] It is of interest that prominent T cell reactivity demonstrated by disease-resistant Wistar Furth rats is directed against a different panel of AChR peptides than disease-susceptable Lewis rats; future studies may reveal that although Wistar Furth rats produce equal amounts of AChR-reactive antibody as Lewis rats, that because of differences in their expressed T cell repertoire the antibody fine-specificities do not contain the disease-causing activity seen in Lewis rats.

C. THE BALANCE OF HELP AND SUPPRESSION IN AUTOIMMUNE DISEASE

The importance of immunoregulatory suppressor T cell activities has over the years stimulated some controversy.[211,212] However, in the face of the clear presence of potential disease-causing specificities in the normal antibody repertoire, the absence of universal immunopathological autoantibody responses leads one to consider that some mechanism must exist for down-regulating detrimental immunoresponsiveness; explanations for the expression of detrimental autoimmunity, therefore, must be based on the premise that although under normal circumstances of health autoreactive helper T cells may slip past the clonal deletion machinery, their ability to stimulate the selective expansion of B cell clones capable of producing disease-causing antibodies is usually suppressed. Studies involving the role of T cell influences on the production of autoantibodies, particularly antibodies with reactivities leading to disease, suggest that there may be a finely tuned balance of T cell-mediated help and suppression created that can be tipped in either direction depending on the circumstances.

Suppressor T cells have been discussed in terms of playing a role in the maintenance of self-tolerance. Down-regulating cells could be valuable for preventing the expansion, to immunopathological levels, of B cell clones producing natural autoantibodies with disease-causing potential. Although mechanisms of antigen recognition and self-nonself discrimination have not been as well defined as for helper T cells, suppressor T cells are likely to be equally capable of specific, antigen driven functions. In a recent example it has been observed that B10.D2 OSN/J mice that are deficient in the fifth component of complement (C5) are capable of producing an antibody response against C5 obtained from MHC-identical, congenic, complement-normal B10.D2 NSN/J mice.[213] The ability of the C5-deficient mice to produce anti-C5 antibodies, however, is inhibited by the neonatal engraftment of the thymus coming from the immunologically tolerant, C5-normal mice. This state of unresponsiveness appears to be "learned" and dependent on the thymus, and, furthermore, the effect can be adoptively transferred from engrafted C5-deficient mice into unengrafted recipients. One explanation for such down-regulation could be the induction of antigen-specific suppressors that interfere with the activity of C5-specific helper T cells and/or B cells. A suppressor T cell repertoire for self-antigens could be generated by antigen-driven or idiotype-driven mechanisms. For example, mechanisms have been previously described in which the production of particular B cell idiotypes may actually drive the expansion of suppressor T cell specificities based on the generation of anti-idiotypic crossreactivities.[214] Although the possible presence of such an idiotypic network has intrigued investigators for many years,[215] exact mechanisms for discrimination of self from non-self in such a system is totally unclear at this time.

The possible importance of suppressor T cell activity and its relationship with helper T cell activity has been mentioned with regard to a variety of autoimmmune disease models. For example, it appears likely that the ability of mice to produce anti-mouse RBC antibodies in response to rat RBC immunization is under the control of $L3T4^-$ $Lyt2^+$ T cells; depending on the state of activation of these suppressive T cells, the expressed mouse antibody repertoire can be shifted in favor of either antibody fine-specificities for non-self rat antigens, or fine-specificities for self mouse antigens.[216,217]

Suppressor cell activity has also been observed in myasthenia gravis systems in which the levels of *in vitro*, helper T cell-dependent stimulation of B cells to produce specific antibody with either antigen (acetylcholine receptor)[218] or mitogen (polk weed)[219] could be modulated depending on the presence of activated $CD8^+$ T cells.

In a rat system of Heymann's nephritis in which injection of renal tubular antigen causes an antibody-mediated membranous glomerulonephritis, it was observed that different rat strains demonstrated different levels of disease severity.[220] It was observed in adoptive transfer experiments that although the antibody response requires helper T cells ($W3/25^+$ cells in the rat), it is also down-regulated by transfer of suppressor T cells ($OX8^+$ cells in the rat). The conclusion reached from these studies must be that both helper and suppressor activities exist, at least in this model of autoimmune disease, and either can be expressed depending on the circumstances; the balance of potential expression of help vs. suppression appears to be tipped by inherited characteristics of the rat strains studied. The suppressive T cells appeared to be short-lived, therefore requiring constant replenishing and the presence of an intact thymus (in contrast to the continued helper T cell activity observed in adult rats following thymectomy); to complicate the picture, the suppressor activity requires an inducer activity derived from the $W3/25^+$ subset, perhaps related to suppressor inducer cells whose lost activity has been implicated in the autoimmunity expressed in lupus patients[221] and multiple sclerosis patients.[222]

Finally, a body of information, coming primarily from studies of SLE patients and lupus-prone mice indicates that disease may be the result of defective immunosuppressive T cells that, when normal, can restrict the production of autoantibodies to specificities with little immunopathological significance. Expression of autoreactivities, and eventually disease symptoms, has been long attributed to the loss of suppressor T cell activity in lupus-prone mice[223-226] as well as in patients with SLE.[221,227] Furthermore, age-associated shifts in class (IgM predominance to IgG predominance) and pathogenicity of anti-DNA antibodies in such models also appear to lead back to changing immunosuppressive activities and resulting changes in the expression of helper activities.[191,228] Various mechanisms can be imagined that result in declining immunoregulatory function that allows usually quiet autoreactivities to be expressed, including inadequate expansion of the suppressor T cell repertoire, inadequate inducer activity,[221] or perhaps a more generalized defect that happens to have great impact on the suppressor cell population such as declining availability of IL-2.[229-231]

V. SUMMARY

The intention of this discussion was to consider the following possibilities: First, that the expression of "forbidden" antibody reactivities against self is frequently a normal and benign occurrence; natural autoantibodies are often of low affinity, IgM in isotype, polyreactive, and are the result of either nonspecific polyclonal B cell activation or antigen-driven activation demonstrating broad crossreactivity. Second, the genetic information necessary for producing natural autoantibodies is encoded in the germline side-by-side with the information for the antibodies against foreign antigens. Third, the potential for producing disease-causing autoantibodies, a potential that is usually buried in the genome and under layers of immunoregulatory safeguards, exists in all individuals. Expression of these harmful auto-reactivities in the form of disease symptoms will occur following some event that selectively expands those B cell clones responsible for the production of antibodies of appropriately high affinity, of the appropriate isotype (often IgG), and of the appropriate binding fine-specificity. This could involve an antigen specific event such as the exposure of the host to a pathogen that contains antigenic determinants that crossreact with an important self antigenic determinant, although this does not exclude mitogen-like stimuli that act to sustain autoreactive B cells initially activated by specific antigen. Genetically determined factors may play a role, such as the mutation of an initially benign VH gene resulting in an autoantibody with higher affinity, or aberrant presentation of self-antigen by products of the MHC. In addition, impaired immunoregulatory networks resulting in increased helper T cell activity or decreased suppressor T cell activity may likely influence the expansion of particular auto-reactivities; these effects may be driven directly by the antigen itself or perhaps indirectly by idiotype-specific mechanisms. Finally, the resulting disease profile will reflect the amount of disease-causing antibody that is in the circulation; the proportion of the total autoantibody that represents the disease-causing subset may be a more meaningful value then the total autoantibody titer (disease-causing plus benign) for purposes of diagnosis and treatment.

REFERENCES

1. **Shoenfeld, Y. and Isenberg, D. A.,** The mosaic of autoimmunity, *Immunol. Today,* 10, 123, 1989.
2. **Strominger, J. L.,** Biology of the human histocompatibility leukocyte antigen (HLA) system and a hypothesis regarding the generation of autoimmune diseases, *J. Clin. Invest.,* 77, 1411, 1986.
3. **Kofler, R., Dixon, F. J., and Theofilopoulos, A. N.,** The genetic origin of autoantibodies, *Immunol. Today,* 8, 375, 1987.
4. **Isenberg, D. A., Shoenfeld, Y., Walport, M., Mackworth-Young, C., Dudeney, C., Todd-Pokropek, A., Brill, S., Weinberger, A., and Pincus, J.,** Detection of crossreactive anti-DNA antibody idiotypes in the serum of systemic lupus erythematosus patients and their relatives, *Arthritis Rheum.,* 28, 999, 1985.
5. **Shoenfeld, Y., Segol, G., Segol, O., Neary, N., Klajman, A., Stollar, B. D., and Isenberg, D. A.,** Detection of antibodies to total histones and their asymptomatic relatives, *Arthritis Rheum.,* 50, 169, 1987.
6. **Barnett, A. H., Eff, C., and Leslie, R. D. G.,** Diabetes in identical twins, *Diabetologia,* 20, 87, 1981.
7. **Block, S. R., Lockshin, M. D., Winfield, J. B., Weksler, M. E., Imamura, M., Winchester, R. J., Mellors, R. C., and Christian, C. L.,** Immunologic observations on nine sets of twins either concordant or discordant for SLE, *Arthritis Rheum.,* 19, 545, 1976.
8. **Propert, D. N.,** *Detection of Immune-Associated Genetic Markers of Human Disease,* Simons, M. and Tait, B., Eds., Livingstone Press, 1984, 70.
9. **Kalka, K., Eskola, J., Granfors, K., Koistinen, J., and Toivanen, A.,** Serum IgA deficiency and anti-IgA antibodies in pernicious anemia, *Clin. Immunol. Immunopathol.,* 46, 55, 1988.
10. **Cassidy, J. T., Burt, A., Petty, R., and Sullivan, D.,** Selective IgA deficiency in connective tissue diseases, *N. Engl. J. Med.,* 280, 275, 1969.

11. **Bluestone, R., Goldberg, L. S., Katz, R. M., Marchesano, J. M., and Calabro, J. J.,** Juvenile rheumatoid arthritis: a serologic survey of 200 consecutive patients, *J. Pediatr.*, 77, 98, 1970.
12. **Yewdall, V., Cameron, J. S., Nathan, A. W., Ogg, C. S., and Williams, D. G.,** Systemic lupus erythematosus and IgA deficiency, *J. Clin. Lab Immunol.*, 10, 13, 1983.
13. **Lachmann, P. J. and Walport, M. J.,** in *Autoimmunity and Autoimmune Disease (Ciba Foundation Symposium 129)* , Evered, D. and Whelan, J., Eds., John Wiley & Sons, New York, 1987, 149.
14. **El-Roeiy, A. and Shoenfeld, Y.,** Autoimmunity and pregnancy, *Am. J. Reprod. Immunol. Microbiol.*, 9, 25, 1985.
15. **Talal, N. and Ahmed, S. A.,** Immunomodulation by hormones: an area of growing importance, *J. Rheumatol.*, 14, 191, 1987.
16. **Ahmed, S. A., Dauphinee, M. J., Montoya, A. I., and Talal, N.,** Estrogen induces normal murine CD5 B cells to produce autoantibodies, *J. Immunol.*, 142, 2647, 1989.
17. **Horowitz, D. S., Borcherding, W., Vishnu Moorthy, A., Cheaney, R., Schulte-Wisserman, H., and Hong, R.,** Induction of suppressor T cells in systemic lupus erythematosus by thymosin and cultured thymic epithelium, *Science,* 197, 999, 1977.
18. **Russel, G. I., Bing, R. F., Jones, A. G., Thurston, H., and Swales, J. D.,** Hydralazine sensitivity: clinical features, autoantibody changes, and HLA-DR phenotype, *Q. J. Med.*, 65, 845, 1987.
19. **Dawkins, R. L., Garlepp, M. J., McDonald, B. L., Williamson, J., Zilko, P. J., and Carrano, J.,** Myasthenia gravis and penicillamine, *J. Rheumatol.*, 7, 169, 1981.
20. **Salama, A. and Mueller-Eckhardt, C.,** On the mechanisms of sensitization and attachment of antibodies to RBC in drug-induced immune hemolytic anemia, *Blood,* 69, 1006, 1987.
21. **Fernandes, G., Friend, P., Yunis, E. J., and Good, R. A.,** Influence of dietary restriction on immunologic function and renal disease in (NZB × NZW)F1 mice, *Proc. Natl. Acad. Sci. U.S.A.*, 75, 1500, 1978.
22. **Morrow, W. J. W., Ohashi, Y., Hallad, J., Pribnow, J., Hirose, S., Shirai, T., and Levy, J. A.,** Dietary fat and immune function. I. Antibody responses, lymphocyte and accessory cell functions in NZB mice, *J. Immunol.*, 135, 3857, 1985.
23. **Shoenfeld, Y. and Cohen, I. R.,** in *The Antigens,* Vol. 7, Sela, M., Ed., Academic Press, New York, 1987, 307.
24. **Oldstone, M. B. A.,** Molecular mimicry and autoimmune disease, *Cell,* 50, 819, 1987.
25. **Srinivasappa, J., Saegusa, J., Prabhakar, B. S., Gentry, M. K., Buchmeier, M. J., Wiktor, T. J., Koprowski, H., Oldstone, M. B., and Notkins, A. L.,** Molecular mimicry: frequency of reactivity of monoclonal antiviral antibodies with normal tissues, *J. Virol.*, 57, 397, 1986.
26. **Schwimmbeck, P. L., Yu, D. T., and Oldstone, M. B.,** Autoantibodies to HLA-B27 in the sera of HLA-B27 patients with ankylosing spondylitis and Reiter's syndrome: molecular mimicry with Klebsiella pneumoniae as potential mechanism of autoimmune disease, *J. Exp. Med.*, 166, 173, 1987.
27. **Shoenfeld, Y.,** A study of anti-polynucleotide antibodies, anti-Klebsiella (K30) antibodies, and anti-DNA antibody idiotypes in ankylosing spondylitis, *Br. J. Rheumatol.*, 26, 168, 1987.
28. **Feldman, M.,** in *Autoimmunity and Autoimmune Disease (Ciba Foundation Symposium 129),* Evered, D. and Whelan, J., Eds., John Wiley & Sons, New York, 1987, 88.
29. **Plotz, P. H.,** Autoantibodies are anti-idiotypic antibodies to anti-viral antibodies, *Lancet,* ii, 824, 1983.
30. **Burnet, F. M.,** *The Clonal Selection Theory of Acquired Immunity,* Cambridge University Press, London, 1959.
31. **Nossal, G. J. V. and Pike, B. L.,** Clonal anergy: persistence in tolerant mice of antigen-binding B lymphocytes incapable of responding to antigen or mitogen, *Proc. Natl. Acad. Sci. U.S.A.*, 77, 1602, 1980.
32. **Nossal, G. J. V.,** Cellular mechanisms of immunologic tolerance, *Annu. Rev. Immunol.*, 1, 33, 1983.
33. **Ternynck, T. and Avrameas, S.,** Murine natural monoclonal autoantibodies: a study of their polyspecificities and their affinities, *Immunol. Rev.*, 94, 99, 1986.
34. **Guilbert, B., Dighiero, G., and Avrameas, S.,** Naturally occurring antibodies against nine common antigens in human sera. I. Detection, isolation and characterization, *J. Immunol.*, 128, 2779, 1982.
35. **Dighiero, G., Lymberi, P., Mazie, J. C., Rouyre, S., Butler-Browne, G. S., Whalen, R. G., and Avrameas, S.,** Murine hybridomas secreting natural monoclonal antibodies reacting with self antigens, *J. Immunol.*, 131, 2267, 1983.
36. **MacKay, I.,** Natural autoantibodies to the fore: forbidden clones to the rear?, *Immunol. Today,* 4, 340, 1983.
37. **Holmberg, D., Forsgren, S., Ivans, F., and Coutinho, A.,** Reactions among IgM antibodies derived from normal neonatal mice, *Eur. J. Immunol.*, 14, 435, 1984.
38. **Dighiero, G., Lymberi, P., Holmberg, D., Lundquist, I., Coutinho, A., and Avrameas, S.,** High frequency of natural autoantibodies in normal new born mice, *J. Immunol.*, 134, 765, 1985.
39. **Holmberg, D. and Coutinho, A.,** Natural antibodies and autoimmunity, *Immunol. Today,* 6, 356, 1985.
40. **Dziarski, R.,** Autoimmunity: polyclonal activation or antigen induction?, *Immunol. Today,* 9, 341, 1988.

41. **Izui, S., McConahey, P. J., and Dixon, F. J.,** Increased spontaneous polyclonal activation of B lymphocytes in mice with spontaneous autoimmune disease, *J. Immunol.,* 121, 2213, 1978.
42. **Dziaski, R.,** Preferential induction of autoantibody secretion in polyclonal activation by peptidoglycan and lipopolysaccharide. I. In vitro studies, *J. Immunol.,* 128, 1018, 1982.
43. **Dziaski, R.,** Preferential induction of autoantibody secretion in polyclonal activation by peptidoglycan and lipopolysaccharide. II. In vivo studies, *J. Immunol.,* 128, 1026, 1982.
44. **Dziaski, R.,** Opposing effects of Xid and Nu mutations on proliferative and polyclonal antibody and autoantibody responses to peptidoglycan, LPS, protein A, and PWM, *Immunology,* 53, 563, 1984.
45. **Dziaski, R.,** Comparison of in vitro and in vivo mitogenic and polyclonal antibody and autoantibody responses to peptidoglycan, LPS, protein A, PWM, PHA, and Con A in normal and autoimmune mice, *J. Clin. Lab Immunol.,* 16, 93, 1985.
46. **Klinman, D. M. and Steinberg, A. D.,** Systemic autoimmune disease arises from polyclonal B cell activation, *J. Exp. Med.,* 165, 1755, 1987.
47. **Ishigatsubo, Y., Steinberg, A. D., and Klinman, D. M.,** Autoantibody production is associated with polyclonal B cell activation in autoimmune mice which express the lpr or gld genes, *Eur. J. Immunol.,* 18, 1089, 1988.
48. **Klinman, D. M., Ishigatsubo, Y., and Steinberg, A. D.,** Acquisition and maturation of expressed B cell repertoires in normal and autoimmune mice, *J. Immunol.,* 141, 801, 1988.
49. **Portanova, J. P., Claman, H. N., and Kotzin, B. L.,** Autoimmunization in murine graft-versus-host disease. I. Selective production of antibodies to histones and DNA, *J. Immunol.,* 135, 3850, 1985.
50. **Fournie, G. F., Lambert, P. H., and Miescher, P. A.,** Release of DNA in circulating blood and induction of anti-DNA antibodies after injection of bacterial lipopolysaccharides, *J. Exp. Med.,* 140, 1109, 1974.
51. **Izui, S., Zaldivar, M., Scher, I., and Lambert, P. H.,** Mechanism for induction of anti-DNA antibodies by bacterial lipopolysaccharides in mice. I. Anti-DNA induction by LPS without significant release of DNA in circulating blood, *J. Immunol.,* 119, 2151, 1977.
52. **Pisetsky, D. S., Klatt, C., Dawson, D., and Roths, J. B.,** The influence of *Yaa* genes on anti-DNA responses in B6-lpr mice, *Clin. Immunol. Immunopathol.,* 37, 369, 1985.
53. **Logtenberg, T., Melissen, P. M. B., Kroon, A., Gmelig-Meyling, F. H. J., and Ballieux, R. E.,** Autoreactive B cells in normal humans: autoantibody production upon lymphocyte stimulation with autoantigen-xenoantigen conjugates, *J. Immunol.,* 140, 446, 1988.
54. **Marion, T. N., Bothwell, A. L., Briles, D. E., and Janeway, C. A.,** IgG anti-DNA autoantibodies within an individual autoimmune mouse are the products of clonal selection, *J. Immunol.,* 142, 4269, 1989.
55. **Bouanani, M., Piechaczyk, M., Pau, B., and Bastide, M.,** Significance of the recognition of certain antigenic regions on the human thyroglobulin molecule by natural autoantibodies from healthy subjects, *J. Immunol.,* 143, 1129, 1989.
56. **Ruf, J., Carayon, P., and Lissitzky, S.,** Various expression of a unique anti-human thyroglobulin antibody repertoire in normal state and autoimmune disease, *Eur. J. Immunol.,* 15, 268, 1985.
57. **Shlomchik, M., Nemazee, D., Van Snick, J., and Weigert, M.,** Variable region sequences of murine IgM anti-IgG monoclonal autoantibodies rheumatoid factors. II. Comparison of hybridomas derived by lipopolysaccharide stimulation and secondary protein immunization, *J. Exp. Med.,* 165, 970, 1987.
58. **Shlomchik, M., Mascelli, M., Shan, H., Radic, M. Z., Pisetsky, D., Marshak-Rothstein, A., and Weigert, M.,** Anti-DNA antibodies from autoimmune mice arise by clonal expansion and somatic mutation, *J. Exp. Med.,* 171, 265, 1990.
59. **Shlomchik, M. J., Marshak-Rothstein, A., Wolfowicz, C. B., Rothstein, T. L., and Weigert, M. G.,** The role of clonal selection and somatic mutation in autoimmunity, *Nature,* 328, 805, 1987.
60. **Sanz, I., Casali, P., Thomas, J. W., Notkins, A. L., and Capra, J. D.,** Nucleotide sequences of eight human natural autoantibody VH regions reveals apparent restricted use of VH families, *J. Immunol.,* 142, 4054, 1989.
61. **Holmberg, D.,** High connectivity, natural antibodies preferentially use 7183 and Q52 VH families, *Eur. J. Immunol.,* 17, 399, 1987.
62. **Sanz, I., Dang, H., Takei, M., Talal, N., and Capra, J. D.,** VH sequence of a human anti-Sm autoantibody: evidence that autoantibodies can be unmutated copies of germline genes, *J. Immunol.,* 142, 883, 1989.
63. **Sanz, I. and Capra, J. D.,** Commentary: the genetic origin of human autoantibodies, *J. Immunol.,* 140, 3283, 1988.
64. **Trepicchio, W., Maruya, A., and Barrett, K. J.,** Heavy chain genes of a lupus anti-DNA autoantibody are encoded in the germline of a non-autoimmune strain of mouse and conserved in strains of mice polymorphic for this gene locus, *J. Immunol.,* 139, 3139, 1987.
65. **Dersimonian, H., Schwartz, R. S., Barrett, K. J., and Stollar, B. D.,** Relationships of human variable region heavy chain germline genes to genes encoding anti-DNA autoantibodies, *J. Immunol.,* 139, 2496, 1987.

66. **Davidson, A., Shefner, R., Livneh, A., and Diamond, B.,** The role of somatic mutation of immunoglobulin genes in autoimmunity, *Annu. Rev. Immunol.*, 5, 85, 1987.
67. **Painter, C., Monestier, M., Bonin, B., and Bona, C. A.,** Functional and molecular studies of V genes expressed in autoantibodies, *Immunol. Rev.*, 94, 75, 1986.
68. **Kofler, R., Noonan, D. J., Levy, D. E., Wilson, M. C., Moller, N. P. H., Dixon, F. J., and Theofilopoulos, A. N.,** Genetic elements used for a murine lupus anti-DNA antibody are closely related to those for antibodies to exogenous antigens, *J. Exp. Med.*, 161, 805, 1985.
69. **Trepicchio, W. and Barrett, K. J.,** The Igh-V locus of MRL mice: restriction fragment length polymorphism in eleven strains of mice as determined with VH and D gene probes, *J. Immunol.*, 134, 2734, 1985.
70. **Kofler, R., Perlmutter, Noonan, D. J., Dixon, F. J., and Theofilopoulos, A. N.,** Immunoglobulin heavy chain variable region gene complex of lupus mice exhibits normal restriction fragment length polymorphism, *J. Exp. Med.*, 162, 346, 1985.
71. **Theofilopoulos, A. N., Kofler, R., Noonan, D. J., Singer, P., and Dixon, F. J.,** Molecular aspects of murine systemic lupus erythematosus, *Springer Semin. Immunopathol.*, 9, 121, 1986.
72. **Monestier, M., Manheimer-Lori, A., Bellon, B., Painter, C., Dang, H., Talal, N., Zanetti, M., Schwartz, R., Pisetsky, D., and Kuppers, R.,** Shared idiotypes and restricted immunoglobulin variable region heavy chain genes characterize murine autoantibodies of various specificities, *J. Clin. Invest.*, 78, 753, 1986.
73. **Trepicchio, W. and Barrett, K. J.,** Eleven MRL-lpr/lpr anti-DNA autoantibodies are encoded by genes from four VH gene families: a potentially biased usage of VH genes, *J. Immunol.*, 138, 2323, 1987.
74. **Bellon, B., Manheimer-Lory, A., Monestier, M., Moran, T., Dimitriu-Bona, A., Alt, F., and Bona, C.,** High frequency of autoantibodies bearing crossreactive idiotopes among hybridomas using VH7183 genes prepared from normal and autoimmune murine strains, *J. Clin. Invest.*, 79, 1044, 1987.
75. **Bona, C. A.,** V genes encoding autoantibodies: molecular and phenotypic characteristics, *Annu. Rev. Immunol.*, 6, 327, 1988.
76. **Komisar, J. L., Leung, K. Y., Crawley, R. R., Talal, N., and Teale, J. M.,** Ig VH gene family repertoire of plasma cells derived from lupus-prone MRL/lpr mice, *J. Immunol.*, 143, 340, 1989.
77. **Monestier, M., Bonin, B., Migliorine, P., Dang, H., Datta, S., Datta, H., Kuppers, R., Rose N., Maurer, P., Talal, N., and Bona, C.,** Autoantibodies of various specificities encoded by genes from the VH J558 family bind to foreign antigens and share idiotopes of antibodies specific for self and foreign antigens, *J. Exp. Med.*, 166, 1109, 1987.
78. **Shoenfeld, Y., Ben-Yehuda, O., Naparstek, Y., Wilner, Y., Frolichman, R., Schattner, A., Lavie, G., Joshua, H., Pinkhas, J., Komeck, R. C., Schwartz, R. S., and Pick, A. I.,** The detection of a common idiotype of anti-DNA antibodies in the sera of patients with monoclonal gammopathies, *J. Clin. Immunol.*, 6, 194, 1986.
79. **Mayus, J. L. and Pisetsky, D. S.,** Expression of a highly conserved anti-DNA idiotype in normal and autoimmune mice, *Clin. Immunol. Immunopathol.*, 34, 366, 1985.
80. **Lymberi, P., Dighiero, G., Ternynck, T., and Avrameas, S.,** A high incidence of crossreactive idiotypes among murine natural autoantibodies, *Eur. J. Immunol.*, 15, 702, 1985.
81. **Portnoi, D., Freitas, A., Holmberg, D., Bandeira, A., and Coutinho, A.,** Immunocompetent autoreactive B lymphocytes are activated cycling cells in normal mice, *J. Exp. Med.*, 164, 25, 1986.
82. **Holmberg, D., Freitas, A. A., Portnoi, D., Jacquemart, F., Avrameas, S., and Coutinho, A.,** Antibody repertoires of normal BALB/c mice: B lymphocyte populations defined by state of activation, *Immunol. Rev.*, 93, 17, 1986.
83. **Portnoi, D., Lundkvist, I., and Coutinho, A.,** Inverse correlation between the utilization of an idiotype in specific immune responses and its representation in pre-immune "natural" antibodies, *Eur. J. Immunol.*, 18, 571, 1988.
84. **Holmberg, D., Wennerstrom, G., Andrade, L., and Coutinho, A.,** The high idiotypic connectivity of "natural" newborn antibodies is not found in adult mitogen-reactive B cell repertoires, *Eur. J. Immunol.*, 16, 82, 1986.
85. **Herzenberg, L. A., Stall, A. M., Lalor, P. A., Sidman, C., Moore, W. A., Parks, D. R., and Herzenberg, L. A.,** The Ly-1 B cell lineage, *Immunol. Rev.*, 93, 81, 1986.
86. **Hardy, R. R. and Hayakawa, K.,** Development and physiology of Ly-1 B and its human homolog, Leu-1 B, *Immunol. Rev.*, 93, 53, 1986.
87. **Gadol, N. and Ault, K. A.,** Phenotypic and functional characterization of human Leu1 (CD5) cells, *Immunol. Rev.*, 93, 23, 1986.
88. **Kipps, T. J. and Vaughan, J. H.,** Genetic influence on the levels of circulating CD5 B lymphocytes, *J. Immunol.*, 139, 1060, 1987.
89. **Kipps, T. J.,** The CD5 B cell, *Adv. Immunol.*, 47, 117, 1989.

90. **Casali, P. and Notkins, A. L.,** CD5+ B lymphocytes, polyreactive antibodies, and the human B cell repertoire, *Immunol. Today,* 10, 364, 1989.
91. **Antin, J. H., Emerson, S. G., Marin, P., Gadol, N., and Ault, K. A.,** Leu1 + (CD5 +) B cells. A major lymphoid subpopulation in fetal spleen: phenotypic and functional studies, *J. Immunol.,* 136, 505, 1986.
92. **Casali, P., Burastero, S. E., Nakamura, M., Inghirami, G., and Notkins, A. L.,** Human lymphocytes making rheumatoid factor and antibody to belong to Leu + B cell subset, *Science,* 236, 77, 1987.
93. **Casali, P. and Notkins, A. L.,** Probing the human B cell repertoire with EBV: polyreactive antibodies and CD5+ B lymphocytes, *Annu. Rev. Immunol.,* 7, 513, 1989.
94. **Pennell, C. A., Arnold, L. W., Haughton, G., and Clarke, S. H.,** Restricted Ig variable gene expression among Ly-1 + B cell lymphomas, *J. Immunol.,* 141, 1788, 1988.
95. **Hayakawa, K. and Hardy, R. R.,** Normal, autoimmune, and malignant CD5+ B cells: the Ly-1 B lineage?, *Annu. Rev. Immunol.,* 6, 197, 1988.
96. **Foerster, I., Gu, H., and Rajewsky, K.,** Germline antibody V regions as determinants of clonal persistence and malignant growth in the B cell compartment, *EMBO J.,* 7, 3693, 1988.
97. **Shlomchik, M. J., Aucoin, A. H., Pisetsky, D. S., and Weigert, M. G.,** Structure and function of anti-DNA autoantibodies derived from a single autoimmune mouse, *Proc. Natl. Acad. Sci. U.S.A.,* 84, 9150, 1987.
98. **Hardin, J. H.,** The lupus autoantigens and the pathogenesis of systemic lupus erythematosus, *Arthritis Rheum.,* 29, 457, 1986.
99. **Tan, E. M., Chan, E. K. L., Sullivan, K. F., and Rubin, R. L.,** Anti-nuclear antibodies: diagnostically specific immune markers and clues toward the understanding of systemic autoimmunity, *Clin. Immunol. Immunopathol.,* 47, 121, 1988.
100. **Tarlington, D., Stall, A. M., and Herzenberg, L. A.,** Repetitive usage of immunoglobulin VH and D gene segments in CD5 B cell clones of (NZB × NZW)F1 mice, *EMBO J.,* 7, 3705, 1988.
101. **Klinman, D. M., Banks, S., Hartman, A., and Steinberg, A. D.,** Natural murine autoantibodies and conventional antibodies exhibit similar degrees of antigenic crossreactivity, *J. Clin. Invest.,* 82, 652, 1988.
102. **Matthes, T., Wolff, A., Soubiran, P., Gros, F., and Dighiero, G.,** 108 Antitubulin antibodies. II. Natural autoantibodies and induced antibodies recognize different epitopes on the tubulin molecule, *J. Immunol.,* 141, 3135, 1988.
103. **Matthes, T. and Dighiero, G.,** Detection of private and recurrent idiotopes on natural anti-tubulin antibodies by monoclonal anti-idiotopic antibodies, *J. Immunol.,* 140, 148, 1988.
104. **Datta, S. K., Stollar, B. D., and Schwartz, R. S.,** Normal mice express idiotypes related to autoantibody idiotypes of lupus mice, *Proc. Natl. Acad. Sci. U.S.A.,* 80, 2723, 1983.
105. **Rauch, J., Murphy, E., Roths, J. B., Stollar, B. D., and Schwartz, R. S.,** A high frequency of idiotype marker of anti-DNA autoantibodies in MRL-lpr/lpr mice, *J. Immunol.,* 129, 236, 1982.
106. **Marion, T. N., Lawton, A. R., Kearney, J. F., and Briles, D. E.,** Anti-DNA autoantibodies in (NZB × NZW)F1 mice are clonally heterogeneous but the majority share a common idiotope, *J. Immunol.,* 128, 668, 1982.
107. **Hahn, B. H. and Ebling, F. M.,** Idiotype restriction in murine lupus: high frequency of three public idiotypes on serum IgG in nephritic NZB mice, *J. Immunol.,* 138, 2110, 1987.
108. **Cowdery, J. S., Jacobi, S. M., Pitts, A. K., and Tyler, T. L.,** Defective B cell clonal regulation and autoantibody production in New Zealand Black mice, *J. Immunol.,* 138, 760, 1987.
109. **Lindstrom, J., Shelton, D., and Fujii, Y.,** Myasthenia gravis, *Adv. Immunol.,* 42, 233, 1988.
110. **Lindstrom, J. M., Lennon, V. A., Seybold, M. E., and Whittingham, S.,** Experimental autoimmune myasthenia gravis and myasthenia gravis: biochemical and immunochemical aspects, *Ann. N.Y. Acad. Sci.,* 274, 254, 1976.
111. **Bartfeld, D. and Fuchs, S.,** Fractionation of antibodies to acetylcholine receptor according to antigenic specificity, *FEBS Lett.,* 105, 303, 1979.
112. **Yeh, T. and Krolick, K. A.,** Clonotypic analysis of anti-acetylcholine receptor antibodies produced against native and denatured antigen, *J. Neuroimmunol.,* 24, 133, 1989.
113. **Plater-Zyberk, C., Maini, R. N., Lam, K., Kennedy, T. D., and Janossy, G.,** A rheumatoid arthritis B cell subset expresses a phenotype similar to that in chronic lymphocytic leukemia, *Arthritis Rheum.,* 28, 971, 1985.
114. **Burastero, S. E., Casali, P., Wilder, R. L., and Notkins, A. L.,** Monoreactive high affinity and polyreactive low affinity rheumatoid factors are produced by CD5 B cells from patients with rheumatoid arthritis, *J. Exp. Med.,* 168, 1979, 1988.
115. **Nakamura, M., Burastero, S. E., Notkins, A. L., and Casali, P.,** Human monoclonal rheumatoid factor-like antibodies from CD5 (Leu-1) B cells are polyreactive, *J. Immunol.,* 140, 4180, 1988.
116. **Hardy, R. R., Hayakawa, K., Shimizu, M., Yamasaki, K., and Kishimoto, T.,** Rheumatoid factor secretion from human Leu-1 B cells, *Science,* 236, 81, 1987.

117. **Nardella, F. A., Teller, D. C., Barber, C. V., and Mannik, M.,** IgG rheumatoid factors and staphylococcal protein A bind to a common molecular site on IgG, *J. Exp. Med.*, 162, 1811, 1985.

118. **Oppliger, I. R., Nardella, F. A., Stone, G. C., and Mannik, M.,** Human rheumatoid factors bear the internal image of the Fc binding region of staphylococcal protein A, *J. Exp. Med.*, 166, 702, 1987.

119. **Weisbart, R. H., Noritake, D. T., Colburn, K. K., and Saxton, R. E.,** Human monoclonal IgG rheumatoid factor has structural homology with bacterial Fc receptor proteins, *J. Immunol.*, 139, 2925, 1987.

120. **Nardella, F. A., Schroder, A. K., Svensson, M., Sjoquist, J., Barber, C., and Christensen, P.,** T15 group A streptococcal Fc receptor binds to the same location on IgG as staphylococcal protein A and IgG rheumatoid factors, *J. Immunol.*, 138, 922, 1987.

121. **Nelson, J. L., Nardella, F. A., and Mannik, M.,** Competition between antigen and anti-idiotypes for rheumatoid factors, *J. Immunol.*, 135, 2357, 1985.

122. **Goldfien, R. D., Chen, P. P., Fong, S., and Carson, D. A.,** Synthetic peptides corresponding to third hypervariable region of human monoclonal IgM rheumatoid factor heavy chains define an immunodominant idiotype, *J. Exp. Med.*, 162, 756, 1985.

123. **Rauch, J., Massicotte, H., and Tannenbaum, H.,** Specific and shared idiotypes found on hybridoma anti-DNA autoantibodies derived from rheumatoid arthritis and systemic lupus erythematosus patients, *J. Immunol.*, 135, 2385, 1985.

124. **Casali, P., Inghirami, G., Nakamura, M., Davies, T. F., and Notkins, A. L.,** Human monoclonals from antigen-specific selection of B lymphocytes and transformed by EBV, *Science*, 234, 476, 1986.

125. **Rauch, J., Tannenbaum, H., Straton, K., Massite, H., and Wild, J.,** Human-human hybridoma autoantibodies with both anti-DNA and rheumatoid factor activities, *J. Clin. Invest.*, 77, 106, 1986.

126. **Darwin, B. S., Grudier, J. P., Klatt, C. L., and Pisetsky, D. S.,** IgG antinuclear antibodies with crossreactive rheumatoid factor activity, *J. Immunol.*, 137, 3796, 1986.

127. **Shoenfeld, Y., Rauch, J., Massicotte, H., Datta, S. K., Andre-Schwartz, J., Stollar, B. D., and Schwartz, R. S.,** Polyspecificity of monoclonal lupus autoantibodies produced by human-human hybridomas, *N. Engl. J. Med.*, 308, 414, 1983.

128. **Karounos, D. G., Grudier, J. P., and Pisetsky, D. S.,** Spontaneous expression of antibodies to DNA of various species origin in sera of normal subjects and patients with systemic lupus erythematosus, *J. Immunol.*, 140, 451, 1988.

129. **Essani, K., Satoh, J., Prabhakar, B. S., McClintock, P. R., and Notkins, A. L.,** Anti-idiotypic antibodies against a human multiple organ-reactive autoantibody: detection of idiotopes in normal individuals and patients with autoimmune diseases, *J. Clin. Invest.*, 76, 1649, 1985.

130. **Uchigata, Y., Prabhakar, B. S., Salata, K. F., Ginsberg-Fellner, F., and Notkins, A. L.,** Human monoclonal multiple organ-reactive autoantibodies distinguished by mouse monoclonal anti-idiotypic antibodies: expression of idiotopes in humans with and without autoimmune diseases, *J. Immunol.*, 138, 4218, 1987.

131. **Isenberg, D. A., Williams, W., Le Page, S., Swana, G., Feldman, R., Addison, I., Bakimer, R., and Shoenfeld, Y.,** A comparison of autoantibodies and common DNA antibody idiotypes in SLE patients and their spouses, *Br. J. Rheumatol.*, 27, 431, 1988.

132. **Isenberg, D. A. and Collins, C.,** Detection of crossreactive anti-DNA antibody idiotypes on renal tissue-bound immunoglobulins from lupus patients, *J. Clin. Invest.*, 76, 287, 1985.

133. **Eliat, D., Fishel, R., and Zlotnick, A.,** A central anti-DNA idiotype in human and murine systemic lupus erythematosus, *Eur. J. Immunol.*, 15, 368, 1985.

134. **Dersimonian, H., McAdam, K. P., Mackworth-Young, C., and Stollar, B. D.,** The recurrent expression of variable region segments in human IgM anti-DNA autoantibodies, *J. Immunol.*, 142, 4027, 1989.

135. **Halpern, R., Davidson, A., Lazo, A., Solomon, G., Lahita, R., and Diamond, B.,** Familial systemic lupus erythematosus: presence of a crossreactive idiotype in healthy family members, *J. Clin. Invest.*, 76, 731, 1985.

136. **Davidson, A., Preud'homme, J. L., Solomon, A., Chang, M. D., Beede, S., and Diamond, B.,** Idiotypic analysis of myeloma proteins: anti-DNA activity of monoclonal immmunoglobulins bearing an SLE idiotype is more common in IgG than IgM antibodies, *J. Immunol.*, 138, 1515, 1987.

137. **Cohen, I. R. and Cooke, A.,** Natural antibodies might prevent autoimmune disease, *Immunol. Today*, 7, 363, 1986.

138. **Kroemer, G. and Wick, G.,** The role of interleukin-2 in autoimmunity, *Immunol. Today*, 10, 246, 1989.

139. **Malkovsky, M. and Medawar, P. B.,** Is immunological tolerance a consequence of interleukin-2 deficit during the recognition of antigen?, *Immunol. Today*, 5, 340, 1984.

140. **DeBoer, R. J. and Hogezeg, P.,** Immunological discrimination between self and non-self by precursor depletion and memory accumulation, *J. Theoret. Biol.*, 124, 343, 1987.

141. **Lehtonen, L., Vainio, O., and Toivanen, P.,** Mechanisms of transplantation tolerance: impairment of tolerance by T cell growth factor, *Transplantation*, 42, 184, 1986.

142. **Malkovsky, M., Medawar, P. B., Thatcher, D. R., Toy, J., Hunt, R., Rayfield, L. S., and Dorie, C.,** Acquired immunological tolerance of foreign cells is impaired by recombinant interleukin-2 or vitamin A acetate, *Proc. Natl. Acad. Sci. U.S.A.*, 82, 536, 1985.
143. **Altman, A., Theofilopoulos, A., Weiner, R., Katz, D. H., and Dixon, F. J.,** Analysis of T cell function in autoimmune murine strains: defects in production and response to interleukin-2, *J. Exp. Med.*, 154, 791, 1981.
144. **Elder, M. E. and MacLaren, N. K.,** Identification of profound peripheral T cell immunodeficiency in the spontaneous diabetic BB rat, *J. Immunol.*, 130, 1723, 1983.
145. **Kuntz, M. M., Innes, J. B., and Weksler, M. E.,** The cellular basis of the impaired autologous mixed lymphocyte reactions in patients with systemic lupus erythematosus, *J. Clin. Invest.*, 63, 151, 1979.
146. **Forre, O., Egemand, T., Doblong, J. H., Kvien, T. K., and Natvig, J. B.,** Autologous mixed lymphocyte reactions in patients with rheumatoid arthritis and juvenile rheumatoid arthritis: both non-T cells and in vivo activated T cells can act as stimulator cells, *Scand. J. Immunol.*, 16, 173, 1982.
147. **Leventhal, B. G., Waldorf, D. S., and Talal, N.,** Impaired lymphocyte transformation and delayed hypersensitivity in Sjogren's syndrome, *J. Clin. Invest.*, 46, 1338, 1967.
148. **Merrill, J. E., Mohlstrom, C., Wittenbogaart, C., and Myers, L. W.,** Response to and production of interleukin-2 by peripheral blood and cerebrospinal fluid lymphocytes of patients with multiple sclerosis, *J. Immunol.*, 133, 1931, 1984.
149. **Huang, Y. P., Perrin, L. H., Miescher, P. A., and Zubler, R. H.,** Correlation of T and B cell activities in vitro and serum IL-2 levels in systemic lupus erythematosus, *J. Immunol.*, 141, 827, 1988.
150. **Lehmann, K. R., Kotzin, B. L., Portanova, J. P., and Santoro, T. J.,** Interleukin-2 is a proliferative signal for B cells from autoimmune mice, *Eur. J. Immunol.*, 16, 1105, 1986.
151. **Rosenberg, Y. J., Steinberg, A. D., and Santoro, T. J.,** T cells from "IL-2 defective" MRL-lpr/lpr mice continue to grow in vitro and produce IL-2 constituitively, *J. Immunol.*, 133, 2545, 1984.
152. **Weston, K. M., Yeh, E. T. H., and Sy, M. S.,** Autoreactivity accelerates the development of autoreactivity and lymphoproliferation in MRL-lpr/lpr, *J. Immunol.*, 139, 734, 1987.
153. **Linker-Israeli, M., Bakke, A. C., Kitridou, R. C., Gendler, S., Gillis, S., and Horowitz, D. A.,** Defective production of interleukin-1 and interleukin-2 in patients with systemic lupus erythematosus, *J. Immunol.*, 130, 2651, 1983.
154. **Ishida, H., Kumagai, S., Umehara, H., Sano, H., Tagaya, Y., Yodoi, J., and Imura, H.,** Impaired expression of high affinity interleukin-2 receptor on activated lymphocytes from patients with systemic lupus erythematosus, *J. Immunol.*, 139, 1070, 1987.
155. **Alcocer-Varela, J. and Alarcan-Segovea, D.,** Decreased production of and response to interleukin-2 by cultured lymphocytes from patients with systemic lupus erythematosus, *J. Clin. Invest.*, 69, 1388, 1982.
156. **Rosenberg, Y. J., Nurse, F., and Begley, C. G.,** IL-2 receptor expression in autoimmune MRL-Ipr/lpr mice: the expanded L3T4$^-$, Lyt2$^-$ population does not express p75 and cannot generate functional high-affinity IL-2 receptors, *J. Immunol.*, 143, 2216, 1989.
157. **Murakawa, Y., Takada, S., Ueda, Y., Suzuki, N., Hoshino, T., and Sakane, T.,** Characterization of T lymphocyte subpopulations responsible for deficient interleukin-2 activity in patients with systemic lupus erythematosus, *J. Immmunol.*, 134, 187, 1985.
158. **Balderas, R. S., Josimovic-Alasevic, O., Diamantstein, T., Dixon, F. J., and Theofilopoulos, A. N.,** Elevated titers of cell-free interleukin-2 receptor in serum of lupus mice, *J. Immunol.*, 139, 1496, 1987.
159. **Wolf, R. E. and Brelsford, W. G.,** Soluble interleukin-2 receptors in systemic lupus erythematosus, *Arthritis Rheum.*, 31, 729, 1988.
160. **Via, C. S. and Shearer, G. M.,** T cell interactions in autoimmunity: insights from a murine model of graft-versus-host disease, *Immunol. Today*, 9, 207, 1988.
161. **Huang, Y. P., Miescher, P. A., and Zubler, R. H.,** The interleukin-2 secretion defect in vitro in systemic lupus erythematosus is reversible in rested cultured T cells, *J. Immunol.*, 137, 3515, 1986.
162. **Santoro, T. J., Lehmann, K. R., Batt, R. A., and Kotzin, B. L.,** The role of L3T4 cells in the pathogenesis of lupus in lpr-bearing mice. I. Defects in the production of interleukins 2 and 3, *Eur. J. Immunol.*, 17, 1131, 1987.
163. **Herron, L. R., Coffman, R. L., Bond, M. W., and Kotzin, B. L.,** Increased autoantibody production by NZB/NZW B cells in response to IL-5, *J. Immunol.*, 141, 842, 1988.
164. **Kitas, G. D., Salmon, M., Farr, M., Gaston, J. S. H., and Bacon, P. A.,** Deficient interleukin-2 production in rheumatoid arthritis: association with active disease and systemic complications, *Clin. Exp. Immunol.*, 73, 242, 1988.
165. **Symons, J. A., Wood, N. C., Di Giovine, F. S., and Duff, G. W.,** Soluble IL-2 receptor in rheumatoid arthritis: correlation with disease activity, IL-1, and IL-2 inhibition, *J. Immunol.*, 141, 2612, 1988.
166. **Keystone, E. C., Snow, K. M., Bombardier, C., Chang, C. H., Nelson, D. L., and Rubin, L. A.,** Elevated soluble interleukin-2 receptor levels in the sera and synovial fluids of patient with rheumatoid arthritis, *Arthritis Rheum.*, 31, 844, 1988.

167. **Buchan, G., Barrett, K., Fujita, T., Taniguchi, T., Maini, R., and Feldman, M.,** Detection of activated T cell products in the rheumatoid joint using cDNA probes to interleukin-2, IL-2 receptor, and IFN-gamma, *Clin. Exp. Immunol.*, 71, 295, 1988.
168. **Eastgate, J. A., Symons, J. A., Wood, N. C., Grinlint, F. M., di Giovine, F. S., and Duff, G. W.,** Correlation of plasma interleukin 1 levels with disease activity in rheumatoid arthritis, *Lancet*, 2(8613), 706, 1988.
169. **Hirano, T., Matsuda, T., Turner, M., Miyasaka, N., Buchan, G., Tang, B., Sato, K., Shimzu, M., Maini, R., and Feldmann, M.,** Excessive production of interleukin 6 (B cell stimulatory factor-2) in rheumatoid arthritis, *Eur. J. Immunol.*, 18, 1797, 1988.
170. **Selmaj, K., Nowak, Z., and Tchorzewsky, H.,** Interleukin-1 and interleukin-2 production by peripheral blood mononuclear cells in multiple sclerosis patients, *J. Neurol. Sci.*, 85, 67, 1988.
171. **Trotter, J. L., Clifford, D. B., Anderson, C. B., van der Veen, R. C., Hicks, B. C., and Banks, G.,** Elevated serum interleukin-2 levels in chronic progressive multiple sclerosis, *N. Engl. J. Med.*, 318, 1206, 1988.
172. **Umehara, H., Kumagai, S., Ishida, H., Suginoshita, T., Maeda, M., and Imura, H.,** Enhanced production of interleukin-2 in patients with progressive systemic sclerosis: hyperactivity of CD4-positive T cells?, *Arthritis Rheum.*, 31, 401, 1988.
173. **Defreitas, A. C., Sandberg-Wollheim, M., Schonely, K., Boufal, M., and Koprowski, H.,** Regulation of interleukin-2 receptors on T cells from multiple sclerosis patients, *Proc. Natl. Acad. Sci. U.S.A.*, 83, 2637, 1986.
174. **Selmaj, K., Plater-Zyberk, C., Rockett, K. A., Maini, R. N., Alam, R., Perkin, G. D., and Rose, F. C.,** Multiple sclerosis: increased expression of interleukin-2 receptors on lymphocytes, *Neurology*, 36, 1392, 1986.
175. **Selmaj, K. W. and Raine, C. S.,** Tumor necrosis factor mediates myelin and oligodendrocyte damage in vitro, *Ann. Neurol.*, 23, 339, 1988.
176. **Brosnan, C. F., Selmaj, K., and Raine, C. S.,** Hypothesis: a role for tumor necrosis factor in immune-mediated demyelination and its relevance to multiple sclerosis, *J. Neuroimmunol.*, 18, 87, 1988.
177. **Schauenstein, K., Kroemer, G., Sundick, R. S., and Wick, G.,** Enhanced response to Con A and production of TCGF by lymphocytes of Obese strain (OS) chickens with autoimmune thyroiditis, *J. Immunol.*, 134, 872, 1985.
178. **Kroemer, G., Schauenstein, K., Dietrich, H., Fassler, R., and Wick, G.,** Mechanisms of T cell hyperactivity in Obese strain (OS) chickens with spontaneous autoimmune thyroiditis, *J. Immunol.*, 138, 2104, 1987.
179. **Hala, K., Schauenstein, K., Neu, N., Kromer, G., Wolf, H., Bock, G., and Wick, G.,** A monoclonal antibody reacting with a membrane determinant expressed on activated chicken lymphocytes, *Eur. J. Immunol.*, 16, 1331, 1986.
180. **Wick, G., Most, J., Schauenstein, K., Kromer, G., Dietrich, H., Ziemiechi, A., Fassler, R., Schwarz, S., Neu, N., and Hala, K.,** Spontaneous autoimmune thyroiditis: a bird's eye view, *Immunol. Today*, 6, 359, 1985.
181. **Kroemer, G., Faessler, R., Hala, G., Boeck, G., Schauenstein, K., Brezinschek, H. P., Neu, N., Dietrich, H., Jacober, R., and Wick, G.,** Genetic analysis of extrathyroidal features of Obese strain (OS) chickens with spontaneous autoimmune thyroiditis, *Eur. J. Immunol.*, 18, 1499, 1988.
182. **Huetz, F., Sciard-Larsson, E., Pereira, P., Portnoi, D., and Coutinho, A.,** T cell dependence of the "natural" autoreactive B cell activation in the spleen of normal mice, *Eur. J. Immunol.*, 18, 1615, 1988.
183. **Huetz, F., Poncet, P., Coutinho, A., and Portnoi, D.,** Ontogenic development of autoantibody repertoires in spleen and peritoneal cavity of normal mice: examples of T cell-dependent and -independent reactivities, *Eur. J. Immunol.*, 19, 1195, 1989.
184. **Logtenberg, T., Kroon, A., Gmelig-Meyling, F., and Ballieux, R. E.,** Antigen-specific activation of autoreactive B cells in normal human individuals, *Eur. J. Immunol.*, 16, 1497, 1986.
185. **Cairns, E., Germain, J. S., and Bell, D. A.,** The in vitro production of anti-DNA antibody by cultured peripheral blood or tonsillar lymphoid cells from normal donors and SLE patients, *J. Immunol.*, 135, 3839, 1985.
186. **Santoro, T. J., Portanova, J. P., and Kotzin, B. L.,** The contribution of L3T4 T cells to lymphoproliferation and autoantibody production in MRL-1pr/1pr mice, *J. Exp. Med.*, 167, 1713, 1988.
187. **Sekigawa, I., Ishida, Y., Hirose, S., Sato, H., and Shirai, T.,** Cellular basis of in vitro anti-DNA antibody production: evidence for T cell dependence of IgG-class anti-DNA antibody synthesis in the (NZB × NZW)F1 hybrid, *J. Immunol.*, 136, 1247, 1986.
188. **Wofsy, D. and Seaman, W. E.,** Reversal of advanced murine lupus in NZB/NZW mice by treatment with monoclonal antibody to L3T4, *J. Immunol.*, 138, 3247, 1987.
189. **Laskin, C. A., Haddad, G., and Soloninka, C. A.,** The regulatory role of NZB T lymphocytes in the production of anti-DNA antibodies in vitro, *J. Immunol.*, 137, 1867, 1986.

190. **Wofsy, D. and Seaman, W. E.,** Successful treatment of autoimmunity in NZB/NZW mice with monoclonal antibody to L3T4, *J. Exp. Med.*, 161, 378, 1985.
191. **Sekigawa, I., Okada, T., Noguchi, K., Ueda, G., Hirose, S., Sato, H., and Shirai, T.,** Class-specific regulation of anti-DNA antibody synthesis and the age-associated changes in (NZB × NZW)F1 hybrid mice, *J. Immunol.*, 138, 2890, 1987.
192. **Moore, J. S. and Hoover, R. G.,** Defective isotype-specific regulation of IgA anti-erythrocyte autoantibody-forming cells in NZB mice, *J. Immunol.*, 142, 4282, 1989.
193. **Ando, D. G., Sercarz, E. E., and Hahn, B. H.,** Mechanisms of T and B cell collaboration in the in vitro production of anti-DNA antibodies in the NZB/NZW F1 murine SLE model, *J. Immunol.*, 138, 3185, 1987.
194. **Hahn, B. H., Ando, D., Ebling, F. M., Panosian-Sahakian, N., Tsao, B., and Kalunian, K. C.,** T cell up-regulation of B cells via their idiotypes contributing to the development of systemic lupus erythematosus: a hypothesis, *Am. J. Med.*, 85, 32, 1988.
195. **Hahn, B. H. and Ebling, F. M.,** A public idiotypic determinant is present on spontaneous cationic IgG antibodies to DNA from mice of unrelated lupus-prone strains, *J. Immunol.*, 133, 3015, 1984.
196. **Saito, T. and Rajewsky, K.,** Helper T cells reacting to idiotype on IgG but not IgM, *J. Exp. Med.*, 162, 1399, 1985.
197. **Hohlfeld, R., Kalies, I., Kohleisen, B., Heinir, K., Conti-Tronconi, B., and Toyka, K. V.,** Myasthenia gravis: stimulation of anti-receptor autoantibodies by autoreactive T cell lines, *Neurology,* 36, 618, 1986.
198. **Hohlfeld, R., Kalies, I., Ernst, M., Keletsen, U., and Wekerle, H.,** T-lymphocytes in experimental autoimmune myasthenia gravis. Isolation of T-helper cell lines, *J. Neurol. Sci.*, 57, 265, 1982.
199. **Hohlfeld, R., Toyka, K. V., Tzartos, S. J., Carson, W. and Conti-Tronconi, B. M.,** Human T helper lymphocytes in myasthenia gravis recognize the nicotinic receptor alpha subunit, *Proc. Natl. Acad. Sci. U.S.A.*, 84, 5379, 1987.
200. **Krolick, K. A. and Urso, O. E.,** Influence of T cell specificity on the antibody response to the acetylcholine receptor, *J. Neuroimmunol.*, 13, 75, 1986.
201. **Krolick, K. A. Urso, O. E.,** Analysis of helper T cell function by acetylcholine receptor-reactive cell lines of defined AChR-subunit specificity, *Cell. Immunol.*, 105, 75, 1987.
202. **Yeh, T. and Krolick, K. A.,** Influence of T cell specificity on the heterogeneity and disease-causing capability of antibody against the acetylcholine receptor, *J. Neuroimmunol.*, 17, 17, 1987.
203. **Fujii, Y. and Lindstrom, J.,** Regulation of antibody production by helper T cell clones in experimental autoimmune myasthenia gravis, *J. Immunol.*, 141, 3361, 1988.
204. **Lennon, V. A., McCormick, D. J., Lambert, E. H., Griesmann, G. E., and Atassi, M. Z.,** Region of peptide 125-147 of acetylcholine receptor alpha subunit is exposed at neuromuscular junction and induces experimental autoimmune myasthenia gravis, T cell immunity, and modulating autoantibodies, *Proc. Natl. Acad. Sci. U.S.A.*, 82, 8805, 1985.
205. **Lindstrom, J. L., Seybold, M. E., Lennon, V. A., Whittingham, S., and Duane, D. D.,** Antibody to acetylcholine receptor in myasthenia gravis: prevalence, clinical correlates, and diagnostic value. *Neurology,* 26, 1054, 1976.
206. **Drachman, D. B., Adams, R. N., Josifek, L. F., and Self, S. G.,** Functional activities of autoantibodies to acetylcholine receptors and the clinical severity of myasthenia gravis, *N. Engl. J. Med.*, 307, 769, 1982.
207. **Hohlfeld, R., Sterz, R., Kalies, I., Wekerle, H., and Peper, K.,** Experimental myasthenia: lack of correlation between the autoantibody titer and the reduction of acetylcholine-controlled ionic channels measured at functioning endplates, *Muscle Nerve,* 6, 160, 1983.
208. **Fujii, Y. and Lindstrom, J.,** Specificity of the T cell immune response to acetylcholine receptor in experimental autoimmune myasthenia gravis: response to subunits and synthetic peptides, *J. Immunol.*, 140, 1830, 1988.
209. **Yeh, T.-M. and Krolick, K. A.,** T cells reactive with a small synthetic peptide of the acetylcholine receptor can provide help for a clonotypically heterogeneous antibody response and subsequently impaired muscle function, *J. Immunol.*, in press.
210. **Zhang, Y., Barkas, T., Juillerat, M., Schwendimann, B., and Wekerle, H.,** T cell epitopes in experimental autoimmune myasthenia gravis of the rat: strain-specific epitopes and cross-reaction between two distinct segments of the a chain of the nicotinic acetylcholine receptor *(Torpedo californica), Eur. J. Immunol.*, 18, 551, 1988.
211. **Moller, G.,** Do suppressor cells exist?, *Scand. J. Immunol.*, 27, 247, 1988.
212. **Tada, T.,** But still it moves! An answer to Professor Goran Moller, *Scand. J. Immunol.*, 27, 623, 1988.
213. **Dugunlewicz, M., Sunshine, G. H., and Borel, Y.,** Role of the thymus in natural tolerance to an autologous protein antigen, *J. Exp. Med.*, 169, 285, 1989.
214. **Hayglass, K. T., Benacerraf, B., and Sy, M.-S.,** The influence of B cell idiotypes on the repertoire of suppressor T cells, *Immunol. Today,* 7, 179, 1986.
215. **Jerne, N. K.,** Toward a network theory of the immune system, *Ann. Immunol.*, 1256, 373, 1974.

216. **Hooper, D. C.,** Self-tolerance for erythrocytes is not maintained by clonal deletion of T helper cells. *Immunol. Today,* 8, 327, 1987.
217. **Walker, K. Z., Gibson, J., and Basten, A.,** Clonal analysis of autoreactive B cells in a murine model of autoimmune hemolytic anemia, *Cell. Immunol.,* 107, 227, 1987.
218. **McIntosh, K. R. and Drachman, D. B.,** Induction of suppressor cells specific for AChR in experimental autoimmune myasthenia gravis, *Science,* 232, 401, 1986.
219. **Lisak, R. P., Laramore, C., Levinson, A. I., Zweiman, B., and Moskovitz, A. R.,** Suppressor T cells in myasthenia gravis and antibodies to acetylcholine receptor, *Ann. Neurol.,* 19, 87, 1986.
220. **Cheng, I. K. P., Dorsch, S. E., and Hall, B. M.,** The regulation of autoantibody production in Heymann's nephritis by T lymphocyte subsets, *Lab. Invest.,* 59, 780, 1988.
221. **Morimoto, C., Steinberg, A. D., Letvin, N. L., Hagen, M., Takeuchi, T., Daley, J., Levine, H., and Schlossman, S. F.,** A defect of immunoregulatory T cell subsets in systemic lupus erythematosus patients demonstrated with anti-2H4 antibody, *J. Clin. Invest.,* 79, 762, 1987.
222. **Chofflon, M., Weiner, H. L., Morimoto, C., and Hafler, D. A.,** Decrease of suppressor inducer ($CD4^+2H4^+$) T cells in multiple sclerosis cerebrospinal fluid, *Ann. Neurol.,* 25, 494, 1989.
223. **Steinberg, A. D.,** Pathogenesis of autoimmunity in New Zealand mice. V. Loss of thymic suppressor function, *Arthritis Rheum.,* 17, 11, 1974.
224. **Krakauer, R. S., Waldmann, T. A., and Strober, W.,** Loss of suppressor T cells in adult NZB/NZW mice, *J. Exp. Med.,* 144, 662, 1976.
225. **DeMoor, J. M., Bell, D. A., and Singhal, S. K.,** Regulation of an in vitro anti-DNA antibody response by thymocytes and T cells from NZB/W and normal mice, *Cell. Immunol.,* 93, 406, 1985.
226. **Miller, R. D. and Calkins, C. E.,** Suppressor T cells and self-tolerance: active suppression required for normal regulation of anti-erythrocyte autoantibody responses in spleen cells from non-autoimmune mice, *J. Immunol.,* 140, 3779, 1988.
227. **Miller, K. B. and Schwartz, R. S.,** Familial abnormalities of suppressor cell function in systemic lupus erythematosus, *N. Engl. J. Med.,* 301, 803, 1979.
228. **Moore, J. S. and Calkins, C. E.,** In vitro regulation of the pathogenic autoantibody response of New Zealand Black mice. I. Loss with age of suppressive activity in T cell populations, *J. Immunol.,* 134, 3838, 1984.
229. **Volk, H. D. and Diamantstein, T.,** IL-2 normalizes defective suppressor T cell function of patients with systemic lupus erythematosus in vitro, *Clin. Exp. Immunol.,* 66, 525, 1986.
230. **Kennedy, M., Wassmer, P., and Erb, P.,** A novel role for interleukin-2 in antibody responses: down regulation of T-helper cell activation, *J. Immunol.,* 139, 110, 1987.
231. **Linker-Israeli, M., Bakke, A. C., Quismorio, F. P., and Horwitz, D. A.,** Correction of interleukin-2 production in patients with systemic lupus erythematosus by removal of spontaneously activated suppressor cells, *J. Clin. Invest.,* 75, 762, 1985.

INDEX

T

V

X